Insights and Advancements in Microfluidics

Special Issue Editors

Weihua Li
Hengdong Xi
Say Hwa Tan

MDPI • Basel • Beijing • Wuhan • Barcelona • Belgrade

MDPI

Special Issue Editors
Weihua Li
University of Wollongong
Australia

Hengdong Xi
Northwestern Polytechnical University
China

Say Hwa Tan
Griffith University
Australia

Editorial Office
MDPI AG
St. Alban-Anlage 66
Basel, Switzerland

This edition is a reprint of the Special Issue published online in the open access journal *Micromachines* (ISSN 2072-666X) from 2016–2017 (available at: http://www.mdpi.com/journal/micromachines/special_issues/insights_advancements_microfluidics).

For citation purposes, cite each article independently as indicated on the article page online and as indicated below:

Author 1; Author 2. Article title. *Journal Name* **Year**, *Article number*, page range.

First Edition 2017

ISBN 978-3-03842-516-8 (Pbk)
ISBN 978-3-03842-517-5 (PDF)

Table of Contents

About the Special Issue Editors

Weihua Li is a Senior Professor and Director of the Advanced Manufacturing Research Strength at the University of Wollongong. He completed his BEng (1992), MEng (1995) at University of Science and Technology of China, and PhD (2001) at Nanyang Technological University. He was with the School of Mechanical and Aerospace Engineering of NTU as a Research Associate/Fellow from 2000 to 2003, before he joined the School of Mechanical, Materials and Mechatronic Engineering as a Lecturer. His research focuses on smart materials and their applications, microfluidics, Lab on a Chip, and intelligent mechatronics. He is serving as editor or editorial board member for more than ten international journals, including IEEE/ASME Transactions on Mechatronics, Smart Materials and Structures, Lab on a Chip, Micromachinese, etc. He has published more than 220 journal papers. He is a recipient of Fellow of Engineers Australia, Fellow of the Institute of Physics (UK), JSPS Invitation Fellowship, Australian Endeavour Fellowship, Vice-chancellor's Award for Excellence in Research Supervision, and numerous Best Paper Awards.

Hengdong Xi is a professor at the School of Aeronautics, Northwestern Polytechnical University, Xi'an, China. He received his Bachelor's degree from Northwest University of China in 2001 and Master and Ph.D degrees from The Chinese University of Hong Kong in 2003 and 2007, respectively. He was a postdoctoral fellow at the Max-Planck Institute for Dynamics and Self-organization in Germany from 2009 to 2012. He was awarded the Humboldt Fellowship in 2010. He got in the "1000-young Talent Program" of China in 2012 and worked as a professor in Shenzhen Graduate School, Harbin Institute of Technology from December 2012. In September 2014, he joined the School of Aeronautics, Northwestern Polytechnical University. His research interests include microfluidics, thermal convection, polymeric drag reduction, and turbulence. His research work has been published in journals such as Physical Review Letters, Journal of Fluid Mechanics, Lab on a Chip, Physics of Fluid and Physical Review E.

Say Hwa Tan is an ARC DECRA fellow with Queensland Micro-and Nanotechnology Centre, Griffith University, Australia. He received his BEng, MEng and PhD degrees from the Nanyang Technological University, Singapore, and the Georg-August-Universität Göttingen/Max Planck Institute for dynamics and self-organization (MPI-DS), Germany, in 2008, 2010 and 2014 respectively. In 2016, he was highlighted as one of the 18 emerging investigators in the journal of Lab on a Chip. Dr Tan has published more than 30 research works in microfluidics. His research has established and pioneered different approaches to manipulate droplets and bubbles using thermal, magnetic, acoustic, pneumatic and electric energy.

micromachines

MDPI

Editorial

Editorial for the Special Issue on the Insights and Advancements in Microfluidics

Say Hwa Tan [1], Heng-Dong Xi [2] and Weihua Li [3,*]

1   Queensland Micro- and Nanotechnology Centre, Griffith University, 170 Kessels Road, Brisbane, QLD 4111, Australia; sayhwa.tan@griffith.edu.au
2   School of Aeronautics, Northwestern Polytechnical University, 127 West Youyi Rd., Xi'an 710072, China; hengdongxi@nwpu.edu.cn
3   School of Mechanical, Materials and Mechatronic Engineering, University of Wollongong, Wollongong, NSW 2522, Australia
*   Correspondence: weihuali@uow.edu.au; Tel.: +61-2-4221-4577

Received: 15 August 2017; Accepted: 15 August 2017; Published: 17 August 2017

We present a total of 19 articles in this special issue of *Micromachines* entitled, "Insights and Advancements in Microfluidics." Among the 19 articles, two perspectives, eight reviews, and nine research articles were solicited from leading researchers, pioneers, and emerging investigators. The topics covered in this issue ranges from biology, chemistry, and physics to the intersection of engineering, optics, and material sciences. As editors for this issue, we are both gratified and extremely thankful for the overwhelming responses and contributions from our fellow colleagues within the field of microfluidics.

The special issue is themed to provide both insights and advancements in microfluidics. This well-timed issue touches on a field which has evolved tremendously in the last few decades. Professor Yanyi Huang from Peking University of China provided his unique insight and perspective on digital polymerase chain reaction (PCR) [1]. In his article, an informative guide was provided on the proper designing rules of digital PCR at the micro-scale. Professor Guoqing Hu from the Chinese Academy of Sciences, Beijing, China, provided his astute insights and perspective on particle manipulation based on hydrodynamic effect [2]. His article summarizes both the progress and fundamental mechanisms in particle manipulation using elasto-inertial microfluidics.

In the eight reviews articles, different branches and sub-branches of microfluidics were presented and comprehensively reviewed. These include microfluidic sensing, liquid handling, optofluidics, the use of microfluidics in cytotoxicity, Janus micro-motors, single-cell impedance cytometry, droplets, and polymer microfluidics. We were extremely fortunate to receive contributions from both Professor Dongqing Li and Professor Nam-Trung Nguyen. Both are leading pioneers and extraordinary visionary leaders in the field of microfluidics. Professor Li et al. [3] reviewed the basic theories in both microfluidic and nanofluidic resistive pulse sensing (RPS). His article focuses on the latest developments in the last six years. Future research direction and challenges in this area are also outlined in the review. Professor Nguyen et al. [4] discussed the recent advances and future perspective on microfluidic liquid handling. The first part of the review covers two main and opposing applications of liquid handling in continuous-flow microfluidics: mixing and separation. The second part focuses on various digital microfluidic strategies based on both droplets and liquid marbles. The applications of the emerging field of liquid-marble-based digital microfluidics are also highlighted in the article. Song et al. [5] provided an overview on the recent development of optofluidics. They discussed the critical challenges that hamper the transformation of optofluidic technologies from lab-based procedures to practical usages and commercialization. Priest et al. [6] reviewed the different microfluidic chips that can used for toxicity screening. Li et al. [7] discussed the self-propulsion of a platinum–silica (Pt–SiO$_2$) spherical Janus micro-motor (JM). Their paper reviews two distinct mechanisms, self-diffusiophoresis and microbubble propulsion, and demonstrates that the physical principles of these mechanisms can be used to fulfill many novel functions. Petchakup et al. [8] discussed the topic of microfluidic-based single-cell impedance cytometry. Their article reviews the recent developments, applications, and discusses the future direction and challenges in this field. Wang et al. [9] re-visited the use of droplet-based microfluidics for the production of both micro and nano particles. Tsao et al. [10] discussed simple, low-cost methods to fabricate polymer-based microfluidics devices. Their paper provides an overview of the different micro-fabrication methods and discusses the current challenges of this research.

The nine research articles provides advancement techniques which can be categorized into two distinct classes: device innovation and fluid dynamics. In device innovation, Shui et al. [11] used the vacuum airbag laminator (VAL) to fabricate large-area and high-throughput PDMS microfluidic devices. The proposed fabrication method can achieve a high bond strength with a maximum breakup pressure of about 739 kPa. Lim and Kim et al. [12] proposed a new method to fabricate an all-glass bifurcation microfluidic chip using an amorphous carbon (AC) mold. The device is then used for blood plasma separation. Koh et al. [13] used a modified xurography method to fabricate a low-cost microfluidic device which can be used for both droplet fission and encapsulation. Lim et al. [14] devised a new bonding method which can be used to enhance the fabrication of thermoplastic microfluidic devices. This method combines an interference fit with a thermal treatment at a low pressure. Sui et al. [15] demonstrated a new microfluidic chip that can be used for rapid capture and analysis of airborne staphylococcus. The whole analysis takes less than 5 h and has a detection limit down to about 27 cells. Choi et al. [16] presented a simple but yet effective approach for facile, on-demand reconfiguration of microfluidic channels using flexible polymer tubing. Both microparticle separation and fluid mixing were successfully implemented by reconfiguring the shape of the tubing.

In fluid dynamics, Tsao et al. [17] used a cyclic olefin copolymer (COC)-based microfluidic device to investigate the flow behavior in a fractured porous medium. His results show that the flow resistance in the main channel with a large radius was higher than that in the surrounding area with small pore channels when the injection or extraction rates were low. When the flow rates were increased, the extraction efficiency of the water and oil in the mainstream channel did not increase monotonically because of the complex two-phase-flow dynamics. Zhao et al. [18] used numerical simulations to investigate the dynamic characteristics of electro-osmosis. His investigation using the finite element method shows that the electro-osmotic flow of power-law fluids under an AC/DC combined driving field is enhanced when compared to a pure DC electric field. Qin et al. [19] used a Y-shaped microfluidic device to study the combined effect of wall shear stress (WSS) and adenosine triphosphate (ATP) signals on intracellular calcium dynamics in vascular endothelial cells (VEC). Both numerical simulation and experimental studies verified the approach. The experimental results also suggest that a combination of WSS and ATP signals (rather than a WSS signal alone) play a more significant role in VEC Ca^{2+} signal transduction induced by blood flow.

Last but not least, we wish to thank all authors who submitted their papers to this special issue. We would also like to acknowledge all the reviewers whose careful and timely reviews ensured the quality of this special issue.

References

1. Liao, P.; Huang, Y. Digital PCR: Endless frontier of 'divide and conquer'. *Micromachines* **2017**, *8*, 231.
2. Liu, C.; Hu, G. High-throughput particle manipulation based on hydrodynamic effects in microchannels. *Micromachines* **2017**, *8*, 73. [CrossRef]
3. Song, Y.; Zhang, J.; Li, D. Microfluidic and nanofluidic resistive pulse sensing: A review. *Micromachines* **2017**, *8*, 204. [CrossRef]
4. Nguyen, N.-T.; Hejazian, M.; Ooi, C.; Kashaninejad, N. Recent advances and future perspectives on microfluidic liquid handling. *Micromachines* **2017**, *8*, 186. [CrossRef]
5. Song, C.; Tan, S. A perspective on the rise of optofluidics and the future. *Micromachines* **2017**, *8*, 152. [CrossRef]

6. McCormick, S.; Kriel, F.; Ivask, A.; Tong, Z.; Lombi, E.; Voelcker, N.; Priest, C. The use of microfluidics in cytotoxicity and nanotoxicity experiments. *Micromachines* **2017**, *8*, 124. [CrossRef]
7. Zhang, J.; Zheng, X.; Cui, H.; Silber-Li, Z. The self-propulsion of the spherical Pt–SiO$_2$ janus micro-motor. *Micromachines* **2017**, *8*, 123. [CrossRef]
8. Petchakup, C.; Li, K.; Hou, H. Advances in single cell impedance cytometry for biomedical applications. *Micromachines* **2017**, *8*, 87. [CrossRef]
9. Wang, J.; Li, Y.; Wang, X.; Wang, J.; Tian, H.; Zhao, P.; Tian, Y.; Gu, Y.; Wang, L.; Wang, C. Droplet microfluidics for the production of microparticles and nanoparticles. *Micromachines* **2017**, *8*, 22. [CrossRef]
10. Tsao, C.-W. Polymer microfluidics: Simple, low-cost fabrication process bridging academic lab research to commercialized production. *Micromachines* **2016**, *7*, 225. [CrossRef]
11. Xie, S.; Wu, J.; Tang, B.; Zhou, G.; Jin, M.; Shui, L. Large-area and high-throughput pdms microfluidic chip fabrication assisted by vacuum airbag laminator. *Micromachines* **2017**, *8*, 218.
12. Jang, H.; Haq, M.; Ju, J.; Kim, Y.; Kim, S.-m.; Lim, J. Fabrication of all glass bifurcation microfluidic chip for blood plasma separation. *Micromachines* **2017**, *8*, 67. [CrossRef]
13. Lim, C.; Koh, K.; Ren, Y.; Chin, J.; Shi, Y.; Yan, Y. Analysis of liquid–liquid droplets fission and encapsulation in single/two layer microfluidic devices fabricated by xurographic method. *Micromachines* **2017**, *8*, 49. [CrossRef]
14. Gong, Y.; Park, J.; Lim, J. An interference-assisted thermal bonding method for the fabrication of thermoplastic microfluidic devices. *Micromachines* **2016**, *7*, 211. [CrossRef]
15. Jiang, X.; Liu, Y.; Liu, Q.; Jing, W.; Qin, K.; Sui, G. Rapid capture and analysis of airborne staphylococcus aureus in the hospital using a microfluidic chip. *Micromachines* **2016**, *7*, 169. [CrossRef]
16. Hahn, Y.; Hong, D.; Kang, J.; Choi, S. A reconfigurable microfluidics platform for microparticle separation and fluid mixing. *Micromachines* **2016**, *7*, 139. [CrossRef]
17. Hsu, S.-Y.; Zhang, Z.-Y.; Tsao, C.-W. Thermoplastic micromodel investigation of two-phase flows in a fractured porous medium. *Micromachines* **2017**, *8*, 38. [CrossRef]
18. Zhao, C.; Zhang, W.; Yang, C. Dynamic electroosmotic flows of power-law fluids in rectangular microchannels. *Micromachines* **2017**, *8*, 34. [CrossRef]
19. Chen, Z.-Z.; Gao, Z.-M.; Zeng, D.-P.; Liu, B.; Luan, Y.; Qin, K.-R. A Y-shaped microfluidic device to study the combined effect of wall shear stress and atp signals on intracellular calcium dynamics in vascular endothelial cells. *Micromachines* **2016**, *7*, 213. [CrossRef]

micromachines

MDPI

Perspective

Digital PCR: Endless Frontier of 'Divide and Conquer'

Peiyu Liao and Yanyi Huang *

Beijing Advanced Innovation Center for Genomics (ICG), Biodynamic Optical Imaging Center (BIOPIC),
College of Engineering, School of Life Sciences, and Peking-Tsinghua Center for Life Sciences,
Peking University, Beijing 100871, China; peiyu@pku.edu.cn
* Correspondence: yanyi@pku.edu.cn; Tel.: +86-010-6275-8323

Received: 28 June 2017; Accepted: 18 July 2017; Published: 25 July 2017

Abstract: Digital polymerase chain reaction (PCR) is becoming ever more recognized amid the overwhelming revolution in DNA quantification, genomics, genetics, and diagnostics led by technologies such as next generation sequencing and studies at the single-cell level. The demand to quantify the amount of DNA and RNA has been driven to the molecular level and digital PCR, with its unprecedented quantification capability, is sure to shine in the coming era. Two decades ago, it emerged as a concept; yet one decade ago, integration with microfluidics invigorated this field. Today, many methods have come to public knowledge and applications surrounding digital PCR is mounting. However, to reach wider accessibility and better practicality, efforts are needed to tackle the remaining problems. This perspective looks back at several inspiring and influential digital PCR approaches in the past and tries to provide a futuristic picture of the trends of digital PCR technologies to come.

Keywords: digital polymerase chain reaction (PCR); microfluidics; emulsion droplet; microwell chip

As the knowledge of molecular genetic scripts accumulates, the quantification of DNA and RNA molecules has become increasingly important. Numerous methods have been developed wittingly or not, to quantify the amount of nucleic acids as precisely as possible. Among the sequence-specific quantification methods, polymerase chain reaction (PCR)-based techniques have always been a favored option. Real-time quantitative PCR is regarded as a routine practice in biomedical laboratories. However, due to its limited counting resolution, this method does not meet the ever more stringent quantification demand, especially when the target concentration is relatively low or PCR inhibitors exist to perturb the exquisite assay. Moreover, as the techniques such as next generation sequencing and single-cell analyses continue to flourish, interest in quantification of nucleic acids has been drawn to the unprecedented single-molecule level. This has given rise to the prosperity of digital PCR technologies.

'Limiting dilution PCR' and 'single-molecule PCR' were the first names used to describe the method, until a much more popular and apposite term—'digital PCR'—was put forward, which soon seized broad attention [1]. The strategy is simple, and the metaphor 'divide and conquer' is a good analogy [2]. DNA samples are separately amplified in independent yet identical partitions, and the all-or-none detection results of each reaction follow Poisson statistics. After counting the sum of positive reactions, by Poisson correction, not only the concentration but the absolute number of target molecules can be obtained.

In early demonstrations [3,4], researchers diluted the samples into a large number (typically hundreds) of reaction vessels in micro-well plates (Figure 1a), the assay then was inevitably laborious and costly, and the partition number was thus limited. The situation changed when microfluidics were employed: reactor multiplication became division, and the total volume of the aqueous sample was therefore reduced to less than 100 μL.

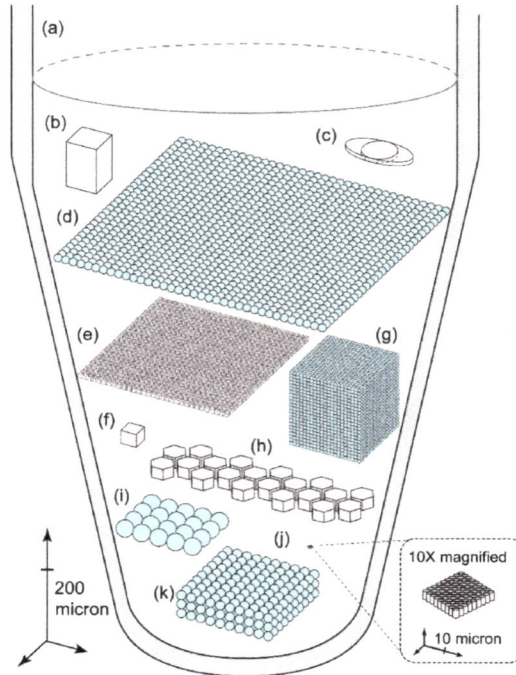

Figure 1. An overview of representative digital PCR approaches. For each partition shown herein represents approximately 1000 partitions in the corresponding methods. Their sizes too roughly reflect those in real-world situations. Micro-chambers in microfluidic chips are shown in pink and emulsion droplets in light blue. (**a**) Digital PCR in well plates [3,4]; (**b**) The first microfluidic chip for digital PCR in 2006 [5]; (**c**) SlipChip in 2010 [6]; (**d**) One-million droplet array in 2011 [7]; (**e**) Megapixel digital PCR in 2011 [8]; (**f**) Digital Array IFC (type: qdPCR 37k) from Fluidigm's Biomark HD (Fluidigm Corporation, South San Francisco, CA, USA); (**g**) RainDrop digital PCR from Raindance Technologies (Lexington, MA, USA); (**h**) QuantStudio 3D from Life Technologies (now Thermo Fisher, Waltham, MA, USA); (**i**) QX200 droplet digital PCR platform from Bio-Rad (Hercules, CA, USA); (**j**) The densest and smallest microwell-based microfluidic digital PCR to date in 2012 [9]; (**k**) Chip-free digital PCR in droplets generated via micro-channel array (MiCA) centrifugation in 2017 [10].

Digital PCR Met Microfluidics

Microfluidic-based digital PCR approaches generally fall into two categories, solid micro-chambers and water-in-oil emulsion. The former typically relies on microfabrication methods to create miniaturized micro-well devices with much smaller wells that are greater in number than our routinely used format (Figure 2). Such devices are commonly made of polydimethylsiloxane (PDMS), while glass and plastics have also been used. Aqueous samples are divided into portions of minute volume and partitioned in the chambers before they undergo PCR thermal cycling. Water-in-oil emulsion digital PCR comes after its counterpart in solid chambers but has developed rapidly since its advent. Flow focusing geometry is most frequently present in droplet generating microfluidic devices (Figure 2a). At the cross-shaped nozzle, the aqueous sample (the disperse phase) is pinched by oil (the continuous phase) and forms droplets. In this fashion, uniform droplets can be continuously generated at high speed under a stable driving force such as pumping pressure. For any methods to partition the water-based sample, volume and number of partitions, uniformity, and robustness are the most essential prerequisites for digital PCR. The reasons are that Poisson distribution has well determined that this method would innately favor smaller partitions in greater number for finer counting resolution and higher dynamic range, and that the thermal cycling procedure would incur considerations such as water evaporation, thermal stability of emulsion or microfluidic chips, and compatibility of thermal cycler, etc. These reasons defined the strategies brought up towards better digital PCR approaches.

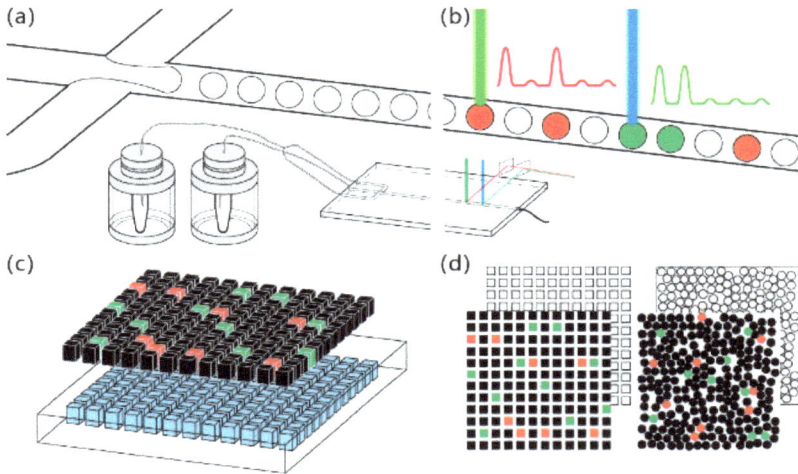

Figure 2. Microfluidic digital PCR partitioning and counting strategies. (**a**) Emulsion droplet generation using flow-focusing microfluidic chip; (**b**) Serial droplet fluorescence reading; (**c**) Minute volumes of PCR sample are partitioned in micro-well microfluidic chip; (**d**) Planar imaging for fluorescence positive chamber counting.

Patterning micron-scale features is no hard job for the prevailing lithography techniques. A notable pioneering integration between digital PCR and microfluidics was reported in 2006 by Ottesen et al. (Figure 1b) [5]. They used a 'Quake valve' structure that was made of multi-layer soft-lithography to actively separate individual reaction chambers. On one chip, 12 identical sections—each of about a thousand 6.5-nL partitions—can be processed simultaneously. This design was further modified in its commercial version. Slightly compromising the partition number per sample, Digital Array IFCs from Fluidigm's Biomark HD made its chip more compact with more samples on one chip, whose most compact type is capable of 48 samples of 770 0.85-nL partitions (Figure 1f).

To enable digital PCR assay in a laboratory with insufficient expertise in microfluidics one may like to devise a method of sample partitioning free of pumps, neither hydraulic nor pneumatic. The SlipChip (Figure 1c) needs no exquisite microfluidic processes after chip fabrication, for its sample loading is just pipetting, and then a simple slipping [6]. SlipChips provide an obvious advantage for their readiness in resource-limited scenarios. Later, an idea called multivolume digital PCR [11,12] came to the researchers as how to minimize the size of the SlipChip and reduce the well number without compromising the performance specifications such as dynamic range. The key point is that the wells in the chip were designed to have different volumes in order to eliminate serial dilution. Simple as the authors claimed, it involves somewhat complicated mathematical derivation in theory and less intuitive interpretation of its results in practice.

The direct approach to wider dynamic range of digital PCR is to augment the amount of partitions. In 2011, the number was boosted to over a million as was described as "megapixel digital PCR" (Figure 1e) [8]. Valve-free partitioning on chip was realized by using fluorinated oil to flush the aqueous sample in the fluorophilic channels. The small volume of the chambers (10 pL) allowed the device's high reactor density (4400 per mm^2). PDMS's permeability of gas and vapor allowed the 'dead-end' design—there were only inlets but no outlets so ambient air in the chambers was squeezed out through the silicon rubber—yet led to the addition of a parylene layer as a vapor barrier to prevent sample evaporation.

Emulsion droplet's capability to reach the same scale of partitioning was validated in the same year. Hatch et al. fabricated a microfluidic chip with 256 droplet generating nozzles that swiftly generated one-million 50-pL droplets (Figure 1d) and captured the fluorescence signals using wide-field imaging [7]. However, PCR that occurs in water-in-oil droplets was not unusual before its application in digital PCR. Emulsion PCR had been widely used in the next generation sequencing field to generate clonally amplified fragments [1]. Raindance Technologies was one of the earliest to use uniform sized water-in-oil droplets for digital PCR [13]. Years later in their commercialized emulsion generator, the droplet volume was reduced to 5 pL, and for a sample of about 70 µL, there were more than 10-million droplets (Figure 1g), which currently marks the record of the highest dynamic range of digital PCR platforms.

In the pursuit of ever smaller partitioning and the extreme of miniaturization of the digital PCR devices, Men et al. fabricated the smallest wells (36 femtoliters) on a PDMS-made microfluidic chip with the densest well arrangement to date (over 20,000 reactors per mm^2) (Figure 1j) [9]. Each device had over 80,000 hexagonally arrayed round-shaped wells of 3.3 µm in diameter and 4.2 µm in depth, using hydraulic pressing valve to actuate the compartmentalization and to prevent the evaporation through PDMS [9]. However, although pursuing limits in science is always fun, such a fine microfluidic device is by no means to be widely used, because for a popular method practicality and accessibility are the priorities.

Proper Design of Digital PCR

Although academia is good at competing over performance specifications, industry seems to know better what is 'just good enough' for a digital PCR solution. The latter, however, has to pay much more attention to the facility's robustness, handiness, cost, and affordability to users, so trade-offs are inevitable. Two representative digital PCR platforms are QuantStudio 3D from Life Technologies (now Thermo Fisher) (Figure 1h) and Bio-Rad's QX200 [14] (Figure 1i). The two coincide in partition volume (about 0.8 nL) and number (about 20,000) but adopt different strategies, the former emulsion droplet and the latter micro-well chip. It seems that their specifications of partition would be most welcomed by biologist users, whose favored sample volume is approximately 20 µL and concentration of interest would fall in this dynamic range in most cases. To the author's knowledge, Bio-Rad may hold the biggest market share in the last two years and it is becoming popular in publications involving digital PCR.

Thanks to marketing propaganda, digital PCR has been well-known to many along with its competency in various applications. Still, apprehensions have kept many watching. Digital PCR seems to be in the similar case that quantitative PCR once was two decades ago, when few could afford to have a fancy machine on their own, but today this machine has become ubiquitous. Price hinders digital PCR's popularity. In the past decade, cost per sample has dramatically decreased to 3 US dollars (Bio-Rad) [2], but the instrument price of the commercial platforms still remains much higher than quantitative PCR machines. To reduce the expenditure on microfluidics, Life Technologies set out to use silicon and Bio-Rad plastic for their partitioning units, respectively. Moreover, the field of microfluidics may not luckily follow Moore's Law, and the cost may linger unless breakthroughs are to come.

Another factor that worries users is that, for emulsion based digital PCR, contamination seems likely to occur. Bio-Rad's whole set contains two separate machines for droplet generation and reading. The droplets of different samples are first transferred by pipetting to a 96-well plate to be thermal cycled, and then droplets are sucked through one syringe nozzle and then serially interrogated at the fluorescent detecting spot. The exposure of amplified DNA products to open experimental environments risks contaminating subsequent assays. It is questionable whether such practice would be welcomed in clinical occasions. Besides, only part of the nucleic acid molecules of interest partake in the quantification assay. For example, in some micro-well partitioning approaches, excess sample was added to make sure all the chambers are taken, and much was removed in the sealing process. Likewise, in the emulsion droplet method involving microfluidic control, fluid waste exists before flow stabilizes to uniformly generate droplets, in addition to droplet breakage and loss amid emulsion transfer and mobile droplet reading. The effective readout rate is therefore limited to around 80% or less. This would not perturb Poisson statistics but is still a nuisance unexpected in stringent assays.

Turning Tides

Late-comers in the digital PCR industry have taken the lessons and made some adaptations. Clarity™ digital PCR (JN Medsys, Singapore) put forward a new method of micro-well partitioning: chip-in-a-tube [15]. A miniaturized chip with 10,000 micro-well that partitions samples but can be placed in the commonly used 200-µL PCR tube, where thermal cycling and signal reading take place. By this means, post-amplification DNA product is sealed throughout the process and normal thermal cyclers can be used. Crystal™ Digital PCR from Stilla Technologies (Villejuif, France) [16] gave another novel droplet generation method other than the conventional ones such as flow focusing, T-junction, or co-flow design. Gradient of confinement was the driving force and gave rise to uniform emulsion droplet generation due to insensitivity to the physical fluid properties [17]. In this method, droplet generation, thermal cycling, and reading happen in a planar flow cell chip in succession, prohibiting carryover contamination by closed processing.

An innovation in monodisperse emulsion generation in 2016 told another story of droplet digital PCR by abandoning the 'chip' format. The device called MiCA has a small glass plate on which there are several through-holes of micron-level diameter [10]. Using centrifugal force to drive the process, aqueous sample is pinched off into monodisperse droplets and jets into surfactant-oil mixture beneath (Figure 1k) [10]. The aqueous sample could be almost completely transformed into droplets while the subsequent reading process rendered the readout rate less perfect, slightly higher than 80% (still higher than many other methods, though) [10]. This centrifugation droplet generation method manifested itself as a new possibility for it is a chip-free, pump-free, and loss-free protocol. Besides, multiple such MiCA tube can work simultaneously in a multiplex way, which is a good sign for high throughput, and the work has also validated that fluorine-free oil phase is compatible of digital PCR, which might help to reduce the oil cost.

In respect to the oil phase used in droplet digital PCR, fluorinated oil has dominated in the past decade. This inert and non-toxic liquid has appreciable performance in microfluidics and has been utilized since the early stage of digital PCR [13]. Its thermal stability, coalescence resistance, low viscosity and gas dissolving capacity contribute to its wide acceptance. Yet some factors keep fluorinated oil from being ideal. One is the high cost, which subject to the relatively tricky fluorine chemistry, would not see considerable decrease. Besides, compatible surfactant for fluorinated oil has to be fluorinated as well and hence is limited to few fluorinated proprietary products. The other, seemingly trivial issue is the high density. Aqueous droplets would gather to the surface of the emulsion and are prone to evaporate and shrink in size due to the large surface-to-volume ratio of emulsion droplets. Hydrocarbon-based emulsion oil has more surfactant options [7,18] and most of them are inexpensive and quite accessible. Also, droplets would not float near the surface but settle at the bottom of the low-density oil, resulting in less evaporation. Given this comparison between the two oil categories, it is still unforeseeable which would be the final winner for digital PCR emulsion, and there remains much room for improvement for the emulsion formulae.

Digital reading or counting of the partitions so far can be generally categorized into two strategies: serial counting and planar imaging (Figure 2b,d, respectively). Regarding the reading scheme, few surprises were brought in the past decade, for these extant strategies seem to have effectively addressed this problem. Micro-well partitioning is a natural fit for planar imaging while its use of serial reading sounds less sensible. For droplet partitioning, most cases opted for serial counting. The aforementioned one-million-droplet digital PCR [7] had to use wide-field optics to image the single layer of close arrayed droplets. That is because if droplets are stacked during imaging, refraction and diffraction induced blurring would interfere with the imaging results and subsequent graphical analysis would be unreliable. There is a dilemma, that droplet digital PCR has a lower cost in partitioning but higher cost in reading, and micro-well digital PCR the other way around. This complex yet interesting problem will have to be overcome by the innovations to come.

In summary, digital PCR has passed its emerging phase and is now trending profoundly, yet it is still not in its peak. Before meeting microfluidics, digital PCR was almost dormant. The past decade has seen how microfluidic techniques boosted biomedical development and digital PCR becomes an ideal example. Current methods have, to a great extent, lowered the cost per sample, which is mainly subject to the consumables. Digital PCR assay's consumables are unlike common molecular biology assays, for they require delicate microfabrication, surface treatment, and special chemistry. To accomplish extraordinary performance specifications would sooner or later be less interesting since many boundaries have been explored. New trends will be to further lower the cost to greatly improve accessibility, and to make the hands-on process as fool-proof and contamination-proof as possible. These trends do not come from nowhere. For example, next generation sequencing has been driving the interest of quantity to a few-cell level, single-cell level, and even few-molecule level. Digital PCR's applications are far reaching: its capability of absolute quantification, single nucleotide polymorphism genotyping, and copy number variation defines its significance in genomics, genetics, and diagnostics. As applications of digital PCR increase, so will the market.

Acknowledgments: This work is supported by National Natural Science Foundation of China (21327808 and 21525521), Ministry of Science and Technology of China (2015AA0200601 and 2016YFC0900100), and Beijing Advanced Innovation Center for Genomics. The authors would like to thank Zitian Chen and Fangli Zhang for the discussion and help on the digital PCR projects.

Conflicts of Interest: The authors declare no conflict of interest.

References

1. Morley, A.A. Digital PCR: A brief history. *Biomol. Detect. Quantif.* **2014**, *1*, 1–2. [CrossRef] [PubMed]
2. Baker, M. Digital PCR hits its stride. *Nat. Methods* **2012**, *9*, 541–544. [CrossRef]
3. Sykes, P.J.; Neoh, S.H.; Brisco, M.J.; Huges, E.; Condon, J.; Morley, A.A. Quantitation of targets for PCR by use of limiting dilution. *Biotechniques* **1992**, *13*, 444–449. [PubMed]
4. Vogelstein, B.; Kinzler, K.W. Digital PCR. *Proc. Natl. Acad. Sci. USA* **1999**, *96*, 9236–9241. [CrossRef]
5. Ottesen, E.A.; Hong, J.W.; Quake, S.R.; Leadbetter, J.R. Microfluidic digital PCR enables multigene analysis of individual environmental bacteria. *Science* **2006**, *314*, 1464–1467. [CrossRef] [PubMed]
6. Shen, F.; Du, W.; Kreutz, J.E.; Fok, A.; Ismagilov, R.F. Digital PCR on a SlipChip. *Lab Chip* **2010**, *10*, 2666–2672. [CrossRef] [PubMed]
7. Hatch, A.C.; Fisher, J.S.; Tovar, A.R.; Hsieh, A.T.; Lin, R.; Pentoney, S.L.; Yang, D.L.; Lee, A.P. 1-million droplet array with wide-field fluorescence imaging for digital PCR. *Lab Chip* **2011**, *11*, 3838–3845. [CrossRef] [PubMed]
8. Heyries, K.A.; Tropini, C.; Vaninsberghe, M.; Doolin, C.; Petriv, O.I.; Singhal, A.; Leung, K.; Hughesman, C.B.; Hansen, C.L. Megapixel digital PCR. *Nat. Methods* **2011**, *8*, 649–651. [CrossRef] [PubMed]
9. Men, Y.; Fu, Y.; Chen, Z.; Sims, P.A.; Greenleaf, W.J.; Huang, Y. Digital polymerase chain reaction in an array of femtoliter polydimethylsiloxane microreactors. *Anal. Chem.* **2012**, *84*, 4262–4266. [CrossRef] [PubMed]
10. Chen, Z.; Liao, P.; Zhang, F.; Jiang, M.; Zhu, Y.; Huang, Y. Centrifugal micro-channel array droplet generation for highly parallel digital PCR. *Lab Chip* **2017**, *17*, 235–240. [CrossRef] [PubMed]
11. Kreutz, J.E.; Munson, T.; Huynh, T.; Shen, F.; Du, W.; Ismagilov, R.F. Theoretical design and analysis of multivolume digital assays with wide dynamic range validated experimentally with microfluidic digital PCR. *Anal. Chem.* **2011**, *83*, 8158–8168. [CrossRef] [PubMed]

12. Shen, F.; Sun, B.; Kreutz, J.E.; Davydova, E.K.; Du, W.; Reddy, P.L.; Joseph, L.J.; Ismagilov, R.F. Multiplexed quantification of nucleic acids with large dynamic range using multivolume digital RT-PCR on a rotational SlipChip tested with HIV and hepatitis C viral load. *J. Am. Chem. Soc.* **2011**, *133*, 17705–17712. [CrossRef] [PubMed]

13. Kiss, M.M.; Otorleva-Donnelly, L.; Beer, N.R.; Warner, J.; Bailey, C.G.; Colston, B.W.; Rothberg, J.M.; Link, D.R.; Leamon, J.H. High-throughput quantitative polymerase chain reaction in picoliter droplets. *Anal. Chem.* **2008**, *80*, 8975–8981. [CrossRef] [PubMed]

14. Hindson, B.J.; Ness, K.D.; Masquelier, D.A.; Belgrader, P.; Heredia, N.J.; Makarewicz, A.J.; Bright, I.J.; Lucero, M.Y.; Hiddessen, A.L.; Legler, T.C.; et al. High-throughput droplet digital PCR system for absolute quantitation of DNA copy number. *Anal. Chem.* **2011**, *83*, 8604–8610. [CrossRef] [PubMed]

15. Low, H.; Chan, S.J.; Soo, G.H.; Ling, B.; Tan, E.L. Clarity digital PCR system: A novel platform for absolute quantification of nucleic acids. *Anal. Bioanal. Chem.* **2017**, *409*, 1869–1875. [CrossRef] [PubMed]

16. Madic, J.; Zocevic, A.; Senlis, V.; Fradet, E.; Andre, B.; Muller, S.; Dangla, R.; Droniou, M.E. Three-color crystal digital PCR. *Biomol. Detect. Quantif.* **2016**, *10*, 34–46. [CrossRef] [PubMed]

17. Dangla, R.; Kayi, S.C.; Baroud, C.N. Droplet microfluidics driven by gradients of confinement. *Proc. Natl. Acad. Sci. USA* **2013**, *110*, 853–858. [CrossRef] [PubMed]

18. Beer, N.B.; Hindson, B.J.; Wheeler, E.K.; Hall, S.B.; Rose, K.A.; Kennedy, I.M.; Colston, B.W. On-chip, real-time, single-copy polymerase chain reaction in picoliter droplets. *Anal. Chem.* **2007**, *79*, 8471–8475. [CrossRef] [PubMed]

micromachines

MDPI

Perspective

High-Throughput Particle Manipulation Based on Hydrodynamic Effects in Microchannels

Chao Liu [1] and Guoqing Hu [2,3,*]

[1] CAS Key Laboratory for Biological Effects of Nanomaterials and Nanosafety, CAS Center for Excellence in Nanoscience, National Center for NanoScience and Technology, Beijing 100190, China; liuchao@imech.ac.cn
[2] State Key Laboratory of Nonlinear Mechanics, Beijing Key Laboratory of Engineered Construction and Mechanobiology, Institute of Mechanics, Chinese Academy of Sciences, Beijing 100190, China
[3] School of Engineering Science, University of Chinese Academy of Sciences, Beijing 100049, China
* Correspondence: guoqing.hu@imech.ac.cn; Tel.: +86-10-8254-4298

Academic Editors: Weihua Li, Hengdong Xi and Say Hwa Tan
Received: 14 January 2017; Accepted: 23 February 2017; Published: 1 March 2017

Abstract: Microfluidic techniques are effective tools for precise manipulation of particles and cells, whose enrichment and separation is crucial for a wide range of applications in biology, medicine, and chemistry. Recently, lateral particle migration induced by the intrinsic hydrodynamic effects in microchannels, such as inertia and elasticity, has shown its promise for high-throughput and label-free particle manipulation. The particle migration can be engineered to realize the controllable focusing and separation of particles based on a difference in size. The widespread use of inertial and viscoelastic microfluidics depends on the understanding of hydrodynamic effects on particle motion. This review will summarize the progress in the fundamental mechanisms and key applications of inertial and viscoelastic particle manipulation.

Keywords: particle manipulation; inertial lift; viscoelastic effects; microfluidics; lab on a chip; high throughput

---

1. Introduction

Cells, bacteria, virus, and biomacromolecules are particles with sizes ranging from tens of micrometers to tens of nanometers (Figure 1). The precise manipulation of these bioparticles is essential to various research and application fields [1,2]. For example, the detection of circulating tumor cells (CTCs) is essential to the early prognostic of cancer and the research on determination of their phenotype and genotype, which can provide better understanding of metastasis process and better guidance of cancer therapy [3–5]. However, CTC detection is challenging due to extremely low CTC concentration (1–100 CTCs/mL of blood) and large blood cell background. Therefore, a prior separation step of CTCs with high-throughput is critical for accurate and sensitive detection.

Conventional separation techniques often rely on immunocapture. Although useful for clinical and research purposes, cells may suffer irreversible damage during labeling. Moreover, the purity and retrieval of CTCs could be affected by the significant variation of the presence of specific biomarkers such as epithelial cell adhesion molecule (EpCAM) or human epidermal growth factor receptor 2 (HER-2) on CTC surface, even for the same tumor type [6,7]. For example, CellSearch® system uses magnetic nanoparticles coated with anti-EpCAM antibodies to capture CTCs from human blood. Although being the only Food and Drug Administration (FDA)-cleared tested technique for capturing and enumerating CTCs, CellSearch® system was recently discontinued due to its high miss rate and low CTC viability. The system fails to identify CTC from 7.5 mL blood samples (the manufacturer's protocol) for nearly half of the 430 tested cancer patients. To achieve a reliable detection, at least 30 mL peripheral blood instead of 7.5 mL has to be collected [8], requiring the ability of high-throughput sample handling [5].

In addition to the immunocapture-based techniques, researchers have tried to enrich and isolate cells and particles based on their physical properties, such as size, shape, deformability, density, compressibility, charge, polarizability, and magnetic susceptibility. Exploiting these biomarkers, extensive methods have been developed, including hydrodynamics-based methods [9], acoustophoresis [10], dielectrophoresis [11], magnetophoresis [12], optophoresis [13], and centrifugation [14] (Table 1). Among them, the manipulation techniques using hydrodynamic effects in microchannels have attracted increasing attention because of the merits from their simple implementation and often high throughput. Specifically, in this review, we will focus on discussing the recent innovations for the hydrodynamic manipulation of particles. To do so, we will first briefly describe the techniques based on particle's following streamlines at low Reynolds number Re ($Re = \rho U D / \eta$, where U is the characteristic velocity, D the characteristic channel dimension, ρ the fluid density, and η the fluid dynamic viscosity). Their advantages and limitations will serve as an introduction and motivation for the more recent techniques based on particle's cross-stream migration caused by hydrodynamic lift forces. The inertial microfluidics is discussed in detail, with emphasis on the fundamental mechanisms and the applications in high-throughput particle manipulation. Examples of specific microfluidic devices will be described and organized regarding their functions and designs, followed by the latest progress in the fundamentals of particle migration that provide better guidelines for the inertial microfluidic communities. We further discuss the more recent innovations in particle manipulation using viscoelastic effects, which enable more flexible working flow conditions and applicable particle sizes.

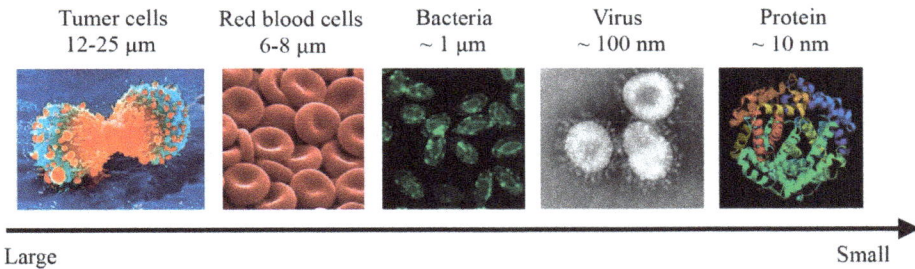

Figure 1. The size ranges of typical types of bioparticles. Cells, bacteria, virus, and biomacromolecules are particles with sizes ranging from tens of micrometers to tens of nanometers.

Table 1. Typical force field types used for microfluidic cell manipulation.

Technique	Separation Marker	Mechanism	Force Scaling with Diameter
Acoustophoresis	a, ρ, α	Ultrasonic sound wave	a^3
Dielectrophoresis	a, ε, σ	Non-uniform electric field	a^3
Magnetophoresis	a, χ	Magnetic field	a^3
Optophoresis	a, n	Optical field	Optical gradient force: a^3 Scattering force: a^6
Centrifugation	a, ρ	Centrifugal force	a^3

Note: a: diameter; ρ: density; α: compressibility; ε: permittivity; σ: electric conductivities; χ: magnetic permeability; n: reflective index.

2. Particle Manipulation Based on Low-Reynolds Number Hydrodynamic Effects

These techniques include pinched flow fractionation (PFF) [15,16], hydrodynamic filtration [17,18], deterministic lateral displacement (DLD) [19,20], and hydrophoresis [21–24], which presume that the particle center will strictly follow the streamline at low *Re* (Figure 2). They are typically based on the fluid and particles interacting with microstructures to separate particles by size, shape, and deformability [25,26]. In PFF, the suspension of particles with different sizes is introduced from a side microchannel into the main microchannel. The flow rates are tuned to pinch particles into a narrow stream and adjacent to the wall, making the centers of all particles located at different streamlines according to their size difference. In a sudden expansion at the downstream, the particles with different sizes are separated with a large lateral distance relative to expansion. With a similar operating principle, hydrodynamic filtration introduces particle suspension from a single inlet of a microchannel with multiple perpendicular branched outlets. The initially dispersed particles are gradually aligned along the side walls by repeatedly withdrawing a small portion of fluid from the main stream through the branched outlets. In the downstream of the microchannel, smaller particles enter into the branched outlets earlier than larger ones because the smaller particles locate closer to the side walls. Therefore, this technique enables particle separation and concentration simultaneously. To determine the size cutoff of the filtered particles, this technique requires precise microchannel fabrication to finely control the velocity profile and flow rate ratio at the branch point. Different from the above two techniques, DLD relies on a micropillar array in which each pillar row is laterally offset with respect to the predecessor row with a finely tuned distance. This design creates different streamline groups that move in the mainstream direction or along the offset sides of the predecessor pillars, depending on the pillar-streamline distance. Due to the steric hindrance of the pillar wall and the streamline following of microparticles, particles smaller than a critical size repeatedly move through the pillar gaps in an average mainstream direction, whereas particles larger than the critical size are laterally "bumped" along the offset direction due to the larger particle-pillar distance. The lateral distance between the large and small particles accumulates after multiple pillar rows, resulting in a final separation. Using these techniques, label-free separation has been achieved for diverse blood cell types [15,18,27] and CTCs [28]. It is worth mentioning that DLD has a very good size resolution, but with a very low throughput. Very recently, size sorting of nanoparticles down to 20 nm and exosomes has been demonstrated using DLD at the flow rates of ~10 nL/h [20]. Hydrophoresis is a separation technique using the particle motion influenced by a microstructure-induced pressure field. It typically uses microfluidic devices containing slanting obstacles to generate a lateral pressure gradient that induces helical recirculation and consequently focuses microparticles at different lateral positions depending on their size or deformability. Hydrophoresis also has a high separation resolution, enabling the discrimination of microparticles with diameter differences as small as 7.3% [21]. PFF and DLD often require sheath flows, which help to obtain high separation resolutions. However, sheath flows might cause several challenges: (1) the branched inlets for sheath flow make poor device parallelizability; (2) the control and operation become complex with sheath flows; and (3) the sample throughputs are limited due to the large sheath–sample flow ratio. Therefore, these techniques are commonly used for handling small volumes of samples.

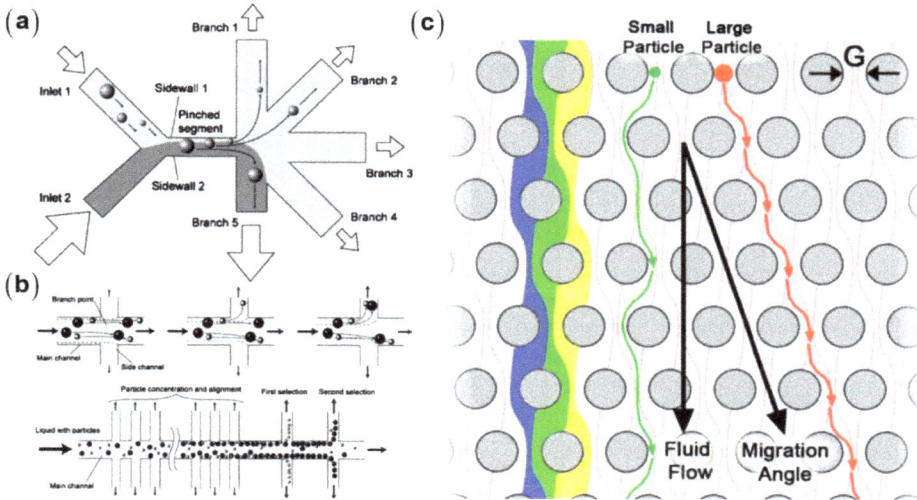

Figure 2. Hydrodynamic particle separation: (**a**) Pinched flow fractionation [15] (Reproduced with permissions from Takagi et al., Lab on a Chip; published by Royal Society of Chemistry, 2005); (**b**) hydrodynamic filtration [17] (Reproduced with permissions from Yamada et al., Lab on a Chip; published by Royal Society of Chemistry, 2005); and (**c**) deterministic lateral displacement [29] (Reproduced with permissions from Inglis et al., Applied Physics Letters; published by American Institute of Physics Publishing, 2004).

3. Inertial Manipulation of Particles

Different from the aforementioned methods, inertial microfluidics works by driving particles cross-stream migration utilizing inertial lift arising from the fluid flow nonlinearity at finite *Re* [30]. Due to its enhanced strength with increasing flow rate, inertial lift is favorable for high-throughput particle manipulation. Suspended in a pressure-driven flow of finite *Re*, particles will laterally migrate due to the acting inertial lift, which can be briefly attributed to the competition between two effects: (1) the shear-gradient-induced lift arising from the curvature of the Poiseuille velocity profile that drives the particle toward the wall; and (2) the wall-induced lift that pushes the particle away from the wall (Figure 3). The magnitude of inertial lift scales as $F_L \propto \rho U^2 a^4$ [30,31]. Although fluid inertia was traditionally thought to be insignificant in microfluidic systems, microchannels are more favorable to realize deterministic particle control than macroscale channels. To produce a sufficient large shear gradient, very high flow speeds are required in macroscale channels, resulting in turbulent flows where the precise control is broken down. By contrast, microchannels can still generate large shear gradients even at relatively low flow speeds.

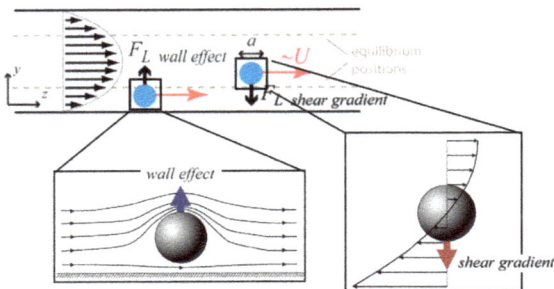

Figure 3. General mechanism of inertial lift: shear-gradient-induced lift arising from the curvature of Poiseuille velocity profile and wall-induced lift arising from the wall repulsion [32,33]. (Reproduced with permissions from Martal et al., Annual Review of Biomedical Engineering; published by Annual Reviews, 2014 and Reproduced with permissions from Amini et al., Lab on a Chip; published by Royal Society of Chemistry, 2014).

3.1. Inertial Particle Focusing in Straight Microchannels

The focusing of cells is essential to their detection and characterization [34–36], whose accuracy and sensitivity highly depend on the focusing quality (the percentage of particles focused at the expected positions). In straight microchannels, the particle focusing depends only on the inertial migration. The shape of microchannel cross-section affects the focusing pattern: particles are focused into a ring in circular pipes [37], focused near the centers of the four channel walls in square channels [9,38], or focused near the centers of the two long channel walls in rectangular channels [39–42]. Therefore, 3D particle focusing cannot be achieved solely by inertial migration in straight microchannels (Figure 4) [39]. An effective solution is to introduce the drag forces of secondary flows to compete with the inertial lift using special structures [43–46] or curved channel shapes [47–52].

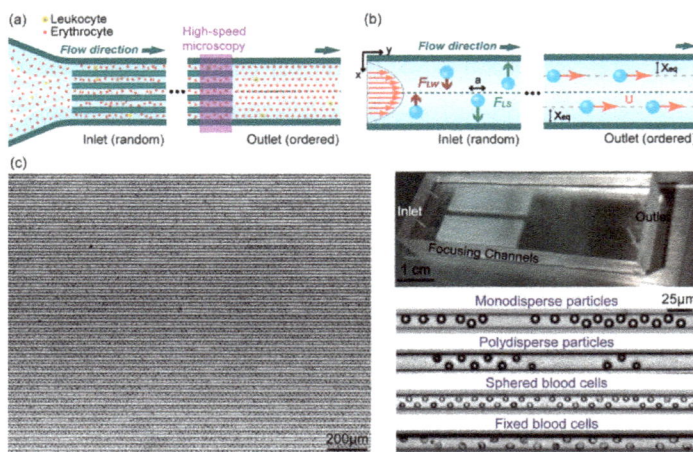

Figure 4. Inertial particle ordering in straight microchannel: (**a**) schematics shows a high-throughput cell ordering in parallel straight microchannels; (**b**) the competition of two opposite lift forces results in particle focusing at specific lateral positions; and (**c**) the massively parallel inertial microfluidic device and zoom-in images of particles and blood cells flowing in a single microchannel [39] (Reproduced with permissions from Hur et al., Lab on a Chip; published by Royal Society of Chemistry, 2010).

3.2. Inertial Particle Separation in Straight Microchannels

Inertial effects in straight microchannels can be used for particle separation (Figure 5) [53,54], which is based on the size-dependent migration velocities resulting from the different scalings between inertial lift (a^4) and viscous drag force (a). Rectangular microchannels are more desirable for particle separation due to the less equilibrium positions [38,55]. Mach et al. used a rectangular straight microchannel with a gradually expanded segment to separate *E. coli* bacteria from human blood samples [53]. The blood cells are focused along the side walls and enter into the branched outlets, whereas bacteria remain dispersed and largely flow into the main outlet. The focusing quality of blood cells degrades at large number densities due to the cell-cell interactions, which need to be minimized via the dilution of blood samples, typically requiring that the length of the equivalent single-cell train is less than 50% of the total microchannel length [33].

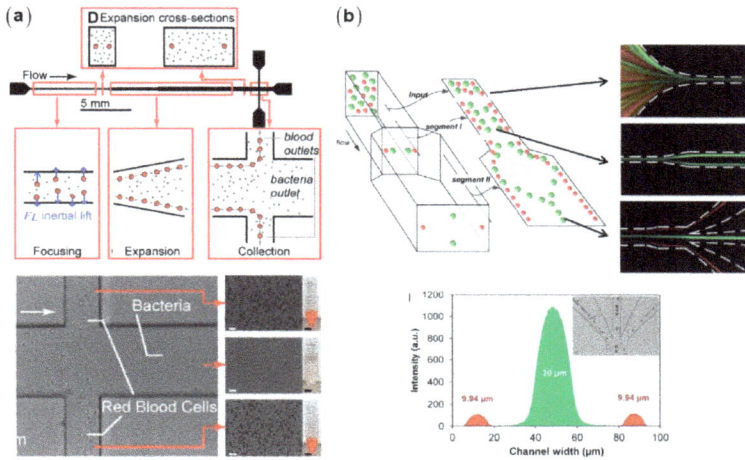

Figure 5. Inertial particle separation in straight microchannels: (**a**) red blood cells and *E. coli* bacteria [53] (Reproduced with permissions from Mach et al., Biotechnology and Bioengineering; published by Wiley Online Library, 2010); and (**b**) polystyrene particles with different diameters [54] (Reproduced with permissions from Zhou et al., Lab on a Chip; published by Royal Society of Chemistry, 2013).

3.3. High-Throughput Particle Transfer and Detection Based on Inertial Microfluidics

The rapid transfer of particles and cells between disparate solutions is important to diverse chemical and biological fields [56,57]. The controllable cell transfer across streams is always challenging as it often requires sophisticated flow control, finely tuned externally applied fields, or precisely manufactured structures. Using a microchannel with shifting aspect ratios ($AR = W/H$, where W is the channel width and H is the height), Gossett et al. realized simple and controllable cell transfer from the side wall to the centerline [42]. The flow rate ratio of transfer fluid to cell fluid is finely tuned to make the transfer fluid occupy the centerline of the main microchannel. Consequently, cells migrate across the interface and enter the transfer fluid at rates exceeding 1000 particles per second. Inertial microfluidics can be also applied to high-speed CTC analysis via cooperation with other high-speed devices [58,59]. Di Carlo's group made a portable CTC clinical detection system using a serpentine microchannel as the high speed cell focuser, achieving a throughput of 10^5 cells per second (Figure 6) [58,59].

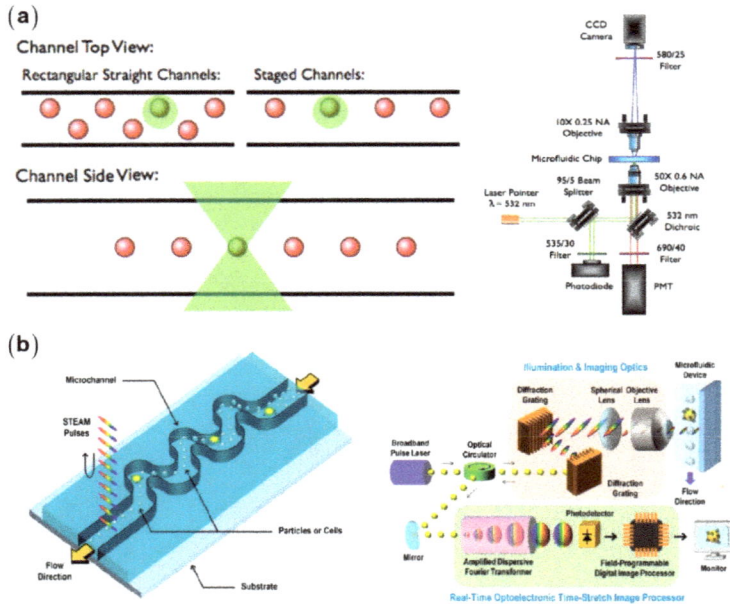

Figure 6. Single-cell imaging flow analyzer based on 3D inertial focusing: (**a**) particle positions relative to the focused laser beam in the cytometer apparatus; and (**b**) a flow analyzer that highlights a microfluidic particle focuser and a real-time imaging system [58,59] (Reproduced with permissions from Goda et al., Proceedings of the National Academy of Sciences of the United States of America; published by National Academy of Sciences of the United States of America, 2012 and Reproduced with permissions from Oakey et al., Analytical Chemistry; published by American Chemical Society, 2010).

3.4. Inertial Particle Separation in Curved Microchannels

Dean flow induced in curved microchannel exerts an additional drag force (Dean drag F_D) on particles, providing a more flexible separation principle. The competition between the inertial lift and the Dean drag results in size-dependent equilibrium positions of particles due to the different force scaling, i.e., $F_L \propto a^4$ and $F_D \propto a$. Dean flows are characterized by Dean number, $De = Re\sqrt{D/R}$, where R is the radius of curvature of the microchannel. The Dean velocity depends on the mainstream flow rate and the curvature of the microchannel $U_{Dean} \propto De^2\eta/(\rho D)$ [60]. Introducing Dean flow changes the particle migration by two aspects: (1) The Dean flow, parallel to the cross-section, can accelerate the migration toward the equilibrium positions, and thus shorten the microchannel length [48]. (2) The ratio of F_L to F_D generally determines particle behaviors: (1) $F_L \gg F_D$, particle migration is dominated solely by inertial lift and the focusing pattern is expected to be the same with that in straight microchannels; (2) $F_L \ll F_D$, particles just follow Dean flows neglecting the inertial lift and therefore no particle focusing can be achieved; and (3) $F_L \sim F_D$, inertial lift and Dean drag synergistically affect the particle migration and lead to different focusing patterns depending on the ratio of F_L to F_D. The applications of the inertial microfluidics lie in the third regime.

There are two types of curved microchannels commonly used for particle separation: spiral (Figure 7) [49–52,61–64] and serpentine microchannels (Figure 8) [9,47,65,66]. Using a single spiral microchannel, Kuntaegowdanahalli et al. successfully separated polystyrene (PS) beads with diameters of 10, 15, and 20 μm with an efficiency of 90% [50]. They further separated two types of tumor cells, SH-SY5Y neuroblastoma cells (average diameter of 15 μm) and C6 glioma cells (average diameter of 8 μm), with an efficiency of 80% and a throughput of 10^6 cells per minute [50]. Using a similar design,

Hou et al. separated CTCs from diluted whole human blood with an efficiency of 85% [67]. All the above devices require sheath flows, leading to limited sample throughputs. Sun et al. separated CTCs from human blood with a high throughput of 2.5×10^8 cells per minute and an efficiency of 90% using a high aspect-ratio microchannel with numerical optimization [68,69]. Bhagat and his collaborators designed a spiral microchannel with a trapezoidal cross-section for particle separation with enhanced resolution and throughput [51]. The trapezoidal design redistributes the Dean flow intensities and inertial lift forces, making the focusing patterns more sensitive to the particle size and the flow rate. The equilibrium positions can sharply shift with a large lateral distance at a size-dependent critical flow rate, leading to a large separation distance. This design is further used for high throughput separation of CTCs at a throughput of 56 mL blood per hour and a recovery of 80% [61].

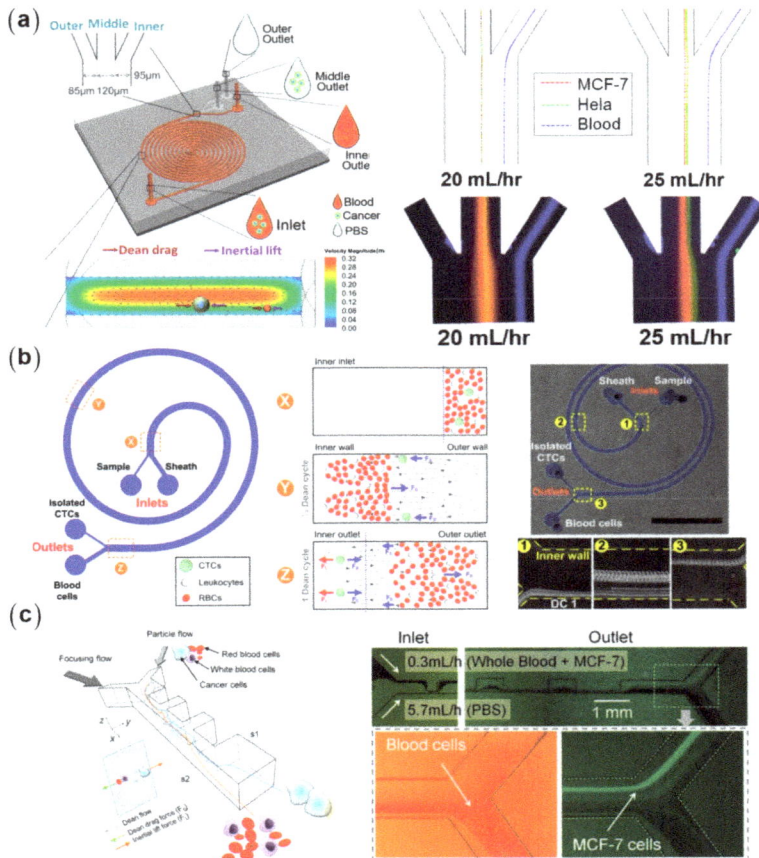

Figure 7. Inertial separation of circulating tumor cells in typical microchannel designs: (**a**) double spiral microchannel [68] (Reproduced with permissions from Sun et al., Lab on a Chip; published by Royal Society of Chemistry, 2012); (**b**) single spiral microchannel [61] (Reproduced with permissions from Warkiani et al., Lab on a Chip; published by Royal Society of Chemistry, 2014); and (**c**) expansion–contraction array microchannel [70] (Reproduced with permissions from Lee et al., Analytical Chemistry; published by American Chemical Society, 2013).

Serpentine microchannels have their curvatures frequently alternated compared with spiral ones, resulting in a more complex competition between the inertial lift and the Dean drag. On the other hand, serpentine microchannels exhibit better parallelizability. Di Carlo et al. adopted asymmetry serpentine microchannel for investigating the inertial focusing behaviors of red blood cells (RBCs) and found that: (1) RBCs suffer no discernable damage during the inertial manipulation; and (2) blockage ratios κ higher than 0.07 are required for successful particle focusing [9]. Based on the asymmetry serpentine design, they successfully separated PS beads with diameters of 9 and 3 μm and isolated platelets from whole human blood [47]. Serpentine microchannels are often designed to be asymmetric to avoid the offset of the counteracting secondary flows in the opposing segments. However, using a symmetry serpentine microchannel, Zhang et al. successfully separated PS beads with diameters of 10 and 3 μm. The 3 μm beads (κ = 0.04) were tightly focused in their microchannels with D of 70 μm, which is inconsistent with the previous claim by Di Carlo group [65,66]. This inconsistency indicates an incomplete understanding of inertial focusing mechanisms.

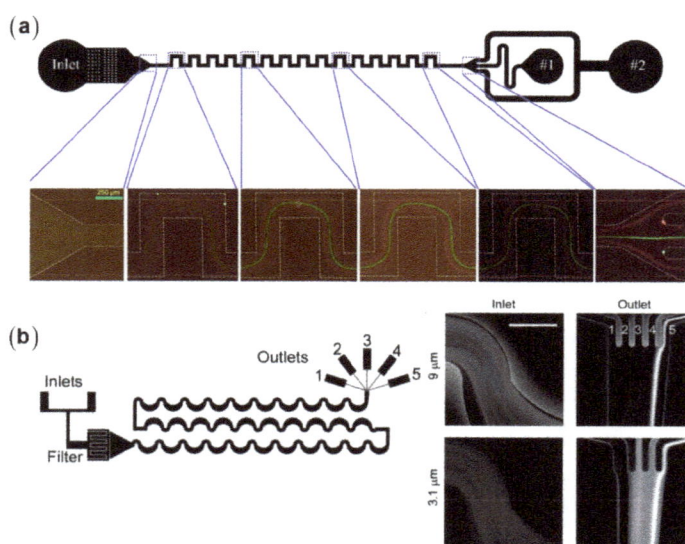

Figure 8. Particle separation in serpentine microchannels: (**a**) the separation of 10 μm and 3 μm polystyrene beads in a symmetry microchannel [65] (Reproduced with permissions from Zhang et al., Scientific Reports; published by Nature Publishing Group, 2014); and (**b**) the separation of 9.3 μm and 3.1 μm polystyrene beads in an asymmetry microchannel [47] (Reproduced with permissions from Di Carlo et al., Analytical Chemistry; published by American Chemical Society, 2008).

3.5. Fundamentals of Inertial Focusing and Recent Development

The optimization of inertial microfluidic devices often requires the evaluation of the focusing pattern of targeted particles, which is determined by the lift force distributions. Since Segre and Silberberg observed the inertial particle focusing in 1961 [37], a plenty of theoretical studies have been proposed to reveal its underlying hydrodynamic mechanism [30,31,71–77]. All these analytical studies were conducted by solving Navier–Stokes equations using the perturbation methods. Investigating the motion of a sphere in a two-dimensional Poiseuille flow, Ho and Leal obtained an explicit formula for the lift force: $F_L = C_L \rho U^2 a^4$, where the lift coefficient C_L is the function of the lateral position and is independent of the detailed undisturbed velocity profile [30]. Their lift formula can successfully explain the inertial focusing patterns in planar or tube Poiseuille flows. However, the restriction of

$Re \ll 1$ and $\kappa \ll 1$ limits its application to practical situations where Re is finite. Using the matched asymptotic perturbation method, Asmolov extended the applicable Re to 3000 (Figure 9) [31]. Asmolov and Matas calculated the lift forces at Re exceeding 1000 with the requisition of $Re_p \ll 1$ and found that with the increasing Re, the equilibrium position shifts closer to the channel wall and the magnitude of C_L decreases, which is consistent with existing experimental and numerical results [31,78]. However, it is still difficult to directly apply these theoretical studies to the realistic cases of finite-sized particles (intermediate κ), where the particles strongly affect the ambient flow field and cause strong nonlinearity, casting challenge on the theoretical analysis.

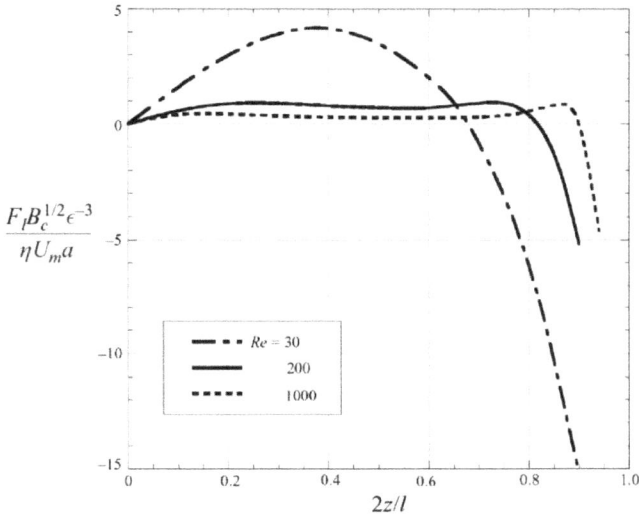

Figure 9. The distributions of inertial lift coefficients at Re ranging from 30 to 1000 showing that the lift profile exhibits a concavity change at higher Re [78] (Reproduced with permissions from Matas et al., Journal of Fluid Mechanics; published by Cambridge University Press, 2004).

Direct numerical simulation (DNS) is able to investigate the motions and acting forces of particles without simplified models [79,80]. In DNS, the acting hydrodynamic force on a particle is calculated by integrating the total stress over its surface. Di Carlo et al. obtained a position-dependent scaling for inertial lift in square channels: $F_L \propto \rho U^2 a^3 / H$ near the channel center and $F_L \propto \rho U^2 a^6 / H^3$ near the channel wall (Figure 10) [41], which is different from the uniform scaling law obtained by theoretical calculations. Using the arbitrary Lagrangian–Eulerian method (ALE), Joseph's group investigated the inertial lift on a sphere in a slit and a circular tube [81] and found that the lift profile exhibits a convexity change at higher Re ($Re > 300$). Here positive lift force directs toward the channel wall while the negative one directs toward the channel center. At low Re ($Re \sim O(10)$), the lift curve concaves downwards with a positive slope near the channel center and a negative slope near the channel wall. At high Re ($Re > 300$), the convexity becomes more complex: two concave-downwards segments occur at the channel center with the equilibrium positions with a maximum in each segment and a concave-upwards segment lies between the two concave-downwards ones. Asmolov's theoretical calculation also obtained a similar convexity change for $Re > 300$ [31]. Matas et al. experimentally observed a distinguished double-ring focusing pattern in a tube at $Re > 760$, which can be explained by the convexity change at high Re [78,82].

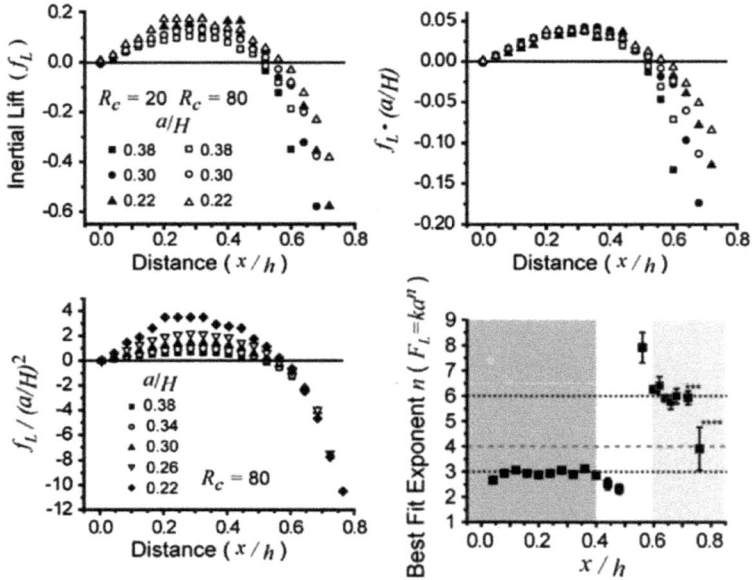

Figure 10. The complex scaling of the inertial lift on finite-sized particles in square microchannels: $F_L \propto \rho U^2 a^3/H$, near the channel center and $F_L \propto \rho U^2 a^6/H^3$ near the channel wall [41] (Reproduced with permissions from Di Carlo et al., Physics Review Letters; published by American Physical Society, 2009).

Microchannels fabricated by the planar soft-lithography methods commonly have square or rectangular cross-sections [83]. In addition to the common four off-center focusing positions near each channel wall, more complex focusing patterns have been observed in square microchannels (Figure 11). Using the lattice Boltzmann method, Chun et al. found eight equilibrium positions in square channels at $Re = 100$ [84]. The similar focusing pattern was also observed in the experiments by Bhagat et al. [85]. The four equilibrium positions near the channel centers disappear at Re exceeding 500 [84]. In rectangular microchannels, particles are typically focused near the centers of the long walls [39–42,53,86,87]. This reduction in equilibrium position makes rectangular microchannels more favorable for particle focusing and separation. However, six or even eight positions have also been observed in rectangular microchannels with similar AR [85,88]. Zhou et al. experimentally showed that the rotation-induced forces play a role in particle migration toward the centers of the long walls [89]. However, the rotation-induced force always directs toward the center of the channel walls, and thus cannot explain the multiple equilibrium positions near the long walls. Gossett et al. numerically investigated the inertial migration of particles at $Re = 80$ in a microchannel with $AR = 2$ [42]. They found two stable equilibrium positions centered at the long walls and two unstable ones centered at the short walls, which is consistent with the typical focusing pattern.

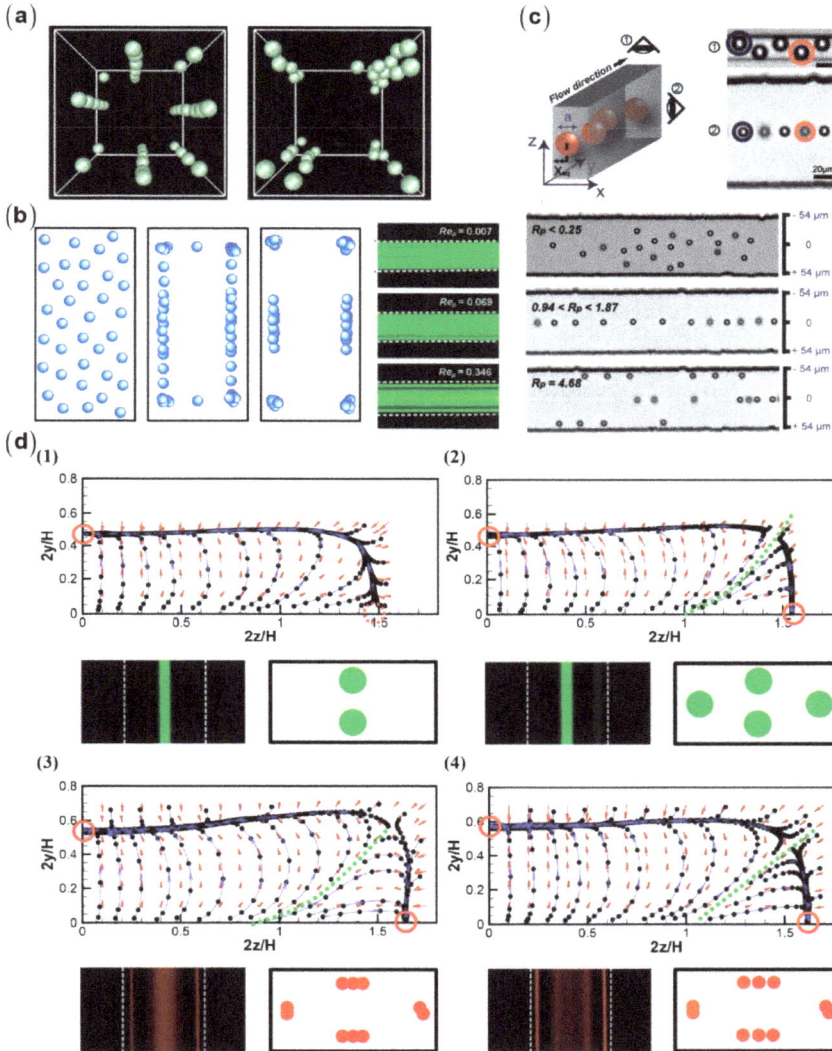

Figure 11. Inertial focusing patterns in rectangular microchannels: (**a**) $AR = 1$, $Re = 100 - 500$, $\kappa = 0.11$ [84] (Reproduced with permissions from Chun et al., Physics of Fluids; published by American Institute of Physics Publishing, 2006); (**b**) $AR = 2$, $Re = 1 - 100$, $\kappa = 0.08$ [85] (Reproduced with permissions from Bhagat et al., Physics of Fluids; published by American Institute of Physics Publishing, 2008); (**c**) $AR = 2$, $Re = 0 - 230$, $\kappa = 0.2$ [39] (Reproduced with permissions from Hur et al., Lab on a Chip; published by Royal Society of Chemistry, 2010); and (**d**) (**1**) $Re = 100$, $\kappa = 0.3$; (**2**) $Re = 200$, $\kappa = 0.3$; (**3**) $Re = 100$, $\kappa = 0.1$; and (**4**) $Re = 200$, $\kappa = 0.1$ [90] (Reproduced with permissions from Liu et al., Lab on a Chip; published by Royal Society of Chemistry, 2015).

Using combination of numerical simulation and experiments, Hu's group systematically investigated the inertial focusing patterns for a wide range of Re, κ, and AR [90]. The typical focusing near the centers of the long walls in rectangular microchannels is obtained at relative low Re. New stable equilibrium positions will emerge at high *Re* due to the stabilization of the sub-stable equilibrium near the centers of the short walls or due to the attractive lift forces near the long walls. The critical *Re* decreases with κ for fixed *AR* and decreases with *AR* for fixed κ. Although it has provided insights for the fundamentals of inertial focusing, DNS is still burdensome when applied to practical long microchannels with complex geometries. Hu's group proposed a fitting formula for the inertial lift on a sphere drawn from DNS data obtained in straight channels [91]. The fitting formula is a function of the parameters of the local flow field, and thus is adaptable to complex microchannels. Being implemented in the Lagrangian particle tracking method, the formula is used to fast predict particle trajectories in some widely used microchannel types (Figure 12).

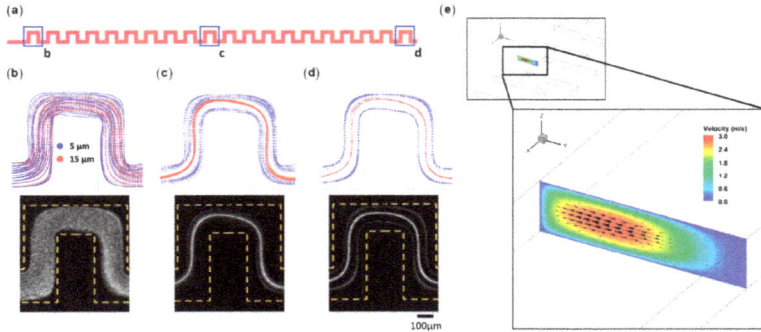

Figure 12. The particle trajectories in a serpentine microchannel calculated by Lagrangian particle tracking based on the explicit lift formula: (**a**) the schematics of the serpentine microchannel; (**b**–**d**) simulation (top row) and experimental observation (bottom row) of particle trajectories are shown at the: (**b**) 1st unit; (**c**) 10th unit; and (**d**) 20th unit; and (**e**) the Dean vortex forming in the zigzag section at the flow rate of 50 mL/h (*Re* ≈ 120) [91] (Reproduced with permissions from Liu et al., *Lab on a Chip*; published by Royal Society of Chemistry, 2016).

4. Viscoelastic Manipulation of Particles

High pressure is needed to generate an inertial flow in a scaled-down microchannel for the manipulation of smaller particles [92]. By contrast, deterministic particle migration can be obtained in viscoelastic fluids even at very low flow speeds, avoiding high pressure drops across the microchannels [93,94]. In addition, the synergetic combination of inertial lift and elastic lift can achieve a real 3D focusing at the microchannel centerline [95]. Most naturally-occurring biochemical samples, such as blood, lymph, saliva, and protein solutions, are viscoelastic [96]. Therefore, the elasto-inertial microfluidic particle manipulation may have wide applications in many biochemical fields.

4.1. Fundamentals of Particle Migration in Viscoelstic Fluids

Viscoelastic microfluidics typically use the aqueous solutions of synthetic or naturally-occurring polymers as carrier mediums. The most commonly used polymers are poly(ethylene oxide) (PEO) and poly(vinylpyrrolidone) (PVP), which have good water solubility. In a sheared or stretched flow, the polymer chains are elongated along the flow direction, causing stress anisotropy, i.e., the non-zero normal stress differences [97]. The first and second normal stress differences are defined as $N_1 = \sigma_{xx} - \sigma_{yy}$ and $N_2 = \sigma_{yy} - \sigma_{zz}$, respectively (here *x*, *y*, and *z* denote the direction of the flow, velocity gradient, and vorticity, respectively). The non-dimensional Weissenberg number (*Wi*) characterizes the fluid elasticity, which is the ratio of the first normal stress difference to the viscous shear stress. *Wi* can be expressed as $\dot{\gamma}\lambda$ using Oldroyd-B constitutive model, where $\dot{\gamma}$ is the shear rate and λ the relaxation time. Using the regular perturbation method, Ho and Leal calculated the lift force on a small sphere (κ ≪ 1) suspended in an inertialess Poiseuille flow of a second-order fluid, showing that the elastic lift stems from the imbalance of normal stress difference over the sphere size [93], i.e., $F_e \sim a^3 \nabla N_1$ (here the effect of N_2 is neglected as $|N_2/N_1| \leq 0.1$ for most polymer solutions [98]). Oldroyd-B model gives the positive value of $N_1 = 2\eta\lambda\dot{\gamma}^2$, indicating compressive normal stresses on the sphere surface that become stronger with increasing local shear rate. In a non-uniformly sheared flow, a net lift force will drive particles toward the positions where $\dot{\gamma}$ have minimums.

4.2. Particle Manipulation in Viscoelastic Microfluidics Devices

In square or rectangular microchannels, particles are focused at the center and four corners [95,99,100]. For *Re* of O(1), particles are only focused along the channel centerline due to synergetic combination of the elastic lift and the wall-induced lift [95,101,102]. This simple focusing pattern can realize particle separation by size difference via sheath flows, curved microchannels, or embedded structures (Figure 13). Using a sheath flow to prefocus different-sized particles along the sidewall, they can be separated due to lateral velocity difference determined by the balance between the elastic lift ($F_e \sim a^3$) and viscous drag ($F_d \sim a$). Separation using sheath flow can work at a relatively wide range of flow rates, but has limited throughputs due to high ratio of sheath flow to sample flow rates. Therefore, whether or not using sheath flows depends on the specific separation task. In addition, inducing secondary flow in curved microchannels to compete with elastic lift can also achieve size-based separation of particles [103,104].

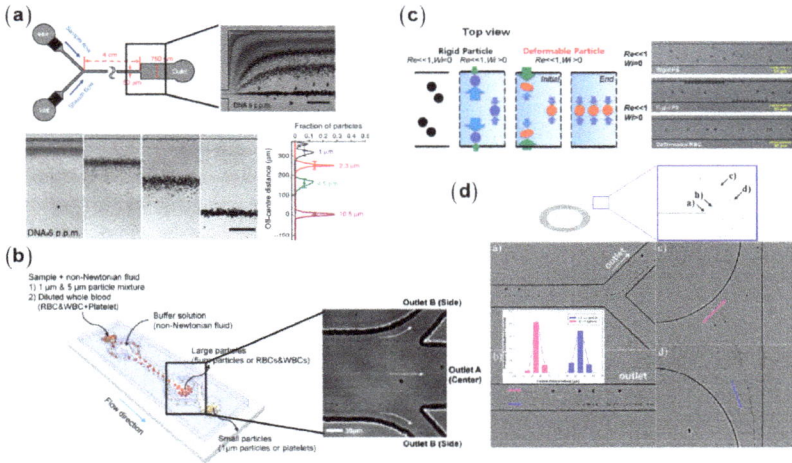

Figure 13. Viscoelastic particle separation: (**a**) separation of PS particles of different diameters with the aid of sheath flow [105] (Reproduced with permissions from Kang et al., Nature Communications; published by Nature Publishing Group, 2013); (**b**) separation of red blood cells and platelets with the aid of sheath flow [101] (Reproduced with permissions from Nam et al., Lab on a Chip; published by Royal Society of Chemistry, 2012); (**c**) separation of rigid and deformable cells [99] (Reproduced with permissions from Yang et al., Soft Matter; published by Royal Society of Chemistry, 2012); and (**d**) separation of PS particles with different diameters in curved microchannel [103] (Reproduced with permissions from Lee et al., Scientific Reports; published by Nature Publishing Group, 2013).

Most viscoelastic particle separations are based on the size-dependent migration velocities. The particle size itself can also affect the focusing pattern. Liu et al. found that 15 µm particles are focused at the both sides of the centerline of a 50-µm-high microchannel at optimized flow rates [106]. By contrast, 5 µm particles are always focused closer to the centerline than 15 µm ones at all investigated flow rates. The mechanism of the off-center focusing relates to the strong coupling between the large particles and the ambient flow field. When a large particle deviates from the channel centerline, the major portion of the viscoelastic fluid chooses to flow through the larger gap between the particle and the channel wall [107]. Therefore, the shear rates and the resultant compressive normal stresses are intensified at the near-center side of the particles, and consequently the particles are driven toward the wall [80]. This off-center focusing pattern occurs for a wide range of *AR* and is determined by the ratio of particle diameter to the narrowest channel dimension. In addition, this off-center focusing can be obtained at a wide range of scales, from macro- to nanoscales, realizing sheathless separations of diverse particles including microparticles, CTCs, RBCs, bacteria, nanoparticles, and biomoleculars (Figure 14) [91,106].

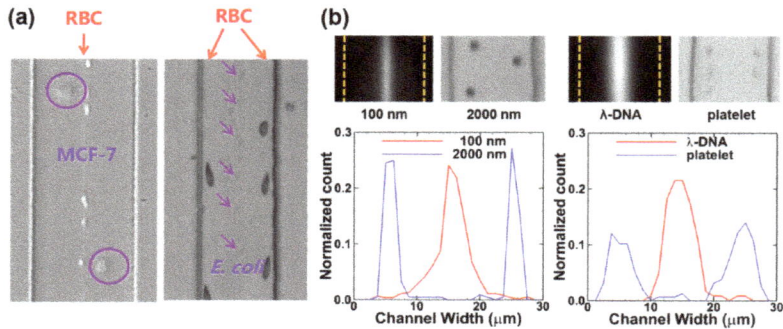

Figure 14. Sheathless separations of diverse binary bioparticle mixtures: (**a**) the separation of the mixture of MCF-7/RBC (left) and RBC/*E. coli* (right) [106] (Reproduced with permissions from Liu et al., *Analytical Chemistry*; published by American Chemical Society, 2015); and (**b**) the separations of smaller binary mixtures of 100 nm/2000 nm polystyrene spheres (left) and λ-DNA/platelet (right) [91] (Reproduced with permissions from Liu et al., *Analytical Chemistry*; published by American Chemical Society, 2016).

Viscoelastic focusing generally works at 1–2 orders of magnitude slower flow rate than inertial focusing due to the focusing degradation at higher flow rates [95]. There are two defocusing mechanisms and each of them corresponds to a solution. One is that the arising shear-gradient-induced lift drives particles away from the channel centerline, indicating that elasticity needs to be enhanced to balance the shear-gradient-induced lift. Kang et al. used a highly elastic medium (5 ppm λ-DNA solution, $\lambda = 0.14$ s) to achieve tight particle focusing over a wide range of flow rates (0.005–2 mL/h) [105]. The other mechanism is that the flow is destabilized when elasticity solely dominates at high *Wi* [108,109]. If fluid inertia and elasticity simultaneously dominate, the flow will keep stable even at the regime in which turbulent flow in the Newtonian fluid is observed. Lim et al. reported particle focusing at extremely high flow rates (1200 mL/h, *Re* = 4630, *Wi* = 566) using a medium with low relaxation time (0.1 *w/v* % hyaluronic acid (HA) solution, $\lambda = 8.7 \times 10^{-4}$ s) [102]. There seems to be contradiction between these two mechanisms, implying an inconclusive understanding of particle migration in viscoelastic medium.

Considering the high price of DNA and HA, existing studies are more focused on optimizing the performance of cheap synthetic polymers, such as PEO and PVP. Liu et al. found that an optimized polymer solution for particle manipulation should have low viscosity, minimized shear shinning, and strong elasticity. The shear thinning is ubiquitous for polymer solutions, especially at high polymer concentration or large molecular weight [106]. Therefore, a trade-off should be properly made between the minimized shear thinning and the necessarily strong elasticity. Liu et al. used naturally denatured PEO solution (4×10^6 Da) to successfully focus 5 μm particles in a 50 μm high microchannel at throughputs one order of magnitude higher than those of newly prepared PEO solutions [106] (Table 2). Compared with its newly prepared counterpart, the denaturized PEO solution has lower viscosities and much weaker shear thinning and still remains highly elastic, eliminating the focusing degradation at higher flow rates.

Table 2. Comparison of recent works on particle focusing and separation using viscoelastic solutions.

	Study	Minimum Particle Size for Successful Manipulation (μm)	Minimum Effective Blockage Ratio	Sample Flow Rate (μL/h)	Focusing Efficiency (%)	Separation Efficiency (%)	Channel Geometry and Footprint	Journal	Manipulation Type
Microparticle	Leshansky et al. [94]	5	0.11	400–2000	>95	N/A	Straight; N/A	Physical Review Letters	Viscoelastic focusing
	Kang et al. [105]	5.8	0.116	5–2000	~100	N/A	Straight; 50 mm long	Nature Communications	Viscoelastic focusing/separation
	Lim et al. [110]	6	0.075	3×10^6	~90	N/A	Straight; 35 mm long	Nature Communications	Elasto-inertial focusing
	Lu et al. [111]	3	0.06	~O(100)	N/A	~100	Straight; 20 mm long	Analytical Chemistry	Elasto-inertial pinched flow fractionation separation
	Liu et al. [106]	1	0.063	10–3000	~100	~100	Straight; 30 mm long	Analytical Chemistry	Elasto-inertial focusing/separation
	Lee et al. [103]	1.5	0.038	~O(100)	N/A	~100	Spiral; 500 mm long	Scientific Reports	Viscoelastic separation
	Yuan et al. [112]	3	0.053	600–4800	~100	~100	Straight; 48 mm long	Lab on a Chip	Elasto-inertial focusing/separation
	Nam et al. [101]	5	0.1	30	N/A	~100	Straight; 25 mm long	Lab on a Chip	Elasto-inertial focusing/separation
	Yang et al. [95]	5.9	0.118	40–320	N/A	>95	Straight; 40 mm long	Lab on a Chip	Elasto-inertial focusing
Nanoparticle	De Santo et al. [113]	0.2	0.04	0.002–0.016	Low: multiple streams	N/A	Straight; 100 mm long	Physical Review Applied	Viscoelastic focusing
	Kim et al. [114]	0.2	0.04	<0.96	85	N/A	Straight; 40 mm long	Lab on a Chip	Viscoelastic focusing
	Liu et al. [91]	0.1	0.014	0.32–2.45	84	>95	Double spiral; >60 mm long; 3×3 mm^2	Analytical Chemistry	Viscoelastic focusing/separation

5. Conclusions

High-throughput particle manipulation is of significance in diverse applications in biological, biomedical, and environmental fields, requiring the ability to handle large volumes of samples. The last decade has seen the rise of inertial microfluidics as a novel tool for high-throughput and label-free particle manipulation utilizing the controllable particle migration driven by inertial lift. However, the pure inertial manipulation achieves its optimal performance often at a narrow flow rate range and faces challenges in handling nanoparticles in down-scaled channels, limiting its widespread usage for the diverse applications. Introducing the viscoelastic effects of carrier medium, the elasto-inertial microfluidic devices can intensively extend the working flow rate range and reduce the applicable particle size. However, there is still a lack of conclusive understanding of particle motion in viscoelastic medium, leading to poor guidelines and difficulties in the design of elasto-inertial microfluidic systems. Systematic studies are urgently needed to elucidate the effects of complex rheological properties, channel geometry, and particle properties on the focusing pattern. Despite the success in laboratories, there are not many reported handling of clinical samples based on the inertial/elasto-inertial concept. To achieve better adaptability to realistic applications, a promising strategy is to couple inertial/elasto-inertial effects with other externally applied physical fields, such as electric, magnetic, and acoustic ones, to further improve the purity and resolution.

Acknowledgments: This work was supported financially by National Natural Science Foundation of China (NSFC) (11572334, 91543125, and 11272321), the CAS Key Research Program of Frontier Sciences (QYZDB-SSW-JSC036), and the CAS Strategic Priority Research Program (XDB22040403) to Guoqing Hu.

Author Contributions: C.L. and G.H. wrote the paper.

Conflicts of Interest: The authors declare no conflict of interest.

References

1. Nolan, J.P.; Sklar, L.A. The emergence of flow cytometry for sensitive, real-time measurements of molecular interactions. *Nat. Biotechnol.* **1998**, *16*, 633–638. [CrossRef] [PubMed]
2. Toner, M.; Irimia, D. Blood-on-a-chip. *Annu. Rev. Biomed. Eng.* **2005**, *7*, 77–103. [CrossRef] [PubMed]
3. Plaks, V.; Koopman, C.D.; Werb, Z. Circulating tumor cells. *Science* **2013**, *341*, 1186–1188. [CrossRef] [PubMed]
4. Yu, M.; Bardia, A.; Aceto, N.; Bersani, F.; Madden, M.W.; Donaldson, M.C.; Desai, R.; Zhu, H.; Comaills, V.; Zheng, Z.; et al. Ex vivo culture of circulating breast tumor cells for individualized testing of drug susceptibility. *Science* **2014**, *345*, 216–220. [CrossRef] [PubMed]
5. Nagrath, S.; Sequist, L.V.; Maheswaran, S.; Bell, D.W.; Irimia, D.; Ulkus, L.; Smith, M.R.; Kwak, E.L.; Digumarthy, S.; Muzikansky, A.; et al. Isolation of rare circulating tumour cells in cancer patients by microchip technology. *Nature* **2007**, *450*, 1235–1239. [CrossRef] [PubMed]
6. Hosokawa, M.; Hayata, T.; Fukuda, Y.; Arakaki, A.; Yoshino, T.; Tanaka, T.; Matsunaga, T. Size-selective microcavity array for rapid and efficient detection of circulating tumor cells. *Anal. Chem.* **2010**, *82*, 6629–6635. [CrossRef] [PubMed]
7. Went, P.T.H.; Lugli, A.; Meier, S.; Bundi, M.; Mirlacher, M.; Sauter, G.; Dirnhofer, S. Frequent epcam protein expression in human carcinomas. *Hum. Pathol.* **2004**, *35*, 122–128. [CrossRef] [PubMed]
8. Lalmahomed, Z.S.; Kraan, J.; Gratama, J.W.; Mostert, B.; Sleijfer, S.; Verhoef, C. Circulating tumor cells and sample size: The more, the better. *J. Clin. Oncol.* **2010**, *28*, e288–e289. [CrossRef] [PubMed]
9. Di Carlo, D.; Irimia, D.; Tompkins, R.G.; Toner, M. Continuous inertial focusing, ordering, and separation of particles in microchannels. *Proc. Natl. Acad. Sci. USA* **2007**, *104*, 18892–18897. [CrossRef] [PubMed]
10. Ding, X.; Peng, Z.; Lin, S.-C.S.; Geri, M.; Li, S.; Li, P.; Chen, Y.; Dao, M.; Suresh, S.; Huang, T.J. Cell separation using tilted-angle standing surface acoustic waves. *Proc. Natl. Acad. Sci. USA* **2014**, *111*, 12992–12997. [CrossRef] [PubMed]
11. Hu, X.; Bessette, P.H.; Qian, J.; Meinhart, C.D.; Daugherty, P.S.; Soh, H.T. Marker-specific sorting of rare cells using dielectrophoresis. *Proc. Natl. Acad. Sci. USA* **2005**, *102*, 15757–15761. [CrossRef] [PubMed]
12. Pamme, N. Magnetism and microfluidics. *Lab Chip* **2006**, *6*, 24–38. [CrossRef] [PubMed]

13. Ashkin, A. Acceleration and trapping of particles by radiation pressure. *Phys. Rev. Lett.* **1970**, *24*, 156–159. [CrossRef]
14. Arosio, P.; Mueller, T.; Mahadevan, L.; Knowles, T.P.J. Density-gradient-free microfluidic centrifugation for analytical and preparative separation of nanoparticles. *Nano Lett.* **2014**, *14*, 2365–2371. [CrossRef] [PubMed]
15. Takagi, J.; Yamada, M.; Yasuda, M.; Seki, M. Continuous particle separation in a microchannel having asymmetrically arranged multiple branches. *Lab Chip* **2005**, *5*, 778–784. [CrossRef] [PubMed]
16. Larson, R.G. The rheology of dilute solutions of flexible polymers: Progress and problems. *J. Rheol.* **2005**, *49*, 1–70. [CrossRef]
17. Yamada, M.; Seki, M. Hydrodynamic filtration for on-chip particle concentration and classification utilizing microfluidics. *Lab Chip* **2005**, *5*, 1233–1239. [CrossRef] [PubMed]
18. Kim, M.; Mo Jung, S.; Lee, K.-H.; Jun Kang, Y.; Yang, S. A microfluidic device for continuous white blood cell separation and lysis from whole blood. *Artif. Organs* **2010**, *34*, 996–1002. [CrossRef] [PubMed]
19. Huang, L.R.; Cox, E.C.; Austin, R.H.; Sturm, J.C. Continuous particle separation through deterministic lateral displacement. *Science* **2004**, *304*, 987–990. [CrossRef] [PubMed]
20. Wunsch, B.H.; Smith, J.T.; Gifford, S.M.; Wang, C.; Brink, M.; Bruce, R.L.; Austin, R.H.; Stolovitzky, G.; Astier, Y. Nanoscale lateral displacement arrays for the separation of exosomes and colloids down to 20 nm. *Nat. Nanotechnol.* **2016**, *11*, 936–940. [CrossRef] [PubMed]
21. Choi, S.; Park, J.-K. Continuous hydrophoretic separation and sizing of microparticles using slanted obstacles in a microchannel. *Lab Chip* **2007**, *7*, 890–897. [CrossRef] [PubMed]
22. Choi, S.; Song, S.; Choi, C.; Park, J.-K. Sheathless focusing of microbeads and blood cells based on hydrophoresis. *Small* **2008**, *4*, 634–641. [CrossRef] [PubMed]
23. Choi, S.; Park, J.-K. Sheathless hydrophoretic particle focusing in a microchannel with exponentially increasing obstacle arrays. *Anal. Chem.* **2008**, *80*, 3035–3039. [CrossRef] [PubMed]
24. Choi, S.; Ku, T.; Song, S.; Choi, C.; Park, J.-K. Hydrophoretic high-throughput selection of platelets in physiological shear-stress range. *Lab Chip* **2011**, *11*, 413–418. [CrossRef] [PubMed]
25. Beech, J.P.; Holm, S.H.; Adolfsson, K.; Tegenfeldt, J.O. Sorting cells by size, shape and deformability. *Lab Chip* **2012**, *12*, 1048–1051. [CrossRef] [PubMed]
26. McGrath, J.; Jimenez, M.; Bridle, H. Deterministic lateral displacement for particle separation: A review. *Lab Chip* **2014**, *14*, 4139–4158. [CrossRef] [PubMed]
27. Davis, J.A.; Inglis, D.W.; Morton, K.J.; Lawrence, D.A.; Huang, L.R.; Chou, S.Y.; Sturm, J.C.; Austin, R.H. Deterministic hydrodynamics: Taking blood apart. *Proc. Natl. Acad. Sci. USA* **2006**, *103*, 14779–14784. [CrossRef] [PubMed]
28. Loutherback, K.; D'Silva, J.; Liu, L.; Wu, A.; Austin, R.H.; Sturm, J.C. Deterministic separation of cancer cells from blood at 10 mL/min. *AIP Adv.* **2012**, *2*, 042107. [CrossRef] [PubMed]
29. Inglis, D.W.; Riehn, R.; Austin, R.H.; Sturm, J.C. Continuous microfluidic immunomagnetic cell separation. *Appl. Phys. Lett.* **2004**, *85*, 5093–5095. [CrossRef]
30. Ho, B.P.; Leal, L.G. Inertial migration of rigid spheres in 2-dimensional unidirectional flows. *J. Fluid Mech.* **1974**, *65*, 365–400. [CrossRef]
31. Asmolov, E.S. The inertial lift on a spherical particle in a plane poiseuille flow at large channel reynolds number. *J. Fluid Mech.* **1999**, *381*, 63–87. [CrossRef]
32. Martel, J.M.; Toner, M. Inertial focusing in microfluidics. *Annu. Rev. Biomed. Eng.* **2014**, *16*, 371–396. [CrossRef] [PubMed]
33. Amini, H.; Lee, W.; Di Carlo, D. Inertial microfluidic physics. *Lab Chip* **2014**, *14*, 2739–2761. [CrossRef] [PubMed]
34. Dongeun, H.; Wei, G.; Yoko, K.; James, B.G.; Shuichi, T. Microfluidics for flow cytometric analysis of cells and particles. *Physiol. Meas.* **2005**, *26*, R73.
35. Chung, T.D.; Kim, H.C. Recent advances in miniaturized microfluidic flow cytometry for clinical use. *Electrophoresis* **2007**, *28*, 4511–4520. [CrossRef] [PubMed]
36. Ateya, D.; Erickson, J.; Howell, P., Jr.; Hilliard, L.; Golden, J.; Ligler, F. The good, the bad, and the tiny: A review of microflow cytometry. *Anal. Bioanal. Chem.* **2008**, *391*, 1485–1498. [CrossRef] [PubMed]
37. Segre, G.; Silberberg, A. Radial particle displacements in poiseuille flow of suspensions. *Nature* **1961**, *189*, 209–210. [CrossRef]

38. Choi, Y.-S.; Seo, K.-W.; Lee, S.-J. Lateral and cross-lateral focusing of spherical particles in a square microchannel. *Lab Chip* **2011**, *11*, 460–465. [CrossRef] [PubMed]
39. Hur, S.C.; Tse, H.T.K.; Di Carlo, D. Sheathless inertial cell ordering for extreme throughput flow cytometry. *Lab Chip* **2010**, *10*, 274–280. [CrossRef] [PubMed]
40. Edd, J.F.; Di Carlo, D.; Humphry, K.J.; Koster, S.; Irimia, D.; Weitz, D.A.; Toner, M. Controlled encapsulation of single-cells into monodisperse picolitre drops. *Lab Chip* **2008**, *8*, 1262–1264. [CrossRef] [PubMed]
41. Di Carlo, D.; Edd, J.; Humphry, K.; Stone, H.; Toner, M. Particle segregation and dynamics in confined flows. *Phys. Rev. Lett.* **2009**, *102*, 094503. [CrossRef] [PubMed]
42. Gossett, D.R.; Tse, H.T.K.; Dudani, J.S.; Goda, K.; Woods, T.A.; Graves, S.W.; Di Carlo, D. Inertial manipulation and transfer of microparticles across laminar fluid streams. *Small* **2012**, *8*, 2757–2764. [CrossRef] [PubMed]
43. Chung, A.J.; Gossett, D.R.; Di Carlo, D. Three dimensional, sheathless, and high-throughput microparticle inertial focusing through geometry-induced secondary flows. *Small* **2013**, *9*, 685–690. [CrossRef] [PubMed]
44. Chung, A.J.; Pulido, D.; Oka, J.C.; Amini, H.; Masaeli, M.; Di Carlo, D. Microstructure-induced helical vortices allow single-stream and long-term inertial focusing. *Lab Chip* **2013**, *13*, 2942–2949. [CrossRef] [PubMed]
45. Lee, M.G.; Choi, S.; Park, J.K. Inertial separation in a contraction-expansion array microchannel. *J. Chromatogr. A* **2011**, *1218*, 4138–4143. [CrossRef] [PubMed]
46. Park, J.S.; Song, S.H.; Jung, H.I. Continuous focusing of microparticles using inertial lift force and vorticity via multi-orifice microfluidic channels. *Lab Chip* **2009**, *9*, 939–948. [CrossRef] [PubMed]
47. Di Carlo, D.; Edd, J.F.; Irimia, D.; Tompkins, R.G.; Toner, M. Equilibrium separation and filtration of particles using differential inertial focusing. *Anal. Chem.* **2008**, *80*, 2204–2211. [CrossRef] [PubMed]
48. Gossett, D.R.; Di Carlo, D. Particle focusing mechanisms in curving confined flows. *Anal. Chem.* **2009**, *81*, 8459–8465. [CrossRef] [PubMed]
49. Bhagat, A.A.S.; Kuntaegowdanahalli, S.S.; Papautsky, I. Continuous particle separation in spiral microchannels using dean flows and differential migration. *Lab Chip* **2008**, *8*, 1906. [CrossRef] [PubMed]
50. Kuntaegowdanahalli, S.S.; Bhagat, A.A.S.; Kumar, G.; Papautsky, I. Inertial microfluidics for continuous particle separation in spiral microchannels. *Lab Chip* **2009**, *9*, 2973–2980. [CrossRef] [PubMed]
51. Guan, G.F.; Wu, L.D.; Bhagat, A.A.S.; Li, Z.R.; Chen, P.C.Y.; Chao, S.Z.; Ong, C.J.; Han, J.Y. Spiral microchannel with rectangular and trapezoidal cross-sections for size based particle separation. *Sci. Rep.* **2013**, *3*, 1475. [CrossRef] [PubMed]
52. Martel, J.M.; Toner, M. Particle focusing in curved microfluidic channels. *Sci. Rep.* **2013**, *3*, 3340. [CrossRef]
53. Mach, A.J.; Di Carlo, D. Continuous scalable blood filtration device using inertial microfluidics. *Biotechnol. Bioeng.* **2010**, *107*, 302–311. [CrossRef] [PubMed]
54. Zhou, J.; Giridhar, P.V.; Kasper, S.; Papautsky, I. Modulation of aspect ratio for complete separation in an inertial microfluidic channel. *Lab Chip* **2013**, *13*, 1919–1929. [CrossRef] [PubMed]
55. Humphry, K.J.; Kulkarni, P.M.; Weitz, D.A.; Morris, J.F.; Stone, H.A. Axial and lateral particle ordering in finite reynolds number channel flows. *Phys. Fluids* **2010**, *22*, 081703. [CrossRef]
56. Duda, D.G.; Cohen, K.S.; Scadden, D.T.; Jain, R.K. A protocol for phenotypic detection and enumeration of circulating endothelial cells and circulating progenitor cells in human blood. *Nat. Protoc.* **2007**, *2*, 805–810. [CrossRef] [PubMed]
57. Pedersen, S.L.; Tofteng, A.P.; Malik, L.; Jensen, K.J. Microwave heating in solid-phase peptide synthesis. *Chem. Soc. Rev.* **2012**, *41*, 1826–1844. [CrossRef] [PubMed]
58. Goda, K.; Ayazi, A.; Gossett, D.R.; Sadasivam, J.; Lonappan, C.K.; Sollier, E.; Fard, A.M.; Hur, S.C.; Adam, J.; Murray, C.; et al. High-throughput single-microparticle imaging flow analyzer. *Proc. Natl. Acad. Sci. USA* **2012**, *109*, 11630–11635. [CrossRef] [PubMed]
59. Oakey, J.; Applegate, R.W.; Arellano, E.; Carlo, D.D.; Graves, S.W.; Toner, M. Particle focusing in staged inertial microfluidic devices for flow cytometry. *Anal. Chem.* **2010**, *82*, 3862–3867. [CrossRef] [PubMed]
60. Squires, T.M.; Quake, S.R. Microfluidics: Fluid physics at the nanoliter scale. *Rev. Mod. Phys.* **2005**, *77*, 977–1026. [CrossRef]
61. Warkiani, M.E.; Guan, G.; Luan, K.B.; Lee, W.C.; Bhagat, A.A.S.; Chaudhuri, P.K.; Tan, D.S.-W.; Lim, W.T.; Lee, S.C.; Chen, P.C.Y.; et al. Slanted spiral microfluidics for the ultra-fast, label-free isolation of circulating tumor cells. *Lab Chip* **2014**, *14*, 128–137. [CrossRef] [PubMed]

62. Seo, J.; Lean, M.H.; Kole, A. Membraneless microseparation by asymmetry in curvilinear laminar flows. *J. Chromatogr. A* **2007**, *1162*, 126–131. [CrossRef] [PubMed]
63. Xiang, N.; Chen, K.; Sun, D.; Wang, S.; Yi, H.; Ni, Z. Quantitative characterization of the focusing process and dynamic behavior of differently sized microparticles in a spiral microchannel. *Microfluid. Nanofluid.* **2013**, *14*, 89–99. [CrossRef]
64. Xiang, N.; Chen, K.; Dai, Q.; Jiang, D.; Sun, D.; Ni, Z. Inertia-induced focusing dynamics of microparticles throughout a curved microfluidic channel. *Microfluid. Nanofluid.* **2015**, *18*, 29–39. [CrossRef]
65. Zhang, J.; Yan, S.; Sluyter, R.; Li, W.; Alici, G.; Nguyen, N.-T. Inertial particle separation by differential equilibrium positions in a symmetrical serpentine micro-channel. *Sci. Rep.* **2014**, *4*, 4527. [CrossRef] [PubMed]
66. Zhang, J.; Yan, S.; Li, W.; Alici, G.; Nguyen, N.-T. High throughput extraction of plasma using a secondary flow-aided inertial microfluidic device. *RSC Adv.* **2014**, *4*, 33149–33159. [CrossRef]
67. Hou, H.W.; Warkiani, M.E.; Khoo, B.L.; Li, Z.R.; Soo, R.A.; Tan, D.S.-W.; Lim, W.-T.; Han, J.; Bhagat, A.A.S.; Lim, C.T. Isolation and retrieval of circulating tumor cells using centrifugal forces. *Sci. Rep.* **2013**, *3*, 1259. [CrossRef] [PubMed]
68. Sun, J.S.; Li, M.M.; Liu, C.; Zhang, Y.; Liu, D.B.; Liu, W.W.; Hu, G.Q.; Jiang, X.Y. Double spiral microchannel for label-free tumor cell separation and enrichment. *Lab Chip* **2012**, *12*, 3952–3960. [CrossRef] [PubMed]
69. Sun, J.S.; Liu, C.; Li, M.M.; Wang, J.D.; Xianyu, Y.L.; Hu, G.Q.; Jiang, X.Y. Size-based hydrodynamic rare tumor cell separation in curved microfluidic channels. *Biomicrofluidics* **2013**, *7*, 011802. [CrossRef] [PubMed]
70. Lee, M.G.; Shin, J.H.; Bae, C.Y.; Choi, S.; Park, J.-K. Label-free cancer cell separation from human whole blood using inertial microfluidics at low shear stress. *Anal. Chem.* **2013**, *85*, 6213–6218. [CrossRef] [PubMed]
71. Bretherton, F.P. The motion of rigid particles in a shear flow at low reynolds number. *J. Fluid Mech.* **1962**, *14*, 284–304. [CrossRef]
72. Saffman, P.G. Lift on a small sphere in a slow shear flow. *J. Fluid Mech.* **1965**, *22*, 385–400. [CrossRef]
73. Cox, R.G.; Brenner, H. The lateral migration of solid particles in poiseuille flow—I theory. *Chem. Eng. Sci.* **1968**, *23*, 147–173. [CrossRef]
74. Vasseur, P.; Cox, R.G. Lateral migration of a spherical-particle in 2-dimensional shear flows. *J. Fluid Mech.* **1976**, *78*, 385–413. [CrossRef]
75. Schonberg, J.A.; Hinch, E.J. Inertial migration of a sphere in poiseuille flow. *J. Fluid Mech.* **1989**, *203*, 517–524. [CrossRef]
76. McLaughlin, J.B. Inertial migration of a small sphere in linear shear flows. *J. Fluid Mech.* **1991**, *224*, 261–274. [CrossRef]
77. McLaughlin, J.B. The lift on a small sphere in wall-bounded linear shear flows. *J. Fluid Mech.* **1993**, *246*, 249–265. [CrossRef]
78. Matas, J.-P.; Morris, J.F.; Guazzelli, É. Inertial migration of rigid spherical particles in poiseuille flow. *J. Fluid Mech.* **2004**, *515*, 171–195. [CrossRef]
79. Hu, H.H.; Joseph, D.D.; Crochet, M.J. Direct simulation of fluid particle motions. *Theor. Comput. Fluid Dyn.* **1992**, *3*, 285–306. [CrossRef]
80. Feng, J.; Hu, H.H.; Joseph, D.D. Direct simulation of initial-value problems for the motion of solid bodies in a newtonian fluid. 2. Couette and poiseuille flows. *J. Fluid Mech.* **1994**, *277*, 271–301. [CrossRef]
81. Yang, B.H.; Wang, J.; Joseph, D.D.; Hu, H.H.; Pan, T.W.; Glowinski, R. Migration of a sphere in tube flow. *J. Fluid Mech.* **2005**, *540*, 109. [CrossRef]
82. Matas, J.-P.; Morris, J.F.; Guazzelli, É. Lateral force on a rigid sphere in large-inertia laminar pipe flow. *J. Fluid Mech.* **2009**, *621*, 59. [CrossRef]
83. Xia, Y.N.; Whitesides, G.M. Soft lithography. *Annu. Rev. Mater. Sci.* **1998**, *28*, 153–184. [CrossRef]
84. Chun, B.; Ladd, A.J.C. Inertial migration of neutrally buoyant particles in a square duct: An investigation of multiple equilibrium positions. *Phys. Fluids* **2006**, *18*, 031704. [CrossRef]
85. Bhagat, A.A.S.; Kuntaegowdanahalli, S.S.; Papautsky, I. Enhanced particle filtration in straight microchannels using shear-modulated inertial migration. *Phys. Fluids* **2008**, *20*, 101702. [CrossRef]
86. Hur, S.C.; Henderson-MacLennan, N.K.; McCabe, E.R.B.; Di Carlo, D. Deformability-based cell classification and enrichment using inertial microfluidics. *Lab Chip* **2011**, *11*, 912–920. [CrossRef] [PubMed]
87. Hur, S.C.; Mach, A.J.; Di Carlo, D. High-throughput size-based rare cell enrichment using microscale vortices. *Biomicrofluidics* **2011**, *5*, 022206. [CrossRef] [PubMed]

88. Bhagat, A.A.S.; Kuntaegowdanahalli, S.S.; Papautsky, I. Inertial microfluidics for continuous particle filtration and extraction. *Microfluid. Nanofluid.* **2009**, *7*, 217–226. [CrossRef]

89. Zhou, J.; Papautsky, I. Fundamentals of inertial focusing in microchannels. *Lab Chip* **2013**, *13*, 1121–1132. [CrossRef] [PubMed]

90. Liu, C.; Hu, G.; Jiang, X.; Sun, J. Inertial focusing of spherical particles in rectangular microchannels over a wide range of reynolds numbers. *Lab Chip* **2015**, *15*, 1168–1177. [CrossRef] [PubMed]

91. Liu, C.; Ding, B.; Xue, C.; Tian, Y.; Hu, G.; Sun, J. Sheathless focusing and separation of diverse nanoparticles in viscoelastic solutions with minimized shear thinning. *Anal. Chem.* **2016**, *88*, 12547–12553. [CrossRef] [PubMed]

92. Ciftlik, A.T.; Ettori, M.; Gijs, M.A.M. High throughput-per-footprint inertial focusing. *Small* **2013**, *9*, 2764–2773. [CrossRef] [PubMed]

93. Ho, B.P.; Leal, L.G. Migration of rigid spheres in a 2-dimensional unidirectional shear-flow of a 2nd-order fluid. *J. Fluid Mech.* **1976**, *76*, 783–799. [CrossRef]

94. Leshansky, A.M.; Bransky, A.; Korin, N.; Dinnar, U. Tunable nonlinear viscoelastic "focusing" in a microfluidic device. *Phys. Rev. Lett.* **2007**, *98*, 234501. [CrossRef] [PubMed]

95. Yang, S.; Kim, J.Y.; Lee, S.J.; Lee, S.S.; Kim, J.M. Sheathless elasto-inertial particle focusing and continuous separation in a straight rectangular microchannel. *Lab Chip* **2011**, *11*, 266–273. [CrossRef] [PubMed]

96. Bhat, P.P.; Appathurai, S.; Harris, M.T.; Pasquali, M.; McKinley, G.H.; Basaran, O.A. Formation of beads-on-a-string structures during break-up of viscoelastic filaments. *Nat Phys* **2010**, *6*, 625–631. [CrossRef]

97. Bird, R.B.; Armstrong, R.C.; Hassager, O.; Curtiss, C.F. *Dynamics of Polymeric Liquids*; Wiley: New York, NY, USA, 1987; Volume 1.

98. Macosko, C.W. *Rheology: Principles, Measurements, and Applications*; VCH: Weinheim, Germany, 1994.

99. Yang, S.; Lee, S.S.; Ahn, S.W.; Kang, K.; Shim, W.; Lee, G.; Hyun, K.; Kim, J.M. Deformability-selective particle entrainment and separation in a rectangular microchannel using medium viscoelasticity. *Soft Matter* **2012**, *8*, 5011–5019. [CrossRef]

100. Cha, S.; Kang, K.; You, J.; Im, S.; Kim, Y.; Kim, J. Hoop stress-assisted three-dimensional particle focusing under viscoelastic flow. *Rheol. Acta* **2014**, *53*, 927–933. [CrossRef]

101. Nam, J.; Lim, H.; Kim, D.; Jung, H.; Shin, S. Continuous separation of microparticles in a microfluidic channel via the elasto-inertial effect of non-newtonian fluid. *Lab Chip* **2012**, *12*, 1347–1354. [CrossRef] [PubMed]

102. Lim, E.J.; Ober, T.J.; Edd, J.F.; Desai, S.P.; Neal, D.; Bong, K.W.; Doyle, P.S.; McKinley, G.H.; Toner, M. Inertio-elastic focusing of bioparticles in microchannels at high throughput. *Nat. Commun.* **2014**, *5*, 4120. [CrossRef] [PubMed]

103. Lee, D.J.; Brenner, H.; Youn, J.R.; Song, Y.S. Multiplex particle focusing via hydrodynamic force in viscoelastic fluids. *Sci. Rep.* **2013**, *19*, 3258. [CrossRef] [PubMed]

104. Xiang, N.; Zhang, X.; Dai, Q.; Cheng, J.; Chen, K.; Ni, Z. Fundamentals of elasto-inertial particle focusing in curved microfluidic channels. *Lab Chip* **2016**, *16*, 2626–2635. [CrossRef] [PubMed]

105. Kang, K.; Lee, S.S.; Hyun, K.; Lee, S.J.; Kim, J.M. DNA-based highly tunable particle focuser. *Nat. Commun.* **2013**, *4*, 2567. [CrossRef] [PubMed]

106. Liu, C.; Xue, C.; Chen, X.; Shan, L.; Tian, Y.; Hu, G. Size-based separation of particles and cells utilizing viscoelastic effects in straight microchannels. *Anal. Chem.* **2015**, *87*, 6041–6048. [CrossRef] [PubMed]

107. Carew, E.O.A.; Townsend, P. Slow visco-elastic flow past a cylinder in a rectangular channel. *Rheol. Acta* **1991**, *30*, 58–64. [CrossRef]

108. Morozov, A.N.; van Saarloos, W. An introductory essay on subcritical instabilities and the transition to turbulence in visco-elastic parallel shear flows. *Phys. Rep.* **2007**, *447*, 112–143. [CrossRef]

109. Meulenbroek, B.; Storm, C.; Morozov, A.N.; van Saarloos, W. Weakly nonlinear subcritical instability of visco-elastic poiseuille flow. *J. Non Newton. Fluid Mech.* **2004**, *116*, 235–268. [CrossRef]

110. Lim, E.J.; Ober, T.J.; Edd, J.F.; McKinley, G.H.; Toner, M. Visualization of microscale particle focusing in diluted and whole blood using particle trajectory analysis. *Lab Chip* **2012**, *12*, 2199–2210. [CrossRef] [PubMed]

111. Lu, X.; Xuan, X. Continuous microfluidic particle separation via elasto-inertial pinched flow fractionation. *Anal. Chem.* **2015**, *87*, 6389–6396. [CrossRef] [PubMed]

112. Yuan, D.; Zhang, J.; Sluyter, R.; Zhao, Q.; Yan, S.; Alici, G.; Li, W. Continuous plasma extraction under viscoelastic fluid in a straight channel with asymmetrical expansion-contraction cavity arrays. *Lab Chip* **2016**, *16*, 3919–3928. [CrossRef] [PubMed]

113. De Santo, I.; D'Avino, G.; Romeo, G.; Greco, F.; Netti, P.A.; Maffettone, P.L. Microfluidic lagrangian trap for brownian particles: Three-dimensional focusing down to the nanoscale. *Phys. Rev. Appl.* **2014**, *2*, 064001. [CrossRef]
114. Kim, J.Y.; Ahn, S.W.; Lee, S.S.; Kim, J.M. Lateral migration and focusing of colloidal particles and DNA molecules under viscoelastic flow. *Lab Chip* **2012**, *12*, 2807–2814. [PubMed]

micromachines

MDPI

Review

Microfluidic and Nanofluidic Resistive Pulse Sensing: A Review

Yongxin Song [1], Junyan Zhang [1] and Dongqing Li [2,*]

[1] Department of Marine Engineering, Dalian Maritime University, Dalian 116026, China;
yongxin@dlmu.edu.cn (Y.S.); junyan@dlmu.edu.cn (J.Z.)

[2] Department of Mechanical and Mechatronics Engineering, University of Waterloo,
Waterloo, ON N2L 3G1, Canada

* Correspondence: dongqing.li@uwaterloo.ca or dongqing@mme.uwaterloo.ca;
Tel.: +1-519-888-4567 (ext. 38682)

Received: 17 April 2017; Accepted: 21 June 2017; Published: 25 June 2017

Abstract: The resistive pulse sensing (RPS) method based on the Coulter principle is a powerful method for particle counting and sizing in electrolyte solutions. With the advancement of micro- and nano-fabrication technologies, microfluidic and nanofluidic resistive pulse sensing technologies and devices have been developed. Due to the unique advantages of microfluidics and nanofluidics, RPS sensors are enabled with more functions with greatly improved sensitivity and throughput and thus have wide applications in fields of biomedical research, clinical diagnosis, and so on. Firstly, this paper reviews some basic theories of particle sizing and counting. Emphasis is then given to the latest development of microfuidic and nanofluidic RPS technologies within the last 6 years, ranging from some new phenomena, methods of improving the sensitivity and throughput, and their applications, to some popular nanopore or nanochannel fabrication techniques. The future research directions and challenges on microfluidic and nanofluidic RPS are also outlined.

Keywords: resistive pulse sensing; particle sizing and counting; microfluidics and nanofluidics; review

1. Introduction

Accurately determining the size and number of particles and cells in electrolyte solutions is an important task in many fields, such as biomedical research [1–6], clinical diagnosis [7–12], and environmental monitoring. Among the methods for particle counting and sizing [13], resistive pulse sensing (also called the Coulter principle) probably is the most popular method. This method was invented by Coulter [14], aiming to replace manual cell counting with an automatic device. In a Coulter counter, a small insulating orifice is immersed into an electrolyte solution with suspended particles. A direct current (DC) voltage is applied through two electrodes placed across the orifice and an electrical current is conducted by the electrolyte solution and the orifice creates a "sensing zone". When a particle passes through the orifice, due to the different resistivity of the particle and electrolyte solution, a temporary electrical resistance change across the orifice is generated. This change is measured in terms of a voltage or a current pulse, whose magnitude is proportional to the volume of the particle for a given orifice. With the invention of Coulter counter, flow cytometry, which is widely used for detecting, counting, sizing cells and particles with a throughput of thousands of cells per second [15–22], became available [23]. The traditional flow cytometer, however, is bulky in size due to bulky pumps, tubes, valves, and other auxiliary components. Furthermore, it is expensive and cannot handle small volumes of samples.

With the advancement of Lab-on-a-chip technologies [24–30], microfluidic and nanofluidic RPS sensors with high sensitivity and accuracy were developed. Besides the basic functions for particle sizing and counting, a nano-RPS sensor can also characterize nanoparticles, DNA, viruses, antigens and so on. Such advancements greatly enrich the powerful abilities of the RPS technology and make the development of a low cost and portable flow cytometer possible.

This paper aims to review the latest development of microfuidic and nanofluidic RPS technologies within the last 6 years, ranging from some new phenomena, nanochannel or nanopore fabrication technologies, methods of improving the sensitivity and throughput, as well as their applications. The paper begins by introducing some basic theories of particle sizing and counting of microfluidic RPS sensors. Then, some interesting phenomena and applications of microfluidic and nanofluidic RPS are reviewed. At the end, the future research directions and challenges on microfluidic and nanofluidic RPS are discussed.

2. Working Principle and New Sensing Phenomena

Microfluidic and nanofluidic RPS employ the principle of the Coulter counter in microfluidic or nanofluidic channels for particle counting and sizing. The working principle and typical system setup is shown in Figure 1. For the system shown in Figure 1, an electrical field is applied across a sensing orifice whose size is much smaller than the main channel which is filled with an electrolyte solution. Each particle passing the sensing orifice will generate a resistive pulse which is processed by the amplification circuit, the data acquisition device, and the computer. Each particle will generate one signal pulse and the magnitude of the signal represents the volume ratio of the particle and the sensing gate. In this way, particle sizing and counting are achieved.

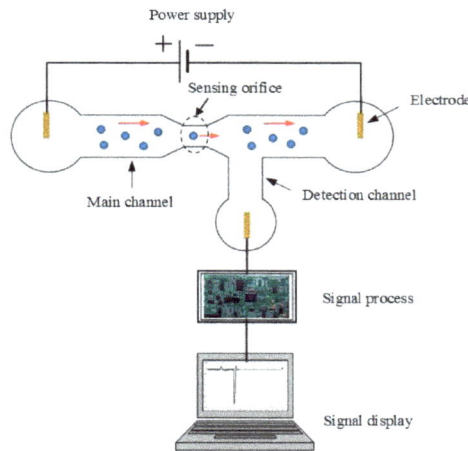

Figure 1. Working principle of a microfluidic resistive pulse sensing (RPS) system.

It should be noted that both DC and alternative current (AC) fields can be applied (named as DC RPS and AC RPS, respectively). For the DC RPS, the system is very simple without using the bulky AC power source and the complicated signal processing circuit. However, the polarization of the electrode in electrolyte solution will increase the electric resistance of the main channel and thus is adverse to the sensitivity of the system. [31–33]. Such polarization effects can be avoided by using an AC electric field which can decrease the polarization resistance by increasing the excitation frequency [33]. In order to minimize the stray capacitance effects [34], however, appropriate frequency must be carefully determined depending on the specific chip design [31,32,35–37].

For particle sizing with the RPS method, the most important question is how to evaluate particle's size based on the detected signal. That is, the relationship between the measured magnitudes of signals and the sizes of the sensing orifice (or sensing channel) and particles must be determined. Such a relationship can be determined by calculating the resistance change caused by a particle entering the sensing channel (ΔR). Table 1 summarizes the relative electrical resistance change caused by spherical particles with different diameters.

Table 1. Relative resistance changes caused by particles passing a sensing orifice. D and L are the diameter and length of the sensing orifice, d is the diameter of the particle.

Relative Particle Diameter (d/D)	Relative Resistance Change ($\Delta R/R$)
Infinite smaller diameter [38]	$\frac{d^3}{D^2L}$
Smaller diameter [39]	$\frac{d^3}{D^2L}\left[\frac{D^2}{2L^2} + \frac{1}{\sqrt{1+(D/L)^2}}\right]F\left(\frac{d^3}{D^3}\right)$
Medium diameter [40,41]	$\frac{d^3}{D^2L} \cdot \frac{1}{1-0.8(d/D)^3}$
Larger diameter [42]	$\frac{D}{L}\left[\frac{\arcsin(d/D)}{\sqrt{1-(d/D)^2}} - \frac{d}{D}\right]$

Note: $F(x) = 1 + 1.264x + 1.347x^2 + 0.648x^3 + 4.167x^4$ [43].

It should be noted that the equations shown in Table 1 were derived based on the assumption that a spherical particle moves along the centerline of a cylindrical microchannel. With the advancement of micro and nanofabrication technology, different types of micro and nanochannels were available and new RPS phenomena in sub-microscale and nanoscale have been demonstrated recently. One interesting phenomenon recently discovered is the generation of a double-peak (resistive-and-conductive peak) when a particle passes through the sensing pore [44,45]. Such a phenomenon is totally different from the classical Coulter theory. Menestrina et al. [45] used a polyethylene terephthalate (PET) pore (12 μm long and 500 nm–1.5 μm in diameter) for RPS detection of particles of several hundred nanometers in KCl solution with different concentrations. They found that double peaks, first a downward resistive peak and then an upward conductive peak, will be generated when the concentration of the KCl solution was below 300 mM. Through numerical simulations and experimental verifications, the authors concluded that the positive peak (conductive) is caused by the modulated ionic concentrations induced by the charged particles entering into the sensing pore.

Weatherall et al. [44] also experimentally investigated the double-peak phenomenon when measuring 200 nm carboxylate polystyrene spherical particles with a tunable RPS method. They found that the onset electrolyte concentration for double-peak formation is about 50 mM. More importantly, the formation of a double-peak is due to the ionic concentration polarization. They also found that the shape of the double-peak depends both on the moving directions of the particle and the applied voltages. For example, a positive voltage bias and a positive hydraulic pressure will only generate a resistive pulse. With a negative voltage bias and a positive pressure, however, a conductive and resistive pulse will be generated.

Considering the 'end effects' of the sensing pore, Willmott et al. [46,47] built up a semi-analytic model to simulate the shapes of the pulses. They found that end effects are prominent when the center of the particle is less than one radius inside the pore entrance. They also found that asymmetry is more obvious for larger and slower moving particles. Such results are valuable for evaluating a particle's size based on the measured pulse because most of the sensing pores are not of the ideal shape with a constant cross-section in practice.

The geometry effects of a nanopore on RPS nanoparticle sensing were also investigated. It was found that the length [48], thickness [49], tip diameter [50], and base ratio [51] of the silicon nitride membrane nanopores can greatly influence the electric field around the ends of the nanopores. More recently, Kaya et al. studied the influence of the cone angle of a conically shaped poly(ethylene terephthalate) (PET) membrane nanopore on DNA sensing. They found that the electric field at the tip of the nanopore increases with the increase in the cone angel. As a result, the magnitude of the pulse will also be increased with the increased cone angel [52].

Beside the geometry of the nanopores, the surface charges at the interface of nanopore–electrolyte solution can also influence the magnitude and duration of a RPS signal pulse. Liu et al. [53] showed that the surface charges can either increase or decrease the magnitude of the signal pulse depending on the positive or negative potential bias applied across the sensing channel. Recently, Weatherall et al. [54] found that particles near the edge of a tunable nanopore entrance generate larger RPS signals. The smallest RPS signal is generated by a particle moving along the center line of the nanopore.

Recent studies also showed that both the shape and material of a particle can influence the detected RPS signals. For example, using a rod-like Au particle (several micrometers in length and several hundred nanometers in diameter) and a tunable elastomeric membrane pore, Platt et al. [55] studied the effects of shape of the particle on RPS signals. They found that the full width at half maximum of the RPS signal can be used to characterize the rod length. More importantly, they showed that for the rod-like particles, the measured current change remains unchanged or slightly increased with the increase in the pore size. However, the current change caused by a spherical particle is decreased.

Lee's study showed that the magnitude of a RPS signal generated by an elongated nanoparticle is larger than that by a spherical nanoparticle of the same volume [56]. Song et al. [57] studied the effect of induced surface charge of metal particles on RPS particle sizing. The experimental results showed that the magnitude of the RPS signals generated by 5 μm magnetic particles is larger than that of a 5 μm polystyrene particle under the same condition, as is shown in Figure 2. Such results are due to the thicker electrical double layer (EDL) formed by the induced charges around the magnetic particles.

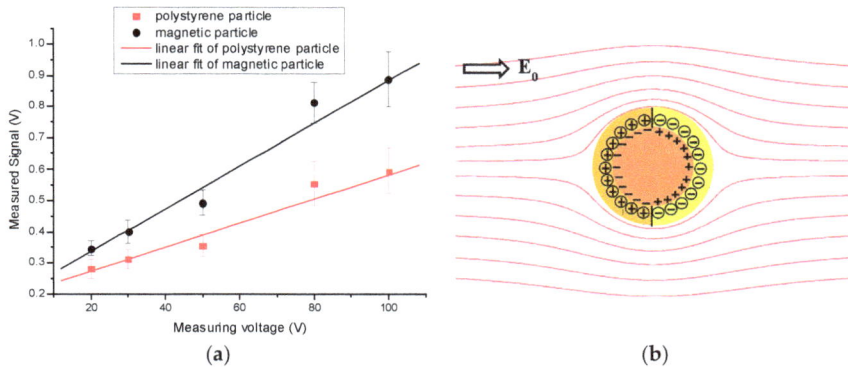

Figure 2. (**a**) Dependence of the magnitudes of measured RPS signals on the applied voltages and (**b**) the induced electrical double layer.

These new phenomena mentioned above are important and not covered in the traditional RPS theory. Such phenomena should be well considered for particle size evaluation with the RPS method.

3. Methods for Throughput and Sensitivity Improvement

3.1. Throughput

One key advantage of microfluidic RPS is its simplicity, because the bulky pumps and valves are no longer needed due to the using of a micro- or nanoscale-sized channel or sensing pore and thus a greatly decreased sample volume. Therefore, microfluidic and nanofluidic RPS is very promising to be developed into portable lab-on-a-chip (LOC) devices. However, the throughput of a portable device is relative low due to the slow sample transportation velocity in a microfluidic channel. The sensitivity is also limited by the limited ability of fabricating a small sensing channel or pore and the simple data processing circuit. For microfluidic and nanofluidic RPS, the throughput and sensitivity are generally coupled with each other. A smaller sensing channel or sensing pore is needed in order to have a higher sensitivity. Achieving this, however, will lower the throughput due to the constraint of using the smaller sensing channel.

To improve the counting throughput, Jagtiani et al. [32] developed a frequency division multiplexing method which can encodes each channel with a specific frequency with a single pair of electrodes and process the signals with a single electronic system. Such a method enables monitoring multi-channels individually and obtaining an 300% improvement of throughput over a single-channel device. For this design, however, each channel needs one specific signal process channel to decode the signals.

Following a similar idea, Liu et al. [58] designed novel coplanar electrodes that can generate orthogonal digital codes when particles pass over the electrodes. The orthogonal digital codes were then decoded by the Code Division Multiple Access (CDMA) principle. Such a design, which can even differentiate the overlapped signals with >90% accuracy, can be easily extended to orthogonally counting particles in an arbitrary number of microfluidic channels and thus provides a very promising approach for solving the throughput problem of a microfluidic flow cytometer.

Song et al. [59] also put forward a novel method to improve the throughput of a DC RPS device. They designed a sensitive differential microfluidic sensor with multiple detecting channels and one common reference channel (Figure 3). With seven detecting channels, an average throughput of 7140 cells/min under a flow rate of 10 µL/min was achieved. Counting throughput can be further increased by increasing the number of the detecting channels. It should be noted that a reference channel makes a good balance between the sensitivity and the throughput.

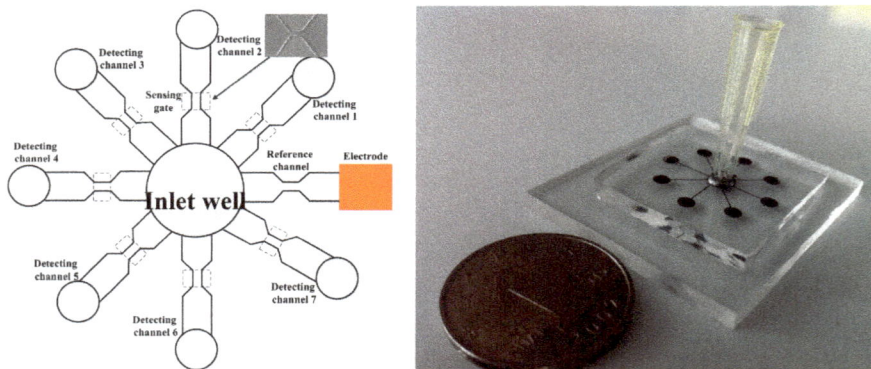

Figure 3. The structure of the high throughput microfluidic chip.

Rather than putting efforts on chip design innovation, Castillo-Fernandez et al. [60] adopted a signal post-treating approach to improve the throughput. They designed an electronic system that allowed a counting throughput of about of 500 counts/s.

3.2. Sensitivity

Theoretically, the sensitivity of the RPS sensor is mainly determined by the volume ratio of the particle and the sensing channel as well as the noise level of the electronic system. Therefore, sensitivity can be improved either by increasing the volume ratio or by using advanced signal processing instruments to cancel the noise.

One way to increase the volume ratio is to decrease the volume of the sensing channel. For smaller particle detection, such as DNA, proteins, and viruses, a nano-sized sensing channel or pore is needed. The details about the methods of fabricating such channels or pores will be reviewed and discussed in Section 5. Here, several other novel methods on decreasing the volume of the sensing channel are reviewed first.

One novel and simple way is to use less conductive focusing solutions to decrease the volume of the sensing channel. Choi et al. [61] used hexadecane as the focusing solution to hydraulically control the width of the passage for the sample solution. The sample solution is focused by the focusing solution (hexadecane) to pass the sensing region. Due to the very low conductivity of hexadecane, the electric field only exists in the sample solution and becomes concentrated in the focused region. In this way, the volume in the sensing region is narrowed and can be flexibly controlled by the flow rate of the focusing solution. Such a RPS sensor has a controllable range of sensitivity and was successfully applied for submicron-sized bacterial detection with a 30 μm wide detection channel.

One problem associated with using oil as the focusing solution is that the large surface tension at the oil–water interface can make the flow unstable. As a result, the sensing volume will also be changing and thus noise level is increased too [62]. To solve this problem, Bernabini et al. added some surfactants into the oil (focusing solution) to stabilize the oil–sample solution interface. Due to the increased stability of the sensing volume by using an oil/surfactant mixture, detection of 1 μm polystyrene particles and discrimination of bacteria and polystyrene particles of similar size with a 200 μm wide sensing channel were achieved [62].

For the above two studies, it should be noted that they used two or four electrodes to apply an AC electric field and the sample solutions were transported by pressure-driven flow. For these designs, the distance between the electrodes defines the length of the sensing region. The focusing solution was only used to decrease the width of the sensing region. More recently, Liu et al. [63] reported a novel electrokinetic-flow-focusing method to narrow the size of the sensing channel and thus improve the sensitivity. This method is particularly useful for a DC RPS system with an elelctrokinetically-driven flow. Figure 4 shows the structure of this novel microfluidic RPS sensor and its working principle. The microfluidic chip has inlet and outlet wells, one main channel, two detecting channels, two focusing channels, and the corresponding wells. The key feature of this sensor is that the focusing solution can flow only from the upstream focusing channel to the downstream focusing channel. In this way, the sensitivity is greatly improved. As a result, detection of 1 μm with a physical sensing gate of 30 μm × 40 μm × 10 μm (width × length × height) was successfully achieved (Figure 5).

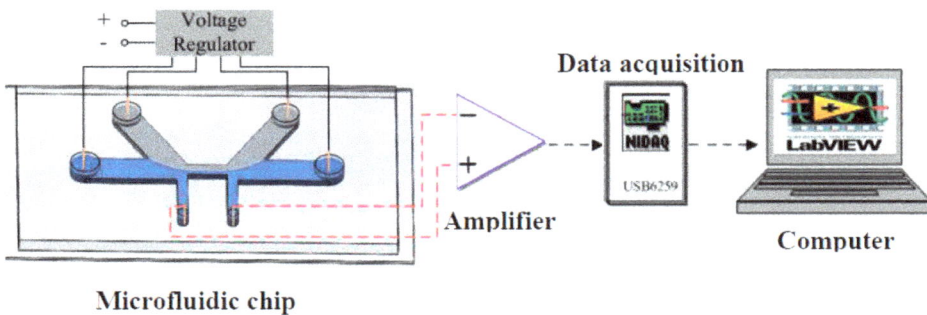

(a)

Figure 4. *Cont.*

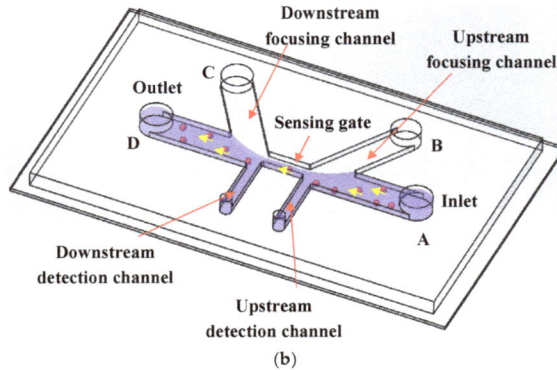

Figure 4. (**a**) A schematic diagram of the system setup, and (**b**) the structure of the microfluidic chip.

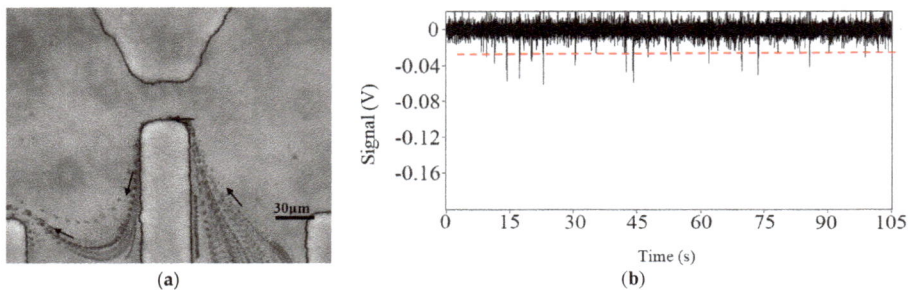

Figure 5. (**a**) Typical flow trajectory, and (**b**) detected RPS signals of 1 μm particles ($V_A = 27$ V, $V_B = 52$ V, $V_C = 0$ V and $V_D = 24$ V).

4. Applications

With the improved RPS sensitivity, especially the ability of fabricating a nano-RPS sensor, label-free particle detection in nanoscale, such as detection of DNA, proteins, viruses, and so on, were possible. With the measured RPS signals, nanoparticle characterization was also reported. In this section, we will review the recent developments on the typical applications of RPS technology in submicro- and nanoscale.

4.1. DNA Detection and Analysis

With the first demonstration of RPS detection of single stranded DNA detection by a biological α-hemolysin pore [64], nano-RPS now is an attractive tool for label-free DNA detection with the distinctive advantages of its simplicity and high sensitivity [65–71]. The commercially available qNano device exemplifies the strength of this technology.

Among the recent investigations, the mostly common used approach of DNA detection with RPS is to detect the change caused by the binding of DNA to a particle. For example, the change of the width and frequency of the RPS signals caused by the binding of thrombin to DNA-coated magnetic beads [67], the change in the full-width half maximum (FWHM) caused by DNA strands hybridized with the probe-grafted magnetic particles [72], have been used for DNA identification.

To increase the small size difference by the adsorption of DNA, an approach of amplification was adopted by Kühnemund [73] and Yang [70]. The key idea is to increase the amount of DNA and thus increase the signal difference and detection sensitivity. Firstly, target DNA, together with padlock probes and capture oligo, is captured by the magnetic particle. After amplification, the magnetic particles are measured with the RPS system. For the particles with target DNA, it will generate signals with larger magnitude and wider FWHM.

In 2013, Traversi et al. [74] fabricated graphene and SiNx nanopores with which the electrokinetic translocation behavior of pNEB DNA and λ-DNA was measured. Since the graphene membrane can be as thin as 0.0335 nm, comparative to the distance between two bases in a DNA chain, it is promising for fabricating a DNA sequencing device. Hernándezainsa et al. [75] fabricated a glass nanocapillary (with a central needle hole) and studied the structures of DNA origami using both visual observation and RPS detection. This demonstrates the possibility of selective detection of double-strand DNA and thus the great potential for DNA sequencing with the RPS method. More recently, Sischka et al. [76] studied the translocation force on DNA molecules in nanopores. It was found that the membrane thickness had a great influence on the translocation of DNA molecules.

4.2. Label-Free Protein Detection

With the improved sensitivity for particle sizing, especially the ability to fabricate a nano-RPS sensor, label-free detection of a specific kind of nanoparticles, such as proteins and DNA, from a mixture solution becomes possible. Rodriguez-Trujilloa et al. [77] measured the specific protein concentrations in a suspension with a microfluidic RPS system. Briefly, bead oligomers were formed when the two functionalized polystyrene beads and the rat IgG serum were mixed together. Such oligomers are larger in size and will generate RPS signals with larger magnitude. By comparing the magnitudes of the detected signals, the presence of the target protein can be easily determined. However, more single bead will be available with the increased protein concentration. The authors attributed such results to the binding and blocking of more sites on the surface of the two beads under high protein concentration.

Following a similar idea, RPS detection of a cancer biomarker [78], human ferritin [79], and even single-molecule proteins [80] were also demonstrated. It should be emphasized that for RPS detection, any specific binding will be reflected by the detected electric signals. Thus, fluorescent labeling is no longer needed. This is much more advantageous and attractive in saving time and simplifying detection.

4.3. Nanoparticle Characterization

The nano-RPS device is a new, yet powerful tool for nanoparticle analysis. Based on a single-particle measurement, much useful information, such as size and concentration distribution, surface charges, and even translocation behavior can be reliably obtained [56,81–86].

Vogel et al. [81] demonstrated the accuracy and reliability of sizing nano-sized particles with a tunable RPS sensor made of polyurethane. The nano-RPS system was fabricated by mounting a resizable elastomeric thermoplastic polyurethane (TPU) membrane on the q-Nano device. The membrane was stretched open to different pore sizes. The experimental results showed that the sizes of different particles, for example polymethyl methacrylate (PMMA) and nonfunctionalized polystyrene particles, adenovirus, and so on, can be accurately evaluated based on the pre-obtained calibration curve. Following a similar calibration methodology, Roberts et al. [82] measured the concentrations of the marine photosynthetic cyanobacterium Prochlorococcus (with a diameter of 600 nm) by using a tunable polyurethane nanopore. A good agreement with the results from microscopy counting was found. Pal et al. [85] measured the size distribution and concentration of engineered nanomaterials in cell culture media with a tunable nanopore RPS method. They found that TRPS technology offers higher resolution and sensitivity compared to the dynamic light scattering (DLS) method. Due to the high accuracy and resolution in sizing nanoparticle, the nanopore RPS sensing technique was applied to characterize the swelling of pH-responsive, polymeric expansile nanoparticles by Colby et al. [83], and to measure the size and the surface charge of silica nanoparticles in serum by Sikora et al. [87]. Compared with the results from other methods, such as transmission electron microscopy, differential centrifugal sedimentation (DCS), and dynamic light scattering (DLS), the tunable RPS sensor showed excellent performance in sensitivity and resolution.

Fraikin et al. [84] developed a polydimethylsiloxane (PDMS) high-throughput nanoparticle RPS sensor that can size and determine the concentration of a nanoparticle suspension and unlabeled bacteriophage T7 in both salt solution and mouse blood plasma with a throughput of about 500,000 particles per second. With this sensor, the authors also firstly discovered some naturally occurring nanoparticles in the native blood plasma. Through both finite-element simulations and experimental verification, Lee investigated the geometry of the pore and the particle on the magnitudes of the detected RPS signals. The results showed that a conical pore can generate larger signals in magnitudes than a cylindrical pore. More importantly, both the size and shape of the nanoparticle can be simultaneously determined based on the magnitude and the y-position of 10% resistive pulse (y 10%) [56].

Besides sizing the bare nanoparticle, estimation of the thickness of the protein layer absorbed onto the nanoparticles in serum was also achieved, using a tunable RPS method by Sikora et al. [87]. Recently, Luo et al. reviewed the advances in the transport motion of a single nanoparticle by the RPS method, with an emphasis on the forces governing the translocation of low-aspect-ratio, non-deformable particles [86].

Determining the zeta potential, an important electrokinetic parameter, of carboxylated polystyrene nanoparticles can also be reliably achieved with the RPS method [88–92]. Eldridge et al. [89] measured the zeta potential of carboxylated polystyrene nanoparticles. The idea is to find the critical pressure at which the drag forces of the pressure-driven flow and electroosmotic flow and the electrophoretic force exerted on the nanoparticle become balanced. Once balanced, the frequency of the signal pulse becomes minimum which can be used to calculate the electrophoretic mobility of the particle (also the zeta potential). By using the same system and the same measuring principle, Somerville et al. [90] measured the zeta potential of a water-in-oil emulsion, demonstrating the ability of measuring zeta potential of a single soft particle. To further evaluate the accuracy of this technique, Eldridge et al. [89] conducted experiments to measure particles with different diameters and surface charges and concluded that the full width half maximum (FWHM) duration of a signal pulse is more reliable in determining zeta potential. More recently, determining the zeta potential of DNA modified particles, discrimination of ssDNA, dsDNA, and small changes in base length for nucleotides were reported by Blundell et al. [69]. All of the above investigations demonstrate the power of the RPS technology to characterize bio-nanoparticles with high sensitivity and resolution.

5. Fabrication of the Nano-Sensing Gate

As reviewed above, there is an increasing interest in nano-RPS technology nowadays, propelled both from the interests in scientific understanding and promising applications in nanoscale. For a nano-RPS device, the most important component is the sensing orifice, which is normally in form of a nanopore or a nanochannel. Therefore, advancement in nanofabrication is vital for the development of a nano-RPS device. In addition to the several well-known review papers [86,93–95], we will review the latest development of fabricating two typical nanosensing orifices: nanopores and PDMS nanochannels.

5.1. Nanopores

The nanopore is one of the most widely used sensing orifices. Several review papers have already summarized the advances in the fabrication of solid-state nanopores and their applications on single-molecule sensing, DNA sequencing, genetics, medical diagnostics, and so on [93–95]. Generally speaking, a nanopore is typically of cylindrical [55,96–98] or conical [99–102] shape for use in RPS. In practice, most of the nanopore is of conical shape due to fabrication limitations. The conical nanopore is much more advantageous in sensitivity and accuracy due to the concentrated electric field at the entrance of a conical nanopore [99]. According to the material of a nanopore, it can be generally divided into two categories: biological nanopores and solid-state nanopores.

5.1.1. Biological Nanopores

α-Hemolysin (α-HL) is one typical biological nanopore used for RPS [103]. The hydrophilic α-HL nanopore can self-assemble into a planar lipid membrane [104]. In 2012, Gopfrich et al. [105] fabricated an α-HL nanopore which is embedded in lipid bilayers. The fabrication procedures can be found elsewhere [106]. Firstly, nanocapillaries of 200 nm diameter were fabricated by using a laser pipet puller. Then, this nanocapillary was put on the baseplate and a charge of Giant Unilamellar Vesicles (GUVs) was introduced. A few seconds later, a vesicle cracked on the nanotip area to create a nanobilayer. The formation of the nanobilayer was detected by measuring the current drop. This was a novel technique for nanobilayer manufacturing. Compared with other conventional bilayer methods, it provided a fast and stable way to make a bilayer.

However, the diameter of α-HL nanopores is only 1.4 nm, which is not suitable for the detection of some large molecules. To deal with this problem, some other biological membrane, such gareolysin [107], FhuA [108–110], and ClyA [111,112] hetero-oligomeric channels formed by NfpA and NfpB [113], were used. For biological nanopores, these are preferable for DNA sensing and sequencing. Although these nanopores achieved some success in a variety of applications, they had several weaknesses that hindered the advance of biological sensing in nanopores, for example, the low mechanical rigidity of the lipid membrane and shorter lifetime, etc. [114].

5.1.2. Solid-State Nanopores

Compared with the biological nanopores, a solid-state nanopore is durable and can be easily fabricated into the desired geometries. Such merits are advantageous for commercial application [115].

(1) Glass

Nanopores made of glass have a high mechanical rigidity and chemical resistance and have been widely used for RPS [114–120]. The most widely used method for glass nanopore fabrication is by the pulling-and-etching method, which is simple and readily available with a relatively high resolution [117,120]. Surface modification techniques can also be applied to the nanopore to achieve some specific purposes. Such a surface charge-modulated RPS sensor was proven to be highly sensitive and selective for target particle detection [118].

He et al. [119] reported a novel surface modification method that can be reliably applied for selectively detecting uric acid (UA). To modify the nanopore, firstly the inner surface of the capillary tip was coated with a poly (L-histidine) (PLH) monolayer, which worked as the substrate for Au assembly. This was followed by in situ chemical reduction of $AuCl_4^-$. An Au nanofilm and 2-thiouracil (2-TU) were generated and coated on the inner surface of the glass nanopore. The binding of the target molecules with the functioned 2-TU will change the ionic current of the RPS system which can be used for quantitative detection of UA.

(2) Polymer Membranes

Polymer membranes were a good material for fabricating the nano-orifice due to their excellent optical properties [121,122]. The track-etched method was a popular method for fabricating a nano-orifice in polymer membranes [52,123] because it is cost-saving, reproducible, and provides easy precision controlling [123]. For example, the authors [52] developed a novel method to accurately control the geometry of the poly(ethylene terephthalate) (PET) nanopore by controlling the proportion of methanol in the etching solution. Their results showed that the pore diameters can be accurately controlled by applying different amount of methanol in the alkali solution. The base diameters of the PET membrane were 512 ± 30 nm and 928 ± 33 nm, with volume proportions of methanol of 0% and 10%, respectively. The maximum deviation of the base diameters was only 5.8%.

(3) Silicon Dioxide and Silicon Nitride

The silicon-based nanopore is another popular nanopore which can be fabricated by using the electron beam writing method [124–126]. The diameter of such a nanopore can be smaller than 10 nm [125]. A chemically etched SiN_x nanopore was recently reported by Kwok [127,128]. The SiN_x membrane was immersed in 3.6 M LiCl buffer solution at pH = 10, where the dielectric breakdown was controlled. The nanopores manufactured by this method showed a very high resolution (2.0 ± 0.5 nm in diameter) and over 95% reproducibility.

Yanagi et al. [129] fabricated a silicon nitride nanopore with a diameter of 1–2 nm, using the technique of multilevel pulse-voltage injection (MPVI). There is a trans chamber and a cis chamber across the 10 nm-thick Si_3N_4 membrane. Two Ag/AgCl electrodes were inserted into the two chambers and connected to an AC power source and an ammeter. The voltages were used to drill a nanopore across the membrane. The nanopore was generated when the measured current in this system exceeds the preset value.

(4) Graphene

Recently, advancements on fabricating graphene nanopores were reported [130–132]. Using electron beam writing, graphene nanopores with diameters less than 5 nm were achieved [131,133]. Besides the traditional electron beam writing method, Deng et al. [134] proposed a technique of fabricating a graphene nanopore of 5 nm in diameter with a helium ion microscope (HIM). For this method, there is a trimer to emit the helium ion beam under a high voltage at the top of an emitter. Under the action of the He^+ ion, a graphene nanopore can be obtained. In that paper, they demonstrated a precise control of nanopore size and shape during the manufacturing process by controlling the dwell time of exposures.

5.2. PDMS Nanochannels

Nanochannels can be fabricated both in polymer and glass substrates. The two major advantages of using these nanochannels are being an integral part in a complex fluidic structure and allowing optical access to objects transported inside the channels. Fabricating a PDMS nanochannel is more attractive due to the distinctive advantage of PDMS in developing microfluidic and nanofluidic devices. Generally, a mold or master is needed for PDMS replica. Such a mold is normally fabricated by methods of the electron beam lithography (EBL) [135–141] and focused ion beam (FIB) [142–149]. For these two methods, the wavelength of the e-beam has a vital influence on the resolution of the nanochannel. However, the equipment is very expensive and not every lab can afford it [150].

To develop a simpler and reliable PDMS nanochannel fabrication method, Peng and Li [151] reported fabricating a single nanocrack on a polymer substrate by using a solvent-induced technique (Figure 6). The minimal dimension of the nanocracks could reach nearly 64 nm in width and 17.4 nm in depth. Using this method, a PDMS nanochannel with a width of about 100 nm was fabricated [152]. The key procedure of this fabrication technique is the making of nanocracks firstly. The next step is to employ the soft lithography technique (SU8 photoresist) to copy this nanochannel mold. At last, the smooth cast slab is solidified and nanoimprinted by pressure gauge.

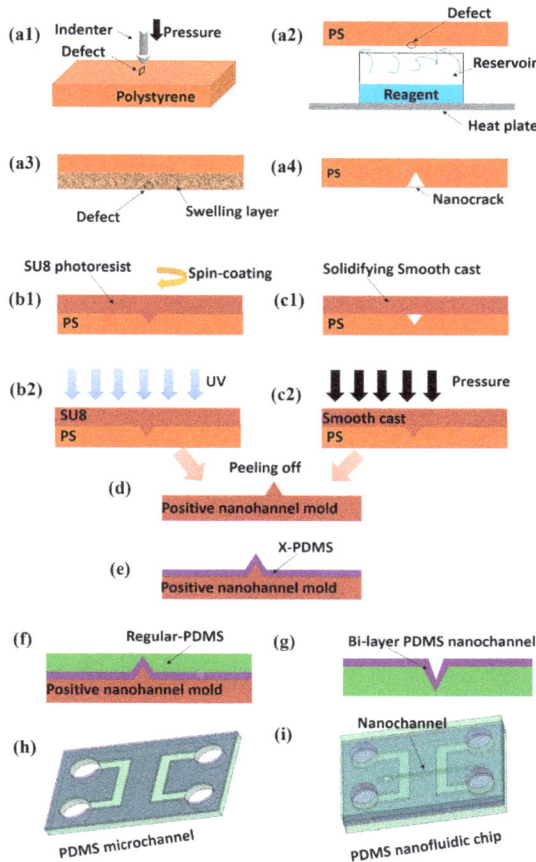

Figure 6. Procedures for polydimethylsiloxane (PDMS) nanochannel fabrication by using a solvent-induced nanocrack. (**a1**) making microdefects on a polystyrene slab by using an indenter of a micro-hardness testing system; (**a2**) absorption of the solvent; (**a3**) swelling of the polystyrene surface and initialization of nanocracks; (**a4**) nanocracks on the polystyrene surface. (**b1**) spin-coating of SU8 photoresist on the polystyrene slab with nanocracks; (**b2**) exposure the SU-8 layer to UV light. (**c1**) fabricating of solidifying smooth cast slab; (**c2**) nanoimprint by using a pressure gauge. (**d**)–(**i**) Show how to make a PDMS micro–nanofluidic chip by using the nanochannel mold: (**d**) nanochannel mold after peeling off process; (**e**) coating of x-PDMS on the nanochannel mold; (**f**) casting another layer of regular PDMS on the x-PDMS; (**g**) bi-layer PDMS nanochannel; (**h**) fabrication of bi-layer microchannel system; (**i**) PDMS micro–nanofluidic chip after bonding.

6. Outlook

The latest development of microfuidic and nanofluidic RPS technologies within the last 6 years is reviewed in this paper. Some new and important phenomena both in microscale and nanoscale have been discovered, which greatly enriches the power of the Coulter principle. Researches on micro- and nano-RPS, both from the aspects of theory and application, still have many challenges and great potential. For example, one big question is the mathematic relationship between the resistance change and the sizes of a particle and a sensing channel at nanoscale, especially when the EDLs of the sensing channel are overlapped. Another challenge which prevents the wide application of the nano-RPS system is the integration of a microchannel with the nanochannel. While there are variety of nanochannel fabrication methods, as reviewed above, difficulties still exist for the integration of a nanochannel with a microchannel. Simple and reliable integration technology can greatly benefit both fundamental and applied research.

Acknowledgments: The authors wish to thank the financial support of National Science Foundation program of China (51679023) and 863 plan (2015AA020409) to Yongxin Song, the Natural Sciences and Engineering Research Council of Canada through a research grant to Dongqing Li, the support from Fundamental Research Funds for the Central Universities (3132016325 and 3132017012) and from the University 111 project of China under Grant No. B08046 is greatly appreciated.

Conflicts of Interest: The authors declare no conflict of interest.

References

1. Wang, Z.; Wang, X.; Liu, S.; Yin, J.; Wang, H. Fluorescently imaged particle counting immunoassay for sensitive detection of DNA modifications. *Anal. Chem.* **2010**, *82*, 9901–9908. [PubMed]
2. Barnard, J.G.; Singh, S.; Randolph, T.W.; Carpenter, J.F. Subvisible particle counting provides a sensitive method of detecting and quantifying aggregation of monoclonal antibody caused by freeze-thawing: Insights into the roles of particles in the protein aggregation pathway. *J. Pharm. Sci.* **2011**, *100*, 492–503. [PubMed]
3. Benech, H.; Théodoro, F.; Herbet, A.; Page, N.; Schlemmer, D.; Pruvost, A. Peripheral blood mononuclear cell counting using a DNA-detection-based method. *Anal. Chem.* **2004**, *330*, 172–174.
4. Huang, B.; Wu, H.K.; Bhaya, D.; Grossman, A.; Granier, S.; Kobilka, B.K.; Zare, R.N. Counting low-copy number proteins in a single cell. *Science* **2007**, *315*, 81–84. [PubMed]
5. Amann, R.I.; Binder, B.J.; Olson, R.J.; Chisholm, S.W.; Devereux, R.; Stahl, D.A. Combination of 16S rRNA-targeted oligonucleotide probes with flow cytometry for analyzing mixed microbial populations. *Appl. Environ. Microbiol.* **1990**, *56*, 1919–1925. [PubMed]
6. Yarnell, J.W.; Baker, I.A.; Sweetnam, P.M.; Bainton, D.; O'Brien, J.R.; Whitehead, P.J.; Elwood, P.C. Fibrinogen, viscosity, and white blood cell count are major risk factors for ischemic heart disease. *Circulation* **1991**, *83*, 836–844. [PubMed]
7. Brüllmann, D.; Pabst, A.; Lehmann, K.; Ziebart, T.; Klein, M.; d'Hoedt, B. Counting touching cell nuclei using fast ellipse detection to assess in vitro cell characteristics: A feasibility study. *Clin. Oral Investig.* **2012**, *16*, 33–38. [PubMed]
8. Braun, S.; Marth, C. Circulating tumor cells in metastatic breast cancer—Toward individualized treatment. *N. Engl. J. Med.* **2004**, *351*, 824–826. [PubMed]
9. Adams, A.A.; Okagbare, P.I.; Feng, J.; Hupert, M.L.; Patterson, D.; Göttert, J.; McCarley, R.L.; Nikitopoulos, D.; Murphy, M.C.; Soper, S.A. Highly efficient circulating tumor cell isolation from whole blood and label-free enumeration using polymer-based microfluidics with an integrated conductivity sensor. *J. Am. Chem. Soc.* **2008**, *30*, 8633–8641.
10. Cristofanilli, M.; Budd, G.T.; Ellis, M.J.; Stopeck, A.; Matera, J.; Miller, M.C.; Reuben, J.M.; Doyle, G.V.; Allard, W.J.; Terstappen, L.W.M.M.; et al. Circulating tumor cells, disease progression, and survival in metastatic breast cancer. *N. Engl. J. Med.* **2004**, *351*, 781–791. [PubMed]
11. Pantel, K.; Riethdorf, S. Are circulating tumor cells predictive of overall survival? *Nat. Rev. Clin. Oncol.* **2009**, *6*, 190–191. [CrossRef] [PubMed]
12. Slade, M.J.; Coombes, R.C. The clinical significance of disseminated tumor cells in breast cancer. *Nat. Rev. Clin. Oncol.* **2007**, *4*, 30–41. [CrossRef] [PubMed]
13. Zhang, H.; Chan, H.C.; Pan, X.; Li, D. Methods for counting particles in microfluidic applications. *Microfluid. Nanofluid.* **2009**, *7*, 739–749. [CrossRef]
14. Wallace, H.C. Means for Counting Particles Suspended in a Fluid. U.S. Patent 2,656,508, 20 October 1953.
15. Saleh, O.A.; Sohn, L.L. Direct Detection of Antibody-Antigen Binding Using an On-Chip Artificial Pore. *Proc. Natl. Acad. Sci. USA* **2003**, *100*, 820–824. [CrossRef] [PubMed]
16. Sam, E.; Mehdi, J.; Robert, W.; Duttonand, R.W.; Davis, J.M.; Dutton, R.W.; Davis, A.R.W. Smart surfaces: Use of electrokinetics for selective modulation of biomolecular affinities. *MRS Online Proc. Libr.* **2012**, *1415*, 153–158.

17. Sam, E.; Mehdi, J.; Dutton, R.W.; Davis, R.W. Microfluidic diagnostic tool for the developing world: Contactless impedance flowcytometry. *Lab Chip* **2012**, *12*, 4499–4507.

18. Sam, E.; Mehdi, J.; Dutton, R.W.; Davis, R.W. Smart surface forelution of protein-protein bound particles: Nanonewton dielectrophoretic forces using atomic layer deposited oxides. *Anal. Chem.* **2012**, *84*, 10793–10801.

19. Mehdi, J.; Sam, E.; Dutton, R.W.; Davis, R.W. Use of negative dielectrophoresis for selective elution of protein-bound particles. *Anal. Chem.* **2012**, *84*, 1432–1438.

20. Mehdi, J.; Davis, R.W. A microfluidic platform for electrical detection of DNA hybridization. *Sens. Actuators B Chem.* **2011**, *154*, 22–27.

21. Davey, H.M; Kell, D.B. Flow cytometry and cell sorting of heterogeneous microbial populations: The importance of single-cell analyses. *Microbiol. Rev.* **1996**, *60*, 641–696. [PubMed]

22. Shapiro, H.M. The evolution of cytometers. *Cytometry Part A* **2004**, *58A*, 13–20. [CrossRef] [PubMed]

23. Moldavan, A. Photo-electric technique for the counting of microscopical cells. *Science* **1934**, *80*, 188–189. [CrossRef] [PubMed]

24. Manz, A.; Harrison, D.J.; Verpoorte, E.M.J.; Fettinger, J.C.; Paulus, A.; Lüdi, H. Planar chips technology for miniaturization and integration of separation techniques into monitoring systems: Capillary electrophoresis on a chip. *J. Chromatogr. A* **1992**, *593*, 253–258. [CrossRef]

25. Whitesides, G.M.; Stroock, A.D. Flexible methods for microfluidics. *Phys. Today* **2001**, *54*, 42–48. [CrossRef]

26. Quake, S. The chips are down—Microfluidic large-scale integration. *TrAC Trends Anal. Chem.* **2002**, *21*, 12–13.

27. Beebe, D.J.; Mensing, G.A.; Walker, G.M. Physics and applications of microfluidics in biology. *Annu. Rev. Biomed. Eng.* **2002**, *4*, 261–286. [CrossRef] [PubMed]

28. Stone, H.A.; Stroock, A.D.; Ajdari, A. Engineering flows in small devices: Microfluidics toward a lab-on-a-chip. *Annu. Rev. Fluid Mech.* **2004**, *36*, 381–411. [CrossRef]

29. Squires, T. Microfluidics: Fluid physics at the nanoliter scale. *Rev. Mod. Phys.* **2005**, *77*, 977–1026. [CrossRef]

30. Whitesides, G.M. The origins and the future of microfluidics. *Nature* **2006**, *442*, 368–373. [CrossRef] [PubMed]

31. Zheng, S.; Liu, M.; Tai, Y.C. Micro coulter counters with platinum black electroplated electrodes for human blood cell sensing. *Biomed. Microdevices* **2008**, *10*, 221–231. [CrossRef] [PubMed]

32. Jagtiani, A.V.; Carletta, J.; Zhe, J. An impedimetric approach for accurate particle sizing using a microfluidic Coulter counter. *J. Micromech. Microeng.* **2011**, *21*, 045036. [CrossRef]

33. Richards, A.L.; Dickey, M.D.; Kennedy, A.S. Design and demonstration of a novel micro-Coulter counter utilizing liquid metal electrodes. *J. Micromech. Microeng.* **2012**, *22*, 115012. [CrossRef]

34. Gawad, S.; Cheung, K.; Seger, U.; Bertsch, A.; Renaud, P. Dielectric spectroscopy in a micromachined flow cytometer: theoretical and practical considerations. *Lab Chip* **2004**, *4*, 241–251. [CrossRef] [PubMed]

35. Gawad, S.; Schild, L.; Renaud, P.H. Micromachined impedance spectroscopy flow cytometer for cell analysis and particle sizing. *Lab Chip* **2001**, *1*, 76–82. [CrossRef] [PubMed]

36. Rodriguez-Trujillo, R.; Mills, C.A.; Samitier, J.; Gomila, G. Low cost micro-coulter counter with hydrodynamic focusing. *Microfluid. Nanofluid.* **2007**, *3*, 171–176. [CrossRef]

37. Jagtiani, A.V.; Carletta, J.; Zhe, J. A microfluidic multichannel resistive pulse sensor using frequency division multiplexing for high throughput counting of micro particles. *J. Micromech. Microeng.* **2011**, *21*, 065004. [CrossRef]

38. Clerk, M.J. *Treatise on Electricity and Magnetism*; Clarendon Press: Oxford, UK, 1904.

39. Deblois, R.W.; Bean, C.P. Counting and sizing of submicron particles by the resistive pulse technique. *Rev. Sci. Instrum.* **1970**, *41*, 909–916. [CrossRef]

40. Smythe, W.R. Flow around a spheroid in a circular tube. *Phys. Fluids* **1964**, *7*, 633–638. [CrossRef]

41. Deblois, R.W.; Bean, C.P.; Wesley, R.K.A. Electrokinetic measurements with submicron particles and pores by the resistive pulse technique. *J. Colloid Interface Sci.* **1977**, *61*, 323–335. [CrossRef]

42. Gregg, E.C.; Steidley, K.D. Electrical counting and sizing of mammalian cells in suspension. *Biophys. J.* **1965**, *5*, 393–405. [CrossRef]

43. Saleh, O.A. A Novel Resistive Pulse Sensor for Biological Measurements. Ph.D. Thesis, Princeton University, Princeton, NJ, USA, January 2003.

44. Weatherall, E.; Willmott, G.R. Conductive and Biphasic Pulses in Tunable Resistive Pulse Sensing. *J. Phys. Chem. B* **2015**, *119*, 5328–5335. [CrossRef] [PubMed]

45. Menestrina, J.; Yang, C.; Schiel, M. Charged Particles Modulate Local Ionic Concentrations and Cause Formation of Positive Peaks in Resistive-Pulse-Based Detection. *J. Phys. Chem. C* **2014**, *118*, 2391–2398. [CrossRef]

46. Stober, G.; Steinbock, L.J.; Keyser, U.F. Modeling of colloidal transport in capillaries. *J. Appl. Phys.* **2009**, *104*, 084702. [CrossRef]

47. Steinbock, L.J.; Stober, G.; Keyser, U.F. Sensing DNA-coatings of microparticles using micropipettes. *Biosens. Bioelectron.* **2009**, *24*, 2423–2427. [CrossRef] [PubMed]

48. Davenport, M.; Healy, K.; Pevarnik, M.; Teslich, N.; Cabrini, S.; Morrison, A.P. The role of pore geometry in single nanoparticle detection. *ACS Nano* **2012**, *6*, 8366–8380. [CrossRef] [PubMed]

49. Tsutsui, M.; Hongo, S.; He, Y.; Taniguchi, M.; Gemma, N.; Kawai, T. Single-nanoparticle detection using a low-aspect-ratio pore. *ACS Nano* **2012**, *6*, 3499–3505. [CrossRef] [PubMed]

50. Carson, S.; Wilson, J.; Aksimentiev, A.; Wanunu, M. Smooth DNA transport through a narrowed pore geometry. *Biophys. J.* **2014**, *107*, 2381–2393. [CrossRef] [PubMed]

51. Sahebi, M.; Azimian, A.R. Effect of some geometrical characteristics of asymmetric nanochannels on acceleration-driven flow. *Microfluid. Nanofluid.* **2015**, *18*, 1–9. [CrossRef]

52. Kaya, D.; Dinler, A.; San, N.; Kececi, K. Effect of Pore Geometry on Resistive-Pulse Sensing of DNA Using Track-Etched PET Nanopore Membrane. *Electrochim. Acta* **2016**, *202*, 157–165. [CrossRef]

53. Liu, J.; Kvetny, M.; Feng, J.; Wang, D.; Wu, B.; Brown, W. Surface charge density determination of single conical nanopores based on normalized ion current rectification. *Langmuir* **2012**, *28*, 1588–1595. [CrossRef] [PubMed]

54. Weatherall, E.; Hauer, P.; Vogel, R.; Willmott, G.R. Pulse Size Distributions in Tunable Resistive Pulse Sensing. *Anal. Chem.* **2016**, *88*, 8648–8656. [CrossRef] [PubMed]

55. Platt, M.; Willmott, G.R.; Lee, G.U. Resistive pulse sensing of analyte-induced multicomponent rod aggregation using tunable pores. *Small* **2012**, *8*, 2436–2444. [CrossRef] [PubMed]

56. Lee, C.; Chen, C. Characterizations of nanospheres and nanorods using resistive-pulse sensing. *Microsyst. Technol.* **2017**, *23*, 299–304. [CrossRef]

57. Song, Y.; Wang, C.; Sun, R. Effect of induced surface charge of metal particles on particle sizing by resistive pulse sensing technique. *J. Colloid Interface Sci.* **2014**, *423*, 20–24. [CrossRef] [PubMed]

58. Liu, R.; Wang, N.; Kamili, F.; Sarioglu, A.F. Microfluidic CODES: A scalable multiplexed electronic sensor for orthogonal detection of particles in microfluidic channels. *Lab Chip* **2016**, *16*, 1350–1357. [CrossRef] [PubMed]

59. Song, Y.; Yang, J.; Pan, X.; Li, D. High-throughput and sensitive particle counting by a novel microfluidic differential resistive pulse sensor with multidetecting channels and a common reference channel. *Electrophoresis* **2015**, *36*, 495–501. [CrossRef] [PubMed]

60. Castillo-Fernandez, O.; Rodriguez-Trujillo, R.; Gomila, G.; Samitier, J. High-speed counting and sizing of cells in an impedance flow microcytometer with compact electronic instrumentation. *Microfluid. Nanofluid.* **2014**, *16*, 91–99. [CrossRef]

61. Choi, H.; Jeon, C.S.; Hwang, I.; Ko, J.; Lee, S.; Choo, J.; Chung, T.D. A flow cytometry-based submicron-sized bacterial detection system using a movable virtual wall. *Lab Chip* **2014**, *14*, 2327–2333. [CrossRef] [PubMed]

62. Bernabini, C.; Holmes, D.; Morgan, H. Micro-impedance cytometry for detection and analysis of micron-sized particles and bacteria. *Lab Chip* **2011**, *11*, 407–412. [CrossRef] [PubMed]

63. Liu, Z.; Li, J.; Yang, J.; Song, Y.; Pan, X.; Li, D. Improving particle detection sensitivity of a microfluidic resistive pulse sensor by a novel electrokinetic flow focusing method. *Microfluid. Nanofluid.* **2017**, *21*, 4. [CrossRef]

64. Kasianowicz, J.J.; Brandin, E.; Branton, D.; Deamer, D.W. Characterization of individual polynucleotide molecules using a membrane channel. *Proc. Natl. Acad. Sci. USA* **1966**, *93*, 13770–13773. [CrossRef]

65. Kozak, D.; Anderson, W.; Vogel, R.; Trau, M. Advances in resistive pulse sensors: Devices bridging the void between molecular and microscopic detection. *Nano Today* **2011**, *6*, 531–545. [CrossRef] [PubMed]

66. Harrer, S.; Kim, S.C.; Schieber, C.; Kannam, S.; Gunn, N.; Moore, S. Label-free screening of single biomolecules through resistive pulse sensing technology for precision medicine applications. *Nanotechnology* **2015**, *26*, 182502. [CrossRef] [PubMed]

67. Billinge, E.R.; Broom, M.; Platt, M. Monitoring aptamer-protein interactions using tunable resistive pulse sensing. *Anal. Chem.* **2014**, *86*, 1030–1037. [CrossRef] [PubMed]

68. Sha, J.; Hasan, T.; Milana, S.; Bertulli, C.; Bell, N.A.; Privitera, G.; Keyser, U.F. Nanotubes complexed with DNA and proteins for resistive-pulse sensing. *ACS Nano* **2013**, *7*, 8857–8869. [CrossRef] [PubMed]

69. Blundell, E.L.; Vogel, R.; Platt, M. Particle charge analysis of DNA-modified nanoparticles using tunable resistive pulse sensing. *Langmuir* **2016**, *32*, 1082–1090. [CrossRef] [PubMed]

70. Yang, A.K.L.; Lu, H.; Wu, S.Y.; Kwok, H.C.; Ho, H.P.; Yu, S.; Kong, S.K. Detection of Panton-Valentine Leukocidin DNA from methicillin-resistant Staphylococcus aureus by resistive pulse sensing and loop-mediated isothermal amplification with gold nanoparticles. *Anal. Chim. Acta* **2013**, *782*, 46–53. [CrossRef] [PubMed]

71. Booth, M.A.; Vogel, R.; Curran, J.M.; Harbison, S.; Travas-Sejdic, J. Detection of target-probe oligonucleotide hybridization using synthetic nanopore resistive pulse sensing. *Biosens. Bioelectron.* **2013**, *45*, 136–140. [CrossRef] [PubMed]

72. Sheng, T.H.; Ling, F.Y.; Li, N.B.; Hong, Q.L. An ultrasensitive and selective fluorescence assay for sudan i and iii against the influence of sudan ii andiv. *Biosens. Bioelectron.* **2013**, *42*, 136–140.

73. Kühnemund, M.; Nilsson, M. Digital quantification of rolling circle amplified single DNA molecules in a resistive pulse sensing nanopore. *Biosens. Bioelectron.* **2015**, *67*, 11–17. [CrossRef] [PubMed]

74. Traversi, F.; Raillon, C.; Benameur, S.M.; Liu, K.; Khlybov, S.; Tosun, M. Detecting the translocation of DNA through a nanopore using graphene nanoribbons. *Nat. Nanotechnol.* **2013**, *8*, 939–945. [CrossRef] [PubMed]

75. Hernándezainsa, S.; Bell, N.A.; Thacker, V.V.; Göpfrich, K.; Misiunas, K.; Fuentesperez, M.E. DNA origami nanopores for controlling DNA translocation. *ACS Nano* **2013**, *7*, 6024–6030. [CrossRef] [PubMed]

76. Sischka, A.; Galla, L.; Meyer, A.J.; Spiering, A.; Knust, S.; Mayer, M. Controlled translocation of DNA through nanopores in carbon nano-, silicon-nitride- and lipid-coated membranes. *Analyst* **2015**, *140*, 4843–4847. [CrossRef] [PubMed]

77. Rodriguez-Trujillo, R.; Ajine, M.A.; Orzan, A.; Mar, M.D.; Larsen, F.; Clausen, C.H. Label-free protein detection using a microfluidic coulter-counter device. *Sens. Actuators B Chem.* **2014**, *190*, 922–927. [CrossRef]

78. Cai, H.; Wang, Y.; Yu, Y.; Mirkin, M.V.; Bhakta, S.; Bishop, G.W.; Rusling, J.F. Resistive-pulse measurements with nanopipettes: Detection of vascular endothelial growth factor C (VEGF-C) using antibody-decorated nanoparticles. *Anal. Chem.* **2015**, *87*, 6403–6410. [CrossRef] [PubMed]

79. Han, Y.; Wu, H.; Liu, F.; Cheng, G.; Zhe, J. Novel quantitative macro biomolecule analysis based on a micro coulter counter. *Anal. Chem.* **2014**, *86*, 9717–9722. [CrossRef] [PubMed]

80. Takakura, T.; Yanagi, I.; Goto, Y.; Ishige, Y.; Kohara, Y. Single-molecule detection of proteins with antigen-antibody interaction using resistive-pulse sensing of submicron latex particles. *Appl. Phys. Lett.* **2016**, *108*, 123701. [CrossRef]

81. Vogel, R.; Willmott, G.; Kozak, D.; Roberts, G.S.; Anderson, W.; Groenewegen, L.; Glossop, B.; Barnett, A.; Turner, A.; Trau, M. Quantitative sizing of nano/microparticles with a tunable elastomeric pore sensor. *Anal. Chem.* **2011**, *83*, 3499–3506. [CrossRef] [PubMed]

82. Roberts, G.S.; Yu, S.; Zeng, Q.; Chan, L.C.L.; Anderson, W.; Colby, A.H.; Grinstaff, M.W.; Reid, S.; Vogel, R. Tunable pores for measuring concentrations of synthetic and biological nanoparticle dispersions. *Biosens. Bioelectron.* **2012**, *31*, 17–25. [CrossRef] [PubMed]

83. Colby, A.H.; Colson, Y.L.; Grinstaff, M.W. Microscopy and tunable resistive pulse sensing characterization of the swelling of pH responsive, polymeric expansile nanoparticles. *Nanoscale* **2013**, *5*, 3496–3504. [PubMed]

84. Fraikin, J.L.; Teesalu, T.; Mckenney, C.M.; Ruoslahti, E.; Cleland, A.N. A high-throughput label-free nanoparticle analyser. *Nat. Nanotechnol.* **2011**, *6*, 308–313. [CrossRef] [PubMed]

85. Pal, A.K.; Aalaei, I.; Gadde, S.; Gaines, P.; Schmidt, D.; Demokritou, P.; Bello, D. High Resolution Characterization of Engineered Nanomaterial Dispersions in Complex Media Using Tunable Resistive Pulse Sensing Technology. *ACS Nano* **2014**, *8*, 9003–9015. [CrossRef] [PubMed]

86. Luo, L.; German, S.; Lan, W.; Holden, D.; Mega, T.; White, H. Resistive-Pulse Analysis of Nanoparticles. *Annu. Rev. Anal. Chem.* **2014**, *7*, 513–535. [CrossRef] [PubMed]

87. Sikora, A.; Shard, A.G.; Minelli, C. Size and ζ-potential measurement of silica nanoparticles in serum using tunable resistive pulse sensing. *Langmuir* **2016**, *32*, 2216–2224. [CrossRef] [PubMed]

88. Arjmandi, N.; Roy, W.V.; Lagae, L.; Borghs, G. Measuring the electric charge and zeta potential of nanometer-sized objects using pyramidal-shaped nanopores. *Anal. Chem.* **2012**, *84*, 8490–8496. [CrossRef] [PubMed]

89. Eldridge, J.A.; Willmott, G.R.; Anderson, W.; Vogel, R. Nanoparticle ζ-potential measurements using tunable resistive pulse sensing with variable pressure. *J. Colloid Interface Sci.* **2014**, *429*, 45–52. [CrossRef] [PubMed]

90. Somerville, J.A.; Willmott, G.R.; Eldridge, J.; Griffiths, M.; McGrath, K.M. Size and charge characterisation of a submicrometre oil-in-water emulsion using resistive pulse sensing with tunable pores. *J. Colloid Interface Sci.* **2013**, *394*, 243–251. [CrossRef] [PubMed]

91. Vogel, R.; Anderson, W.; Eldridge, J.; Glossop, B.; Willmott, G. A variable pressure method for characterizing nanoparticle surface charge using pore sensors. *Anal. Chem.* **2012**, *84*, 3125–3131. [CrossRef] [PubMed]

92. Kozak, D.; Anderson, W.; Vogel, R.; Chen, S.; Antaw, F.; Trau, M. Simultaneous Size and zeta-Potential Measurements of Individual Nanoparticles in Dispersion Using Size-Tunable Pore Sensors. *ACS Nano* **2012**, *6*, 6990–6997. [CrossRef] [PubMed]

93. Dekker, C. Solid-state nanopores. *Nat. Nanotechnol.* **2007**, *2*, 209–215. [CrossRef] [PubMed]

94. Howorka, S.; Siwy, Z. Nanopore analytics: Sensing of single molecules. *Chem. Soc. Rev.* **2009**, *38*, 2360–2384. [CrossRef] [PubMed]

95. Venkatesan, B.; Bashir, R. Nanopore sensors for nucleic and acid analysis. *Nat. Nanotechnol.* **2011**, *6*, 615–624. [CrossRef] [PubMed]

96. Chen, J.T.; Wei, T.H.; Chang, C.W.; Ko, H.W.; Chu, C.W.; Chi, M.H.; Tsai, C.C. Fabrication of polymer nanopeapods in the nanopores of anodic aluminum oxide templates using a double-solution wetting method. *Macromolecules* **2014**, *47*, 5227–5235. [CrossRef]

97. Nasir, S.; Ali, M.; Ramirez, P.; Gómez, V.; Oschmann, B.; Muench, F.; Ensinger, W. Fabrication of Single Cylindrical Au-Coated Nanopores with Non-Homogeneous Fixed Charge Distribution Exhibiting High Current Rectifications. *ACS Appl. Mater. Interfaces* **2014**, *6*, 12486–12494. [CrossRef] [PubMed]

98. Bandara, Y.N.D.; Karawdeniya, B.I.; Dwyer, J.R. Real-Time Profiling of Solid-State Nanopores During Solution-Phase Nanofabrication. *ACS Appl. Mater. Interfaces* **2016**, *8*, 30583–30589. [CrossRef] [PubMed]

99. Lan, W.J.; Holden, D.A.; Zhang, B.; White, H.S. Nanoparticle transport in conical-shaped nanopores. *Anal. Chem.* **2011**, *83*, 3840–3847. [CrossRef] [PubMed]

100. Li, J.; Li, C.; Gao, X. Structural evolution of self-ordered alumina tapered nanopores with 100 nm interpore distance. *Appl. Surf. Sci.* **2011**, *257*, 10390–10394. [CrossRef]

101. Wei, R.; Pedone, D.; Zürner, A.; Döblinger, M.; Rant, U. Fabrication of metallized nanopores in silicon nitride membranes for single-molecule sensing. *Small* **2010**, *6*, 1406–1414. [CrossRef] [PubMed]

102. Sheng, Q.; Wang, L.; Wang, C.; Wang, X.; Xue, J. Fabrication of nanofluidic diodes with polymer nanopores modified by atomic layer deposition. *Biomicrofluidics* **2014**, *8*, 052111. [CrossRef] [PubMed]

103. Song, L.; Hobaugh, M.R.; Shustak, C.; Cheley, S.; Bayley, H.; Gouaux, J.E. Structure of staphylococcal α-hemolysin, a heptameric transmembrane pore. *Science* **1996**, *274*, 1859–1865. [CrossRef] [PubMed]

104. Yusko, E.C.; Johnson, J.M.; Majd, S.; Prangkio, P.; Rollings, R.C.; Li, J.; Mayer, M. Controlling protein translocation through nanopores with bio-inspired fluid walls. *Nat. Nanotechnol.* **2011**, *6*, 253–260. [CrossRef] [PubMed]

105. Göpfrich, K.; Kulkarni, C.V.; Pambos, O.J.; Keyser, U.F. Lipid nanobilayers to host biological nanopores for DNA translocations. *Langmuir* **2012**, *29*, 355–364. [CrossRef] [PubMed]

106. Hemmler, R.; Böse, G.; Wagner, R.; Peters, R. Nanopore unitary permeability measured by electrochemical and optical single transporter recording. *Biophys. J.* **2005**, *88*, 4000–4007. [CrossRef] [PubMed]

107. Pastoriza-Gallego, M.; Rabah, L.; Gibrat, G.; Thiebot, B.; van der Goot, F.G.; Auvray, L.; Pelta, J. Dynamics of unfolded protein transport through an aerolysin pore. *J. Am. Chem. Soc.* **2011**, *133*, 2923–2931. [CrossRef] [PubMed]

108. Yilun, Y.; Xing, Z.; Yu, L.; Mengzhu, X.; Honglin, L.; Yitao, L. Single molecule study of the weak biological interactions between p53 and DNA. *Acta Chim. Sin.* **2013**, *71*, 44–50.

109. Mohammad, M.M.; Raghuvaran Iyer, K.R.H.; McPike, M.P.; Borer, P.N.; Movileanu, L. Engineering a rigid protein tunnel for biomolecular detection. *J. Am. Chem. Soc.* **2012**, *134*, 9521–9531. [CrossRef] [PubMed]

110. Niedzwiecki, D.J.; Mohammad, M.M.; Movileanu, L. Inspection of the engineered fhua δc/δ4l protein nanopore by polymer exclusion. *Biophys. J.* **2012**, *103*, 2115–2124. [CrossRef] [PubMed]

111. Soskine, M.; Biesemans, A.; Moeyaert, B.; Cheley, S.; Bayley, H.; Maglia, G. An engineered ClyA nanopore detects folded target proteins by selective external association and pore entry. *Nano Lett.* **2012**, *12*, 4895–4900. [CrossRef] [PubMed]

112. Ying, Y.L.; Li, D.W.; Liu, Y.; Dey, S.K.; Kraatz, H.B.; Long, Y.T. Recognizing the translocation signals of individual peptide–oligonucleotide conjugates using an α-hemolysin nanopore. *Chem. Commun.* **2012**, *48*, 8784–8786. [CrossRef] [PubMed]

113. Singh, P.R.; Bárcena-Uribarri, I.; Modi, N.; Kleinekathöfer, U.; Benz, R.; Winterhalter, M.; Mahendran, K.R. Pulling peptides across nanochannels: Resolving peptide binding and translocation through the hetero-oligomeric channel from nocardia farcinica. *ACS Nano* **2012**, *6*, 10699–10707. [CrossRef] [PubMed]

114. Kudr, J.; Skalickova, S.; Nejdl, L.; Moulick, A.; Ruttkay–Nedecky, B.; Adam, V.; Kizek, R. Fabrication of solid-state nanopores and its perspectives. *Electrophoresis* **2015**, *36*, 2367–2379. [CrossRef] [PubMed]

115. Sha, J.; Si, W.; Xu, W.; Zou, Y.; Chen, Y. Glass capillary nanopore for single molecule detection. *Sci. China Technol. Sci.* **2015**, *58*, 803–812. [CrossRef]

116. Steinbock, L.J.; Otto, O.; Chimerel, C.; Gornall, J.; Keyser, U.F. Detecting DNA folding with nanocapillaries. *Nano Lett.* **2010**, *10*, 2493–2497. [CrossRef] [PubMed]

117. Steinbock, L.J.; Bulushev, R.D.; Krishnan, S.; Raillon, C.; Radenovic, A. DNA translocation through low-noise glass nanopores. *ACS Nano* **2013**, *7*, 11255–11262. [CrossRef] [PubMed]

118. Cai, S.L.; Cao, S.H.; Zheng, Y.B.; Zhao, S.; Yang, J.L.; Li, Y.Q. Surface charge modulated aptasensor in a single glass conical nanopore. *Biosens. Bioelectron.* **2015**, *71*, 37–43. [CrossRef] [PubMed]

119. He, H.; Xu, X.; Wang, P.; Chen, L.; Jin, Y. The facile surface chemical modification of a single glass nanopore and its use in the nonenzymatic detection of uric acid. *Chem. Commun.* **2015**, *51*, 1914–1917. [CrossRef] [PubMed]

120. Bafna, J.A.; Soni, G.V. Fabrication of low noise borosilicate glass nanopores for single molecule sensing. *PLoS ONE* **2016**, *11*, e0157399. [CrossRef] [PubMed]

121. Ali, M.; Yameen, B.; Neumann, R.; Ensinger, W.; Knoll, W.; Azzaroni, O. Biosensing and supramolecular bioconjugation in single conical polymer nanochannels. Facile incorporation of biorecognition elements into nanoconfined geometries. *J. Am. Chem. Soc.* **2008**, *130*, 16351–16357. [CrossRef] [PubMed]

122. Wang, J.; Martin, C.R. A new drug-sensing paradigm based on ion-current rectification in a conically shaped nanopore. *Nanomedicine* **2008**, *3*, 13–20. [CrossRef] [PubMed]

123. Chavan, V.; Agarwal, C.; Pandey, A.K.; Nair, J.P.; Surendran, P.; Kalsi, P.C.; Goswami, A. Controlled development of pores in polyethylene terepthalate sheet by room temperature chemical etching method. *J. Membr. Sci.* **2014**, *471*, 185–191. [CrossRef]

124. Venta, K.; Shemer, G.; Puster, M.; Rodriguez-Manzo, J.A.; Balan, A.; Rosenstein, J.K.; Drndić, M. Differentiation of short single-stranded DNA homopolymers in solid-state nanopores. *ACS Nano* **2013**, *7*, 4629–4636. [CrossRef] [PubMed]

125. Zhang, M.; Schmidt, T.; Sangghaleh, F.; Roxhed, N.; Sychugov, I.; Linnros, J. Oxidation of nanopores in a silicon membrane: Self-limiting formation of sub-10 nm circular openings. *Nanotechnology* **2014**, *25*, 355302. [CrossRef] [PubMed]

126. Balme, S.; Coulon, P.E.; Lepoitevin, M.; Charlot, B.; Yandrapalli, N.; Favard, C.; Janot, J.M. Influence of Adsorption on Proteins and Amyloid Detection by Silicon Nitride Nanopore. *Langmuir* **2016**, *32*, 8916–8925. [CrossRef] [PubMed]

127. Kwok, H.; Briggs, K.; Tabard-Cossa, V. Nanopore fabrication by controlled dielectric breakdown. *PLoS ONE* **2014**, *9*, e92880. [CrossRef] [PubMed]

128. Briggs, K.; Kwok, H.; Tabard-Cossa, V. Automated Fabrication of 2-nm Solid-State Nanopores for Nucleic Acid Analysis. *Small* **2014**, *10*, 2077–2086. [CrossRef] [PubMed]

129. Yanagi, I.; Akahori, R.; Hatano, T.; Takeda, K.I. Fabricating nanopores with diameters of sub-1 nm to 3 nm using multilevel pulse-voltage injection. *Sci. Rep.* **2014**, *4*, 5000. [CrossRef] [PubMed]

130. Kumar, A.; Park, K.B.; Kim, H.M.; Kim, K.B. Noise and its reduction in graphene based nanopore devices. *Nanotechnology* **2013**, *24*, 495503. [CrossRef] [PubMed]

131. Fox, D.S.; Maguire, P.; Zhou, Y.; Rodenburg, C.; O'Neill, A.; Coleman, J.N.; Zhang, H. Sub-5 nm graphene nanopore fabrication by nitrogen ion etching induced by a low-energy electron beam. *Nanotechnology* **2016**, *27*, 195302. [CrossRef] [PubMed]

132. Jung, W.; Kim, J.; Kim, S.; Park, H.G.; Jung, Y.; Han, C.S. A Novel Fabrication of 3.6 nm High Graphene Nanochannels for Ultrafast Ion Transport. *Adv. Mater.* **2017**, *29*, 1605854. [CrossRef] [PubMed]

133. Goyal, G.; Lee, Y.B.; Darvish, A.; Ahn, C.W.; Kim, M.J. Hydrophilic and size-controlled graphene nanopores for protein detection. *Nanotechnology* **2016**, *27*, 495301. [CrossRef] [PubMed]

134. Deng, Y.; Huang, Q.; Zhao, Y.; Zhou, D.; Ying, C.; Wang, D. Precise fabrication of a 5 nm graphene nanopore with a helium ion microscope for biomolecule detection. *Nanotechnology* **2016**, *28*, 045302. [CrossRef] [PubMed]

135. Perry, J.M.; Zhou, K.; Harms, Z.D.; Jacobson, S.C. Ion transport in nanofluidic funnels. *ACS Nano* **2010**, *4*, 3897–3902. [CrossRef] [PubMed]

136. Kim, S.H.; Cui, Y.; Lee, M.J.; Nam, S.W.; Oh, D.; Kang, S.H.; Park, S. Simple fabrication of hydrophilic nanochannels using the chemical bonding between activated ultrathin PDMS layer and cover glass by oxygen plasma. *Lab Chip* **2011**, *11*, 348–353. [CrossRef] [PubMed]

137. Harms, Z.D.; Mogensen, K.B.; Nunes, P.S.; Zhou, K.; Hildenbrand, B.W.; Mitra, I.; Jacobson, S.C. Nanofluidic devices with two pores in series for resistive-pulse sensing of single virus capsids. *Anal. Chem.* **2011**, *83*, 9573–9578. [CrossRef] [PubMed]

138. Nam, S.W.; Lee, M.H.; Lee, S.H.; Lee, D.J.; Rossnagel, S.M.; Kim, K.B. Sub-10-nm nanochannels by self-sealing and self-limiting atomic layer deposition. *Nano Lett.* **2010**, *10*, 3324–3329. [CrossRef] [PubMed]

139. Fouad, M.; Yavuz, M.; Cui, B. Nanofluidic channels fabricated by e-beam lithography and polymer reflow sealing. *J. Vac. Sci. Technol. B* **2010**, *28*, C6I11–C6I13. [CrossRef]

140. Chen, Y. Nanofabrication by electron beam lithography and its applications: A review. *Microelectron. Eng.* **2015**, *135*, 57–72. [CrossRef]

141. Williams, C.; Bartholomew, R.; Rughoobur, G.; Gordon, G.S.; Flewitt, A.J.; Wilkinson, T.D. Fabrication of nanostructured transmissive optical devices on ITO-glass with UV1116 photoresist using high-energy electron beam lithography. *Nanotechnology* **2016**, *27*, 485301. [CrossRef] [PubMed]

142. Mussi, V.; Fanzio, P.; Repetto, L.; Firpo, G.; Scaruffi, P.; Stigliani, S.; Valbusa, U. DNA-functionalized solid state nanopore for biosensing. *Nanotechnology* **2010**, *21*, 145102. [CrossRef] [PubMed]

143. Tian, Z.P.; Lu, K.; Chen, B. Unique nanopore pattern formation by focused ion beam guided anodization. *Nanotechnology* **2010**, *21*, 405301. [CrossRef] [PubMed]

144. Yamamoto, T.; Fujii, T. Nanofluidic single-molecule sorting of DNA: A new concept in separation and analysis of biomolecules towards ultimate level performance. *Nanotechnology* **2010**, *21*, 395502. [CrossRef] [PubMed]

145. Menard, L.D.; Ramsey, J.M. The fabrication of sub-5-nm nanochannels in insulating substrates using focused ion beam milling. *Nano Lett.* **2011**, *11*, 512–517. [CrossRef] [PubMed]

146. Angeli, E.; Manneschi, C.; Repetto, L.; Firpo, G.; Valbusa, U. DNA manipulation with elastomeric nanostructures fabricated by soft-moulding of a FIB-patterned stamp. *Lab Chip* **2011**, *11*, 2625–2629. [CrossRef] [PubMed]

147. Fanzio, P.; Mussi, V.; Manneschi, C.; Angeli, E.; Firpo, G.; Repetto, L.; Valbusa, U. DNA detection with a polymeric nanochannel device. *Lab Chip* **2011**, *11*, 2961–2966. [CrossRef] [PubMed]

148. Wu, J.; Chantiwas, R.; Amirsadeghi, A.; Soper, S.A.; Park, S. Complete plastic nanofluidic devices for DNA analysis via direct imprinting with polymer stamps. *Lab Chip* **2011**, *11*, 2984–2989. [CrossRef] [PubMed]

149. La Ferrara, V.; Aneesh, P.M.; Veneri, P.D.; Mercaldo, L.V.; Usatii, I.; Polichetti, T.; Cusano, A. Focused ion beam strategy for nanostructure milling in doped silicon oxide layer for light trapping applications. *Vacuum* **2014**, *99*, 135–142. [CrossRef]

150. Duan, C.; Wang, W.; Xie, Q. Review article: Fabrication of nanofluidic devices. *Biomicrofluidics* **2013**, *7*, 026501. [CrossRef] [PubMed]

151. Peng, R.; Li, D. Fabrication of nanochannels on polystyrene surface. *Biomicrofluidics* **2015**, *9*, 024117. [CrossRef] [PubMed]

152. Peng, R.; Li, D. Fabrication of polydimethylsiloxane (PDMS) nanofluidic chips with controllable channel size and spacing. *Lab Chip* **2016**, *16*, 3767–3776. [CrossRef] [PubMed]

micromachines

MDPI

Review

Recent Advances and Future Perspectives on Microfluidic Liquid Handling

Nam-Trung Nguyen *, Majid Hejazian, Chin Hong Ooi and Navid Kashaninejad

Queensland Micro- and Nanotechnology Centre, Nathan Campus, Griffith University, 170 Kessels Road, Brisbane, QLD 4111, Australia; majid.hejazian@griffithuni.edu.au (M.H.); chinhong.ooi@griffithuni.edu.au (C.H.O.); n.kashaninejad@griffith.edu.au (N.K.)
* Correspondence: nam-trung.nguyen@griffith.edu.au; Tel.: +61-(0)7373-53921

Academic Editor: Shih-Kang Fan
Received: 23 May 2017; Accepted: 8 June 2017; Published: 12 June 2017

Abstract: The interdisciplinary research field of microfluidics has the potential to revolutionize current technologies that require the handling of a small amount of fluid, a fast response, low costs and automation. Microfluidic platforms that handle small amounts of liquid have been categorised as continuous-flow microfluidics and digital microfluidics. The first part of this paper discusses the recent advances of the two main and opposing applications of liquid handling in continuous-flow microfluidics: mixing and separation. Mixing and separation are essential steps in most lab-on-a-chip platforms, as sample preparation and detection are required for a variety of biological and chemical assays. The second part discusses the various digital microfluidic strategies, based on droplets and liquid marbles, for the manipulation of discrete microdroplets. More advanced digital microfluidic devices combining electrowetting with other techniques are also introduced. The applications of the emerging field of liquid-marble-based digital microfluidics are also highlighted. Finally, future perspectives on microfluidic liquid handling are discussed.

Keywords: continuous microfluidics; micromixers; cell separation; digital microfluidics; liquid marbles; electrowetting-on-dielectric (EWOD); microfluidic liquid handling

---

1. Introduction

In recent years, the technology of microfluidics has progressed rapidly and become an integral part in many engineering and biomedical applications [1]. Microfluidics has been regarded as the main driver for the paradigm shift in four main areas: molecular analysis, biodefence, molecular biology and microelectronics [2]. The integration of microfluidic components into a single chip led to the advent of lab-on-a-chip (LOC) [3], micro total analysis system (μTAS) [4] and point-of-care (POC) diagnostic devices [5]. In such devices, as well as most biological processes, liquid handling is of great importance, as its quality can significantly affect the end results. According to the way a small liquid amount is handled and manipulated, the field of microfluidics is further classified as continuous-flow microfluidics and digital (droplet-based) microfluidics.

Continuous-flow microfluidics requires an external means to deliver the continuous flow of a single liquid phase or multiple phases through microchannels [6]. The two major and opposing fluid handling tasks of continuous-flow microfluidics are mixing and separation. In particular, mixing of reactants is required to initiate the interactions involved in biological processes such as protein folding and enzyme reactions [7]. For instance, in tumor-on-a-chip microfluidic platforms [8], mixing and delivery of a combination of drugs are necessary. Separation also plays an important role in sample preparation for both analytical chemistry and biological applications [9]. Additionally, cell sorting and separation need to be carried out precisely to develop microfluidic disease models and POC diagnostic tools. Yet, using continuous flow microfluidic technology for mixing and separation seems paradoxical. On the one hand, the high surface-to-volume ratio in microfluidics reduces the required sample, and is ideal for biological, biochemical and pharmaceutical applications. On the other hand, the dominant laminar and low-Reynolds-number flow regime delays the mixing and separation process, and requires a larger mixing and separation length scale. This problem indicates the need for innovative mixing/separation methods, especially for LOC applications, where a number of components need to be integrated on a single chip. Traditionally, these methods are categorised as passive (without external energy) and active (in the presence of external energy) techniques [7].

The advantages and disadvantages of passive methods, which utilise chaotic advection to reduce the mixing time, were extensively reviewed by Suh and Kang [10]. The operation principles and mixing capabilities of a broad range of predominantly used micromixers were reviewed by Lee et al. [11]. Ward and Fan [12] categorised and discussed a variety of basic passive microfluidic mixing enhancement techniques, such as slanted wells/pillars, multiphase mixing enhancement and active enhancement techniques, such as thermal enhancement, acoustic waves and flow pulsation. A number of review articles addressed the current state of microfluidic separation techniques. For instance, Sajeesh and Sen presented a comprehensive review on different microfluidic passive and active techniques for particle separation and sorting [9]. In cell biology, microfluidic methods that do not require biochemical labels to isolate and identify cells are referred to as label-free techniques, and have attracted a great deal of attention. Gosset et al. reviewed label-free microfluidic techniques that use the intrinsic properties of the cell, such as its size and other physical signatures [13]. Microfluidic techniques can also be used for detection and separation of cancer cells. Chen et al. [14] discussed high-throughput microfluidic techniques, such as cell-affinity micro-chromatography and magnetically activated sorting. Shields et al. presented recent advances in microfluidic cell separation, along with the challenges in the commercialisation of such devices for practical clinical applications [15].

Combining microfluidics with the science of emulsion, digital microfluidics (DFM) has been developed as a technology dealing with the manipulation of individual droplets, rather than continuous streams of liquid [16]. This field has numerous applications and has the potential to revolutionise various biochemical and biomedical protocols, as well as cell-based assays [17]. DMF has numerous advantages, such as minimum reagent requirement, fast response rates, and more importantly, the capability of performing several parallel procedures [18]. These advantages make DMF an ideal candidate for practical LOC and POC diagnostic devices in clinical use [19]. However, there are still many challenges that need to be addressed in this field, such as droplet evaporation, droplet handling techniques, material selection, etc. [20]. A few recent review articles exist in this emerging field. Samiei et al. [21] reviewed the recent advances in DMF regarding fabrication technology, handling of biological reagents, packaging and portability. Using magnetic actuations to handle the individual droplet is also of great interest. Possibilities and challenges of magnetic digital microfluidics were recently reviewed by Zhang and Nguyen [22].

The scope of the present review paper is summarised in Figure 1. The first part of this paper discusses recent advances of continuous-flow microfluidics in liquid handling, i.e., mixing and separation. In particular, recent progress regarding two key mixing enhancement techniques, namely external forces and complex geometry, are revisited. Subsequently, the paper discusses continuous-flow microfluidic separation techniques such as magnetofluidics, inertial microfluidics, acoustofludics, dielectrophoretics and optofludics. The second part of this paper mainly deals with the advances in both droplet-based DMF and liquid-marble-based DMF. This part discusses the most common methods of droplet-based DMF, such as electrowetting-on-dielectric (EWOD), dielectrophoresis, and magnetic techniques to dispense, move or mix droplets. Finally, the promising field of liquid-marble-based DMF, along with its application as a microbioreactor to culture three-dimensional tissues, will be highlighted.

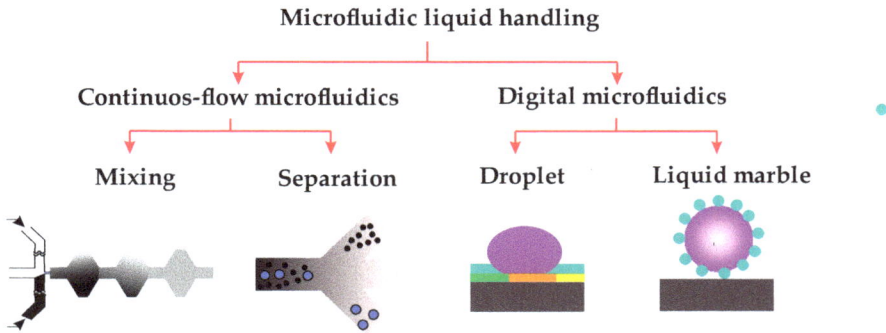

Figure 1. Scope of the present review. Microfluidic liquid handling are important parts of biological processes that can be divided into continuous-flow microfluidics and digital microfluidics. Mixing and separation are two common liquid handling techniques in continuous-flow microfluidics, whereas droplet-based and liquid-marble-based digital microfluidic technologies are used for manipulating discrete droplets.

2. Continuous Flow Microfluidics

2.1. Mixing

Mixing is an essential step in most lab-on-a-chip platforms, as sample preparation is required for a variety of biological and chemical assays. Diffusion-based mixing techniques fail to satisfy the recent demand for rapid and homogeneous mixing. Various strategies have been implemented to enhance the efficiency of continuous-flow microfluidic mixing. In this section, we present the recent advances in continuous mixing with microfluidics.

2.1.1. Mixing with External Energy Sources

One of the strategies for increasing mixing efficiency is employing external energy sources to create disturbances, such as acoustic, magnetic, electrostatic. Mass transport of a species in a superparamagnetic solution can be enhanced with an external magnetic field [23]. Utilising embedded electromagnets for magnetofluidic actuation, Mao and Koser [24] demonstrated that the mixing of two streams can be significantly improved. Hejazian and Nguyen [25] proposed a rapid and efficient micromixer using a permanent magnet and a magnetic fluid. The permanent magnet induces a non-uniform magnetic field, and correspondingly, a secondary flow, that mixes a non-magnetic stream with another stream containing diluted ferrofluid. Workamp et al. [26] presented a microfluidic suspension-based mixer with low pressure drop. The mixer consists of a chamber where particles are driven by a moving magnet. Peng et al. [27] proposed a micromixer based on parallel manipulation of individual magnetic microbeads. Rotating magnets generate a circular motion of magnetic beads. As a result, local vortices are created across the microchannel, leading to efficient mixing. Venancio-Marques et al. [28] demonstrated optofluidic mixing in a microfluidic device. As shown schematically in Figure 2, the system consists of three streams, a photosensitive water stream sandwiched between two oil phases. Without light illumination, the flow system is a typical flow-focusing configuration [29]. Light illumination generates water micro-droplets that stir and mix the two continuous oil streams.

Ober et al. [30] examined a rational framework for designing microfluidic active mixers, Figure 3. The micromixers were 3D printed and integrated with a rotating impeller. The capability of continuous mixing of complex fluids was demonstrated. Furthermore, the relationships between mixer dimensions and operating conditions were verified experimentally.

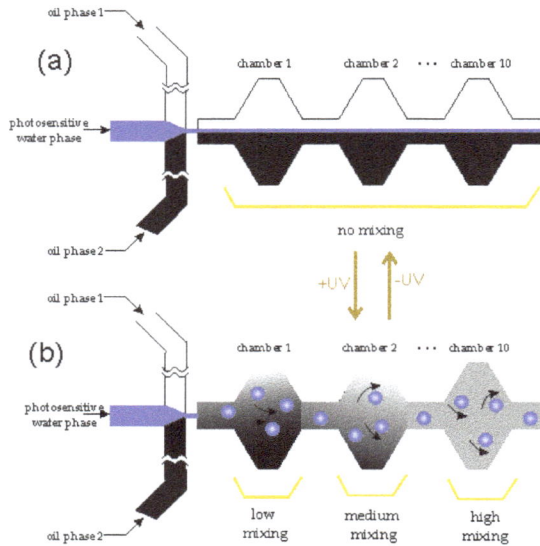

Figure 2. Schematic of the reversible optofluidic mixer developed by Venancio-Marques et al. [28]: (a) when the ultraviolet (UV) is off, the two oil phases are not mixed together; (b) at the presence of UV, the photosensitive water turns into the droplets, causing the mixing between two oil phases. Adapted from [28].

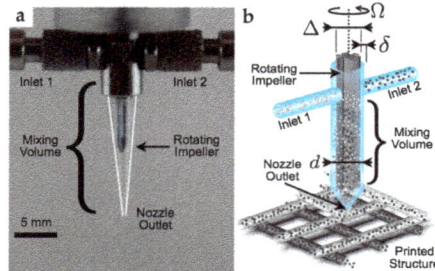

Figure 3. Impeller-based active mixer developed by Ober et al. [30]: (a) optical image of the mixer; (b) representation of the mixing nozzle. Reproduced with permission (granted by PNAS for non-commercial purposes) from [30].

Cui et al. [31] proposed a microfluidic mixer based on acoustically induced vortices created by localized ultrahigh frequency (UHF) acoustic fields. A UHF piezoelectric resonator (SMR) was capable of generating powerful acoustic streaming vortices, resulting in efficient mixing. The authors reported homogeneous mixing, with 87% mixing efficiency at a Peclet number of 35,520, within just 1 ms. Fang et al. [32] proposed a micromixer with a streamline herringbone structure, based on total glass. High direct current (DC) voltage-activated migration condition was applied to the microfluidic device as well, and the performance of the mixer was investigated. They reported an efficiency of over 90% in 20 mm, in a mixing channel of only 300 nL. Shang et al. [33] explored a vortex generated by an acoustic actuator within a circular chamber to improve mixing. The strength of the vortex was tuned by the applied voltage. Their research thus showed that mixing efficiency can be increased by adjusting the voltage.

2.1.2. Mixing with Complex Geometries

Using external actuations to increase the mixing efficiency could be expensive and challenging [29]. Another alternative technique for increasing the mixing efficiency is utilizing relatively complex geometries for chaotic advection. As the flow regime in most microfluidic systems is laminar, the quality of mixing is highly dependent on chaotic advection induced by the geometry of the microchannel. Wu and Nguyen [29] evaluated, both analytically and experimentally, the mixing efficacy of a rectangular microchannel using two-phase hydraulic focusing. To that end, two streams of sheath flow were used to hydraulically focus two streams of sample flow. Their results showed that the focusing ratio was a function of both viscosity ratio and flow rate of sheath and sample flows. To further enhance the mixing efficiency, Nguyen and Huang [34] combined the hydrodynamic focusing technique with time-interleaved segmentation. The results of the paper revealed that, while hydrodynamic focusing could reduce the transversal mixing path, sequential segmentation could also be used to decrease the axial mixing path. It was found that switching frequency and average flow velocity also affected the mixing quality. Cortelezzi et al. [35] proposed a geometrically scalable micromixer capable of achieving fast mixing over a wide range of operating conditions. As shown in Figure 4, the mixer consists of a cylindrical mixing chamber and a cylindrical obstacle. With alternate switching of the inlets to create time-interleaved segmentation, the mixer could reach an efficiency of about 90.8%.

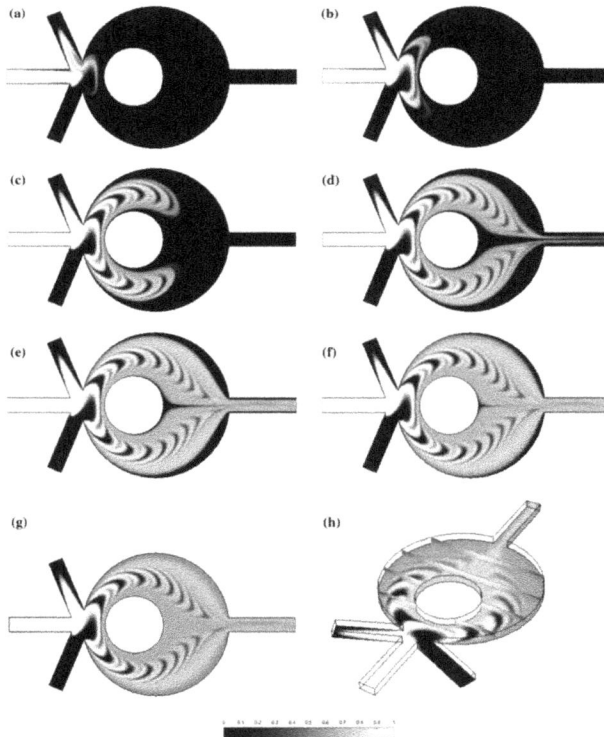

Figure 4. Fast response and geometrically scalable micromixer proposed by Cortelezzi et al. [35]: (**a**–**g**) two-dimensional representation of concentration distribution when time evolves form 2.6, 3.6, 7.6, 11.6, 19.6, 39.6 to 199.6 s, respectively; (**h**) three-dimensional representation of the concentration distribution at 199.6 s. Reproduced with permission from the original in study [35].

Kwak et al. [36] proposed the use of a positive repeated pattern of a staggered herringbone mixer (SHM) in a microchannel to improve mixing efficiency, and compared the results with those obtained from the negative pattern of SHM. It was found that the mixing efficacy would be higher if positive SHM and/or forward flow were used. In particular, a positive pattern SHM could reach completed mixing after two cycles with both forward and reverse flows, while four and five cycles were needed for complete mixing in the negative pattern SHM with forward and reverse flow directions, respectively, Figure 5.

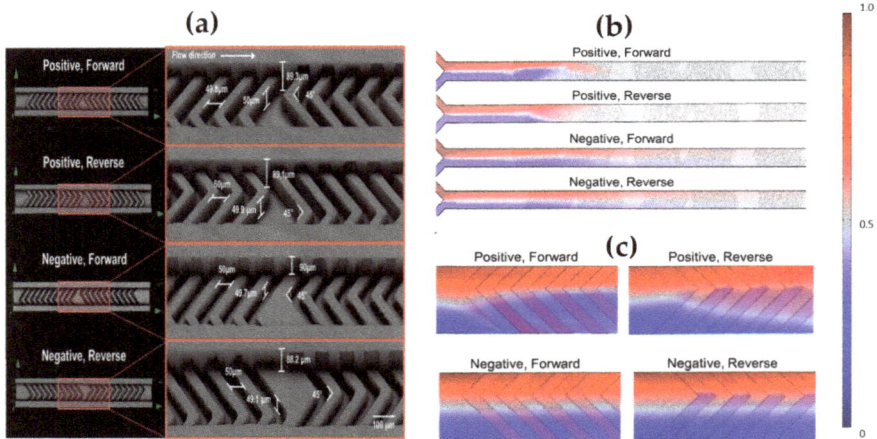

Figure 5. The staggered herringbone mixer (SHM) created by Kwak et al. [36]. (**a**) Detailed pattern structures and flow directions; (**b**) Mixing quality after 2.5 cycles in positive and negative SHM structures subject to both forward and reverse flow directions; (**c**) Top view images of four different SHMs indicating the mixing efficiency at the beginning of the first cycle. Red and blue colors correspond to fluorescence dye and water, respectively, while white color indicates complete mixing. Reproduced with permission (under Creative Commons Attribution (CC BY) license) from [36].

Salieb-Beugelaar et al. [37] presented microfluidic 3D helix mixers for controlled chemical reactions. The authors created the complex channel geometry with thread embedded in polydimethylsiloxane (PDMS). The threads created double helix and triple helix structures in the same device.

Adam and Hashim [38] reported the design and the fabrication of a micromixer with short turns and showed that it could reach a mixing efficiency of 98% at Reynolds number less than 2. Sivashankar et al. [39] proposed a micromixer with a twisted structure to enhance mixing. The 3D microfluidic mixer was fabricated by laser micromachining. The results showed that good mixing can be achieved with more than three mixing units. Wang et al. [40] used triangle baffles embedded in a microchannel to enhance mixing. The simulation results show that mixing efficiency can be improved by increasing the apical angle of the triangles from 30° to 150°. Lehmann et al. [41] performed continuous recalcification of citrated whole blood using a microfluidic herringbone mixer. A herringbone structure was fabricated on top of the channel to generate transverse flows within the microfluidic channel.

Plevniak et al. [42] demonstrated a 3D printed microfluidic mixer for fast mixing of reagents with blood through capillary force. The device was integrated with a smartphone for the point-of-care diagnosis of anemia from a finger-prick blood sample. The results obtained with the device are in line with clinical measurements. Li et al. [43] proposed a microfluidic mixer consisting of an irregular Y junction followed by an observation channel. The mixer was ultra-rapid, as complete mixing was achieved with a mixing time of just 5.5 μs. The authors interrogated the hairpin formation in the early folding process of human telomere G–quadruplex.

2.2. Separation

In the last two decades, significant advances have been made in the development of continuous-flow microfluidic separation. With continuous injection and collection of samples, a high separation throughput can be achieved. Moreover, continuous-flow microfluidic separation also has the benefit of real-time monitoring, and the potential for the integration with other continuous-flow processes [44]. Based on the unique signature of the sample components, a suitable external force can be chosen for the separation process. The separation of particles and cells can employ a variety of external forces such as hydrodynamic, electrophoretic, dielectrophoretic, magnetophoretic, acoustic, and inertial force [45]. In this section, we explore the current range of continuous separation methods.

2.2.1. Magnetofluidic Separation

Continuous-flow magnetofluidic separation has recently gained considerable interest from the research community. Due to the contactless nature of magnetic force, magnetofluidic methods do not alter the pH level or the temperature of the sample, and as a result, it has no negative effect on the viability of cells [45–47]. Magnetofluidic separation of cells and particles is categorised into two main concepts: positive and negative magnetophoresis. If the magnetic susceptibility of the medium fluid is higher than that of the particles, negative magnetophoresis occurs, and vice versa. Over the last decade, a number of reviews have been published on magnetofluidics, reporting a diverse range of techniques for separation of particles and cells, based on negative and positive magnetophoresis [47–51]. Superparamagnetic carrier fluids, such as ferrofluid, create a secondary flow towards the source of a magnetic field. This phenomenon is called magnetoconvection [23,52]. Exploiting magnetoconvection, a highly size-sensitive separation of microparticles was achieved within a microchannel [53]. Using two arrays of attracting magnets, non-magnetic polystyrene micro-particles were captured in different locations along a straight microchannel. Applying a similar concept, Zhou et al. [54] introduced a platform for simultaneous capture of non-magnetic and magnetic particles. For this purpose, an external magnetic field was generated with a permanent magnet positioned next to a T–junction in the microchannel.

Particle focusing with magnetofluidics has been reported using two sets of repelling magnets [49,51,55]. Liang and Xuan [56] reported sheathless focusing of non-magnetic particles. A T–microchannel, a single permanent magnet, and diluted ferrofluid as the superparamagnetic carrier fluid, were used for this purpose. A relatively strong magnetic field gradient should be implemented to achieve high efficiency and size sensitivity. For instance, decreasing the distance between the external magnetic field source and the fluidic channel is a solution for increasing the magnetic field gradient. Zhou and Wang [57] introduced a convenient and low-cost technique for the enhancement of magnetic field gradient. For this purpose, a prefabricated channel was formed next to the microfluidic channel. A mixture of iron powder and polydimethylsiloxane (PDMS) was injected into the channel. The iron–PDMS structures were placed just a few microns from the microchannel. Separation of nanoparticles with magnetofluidics has also recently gained attention. Wu et al. [58] proposed an efficient method for size-selective separation of magnetic nanospheres using a magnetofluidic device. Two monodisperse nanosphere samples (90 nm and 160 nm) were successfully separated from the polydispersing particles solution, with varied particle diameters from 40 to 280 nm.

2.2.2. Inertial Microfluidics

Inertial microfluidics is another emerging field of continuous-flow particle separation. Inertial microfluidics is a suitable method for rare cell sorting, due to such various advantages as high throughput, simplicity, precise manipulation and low cost [59,60]. A number of reviews have summarised the existing techniques and designs of inertial microfluidics [59–63]. The inertial force is often combined with other forces, such as hydraulic, magnetic, centrifugal, or hydrodynamic forces, in order to obtain a higher separation efficiency. Ahn et al. [64] designed a sheathless elasto-inertial focusing microfluidic separator, and performed a systematic study evaluating the parameters affecting the performance of a microfluidic separator based on inertial microfluidics. The schematic illustration along with the working principles of their fabricated microfluidic separator is shown in Figure 6.

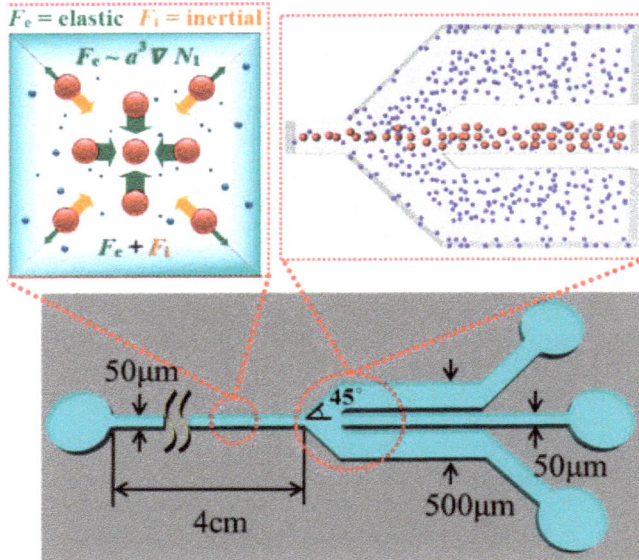

Figure 6. Microfluidic separator based on inertial microfluidics with its working principles developed by Ahn et al. [64]. Reproduced with permission from [64].

Optimisation parameters, such as particle concentration and flow rate, as well as the effect of particle–particle interaction in the separation process, were determined [65]. Combining lift forces and Dean flow drag forces, algae species were separated, based on their shape and size, in a spiral microchannel. Monoraphidium species was successfully separated from the differently shaped Cyanothece, with 77% separation efficiency.

Zhou et al. [66] demonstrated a hybrid method based on the combination of inertial microfluidics and magnetofluidics for size-selective separation of micro-particles. Spherical diamagnetic polystyrene particles of 10 μm and 20 μm were successfully separated using this technique. Clime et al. [67] furthermore demonstrated filtration and extraction of pathogens from food samples, utilising hydrodynamic focusing and inertial lateral migration. The microfluidic platform was capable of removing up to 50% of debris from ground beef samples.

2.2.3. Acoustofluidic Separation

The use of acoustic waves is another technique that has been used for continuous particle separation with microfluidics. Because of such advantages as simplicity of design, low-cost, and biocompatibility due to its contactless nature, acoustic wave devices have been integrated with microfluidic devices. A number of recent reviews reported on the different configurations of acoustofluidic devices [68–73]. Mathew et al. [74] developed a two-dimensional dynamic model for tracing the path of microparticles in continuous-flow microfluidics employing acoustic waves. The effect of parameters, such as acoustic energy density and initial vertical location, on the displacement of microparticles were examined with this model. Shields et al. [75] designed a multi-stage microfluidic platform for separation of cancer cells from blood. In the first module, the acoustic standing wave is exploited for immediate alignment of cells. Magnetic separation techniques then purify and capture individual cells for on-chip analyses, in the next two steps. Ng et al. [76] designed a flow-rate-insensitive device for continuous particle sorting, Figure 7. The device uses surface acoustic waves that combine both standing and travelling wave components to create pressure nodes. The particles were trapped in locations with a stable pressure based on their size, and separated through a distinct exit.

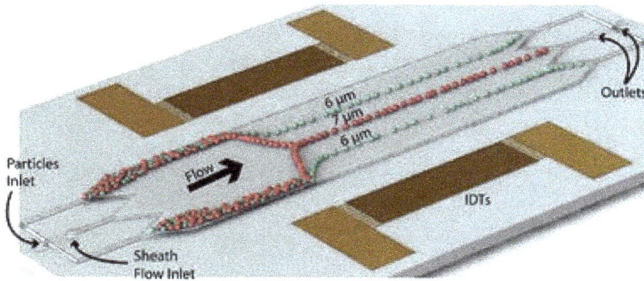

Figure 7. Schematic illustration of flow-rate-insensitive device for continuous particle sorting, designed by Ng et al. [76]. Reproduced with permission from [64,76].

2.2.4. Dielectrophoretic Separation

Dielectrophoretic method has been another area of interest for continuous particle separation with microfluidics in recent years. The use of dielectrophoretic force with microfluidics for continuous particle separation has advantages such as low cost, rapidity, size sensitivity, and selectivity. Previously published reviews discuss a variety of techniques used for dielectrophoretics-based cell and particle separation [77–81]. Cui et al. [82] proposed a dielectrophoresis (DEP)-based method for size-based particle separation. The authors demonstrated the extraction of larger particles, retaining small particles, and also eluting mid-size particles using pulsed dielectrophoresis. Kim et al. [83] proposed an integrated Dielectrophoretic–Magnetic Activated Cell Sorter (iDMACS). The target cell types were sorted based on surface markers, via specific receptor–ligand binding to either DEP or magnetic tags. The device could achieve 900-fold enrichment of multiple bacterial target cell types, with over 95% purity after a single round of separation. Yang et al. [84] examined dielectrophoresis (DEP)-active hydrophoresis for sorting particles and cells. The device consists of prefocusing and sorting steps, and achieved highly efficient and pure separation of both viable and nonviable Chinese Hamster Ovary (CHO) cells from medium fluid.

2.2.5. Optofluidic Separation

Kotari et al. [85] exploited optical radiation pressure for particle separation in a microfluidic device. Figure 8 illustrates the experimental setup for lateral particle sorting which uses SU–8 as a waveguide to irradiate a near-infrared (NIR) laser beam to facilitate the observation of particle distribution. Using scattering force, particles are manipulated corresponding to the amount of light received by them. Polystyrene beads were successfully transported by the optical scattering force with an energy density of less than $10\ mW/mm^2$.

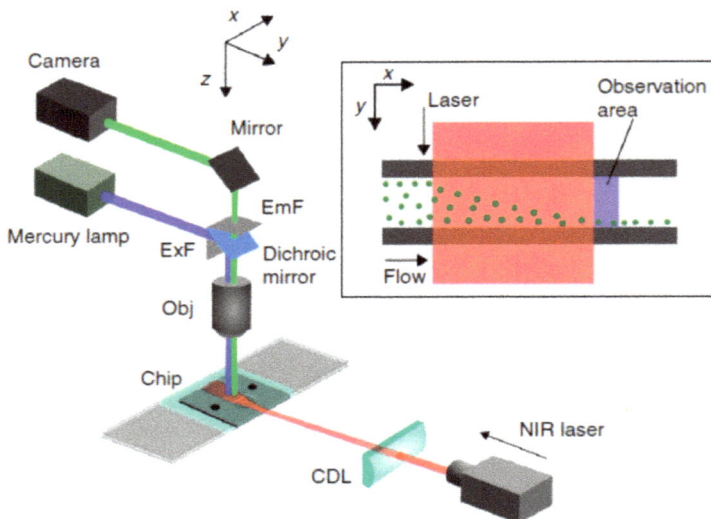

Figure 8. The experimental test rig of microfluidic optical radiation pressure for particle separation developed by Kotari et al. [85]. The SU–8 layer on the microchip guides the near-infrared (NIR) laser beam through the lens. Reproduced with permission from original in study [85].

2.3. Advanced Continuous-Flow Microfluidics with Combined Mixing and Separation

For many biological and chemical analyses, mixing of reagents, and subsequent separation from the remaining sample and vice versa, are the main reasons for making these analyses labour-intensive, time-consuming, expensive and cumbersome. The unique feature of microfluidics is that it allows for the integration of both mixing and separating components on a single chip. In addition, incorporating gas-permeable PDMS membranes into such microfluidic platforms allows for the fabrication of advanced microbioreactors, capable of performing a variety of chemical and biological processes. Specialised POC diagnostic platforms, such as lab-on-a-disc, show great promise for fast, reliable and cost-effective immunoassay tools. For example, the lab-on-a-disc platform developed by Kuo and Li [86] allowed for the separation of plasma from whole blood in only six seconds. Subsequently, the plasma-free blood was able to be mixed with related reagents for other diagnostic tests. The microfluidic device for the prothrombin time (PT) test was 15 times faster than the conventional bench-top counterpart. For both diagnostic and therapeutic purposes, high-throughput label-free microfluidic cell sorters are in great demand. Using the passive hydrodynamic approach, Tallapragada et al. [87] proposed a scale-independent method to separate and encapsulate inertial particles, specifically pancreatic islets, in serpentine microchannels. Finally, microfluidic chromatographic platforms have also opened up new avenues for separation chemistry, especially for protein purification [88].

3. Digital Microfluidics

Digital microfluidics (DMF) involves the manipulation of small, discrete droplets, usually in the microlitre scale or smaller. The main tasks of DMF involve dispensing droplets, moving droplets, merging droplets or mixing contents within a droplet. Numerous techniques have been developed to perform these tasks, as elaborated on in extensive recent reviews [20–22,89–91].

3.1. Droplet-Based DMF

DMF devices can have a basic open planar form, where the droplet is placed on a solid planar surface. The plate is usually engineered to provide an energy gradient to drive the droplet. In some cases, a top plate is added to facilitate control of the sandwiched droplet. With proper design, droplets can be moved across the plate in two dimensions. However, the droplet can also be further controlled by constructing channels between the plates, thus restricting the droplet to a one-dimensional movement. The immiscible fluid surrounding the droplet maintains the separation of droplets. Specially treated surfaces in contact with the droplet minimises loss of liquid during transport.

3.1.1. Electrowetting-on-Dielectric (EWOD) Technique

One of the most popular techniques in DMF is electrowetting-on-dielectric (EWOD). A droplet is placed between two plates, one of which contains a dielectric layer. A voltage difference across the droplet generates asymmetric droplet contact angles, thus creating a driving force. Switching the voltage difference in a timely manner moves the droplet [92–94]. Optoelectrowetting is a modified version of the EWOD technique, where the voltage switching is accomplished optically [95,96]. Recently, Geng et al. [93] reported a pioneering work regarding the use of the dielectrowetting [97] instead of EWOD to manipulate both conductive or non-conductive droplets. This concept removes the need for a top plate and provides easy access to the droplets. The principle operation of such a technique is shown in Figure 9.

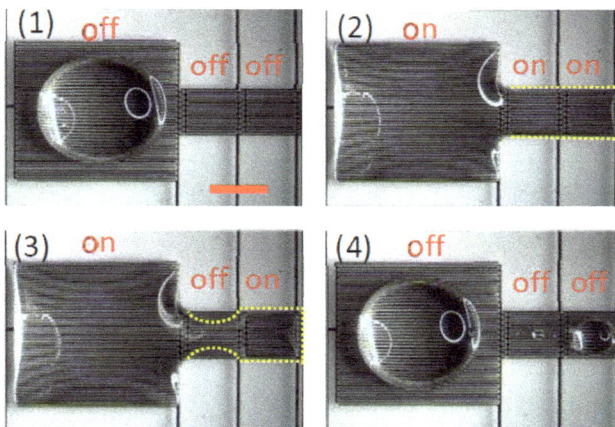

Figure 9. Droplet dispensing using dielectrowetting. (**1**) A 22-μL sessile droplet of propylene carbonate on the electrode pads. (**2**) The electrode pads are turned on, spreading the droplet. (**3**) The middle pad is turned off to pinch off the droplet. (**4**) All the pads are turned off and droplet separation is complete. Reproduced with permission from the original in study [93].

3.1.2. Dielectrophoretic Technique

The dielectrophoresis technique similarly uses electrostatic force, but the droplet itself acts as a dielectric [98,99]. In one of the most recent works, Iwai et al. [100] combine "finger-powered" microfluidics with piezoelectric elements to achieve dielectrophoretic droplet manipulation. The device harnesses the user's mechanical input and converts it into electrostatic energy, which is then used to move the droplets suspended in fluids, Figure 10.

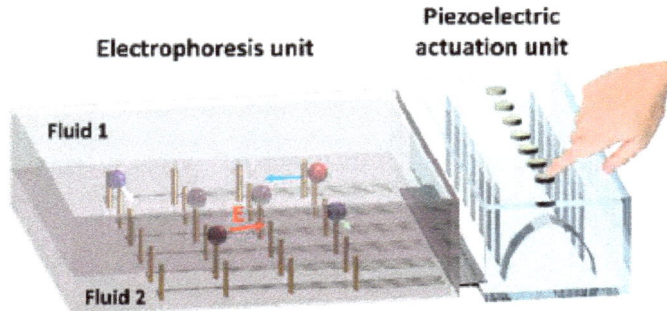

Figure 10. Schematic showing the finger-powered electrophoresis unit. The electrodes are connected to the piezoelectric actuation unit, which can be actuated by pressing on it. The electrophoresis unit contains both the electrode array and the droplets, suspended in a binary fluid. Reproduced with permission from the original in study [100].

3.1.3. Magnetic-Based Techniques

Instead of an electric field, a magnetic field can be applied to move droplets containing magnetite via magnetowetting [101]. The magnetic field generates a body force throughout the entire droplet. Displacing a permanent magnet under a ferrofluid droplet creates asymmetric contact angles and moves the droplet.

3.1.4. Other Techniques

Droplets can be manipulated using other means such as surface acoustic waves (SAW) [102–104] or thermocapillary forces [105,106]. Acoustic energy is generated using a piezoelectric element and transferred to the droplet. As the SAW hit the droplet, energy is transferred onto the droplet, which causes it to de-pin from the surface and move. More energetic SAW can even cause droplets to nebulise. Unlike EWOD, most SAW devices need only one plate. On the other hand, a thermocapillary-based DMF device moves a droplet using capillary forces generated by surface tension gradients which arise from temperature differentials. Nguyen and Huang [107] evaluated the manipulation of droplets in long capillaries under a variable temperature field. In particular, they evaluated the initial behaviour of liquid motion under a transient temperature gradient, both analytically and experimentally.

3.2. Liquid-Marble-Based DMF

Another growing field in DMF is the use of liquid marble (LM) as the discrete platform. The LM is a small droplet encapsulated by a hydrophobic coating, which consists of a porous particle layer [108–111]. The hydrophobic and porous shell removes the need for surface treatment, as the droplet is physically isolated from its surroundings. An added benefit is that a LM is able to float on a liquid surface [112,113] and seemingly skid around with low friction [114,115]. As discussed by Ooi and Nguyen [116] in a comprehensive review paper, numerous techniques to manipulate the LM have been derived. Among the most popular techniques is manipulating a LM containing magnetite using a permanent magnet [117–121]. Zhao et al. [120] used an encapsulated LM driven by a permanent magnet as a bioreactor, as illustrated in Figure 11. Furthermore, a LM can be driven by thermo- [122,123] or soluto-capillary forces, and even carry its own propellant whilst doing so [124–126].

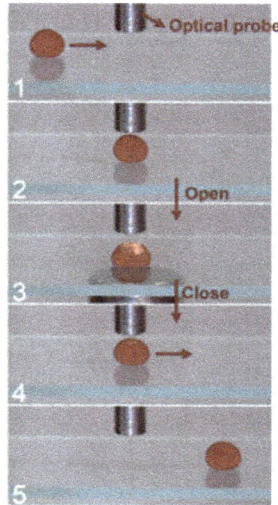

Figure 11. A magnetite-covered LM used as a miniature bioreactor. (**1**,**2**) The LM containing the reactants is moved towards the optical probe using a permanent magnet. (**3**) The coating of the LM can be "opened" to reveal its contents by increasing the magnetic field. (**4**,**5**) The coating opening process is reversed and the LM is moved away from the probe. Reproduced with permission (CC BY license) from original in study [120].

3.3. Advanced Digital Microfluidic Platforms

Recent advances in manipulating microdroplets predominantly involve EWOD-based devices. Researchers have pioneered the use of DMF in the immunoprecipitation process [127]. This concept was accomplished using an existing DMF device which combines both EWOD and magnetic manipulation of the droplet [128,129]. DMF has also been used for the first time in solid-phase micro extraction [130], as well as in high field nuclear magnetic resonance spectroscopy [131]. Nanostructure initiator mass spectrometry (NIMS) arrays can be integrated into an EWOD device to conduct enzyme screening, which potentially increases the throughput of the process [132]. On the cost-reduction front, a specially designed EWOD system has been manipulated using a smart phone to conduct chemiluminescence sensing [133].

However, liquid marble has recently shown its potential as an emerging digital microfluidic platform, especially for biological applications. The most prominent application of liquid marble has been cell culture and the ability to form three-dimensional spheroids due to its respirable and non-adhesive coating [134–137]. Liquid marble can be dehydrated to form hollow shells [138], which then is used for drug encapsulation and release [139]. Liquid marble can also be used as a microbioreactor, as it can accommodate liquid volumes across several orders of magnitude and still can be easily handled [118,140]. Recently, a spinning liquid marble has been used to improve mixing [141].

4. Conclusions and Perspectives

This paper summarises the most recent and advantageous advances in liquid handling modalities, using both continuous-flow and digital microfluidics. Due to the importance of mixing and separation in biological and chemical procedures, we confined the scope of continuous-flow microfluidics to these two topics. Mixing is an essential step in most lab-on-a-chip platforms, as sample preparation is required for a variety of biological and chemical assays. Diffusion-based mixing techniques fail to satisfy the recent demand for rapid and homogeneous mixing. Advances in two major mixing enhancement strategies, i.e., mixing with external energy sources, as well as complex channels geometry, were reviewed. Continuous-flow microfluidic separation also has the benefit of real-time monitoring and the potential for the integration with other continuous-flow processes. Based on the unique signature of the sample components, a suitable external force can be chosen for the separation process. Cutting-edge advances in continuous-flow microfluidic separation techniques, including magnetofluidics, inertial microfluidics, acoustofludics, dielectrophoretics and optofludics, were reviewed and discussed. Emerging applications of combined continuous-flow separation and mixing technologies for more advanced microfluidic platforms, such as diagnostic and therapeutic microbioreactors, lab-on-a-disc and microfluidic chromatography for protein purification, were introduced.

The second part of this paper was dedicated to digital microfluidics for handling microdroplets and liquid marbles. Droplet-based DMF techniques, such as electrowetting-on-dielectric (EWOD), dielectrophoresis, and magnetic methods were discussed. The applications of more advanced combinatorial DMF devices were also introduced. In addition, manipulation techniques for liquid marble as a microbioreactor were presented.

Recent advances in microfluidics indicate that more complex microfluidic structures, especially for mixing applications, could be fabricated with 3D printing. The design freedom provided by 3D printing will allow for novel designs, which to date cannot be obtained with planar micromachining techniques, such as soft lithography with PDMS. Microfluidic cell culture can be considered as the next-generation technique for biomedical and pharmaceutical applications. Liquid marble has emerged as a promising digital microfluidics platform. Continuous-flow microfluidics will continue to be used for applications that require high throughput. However, the problem of bulky external liquid delivery and the need of optical microscopy for characterisation makes continuous-flow microfluidics less suitable for applications with limited sample size. Digital microfluidics with droplets and liquid marbles is the solution for the problem of bulky external systems, as well as the relatively large sample volume. In the near future, we could expect more reports on this unique research area. As most recent works are only on the proof-of-concept of liquid-marble-based digital microfluidics, automated systems for creating liquid marble and the controlled manipulation of liquid marble, such as coalescence and splitting, are areas of interest for bringing this platform closer to practical use.

Acknowledgments: We acknowledge the Australian Research Council for the grant support DP170100277.

Author Contributions: N.-T.N. developed the structure of the paper. Other authors collected and analysed the literature. All authors wrote the paper.

Conflicts of Interest: The authors declare no conflict of interest.

References

1. Nguyen, N.-T.; Wereley, S.T. *Fundamentals and Applications of Microfluidics*; Artech House: London, UK, 2002.
2. Whitesides, G.M. The origins and the future of microfluidics. *Nature* **2006**, *442*, 368–373. [CrossRef] [PubMed]
3. Nguyen, N.-T.; Shaegh, S.A.M.; Kashaninejad, N.; Phan, D.-T. Design, fabrication and characterization of drug delivery systems based on lab-on-a-chip technology. *Adv. Drug Deliv. Rev.* **2013**, *65*, 1403–1419. [CrossRef] [PubMed]
4. Reyes, D.R.; Iossifidis, D.; Auroux, P.-A.; Manz, A. Micro total analysis systems. 1. Introduction, theory, and technology. *Anal. Chem.* **2002**, *74*, 2623–2636. [PubMed]
5. Jung, W.; Han, J.; Choi, J.-W.; Ahn, C.H. Point-of-care testing (POCT) diagnostic systems using microfluidic lab-on-a-chip technologies. *Microelectron. Eng.* **2015**, *132*, 46–57. [CrossRef]
6. Kashaninejad, N.; Chan, W.K.; Nguyen, N.-T. Fluid mechanics of flow through rectangular hydrophobic microchannels. In Proceedings of the ASME 2011 9th International Conference on Nanochannels, Microchannels, and Minichannels, Edmonton, AL, Canada, 19–22 June 2011; pp. 647–655.
7. Nguyen, N.-T.; Wu, Z. Micromixers—A review. *J. Micromech. Microeng.* **2004**, *15*, R1. [CrossRef]
8. Kashaninejad, N.; Nikmaneshi, M.R.; Moghadas, H.; Kiyoumarsi Oskouei, A.; Rismanian, M.; Barisam, M.; Saidi, M.S.; Firoozabadi, B. Organ-Tumor-on-a-Chip for Chemosensitivity Assay: A Critical Review. *Micromachines* **2016**, *7*, 130. [CrossRef]

9.   Sajeesh, P.; Sen, A.K. Particle separation and sorting in microfluidic devices: A review. *Microfluid. Nanofluid.* **2014**, *17*, 1–52. [CrossRef]

10.   Suh, Y.K.; Kang, S. A Review on Mixing in Microfluidics. *Micromachines* **2010**, *1*, 82–111. [CrossRef]

11.   Lee, C.-Y.; Chang, C.-L.; Wang, Y.-N.; Fu, L.-M. Microfluidic Mixing: A Review. *Int. J. Mol. Sci.* **2011**, *12*, 3263–3287. [CrossRef] [PubMed]

12.   Ward, K.; Fan, Z.H. Mixing in microfluidic devices and enhancement methods. *J. Micromech. Microeng.* **2015**, *25*, 094001. [CrossRef] [PubMed]

13.   Gossett, D.R.; Weaver, W.M.; Mach, A.J.; Hur, S.C.; Tse, H.T.K.; Lee, W.; Amini, H.; Di Carlo, D. Label-free cell separation and sorting in microfluidic systems. *Anal. Bioanal. Chem.* **2010**, *397*, 3249–3267. [CrossRef] [PubMed]

14.   Chen, J.; Li, J.; Sun, Y. Microfluidic approaches for cancer cell detection, characterization, and separation. *Lab Chip* **2012**, *12*, 1753–1767. [CrossRef] [PubMed]

15.   Shields, C.W.; Ohiri, K.A.; Szott, L.M.; López, G.P. Translating microfluidics: Cell separation technologies and their barriers to commercialization. *Cytom. B Clin. Cytom.* **2017**, *92*, 115–125. [CrossRef] [PubMed]

16.   Choi, K.; Ng, A.H.; Fobel, R.; Wheeler, A.R. Digital microfluidics. *Annu. Rev. Anal. Chem.* **2012**, *5*, 413–440. [CrossRef] [PubMed]

17.   Barbulovic-Nad, I.; Yang, H.; Park, P.S.; Wheeler, A.R. Digital microfluidics for cell-based assays. *Lab Chip* **2008**, *8*, 519–526. [CrossRef] [PubMed]

18.   Xu, T.; Chakrabarty, K. Parallel scan-like test and multiple-defect diagnosis for digital microfluidic biochips. *IEEE Trans. Biomed. Circuits Syst.* **2007**, *1*, 148–158. [CrossRef] [PubMed]

19.   Fair, R.B. Digital microfluidics: Is a true lab-on-a-chip possible? *Microfluid. Nanofluid.* **2007**, *3*, 245–281. [CrossRef]

20.   Jebrail, M.J.; Bartsch, M.S.; Patel, K.D. Digital microfluidics: A versatile tool for applications in chemistry, biology and medicine. *Lab Chip* **2012**, *12*, 2452–2463. [CrossRef] [PubMed]

21.   Samiei, E.; Tabrizian, M.; Hoorfar, M. A review of digital microfluidics as portable platforms for lab-on-a-chip applications. *Lab Chip* **2016**, *16*, 2376–2396. [CrossRef] [PubMed]

22.   Zhang, Y.; Nguyen, N.-T. Magnetic digital microfluidics—A review. *Lab Chip* **2017**, *17*, 994–1008. [CrossRef] [PubMed]

23.   Hejazian, M.; Phan, D.T.; Nguyen, N.T. Mass transport improvement in microscale using diluted ferrofluid and a non-uniform magnetic field. *RSC Adv.* **2016**, *6*, 62439–62444. [CrossRef]

24.   Mao, L.; Koser, H. Overcoming the Diffusion Barrier: Ultra-Fast Micro-Scale Mixing Via Ferrofluids. In Proceedings of the 2007 International Solid-State Sensors, Actuators and Microsystems Conference (TRANSDUCERS 2007), Lyon, France, 10–14 June 2007; pp. 1829–1832.

25.   Hejazian, M.; Nguyen, N.-T. A rapid magnetofluidic micromixer using diluted ferrofluid. *Micromachines* **2017**, *8*, 37. [CrossRef]

26.   Marcel, W.; Vittorio, S.; Joshua, A.D. A simple low pressure drop suspension-based microfluidic mixer. *J. Micromech. Microeng.* **2015**, *25*, 094003.

27.   Peng, Z.C.; Hesketh, P.; Mao, W.; Alexeev, A.; Lam, W. A microfluidic mixer based on parallel, high-speed circular motion of individual microbeads in a rotating magnetic field. In Proceedings of the 2011 16th International Solid-State Sensors, Actuators and Microsystems Conference, TRANSDUCERS'11, Beijing, China, 5–9 June 2011; pp. 1292–1295.

28.   Venancio-Marques, A.; Barbaud, F.; Baigl, D. Microfluidic mixing triggered by an external LED illumination. *J. Am. Chem. Soc.* **2013**, *135*, 3218–3223. [CrossRef] [PubMed]

29.   Wu, Z.; Nguyen, N.-T. Rapid Mixing Using Two-Phase Hydraulic Focusing in Microchannels. *Biomed. Microdevices* **2005**, *7*, 13–20. [CrossRef] [PubMed]

30.   Ober, T.J.; Foresti, D.; Lewis, J.A. Active mixing of complex fluids at the microscale. *Proc. Natl. Acad. Sci. USA* **2015**, *112*, 12293–12298. [CrossRef] [PubMed]

31.   Cui, W.; Zhang, H.; Zhang, H.; Yang, Y.; He, M.; Qu, H.; Pang, W.; Zhang, D.; Duan, X. Localized ultrahigh frequency acoustic fields induced micro-vortices for submilliseconds microfluidic mixing. *Appl. Phys. Lett.* **2016**, *109*, 253503. [CrossRef]

32.   Fang, F.; Zhang, N.; Liu, K.; Wu, Z.-Y. Hydrodynamic and electrodynamic flow mixing in a novel total glass chip mixer with streamline herringbone pattern. *Microfluid. Nanofluid.* **2015**, *18*, 887–895. [CrossRef]

33. Shang, X.; Huang, X.; Yang, C. Mixing enhancement by the vortex in a microfluidic mixer with actuation. *Exp. Therm. Fluid Sci.* **2015**, *67*, 57–61. [CrossRef]

34. Nguyen, N.-T.; Huang, X. Mixing in microchannels based on hydrodynamic focusing and time-interleaved segmentation: Modelling and experiment. *Lab Chip* **2005**, *5*, 1320–1326. [CrossRef] [PubMed]

35. Cortelezzi, L.; Ferrari, S.; Dubini, G. A scalable active micro-mixer for biomedical applications. *Microfluid. Nanofluid.* **2017**, *21*, 31. [CrossRef]

36. Kwak, T.J.; Nam, Y.G.; Najera, M.A.; Lee, S.W.; Strickler, J.R.; Chang, W.-J. Convex Grooves in Staggered Herringbone Mixer Improve Mixing Efficiency of Laminar Flow in Microchannel. *PLoS ONE* **2016**, *11*, e0166068. [CrossRef] [PubMed]

37. Salieb-Beugelaar, B.G.; Gonçalves, D.; Wolf, P.M.; Hunziker, P. Microfluidic 3D Helix Mixers. *Micromachines* **2016**, *7*, 189. [CrossRef]

38. Adam, T.; Hashim, U. Design and fabrication of micro-mixer with short turns angles for self-generated turbulent structures. *Microsyst. Technol.* **2016**, *22*, 433–440. [CrossRef]

39. Sivashankar, S.; Agambayev, S.; Mashraei, Y.; Li, E.Q.; Thoroddsen, S.T.; Salama, K.N. A "twisted" microfluidic mixer suitable for a wide range of flow rate applications. *Biomicrofluidics* **2016**, *10*, 034120. [CrossRef] [PubMed]

40. Wang, L.; Ma, S.; Wang, X.; Bi, H.; Han, X. Mixing enhancement of a passive microfluidic mixer containing triangle baffles. *Asia Pac. J. Chem. Eng.* **2014**, *9*, 877–885. [CrossRef]

41. Lehmann, M.; Wallbank, A.M.; Dennis, K.A.; Wufsus, A.R.; Davis, K.M.; Rana, K.; Neeves, K.B. On-chip recalcification of citrated whole blood using a microfluidic herringbone mixer. *Biomicrofluidics* **2015**, *9*, 064106. [CrossRef] [PubMed]

42. Plevniak, K.; Campbell, M.; He, M. 3D printed microfluidic mixer for point-of-care diagnosis of anemia. In Proceedings of the 2016 38th Annual International Conference of the IEEE Engineering in Medicine and Biology Society (EMBC), Orlando, FL, USA, 16–20 August 2016; pp. 267–270.

43. Li, Y.; Liu, C.; Feng, X.; Xu, Y.; Liu, B.F. Ultrafast microfluidic mixer for tracking the early folding kinetics of human telomere G-quadruplex. *Anal. Chem.* **2014**, *86*, 4333–4339. [CrossRef] [PubMed]

44. Pamme, N. Continuous flow separations in microfluidic devices. *Lab Chip* **2007**, *7*, 1644–1659. [CrossRef] [PubMed]

45. Chen, Y.; Li, P.; Huang, P.-H.; Xie, Y.; Mai, J.D.; Wang, L.; Nguyen, N.-T.; Huang, T.J. Rare cell isolation and analysis in microfluidics. *Lab Chip* **2014**, *14*, 626–645. [CrossRef] [PubMed]

46. Karle, M.; Vashist, S.K.; Zengerle, R.; Stetten, F.V. Microfluidic solutions enabling continuous processing and monitoring of biological samples: A review. *Anal. Chim. Acta* **2016**, *929*, 1–22. [CrossRef] [PubMed]

47. Nguyen, N.T. Micro-magnetofluidics: Interactions between magnetism and fluid flow on the microscale. *Microfluid. Nanofluid.* **2012**, *12*, 1–16. [CrossRef]

48. Hejazian, M.; Li, W.; Nguyen, N.T. Lab on a chip for continuous-flow magnetic cell separation. *Lab Chip* **2015**, *15*, 959–970. [CrossRef] [PubMed]

49. Gijs, M.A.M. Magnetic bead handling on-chip: New opportunities for analytical applications. *Microfluid. Nanofluid.* **2004**, *1*, 22–40. [CrossRef]

50. Gijs, M.A.M.; Lacharme, F.; Lehmann, U. Microfluidic Applications of Magnetic Particles for Biological Analysis and Catalysis. *Chem. Rev.* **2010**, *110*, 1518–1563. [CrossRef] [PubMed]

51. Pamme, N. Magnetism and microfluidics. *Lab Chip* **2006**, *6*, 24–38. [CrossRef] [PubMed]

52. Hejazian, M.; Nguyen, N.-T. Negative magnetophoresis in diluted ferrofluid flow. *Lab Chip* **2015**, *15*, 2998–3005. [CrossRef] [PubMed]

53. Hejazian, M.; Nguyen, N.-T. Magnetofluidic concentration and separation of non-magnetic particles using two magnet arrays. *Biomicrofluidics* **2016**, *10*, 044103. [CrossRef] [PubMed]

54. Zhou, Y.; Kumar, D.T.; Lu, X.; Kale, A.; DuBose, J.; Song, Y.; Wang, J.; Li, D.; Xuan, X. Simultaneous diamagnetic and magnetic particle trapping in ferrofluid microflows via a single permanent magnet. *Biomicrofluidics* **2015**, *9*, 044102. [CrossRef] [PubMed]

55. Afshar, R.; Moser, Y.; Lehnert, T.; Gijs, M.A.M. Three-dimensional magnetic focusing of superparamagnetic beads for on-chip agglutination assays. *Anal. Chem.* **2011**, *83*, 1022–1029. [CrossRef] [PubMed]

56. Liang, L.; Xuan, X. Continuous sheath-free magnetic separation of particles in a U-shaped microchannel. *Biomicrofluidics* **2012**, *6*, 044106. [CrossRef] [PubMed]

57. Zhou, R.; Wang, C. Microfluidic separation of magnetic particles with soft magnetic microstructures. *Microfluid. Nanofluid.* **2016**, *20*, 48. [CrossRef]

58. Wu, J.; Yan, Q.; Xuan, S.; Gong, X. Size-selective separation of magnetic nanospheres in a microfluidic channel. *Microfluid. Nanofluid.* **2017**, *21*, 47. [CrossRef]

59. Zhang, J.; Yan, S.; Yuan, D.; Alici, G.; Nguyen, N.T.; Ebrahimi Warkiani, M.; Li, W. Fundamentals and applications of inertial microfluidics: A review. *Lab Chip* **2016**, *16*, 10–34. [CrossRef] [PubMed]

60. Di Carlo, D. Inertial microfluidics. *Lab Chip* **2009**, *9*, 3038–3046. [CrossRef] [PubMed]

61. McGrath, J.; Jimenez, M.; Bridle, H. Deterministic lateral displacement for particle separation: A review. *Lab Chip* **2014**, *14*, 4139–4158. [CrossRef] [PubMed]

62. Martel, J.M.; Toner, M. Inertial focusing in microfluidics. *Annu. Rev. Biomed. Eng.* **2014**, *16*, 371–396. [CrossRef] [PubMed]

63. Nan, X.; Zhu, X.; Ni, Z. Application of inertial effect in microfluidic chips. *Prog. Chem.* **2011**, *23*, 1945–1958.

64. Ahn, S.W.; Lee, S.S.; Lee, S.J.; Kim, J.M. Microfluidic particle separator utilizing sheathless elasto-inertial focusing. *Chem. Eng. Sci.* **2015**, *126*, 237–243. [CrossRef]

65. Schaap, A.; Dumon, J.; Toonder, J. Sorting algal cells by morphology in spiral microchannels using inertial microfluidics. *Microfluid. Nanofluid.* **2016**, *20*, 125. [CrossRef]

66. Zhou, Y.; Song, L.; Yu, L.; Xuan, X. Inertially focused diamagnetic particle separation in ferrofluids. *Microfluid. Nanofluid.* **2017**, *21*, 14. [CrossRef]

67. Clime, L.; Hoa, X.D.; Corneau, N.; Morton, K.J.; Luebbert, C.; Mounier, M.; Brassard, D.; Geissler, M.; Bidawid, S.; Farber, J.; et al. Microfluidic filtration and extraction of pathogens from food samples by hydrodynamic focusing and inertial lateral migration. *Biomed. Microdevices* **2015**, *17*, 17. [CrossRef] [PubMed]

68. Barani, A.; Paktinat, H.; Janmaleki, M.; Mohammadi, A.; Mosaddegh, P.; Fadaei-Tehrani, A.; Sanati-Nezhad, A. Microfluidic integrated acoustic waving for manipulation of cells and molecules. *Biosens. Bioelectron.* **2016**, *85*, 714–725. [CrossRef] [PubMed]

69. Sadhal, S.S. Acoustofluidics 15: Streaming with sound waves interacting with solid particles. *Lab Chip* **2012**, *12*, 2600–2611. [CrossRef] [PubMed]

70. Wiklund, M.; Green, R.; Ohlin, M. Acoustofluidics 14: Applications of acoustic streaming in microfluidic devices. *Lab Chip* **2012**, *12*, 2438–2451. [CrossRef] [PubMed]

71. Voiculescu, I.; Nordin, A.N. Acoustic wave based MEMS devices for biosensing applications. *Biosens. Bioelectron.* **2012**, *33*, 1–9. [CrossRef] [PubMed]

72. Nam, J.; Lim, H.; Shin, S. Manipulation of microparticles using surface acoustic wave in microfluidic systems: A brief review. *Korea Aust. Rheol. J.* **2011**, *23*, 255–267. [CrossRef]

73. Länge, K.; Rapp, B.E.; Rapp, M. Surface acoustic wave biosensors: A review. *Anal. Bioanal. Chem.* **2008**, *391*, 1509–1519. [CrossRef] [PubMed]

74. Mathew, B.; Alazzam, A.; El-Khasawneh, B.; Maalouf, M.; Destgeer, G.; Sung, H.J. Model for tracing the path of microparticles in continuous flow microfluidic devices for 2D focusing via standing acoustic waves. *Sep. Purif. Technol.* **2015**, *153*, 99–107. [CrossRef]

75. Shields, C.W.I.; Wang, J.L.; Ohiri, K.A.; Essoyan, E.D.; Yellen, B.B.; Armstrong, A.J.; López, G.P. Magnetic separation of acoustically focused cancer cells from blood for magnetographic templating and analysis. *Lab Chip* **2016**, *16*, 3833–3844. [CrossRef] [PubMed]

76. Ng, J.W.; Collins, D.J.; Devendran, C.; Ai, Y.; Neild, A. Flow-rate-insensitive deterministic particle sorting using a combination of travelling and standing surface acoustic waves. *Microfluid. Nanofluid.* **2016**, *20*. [CrossRef]

77. Devi, U.V.; Puri, P.; Sharma, N.N.; Ananthasubramanian, M. Electrokinetics of Cells in Dielectrophoretic Separation: A Biological Perspective. *BioNanoScience* **2014**, *4*, 276–287. [CrossRef]

78. Jubery, T.Z.; Srivastava, S.K.; Dutta, P. Dielectrophoretic separation of bioparticles in microdevices: A review. *Electrophoresis* **2014**, *35*, 691–713. [CrossRef] [PubMed]

79. Dash, S.; Mohanty, S. Dielectrophoretic separation of micron and submicron particles: A review. *Electrophoresis* **2014**, *35*, 2656–2672. [CrossRef] [PubMed]

80. Khoshmanesh, K.; Nahavandi, S.; Baratchi, S.; Mitchell, A.; Kalantar-zadeh, K. Dielectrophoretic platforms for bio-microfluidic systems. *Biosens. Bioelectron.* **2011**, *26*, 1800–1814. [CrossRef] [PubMed]

81. Zhang, C.; Khoshmanesh, K.; Mitchell, A.; Kalantar-Zadeh, K. Dielectrophoresis for manipulation of micro/nano particles in microfluidic systems. *Anal. Bioanal. Chem.* **2010**, *396*, 401–420. [CrossRef] [PubMed]

82. Cui, H.H.; Voldman, J.; He, X.F.; Lim, K.M. Separation of particles by pulsed dielectrophoresis. *Lab Chip* **2009**, *9*, 2306–2312. [CrossRef] [PubMed]

83. Kim, U.; Soh, H.T. Simultaneous sorting of multiple bacterial targets using integrated Dielectrophoretic-Magnetic Activated Cell Sorter. *Lab Chip* **2009**, *9*, 2313–2318. [CrossRef] [PubMed]

84. Yan, S.; Zhang, J.; Yuan, Y.; Lovrecz, G.; Alici, G.; Du, H.; Zhu, Y.; Li, W. A hybrid dielectrophoretic and hydrophoretic microchip for particle sorting using integrated prefocusing and sorting steps. *Electrophoresis* **2015**, *36*, 284–291. [CrossRef] [PubMed]

85. Kotari, H.; Motosuke, M. Simple applications of microparticle transportation by tender optical scattering force. *Microfluid. Nanofluid.* **2015**, *18*, 549–558. [CrossRef]

86. Kuo, J.-N.; Li, B.-S. Lab-on-CD microfluidic platform for rapid separation and mixing of plasma from whole blood. *Biomed. Microdevices* **2014**, *16*, 549–558. [CrossRef] [PubMed]

87. Tallapragada, P.; Hasabnis, N.; Katuri, K.; Sudarsanam, S.; Joshi, K.; Ramasubramanian, M. Scale invariant hydrodynamic focusing and sorting of inertial particles by size in spiral micro channels. *J. Micromech. Microeng.* **2015**, *25*, 084013. [CrossRef]

88. Millet, L.J.; Lucheon, J.D.; Standaert, R.F.; Retterer, S.T.; Doktycz, M.J. Modular microfluidics for point-of-care protein purifications. *Lab Chip* **2015**, *15*, 1799–1811. [CrossRef] [PubMed]

89. Freire, S.L.S. Perspectives on digital microfluidics. *Sens. Actuators A Phys.* **2016**, *250*, 15–28. [CrossRef]

90. Mashaghi, S.; Abbaspourrad, A.; Weitz, D.A.; van Oijen, A.M. Droplet microfluidics: A tool for biology, chemistry and nanotechnology. *TrAC Trends Anal. Chem.* **2016**, *82*, 118–125. [CrossRef]

91. Shembekar, N.; Chaipan, C.; Utharala, R.; Merten, C.A. Droplet-based microfluidics in drug discovery, transcriptomics and high-throughput molecular genetics. *Lab Chip* **2016**, *16*, 1314–1331. [CrossRef] [PubMed]

92. Pollack, M.G.; Shenderov, A.D.; Fair, R.B. Electrowetting-based actuation of droplets for integrated microfluidics. *Lab Chip* **2002**, *2*, 96–101. [CrossRef] [PubMed]

93. Geng, H.; Feng, J.; Stabryla, L.M.; Cho, S.K. Dielectrowetting manipulation for digital microfluidics: Creating, transporting, splitting, and merging of droplets. *Lab Chip* **2017**, *17*, 1060–1068. [CrossRef] [PubMed]

94. Aijian, A.P.; Garrell, R.L. Digital Microfluidics for Automated Hanging Drop Cell Spheroid Culture. *J. Lab. Autom.* **2015**, *20*, 283–295. [CrossRef] [PubMed]

95. Pei, S.N.; Valley, J.K.; Neale, S.L.; Jamshidi, A.; Hsu, H.Y.; Wu, M.C. Light-actuated digital microfluidics for large-scale, parallel manipulation of arbitrarily sized droplets. In Proceedings of the 2010 IEEE 23rd International Conference on Micro Electro Mechanical Systems (MEMS), Hong Kong, China, 24–28 January 2010; pp. 252–255.

96. Chiou, P.Y.; Moon, H.; Toshiyoshi, H.; Kim, C.-J.; Wu, M.C. Light actuation of liquid by optoelectrowetting. *Sens. Actuators A Phys.* **2003**, *104*, 222–228. [CrossRef]

97. McHale, G.; Brown, C.V.; Newton, M.I.; Wells, G.G.; Sampara, N. Dielectrowetting Driven Spreading of Droplets. *Phys. Rev. Lett.* **2011**, *107*, 186101. [CrossRef] [PubMed]

98. Peng, C.; Wang, Y.; Sungtaek Ju, Y. Finger-powered electrophoretic transport of discrete droplets for portable digital microfluidics. *Lab Chip* **2016**, *16*, 2521–2531. [CrossRef] [PubMed]

99. Hunt, T.P.; Issadore, D.; Westervelt, R.M. Integrated circuit/microfluidic chip to programmably trap and move cells and droplets with dielectrophoresis. *Lab Chip* **2008**, *8*, 81–87. [CrossRef] [PubMed]

100. Iwai, K.; Shih, K.C.; Lin, X.; Brubaker, T.A.; Sochol, R.D.; Lin, L. Finger-powered microfluidic systems using multilayer soft lithography and injection molding processes. *Lab Chip* **2014**, *14*, 3790–3799. [CrossRef] [PubMed]

101. Nguyen, N.T.; Zhu, G.P.; Chua, Y.C.; Phan, V.N.; Tan, S.H. Magnetowetting and Sliding Motion of a Sessile Ferrofluid Droplet in the Presence of a Permanent Magnet. *Langmuir* **2010**, *26*, 12553–12559. [CrossRef] [PubMed]

102. Monkkonen, L.; Edgar, J.S.; Winters, D.; Heron, S.R.; Mackay, C.L.; Masselon, C.D.; Stokes, A.A.; Langridge-Smith, P.R.R.; Goodlett, D.R. Screen-printed digital microfluidics combined with surface acoustic wave nebulization for hydrogen-deuterium exchange measurements. *J. Chromatogr. A* **2016**, *1439*, 161–166. [CrossRef] [PubMed]

103. Renaudin, A.; Tabourier, P.; Camart, J.-C.; Druon, C. Surface acoustic wave two-dimensional transport and location of microdroplets using echo signal. *J. Appl. Phys.* **2006**, *100*, 116101. [CrossRef]

104. Guttenberg, Z.; Muller, H.; Habermuller, H.; Geisbauer, A.; Pipper, J.; Felbel, J.; Kielpinski, M.; Scriba, J.; Wixforth, A. Planar chip device for PCR and hybridization with surface acoustic wave pump. *Lab Chip* **2005**, *5*, 308–317. [CrossRef] [PubMed]

105. Chen, J.Z.; Troian, S.M.; Darhuber, A.A.; Wagner, S. Effect of contact angle hysteresis on thermocapillary droplet actuation. *J. Appl. Phys.* **2005**, *97*, 014906. [CrossRef]
106. Darhuber, A.A.; Valentino, J.P.; Troian, S.M.; Wagner, S. Thermocapillary actuation of droplets on chemically patterned surfaces by programmable microheater arrays. *J. Microelectromech. Syst.* **2003**, *12*, 873–879. [CrossRef]
107. Nguyen, N.-T.; Huang, X. Thermocapillary effect of a liquid plug in transient temperature fields. *Jpn. J. Appl. Phys.* **2005**, *44*, 1139–1142. [CrossRef]
108. Aussillous, P.; Quere, D. Properties of liquid marbles. *Proc. R. Soc. A Math. Phys. Eng. Sci.* **2006**, *462*, 973–999. [CrossRef]
109. McHale, G.; Newton, M.I. Liquid marbles: Principles and applications. *Soft Matter* **2011**, *7*, 5473–5481. [CrossRef]
110. McHale, G.; Newton, M.I. Liquid marbles: Topical context within soft matter and recent progress. *Soft Matter* **2015**, *11*, 2530–2546. [CrossRef] [PubMed]
111. Bormashenko, E. New insights into liquid marbles. *Soft Matter* **2012**, *8*, 11018–11021. [CrossRef]
112. Cengiz, U.; Erbil, H.Y. The lifetime of floating liquid marbles: The influence of particle size and effective surface tension. *Soft Matter* **2013**, *9*, 8980–8991. [CrossRef]
113. Bormashenko, E.; Pogreb, R.; Musin, A. Stable water and glycerol marbles immersed in organic liquids: From liquid marbles to Pickering-like emulsions. *J. Colloid Interface Sci.* **2012**, *366*, 196–199. [CrossRef] [PubMed]
114. Ooi, C.H.; Nguyen, A.V.; Evans, G.M.; Dao, D.V.; Nguyen, N.T. Measuring the Coefficient of Friction of a Small Floating Liquid Marble. *Sci. Rep.* **2016**, *6*, 38346. [CrossRef] [PubMed]
115. Ooi, C.H.; Plackowski, C.; Nguyen, A.V.; Vadivelu, R.K.; John, J.A.S.; Dao, D.V.; Nguyen, N.-T. Floating mechanism of a small liquid marble. *Sci. Rep.* **2016**, *6*, 21777. [CrossRef] [PubMed]
116. Ooi, C.H.; Nguyen, N.-T. Manipulation of liquid marbles. *Microfluid. Nanofluid.* **2015**, *19*, 483–495. [CrossRef]
117. Khaw, M.K.; Ooi, C.H.; Mohd-Yasin, F.; Vadivelu, R.; John, J.S.; Nguyen, N.-T. Digital microfluidics with a magnetically actuated floating liquid marble. *Lab Chip* **2016**, *16*, 2211–2218. [CrossRef] [PubMed]
118. Xue, Y.; Wang, H.; Zhao, Y.; Dai, L.; Feng, L.; Wang, X.; Lin, T. Magnetic Liquid Marbles: A "Precise" Miniature Reactor. *Adv. Mater.* **2010**, *22*, 4814–4818. [CrossRef] [PubMed]
119. Zhao, Y.; Fang, J.; Wang, H.; Wang, X.; Lin, T. Magnetic Liquid Marbles: Manipulation of Liquid Droplets Using Highly Hydrophobic Fe$_3$O$_4$ Nanoparticles. *Adv. Mater.* **2010**, *22*, 707–710. [CrossRef] [PubMed]
120. Zhao, Y.; Xu, Z.G.; Parhizkar, M.; Fang, J.; Wang, X.G.; Lin, T. Magnetic liquid marbles, their manipulation and application in optical probing. *Microfluid. Nanofluid.* **2012**, *13*, 555–564. [CrossRef]
121. Zhang, L.; Cha, D.; Wang, P. Remotely Controllable Liquid Marbles. *Adv. Mater.* **2012**, *24*, 4756–4760. [CrossRef] [PubMed]
122. Paven, M.; Mayama, H.; Sekido, T.; Butt, H.-J.; Nakamura, Y.; Fujii, S. Liquid Marbles: Light-Driven Delivery and Release of Materials Using Liquid Marbles (Adv Funct. Mater. 19/2016). *Adv. Funct. Mater.* **2016**, *26*, 3372. [CrossRef]
123. Kavokine, N.; Anyfantakis, M.; Morel, M.; Rudiuk, S.; Bickel, T.; Baigl, D. Light-Driven Transport of a Liquid Marble with and against Surface Flows. *Angew. Chem. Int. Ed.* **2016**, *55*, 11183–11187. [CrossRef] [PubMed]
124. Bormashenko, E.; Bormashenko, Y.; Grynyov, R.; Aharoni, H.; Whyman, G.; Binks, B.P. Self-Propulsion of Liquid Marbles: Leidenfrost-like Levitation Driven by Marangoni Flow. *J. Phys. Chem. C* **2015**, *119*, 9910–9915. [CrossRef]
125. Ooi, C.H.; Nguyen, A.V.; Evans, G.M.; Gendelman, O.; Bormashenko, E.; Nguyen, N.-T. A floating self-propelling liquid marble containing aqueous ethanol solutions. *RSC Adv.* **2015**, *5*, 101006–101012. [CrossRef]
126. Bormashenko, E. Liquid Marbles, Elastic Nonstick Droplets: From Minireactors to Self-Propulsion. *Langmuir* **2017**, *33*, 663–669. [CrossRef] [PubMed]
127. Seale, B.; Lam, C.; Rackus, D.G.; Chamberlain, M.D.; Liu, C.; Wheeler, A.R. Digital Microfluidics for Immunoprecipitation. *Anal. Chem.* **2016**, *88*, 10223–10230. [CrossRef] [PubMed]
128. Mei, N.; Seale, B.; Ng, A.H.C.; Wheeler, A.R.; Oleschuk, R. Digital microfluidic platform for human plasma protein depletion. *Anal. Chem.* **2014**, *86*, 8466–8472. [CrossRef] [PubMed]
129. Ng, A.H.C.; Choi, K.; Luoma, R.P.; Robinson, J.M.; Wheeler, A.R. Digital microfluidic magnetic separation for particle-based immunoassays. *Anal. Chem.* **2012**, *84*, 8805–8812. [CrossRef] [PubMed]

130. Choi, K.; Boyacı, E.; Kim, J.; Seale, B.; Barrera-Arbelaez, L.; Pawliszyn, J.; Wheeler, A.R. A digital microfluidic interface between solid-phase microextraction and liquid chromatography–mass spectrometry. *J. Chromatogr. A* **2016**, *1444*, 1–7. [CrossRef] [PubMed]

131. Swyer, I.; Soong, R.; Dryden, M.D.M.; Fey, M.; Maas, W.E.; Simpson, A.; Wheeler, A.R. Interfacing digital microfluidics with high-field nuclear magnetic resonance spectroscopy. *Lab Chip* **2016**, *16*, 4424–4435. [CrossRef] [PubMed]

132. Heinemann, J.; Deng, K.; Shih, S.C.C.; Gao, J.; Adams, P.D.; Singh, A.K.; Northen, T.R. On-chip integration of droplet microfluidics and nanostructure-initiator mass spectrometry for enzyme screening. *Lab Chip* **2017**, *17*, 323–331. [CrossRef] [PubMed]

133. Zeng, Z.; Zhang, K.; Wang, W.; Xu, W.; Zhou, J. Portable electrowetting digital microfluidics analysis platform for chemiluminescence sensing. *IEEE Sens. J.* **2016**, *16*, 4531–4536. [CrossRef]

134. Arbatan, T.; Al-Abboodi, A.; Sarvi, F.; Chan, P.P.; Shen, W. Tumor inside a pearl drop. *Adv. Healthc. Mater.* **2012**, *1*, 467–469. [CrossRef] [PubMed]

135. Sarvi, F.; Jain, K.; Arbatan, T.; Verma, P.J.; Hourigan, K.; Thompson, M.C.; Shen, W.; Chan, P.P. Cardiogenesis of embryonic stem cells with liquid marble micro-bioreactor. *Adv. Healthc. Mater.* **2015**, *4*, 77–86. [CrossRef] [PubMed]

136. Vadivelu, R.K.; Ooi, C.H.; Yao, R.-Q.; Tello Velasquez, J.; Pastrana, E.; Diaz-Nido, J.; Lim, F.; Ekberg, J.A.K.; Nguyen, N.-T.; St John, J.A. Generation of three-dimensional multiple spheroid model of olfactory ensheathing cells using floating liquid marbles. *Sci. Rep.* **2015**, *5*, 15083. [CrossRef] [PubMed]

137. Tian, J.; Fu, N.; Chen, X.D.; Shen, W. Respirable liquid marble for the cultivation of microorganisms. *Colloids Surf. B* **2013**, *106*, 187–190. [CrossRef] [PubMed]

138. Eshtiaghi, N.; Hapgood, K.P. A quantitative framework for the formation of liquid marbles and hollow granules from hydrophobic powders. *Powder Technol.* **2012**, *223*, 65–76. [CrossRef]

139. Zhang, G.; Wang, C. Pickering emulsion-based marbles for cellular capsules. *Materials* **2016**, *9*, 572. [CrossRef]

140. Arbatan, T.; Li, L.; Tian, J.; Shen, W. Liquid marbles as micro-bioreactors for rapid blood typing. *Adv. Healthc. Mater.* **2012**, *1*, 80–83. [CrossRef] [PubMed]

141. Han, X.; Lee, H.K.; Lim, W.C.; Lee, Y.H.; Phan-Quang, G.C.; Phang, I.Y.; Ling, X.Y. Spinning Liquid Marble and Its Dual Applications as Microcentrifuge and Miniature Localized Viscometer. *ACS Appl. Mater. Interfaces* **2016**, *8*, 23941–23946. [CrossRef] [PubMed]

micromachines

MDPI

Review

A Perspective on the Rise of Optofluidics and the Future

Chaolong Song [1],* and Say Hwa Tan [2]

[1] School of Mechanical Engineering and Electronic Information, China University of Geosciences (Wuhan), Wuhan 430074, China
[2] Queensland Micro- and Nanotechnology Centre, Griffith University, 170 Kessels Road, Brisbane, QLD 4111, Australia; sayhwa.tan@griffith.edu.au
* Correspondence: songcl@cug.edu.cn; Tel.: +86-27-6788-3273

Academic Editors: Weihua Li, Hengdong Xi and Nam-Trung Nguyen
Received: 13 February 2017; Accepted: 2 May 2017; Published: 8 May 2017

Abstract: In the recent past, the field of optofluidics has thrived from the immense efforts of researchers from diverse communities. The concept of optofluidics combines optics and microfluidics to exploit novel properties and functionalities. In the very beginning, the unique properties of liquid, such as mobility, fungibility and deformability, initiated the motivation to develop optical elements or functions using fluid interfaces. Later on, the advancements of microelectromechanical system (MEMS) and microfluidic technologies enabled the realization of optofluidic components through the precise manipulation of fluids at microscale thus making it possible to streamline complex fabrication processes. The optofluidic system aims to fully integrate optical functions on a single chip instead of using external bulky optics, which can consequently lower the cost of system, downsize the system and make it promising for point-of-care diagnosis. This perspective gives an overview of the recent developments in the field of optofluidics. Firstly, the fundamental optofluidic components will be discussed and are categorized according to their basic working mechanisms, followed by the discussions on the functional instrumentations of the optofluidic components, as well as the current commercialization aspects of optofluidics. The paper concludes with the critical challenges that might hamper the transformation of optofluidic technologies from lab-based procedures to practical usages and commercialization.

Keywords: optofluidics; microfluidics; adaptive optics; optofluidic components; lab-on-chip applications

---

1. The Origin of Optofluidics and Its Motivation

Optofluidics is referred to as the sciences and technologies which explore novel functions and the related physics through the marriage between light and fluid [1]. Optofluidics mainly aims at using microfluidic technologies to enable various optical functions or using optics to manipulate fluids at microscale. In recent years, growing interests and intense efforts have been directed to this field. Specifically, researchers have been developing reconfigurable optics using deformable fluids and in turn taking advantages of tunable optics to control or sense the fluid sample. The concept and implementation of using fluid to fabricate optical components and realize optical functions is not new. Stemming from the early 1900s, a deformable mirror was proposed and demonstrated by spinning fluid in a circular chamber to form a paraboloid shape [2]. The curvature of the mirror interface can be controlled by the spinning speed to construct a mirror with tunable focus. Another example of fluid-based optics is the commonly used oil-immersion microscope for enhanced magnification. These are good examples considered as rudimental development of adaptive optics. However, the most popular way to manufacture optics for centuries has been shaping and polishing solid materials, and the methodology to research and develop optofluidics has not been systematic until the recent maturity of MEMS and microfluidic technologies [3].

Figure 1 shows the evolution from conventional optics to tunable, adaptive and fluid-based optics. Since the first paper on light-steering using a micro-mirror published in 1980 by Petersen et al. [4], the MEMS technology has enabled the development of miniaturized solid-based optical devices mainly driven by the demands from digital display market, and subsequently helped to generate a new field-optical MEMS. Meanwhile, the advancements of MEMS technologies have also enabled the fabrication of microfluidic architectures with precisions down to nanoscale, including the fluid transfer in microchannels, the integrated mechatronic fluid control/monitoring units and fluid pumping/recycling sub-systems. Later on, the study and understanding of microfluidic mechanics opened up a platform to realize precise manipulation of fluid at sub-micron scale. All these developments and achievements have slowly paved the way to the point that it is feasible to fabricate and integrate the optical elements on microfluidic chips. This perspective will be mainly focused on the marriage between microfluidics and optics to develop tunable optical instruments and their applications for microfluidics.

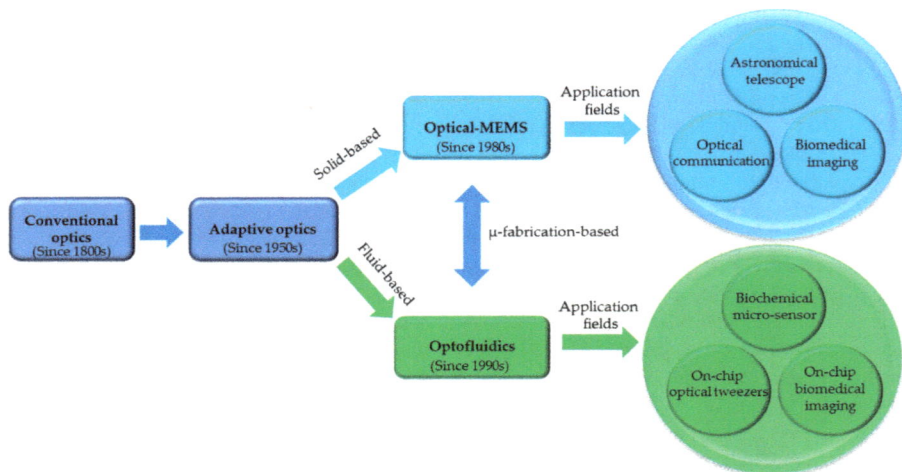

Figure 1. The evolvement of optofluidics from conventional optics with its applications in the fields of biochemistry, biomedical engineering, analytical chemistry, etc.

From a physical point of view, fluids can offer unique properties to realize novel optical functions: (1) replacing a liquid with another liquid inside an optical cavity can easily tune optical performance or function of the optical cavity; (2) refractive index (RI) gradients can be created by the control of diffusion between two miscible fluids; (3) an optically smooth interface can be formed between two immiscible liquids; and (4) light and sample can transfer along the same channel which allows long light–matter interaction length. These novel properties show promise to equip adaptive optical systems with more flexibilities and functionalities. In 2006, the review paper by Psaltis et al. summarized and categorized the early development of reconfigurable optical systems synergistically using optics and fluidics, and conceptualized the field of optofluidics [5]. Since then, optofluidics has intensively summoned research interests from the people working in multiple disciplines, such as biology, biomedical engineering, analytical chemistry, etc.

A good number of review papers have been published with emphasis on a specific area dealing with applications using optofluidic technologies, such as whispering gallery mode sensor [6,7], label free bio-chemical sensor [8], microfluidic cytometry [9], energy production and harvesting [10], on-chip interferometry [11], on-chip treatment and manipulation of bio-cells [12,13], point-of-care genetic analysis [14], Raman spectroscopy [15], etc. There are also papers summarizing a group of specific optofluidic components, such as waveguides [16,17] and lenses [18,19], or discussing various optofluidic components operated by a specific working mechanism, for example fluid interface actuation [20]. However, few papers were dedicated to systematically discuss the development of fundamental optofluidic components and categorize them in terms of their working mechanisms. In this paper, we aim to comprehensively capture the miniaturized optical components using optofluidic technologies and classify them in terms of three working mechanisms: (1) liquid replacing; (2) refractive index gradient control; and (3) fluid interface deformation. Moreover, we discuss the basic optofluidic instrumentations by the utilization of those fundamental components. Finally, we look into the commercialization aspect of optofluidics and envision its future.

2. The Inventions of Optofluidic Components

Figure 2 illustrates the concept of an ideal optofluidic system with light emitter, light maneuvering/delivering channel, light–matter interaction cavity, and the decoding unit for the interpretation of the modulated light after interaction. One critical issue for the integrated optics is the feasibility and convenience of optical alignment. Different from free-space optics, which can be easily aligned by adjusting its physical positions, the integrated optics does not have that luxury of position adjustment after fabrication. Hence, self-tunability is of great importance for lab-on-a-chip optofluidic components. To manipulate and deliver the light with high precision and convenient alignment, a number of fundamental optofluidic components have been proposed and demonstrated. In the following, they will be discussed and categorized in terms of main three working mechanisms to achieve their tunability: (1) liquid replacing; (2) RI gradient control; and (3) fluid interface control.

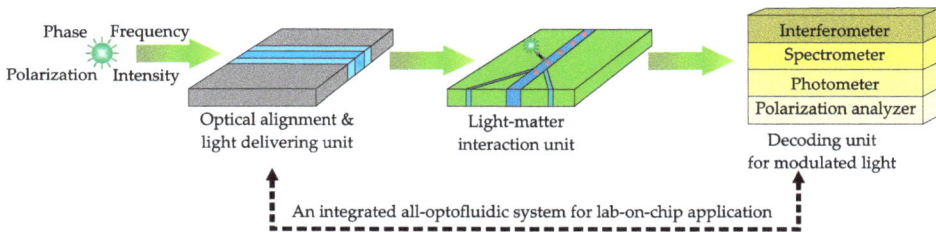

Phase   Frequency

Polarization   Intensity

Interferometer
Spectrometer
Photometer
Polarization analyzer

Optical alignment &
light delivering unit

Light-matter
interaction unit

Decoding unit
for modulated light

An integrated all-optofluidic system for lab-on-chip application

Figure 2. Schematic of an ideal all-optofluidic system for lab-on-chip applications.

2.1. Liquid Replacing

The function of most solid optical components is permanently fixed. Recently, the infiltration or injection of liquid into hollow structures has brought unique light–liquid interactions and excellent tunability of the optical components. Examples are infiltrated photonic crystal devices, such as photonic crystal fibers or planar photonic crystals. Thermal actuation [21], manual use of micropipette [22], laser-induced liquid recondensation [23] and soft lithography nanofluidic technique [24] have been utilized to inject liquid into photonic crystal structures. In porous photonic crystal devices, the selectively introduced liquids can significantly tune the transmission behavior of the device, such as emission wavelength [25], propagation direction [26] and speed [27] or spectral response [28,29].

Another example of replacing liquid inside an optical component is the optical switch based on the phenomenon of total internal reflection [30]. A mirror channel was filled with water (low refractive index) or salt solution (high refractive index) to determine whether total internal reflection happens or not, and therefore to allow whether the light beam can propagate through the mirror channel. Groisman et al. reported another optical switch using blazed diffractive grating [31]. For such a grating, diffraction maximum occurs when the optical path difference over one period of the grating is equal to an integer number of the wavelength of the incidence light. The fluid medium above the grating structure can be replaced by liquids with different RIs, and therefore the optical path difference can be changed accordingly to diffract the transmitted light beam with different propagation directions.

Waveguide is one of the key elements in optical systems for confinement and transmission of light, equivalent to channel in terms of fluid transportation [16]. The integration and functionality of waveguide in microfluidic systems are of great interest in the research community. A tunable and reconfigurable waveguide can be used as sensors for biochemical analysis, which can offer longer light–matter interaction length and thus enhanced sensitivity [32]. A simple method to realize a tunable waveguide is to make a hollow channel filled with higher RI liquid embraced by lower RI solid material (Figure 3a) [33,34]. The optical characteristics can be easily adjusted by replacing the liquid inside the channel.

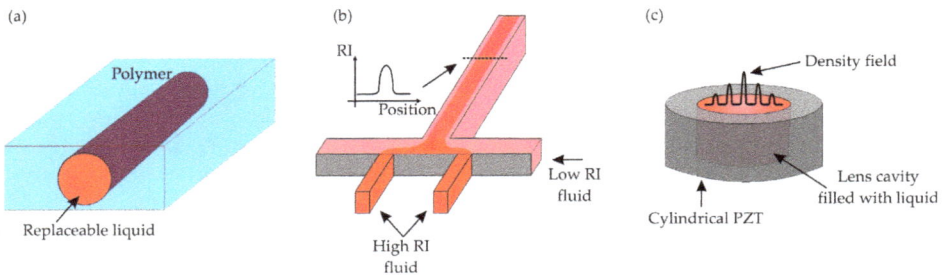

Figure 3. (**a**) A liquid core waveguide which can be tuned by replacing the liquid in the hollow channel; (**b**) an optofluidic GRIN lens configured by two miscible fluids; and (**c**) an acoustics-driven gradient refractive index lens.

Llobera et al. reported a hollow prism built in a microfluidic system for optical sensing [35,36]. The analytes can be injected into the hollow prism chamber, and the interrogation light can be delivered by optical fiber inserted into the microfluidic platform. The RI shift of the analytes can be monitored by the measurement of deviation angle of the interrogation light. The sensing performance and efficiency were investigated using solutions of different concentrations of fluorescein. The devices were found to have good sensitivity in the measurement of absorption and shift of refractive index of the analytes.

2.2. Liquid RI Gradient Control

Instead of replacing the liquid inside an optical component, tuning refractive index of liquid by subjecting it to a flow field, temperature field, electrical field, acoustic field or mechanical strain are the alternative methods to manipulate the optical properties of light propagating in the optical material such as the optical path, phase change and polarization [37]. The generation of refractive index gradient is the most common method to manipulate light properties. The concept of refractive index gradient was already widely used in traditional fiber optics [38–41]. Light rays follow sinusoidal paths in the gradient-index fiber, with the advantage of decreasing the modal dispersion compared with multi-mode step-index fiber. Gradient refractive index (GRIN) also lends its strength to lens optics [42–45]. A proper distribution of index gradient in the direction perpendicular to the optical axis of the lens can help to eliminate the aberrations caused by traditional spherical lenses without varying the shape of the lens. The problem with solid gradient-index optics is the complete lack of tunability. The distribution of the gradient-index is fixed. Moreover, the fabrication process to establish such an index gradient is often complex, requiring field-assisted ion-exchange, micro-controlled dip-coating or vapor deposition.

The refractive index gradient can be controlled through the variation in concentration of a solution, which could be well defined by diffusion of microfluidic laminar flows. Based on this motivation, two-dimensional (in-plane) [46] and three-dimensional (out-of-plane) [47] optofluidic gradient refractive index lenses were proposed and demonstrated (Figure 3b). The laminar flow in microscale gives relative stability of optical performance to these devices, and it is found that the focal length of the lens can be tuned by the diffusion process, which can be well defined by setting a proper flow rate. Recently, Le et al. improved the in-plane gradient index lens by designing a micro-channel of cylindrical geometry with two-dimensional beam focusing capability, and carried out a parametric study to investigate the influences from mass fraction of core solution, flow rate and diffusion coefficient [48]. It is found that focal length can be effectively tuned by adjusting the mass fraction and the flow rate, and the beam spot size can be tuned by controlling the relative slip between the flows at core and cladding inlets. Another research shows that controlling the mass fraction can also be used to reduce aberrations of the gradient index lens [49].

Another mechanism to produce the refractive index gradient is by placing the liquid in a temperature gradient field [50]. In this work, a liquid core/liquid cladding configuration is employed to build an optofluidic waveguide. Two cladding streams at higher temperature sandwich a core stream at lower temperature, which creates a temperature gradient field. The refractive index and many other physical properties are temperature dependent. Therefore the temperature gradient field generates a distribution of refractive index with higher value at the center of the channel. This refractive index gradient enables the liquid core/liquid cladding flow a waveguide function. However, the thermo-optical coefficient dn/dT of most liquids is relatively small. Thus, a large temperature contrast is required to enable an effective waveguide function. Another drawback is the decay of temperature contrast along the flow direction due to dissipation of heat, which leads to unevenly distributed refractive index along the optical axis. This phenomenon causes a lot of optical energy loss during the propagation of light. To improve these drawbacks, Chen et al. proposed to integrate two heat sources into the lens-development chamber, and built up a high temperature gradient, which enabled strongly bending of light to realize lensing effect [51].

Acoustics as an alternative method to enable liquid GRIN lens can offer faster response than the temperature and diffusion driven mechanisms. It was found through theoretical studies and experimentations that acoustic standing wave created inside a cylindrical cavity by a circular piezoelectric transducer can generate a refractive index gradient and consequently result in lensing effect (illustrated in Figure 3c) [52,53]. In particular, the local refractive index of liquid was found to be dependent on a time-invariant density field superposed on the linear motion of the acoustic standing wave, and the time-invariant density field can be approximately described by a combination of Bessel functions. Later, generation and optimization of tunable Bessel beam was demonstrated using such acoustics-driven GRIN lens by McLeod et al. [54,55] Due to the convenience to incorporate the acoustics-driven GRIN lens into conventional optical systems, the tunable Bessel beam has quickly found its applications in bulk imaging [56], two-photon microscopy [57], three-dimensional microscopy [58], etc.

2.3. Fluid Interface Deformation

Recently, a number of optofluidic components based on fluid interface deformation have been proposed and demonstrated, including lenses, waveguides, prisms, apertures, etc. According to the light propagation direction relative to the plane of the substrate, which normally carries the test sample, those components can be divided into two categories: in-plane and out-of-plane types. The in-plane components manipulate beam shape in the plane of the substrate, whereas the out-of-plane ones manipulate the incident light perpendicular to the plane of the substrate.

A key advantage of the out-of-plane components is the compatibility with conventional solid-state optical system and thus the feasibility to replace the conventional ones with tunable optofluidic components in applications such as digital cameras or cell phones. Among the optofluidic components, optofluidic lenses have drawn considerable attention and have been intensively investigated due to their wide applications. A popular method to develop an out-of-plane optofluidic lens is to fill functioning fluid into a circular chamber sealed with a deformable membrane (Figure 4a). Inspired by a human eye's crystalline lens, Ahn et al. proposed to use a deformable glass diaphragm with a thickness of 40 μm to enable a variable focal length [59]. However, the problem with this configuration is the high Young's modulus of the glass diaphragm which limits the tuning range. Only up to 50 μm of deformation for a 10 mm lens can be achieved. To solve this problem, several groups proposed to use transparent elastic polymer (PDMS) as the diaphragm material to achieve a higher diaphragm deformation range [60–62]. Instead of using pneumatic tuning method to control the diaphragm, other working mechanisms such as acoustic [63], electromagnetic [64,65], electro-wetting [66] and electrochemical [67] actuations can be used to modulate the pressure in the liquid lens chamber and thus to tune the focus of the lens. Due to the inherently spherical shape, those lenses suffer from substantial spherical aberrations. Mishra et al. demonstrated to minimize the spherical aberration synthetically using electric field and hydrostatic pressure to manipulate the local curvature of the lens interface [68].

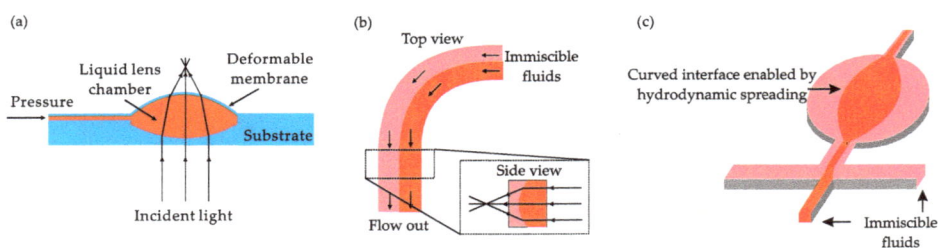

Figure 4. Illustration of optofluidics lenses: (**a**) out-of-plane lens tuned by applying pressure on a deformable membrane; (**b**) in-plane lens enabled by Dean flow; and (**c**) in-plane lens configured by liquid core liquid cladding structure.

The advantage of in-plane components over out-of-plane ones is the possibility of integrating them into microfluidic networks which perform the sample analysis. A high level of integration helps to avoid the manual alignment of optical components as well as to reduce the cost of external bulky components. Different configurations of in-plane liquid/liquid or liquid/air interface optical components have been reported recently. They can be generally categorized into two groups: (a) interfacial intension based tuning; and (b) hydrodynamic tuning. The balance between the applied pressure and the interfacial tension can be used to tune the curvature of the liquid lens interface. Lien et al. fabricated a spherical micro-lens by injecting liquid-phase PDMS into the microfluidic channel [69]. One drawback of this fabrication method is the difficulty to pump highly viscous PDMS into a micro-channel and change the curvature of the PDMS–air interface within a short period of time. To improve this technique, several groups used liquids with low viscosity, such as water, to form the spherical interface driven by capillary force inside the microfluidic networks [70–72]. Since these naturally formed interfaces always present spherical shape, they can be only used for lens configuration.

Hydrodynamic tuning is a more flexible method to manipulate the liquid/liquid interface and realize different functional structures, such as waveguides, lenses, prisms, etc. In such configurations, multiple flows propagate in the channels with specific geometries. In a straight-line channel, a core flow with higher RI sandwiched by two cladding flows with lower RI can form a typical waveguide structure [73]. The width ratio between core and cladding flows can be changed by the pumping flow rates to tune the propagation mode of the light wave. By arranging the flows in a 90° curved channel (Figure 4b), the interface between the flows can bend along perpendicular direction to the substrate to realize a lensing function [74]. If the flow with a liquid core liquid cladding structure enters a shallow cavity (Figure 4c), the interfaces can deform according the geometry of the cavity boundary. Triangular [75,76], rectangular [77–79], hexagonal [80] and circular [81–84] cavities have been proposed to generate a curved interface which can bend the in-plane light according to the RI difference between the core and cladding flows and the shapes of the interfaces. Combining the concepts of liquid replacing and interface deformation, Song et al. designed a rectangular cavity with five inlet channels which allows the switching of fluids inside the cavity, and realization of a lens with both light converging and diverging capabilities [85].

The categorization of optofluidic components is illustrated in Figure 5. In terms of working mechanisms, they can be divided into three groups: (1) liquid replacing; (2) RI gradient control; and (3) fluid interface deformation. Based on these working mechanisms, numerous components have been proposed and demonstrated, such as lenses, waveguides, prisms, apertures, optical cavities, etc. According to the direction of light manipulation, these components can be grouped as out-of-plane and in-plane components. The out-of-plane optofluidic components do not need to be necessarily confined in micro-channels, and thus they can be conveniently tuned by external forces generated by either separate devices or integrated functioning structures. For example, the droplet lens confined in a tube can be tuned by acoustic vibration from a microphone [63]; a liquid prism confined in a cylinder with the wall of the cylinder coated with metal electrodes can be tuned by electro-wetting effect [86]. However, for those in-plane components confined in microfluidic channels for lab-on-chip applications, using non-fluidic driven forces might require integration of complex functional structures, which needs complicated fabrication procedures, and therefore the entire cost of devices rises. To exploit fabrication-convenient and cost-effective solutions, several ways of tuning the optofluidic components have been intensively investigated, such as hydrostatic tuning, RI gradient tuning, and hydrodynamic tuning. Owing to the promising future of optofluidics for lab-on-chip applications, the following section will be more focused on the discussion of using in-plane optofluidic components to develop the optical instruments for bio-chemical analysis and treatment.

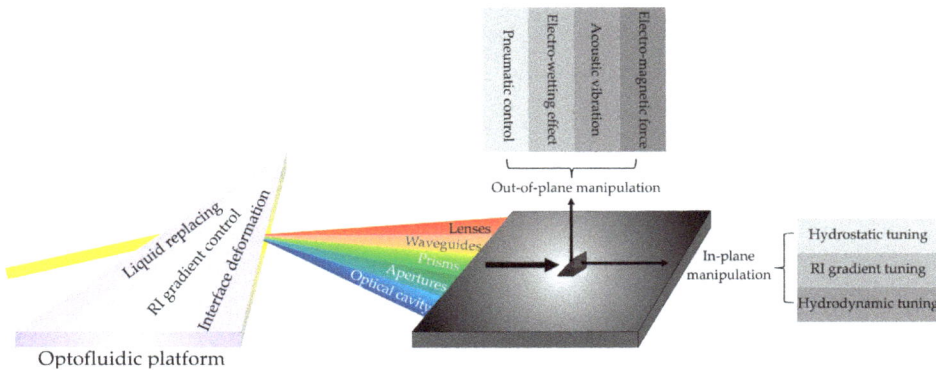

Figure 5. The categorization of optofluidic components in terms of their working mechanisms.

3. Instrumentation of Optofluidics

Based on the abovementioned optofluidic components, a variety of detection and analytical tools have been developed, which have applications in biomedical research, biochemistry, pharmaceuticals, healthcare, and environmental monitoring. The integration of optofluidic components on a microchip can greatly reduce the fabrication cost and significantly downsize the entire systems. It is promising and possible to make the system portable for field test even in remote regions that cannot afford high-cost equipment.

Fundamentally, light carries the information encoded with its basic properties, such as phase, electromagnetic frequency, intensity or polarization. The interaction of light with the inspected sample can modulate its properties, which need to be decoded by specific tools such as interferometer, spectrometer, photometer and polarization analyzer (Figure 2). This section will discuss the instrumentation of optofluidics under such categories.

3.1. Optofluidic Interferometry

Thanks to the realization of on-chip waveguide, the Mach–Zehnder interferometer (MZI) can be easily developed with both sensing and reference arms aligned on an optofluidic chip (Figure 6a). The sensing arm passes through a detection window with its evanescent field penetrating into the fluid analyte in the detection window. Any subtle RI variation in the analyte can result in a change in the optical path. The change in optical path changes the intensity from the output end. By measuring the signal change, the MZI can precisely measure the phase difference between the light waves propagating through the two arms.

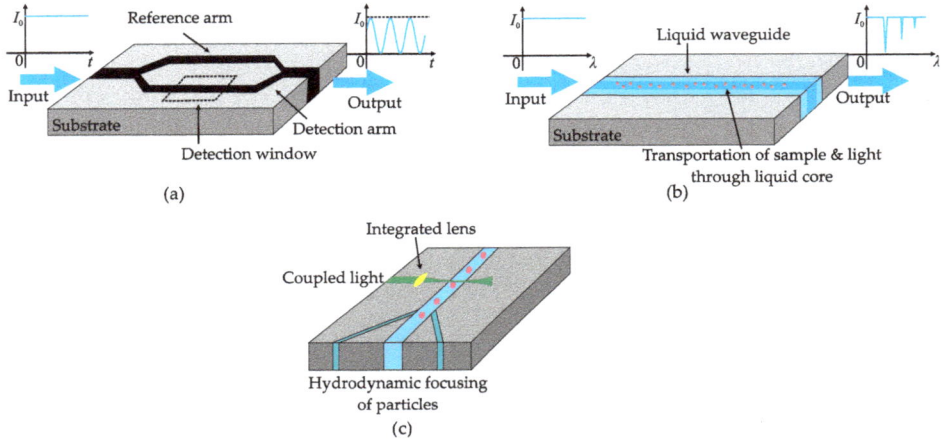

Figure 6. Schematics of functional optofluidic devices: (**a**) integrated Mach–Zehnder interferometer for label-free detection of biochemical sample; (**b**) optofluidic enhanced Raman spectroscopy for analytical chemistry; and (**c**) a typical configuration for optical detection or manipulation of micro-particles.

The first on-chip MZI was realized by growing Si_3N_4 on a silicon substrate to configure the two waveguides arms, and successfully measure the concentration of biomolecules [87–89]. Following works involved the improvements in the waveguide material and configuration of the interferometer [90]. It has been demonstrated that the environmental noise can be canceled by proper design of the reference arm [91]. To improve the precision in signal measurement, a phase modulator can be incorporated in the reference arm to track the quadrature points and enable linearity in signal changing [92]. However, the transduction signal of those configurations relies on the interaction between evanescent field of guided mode and the analyte, which requires sufficiently long interaction length in order to achieve a decent sensitivity. To improve the sensitivity, an extrinsic configuration with the sensing arm orthogonally crossing the analyte flowing channel has been proposed to increase the light-analyte interaction length [93,94]. The drawback of this configuration is that the separation between the two waveguides across the detection window might cause severe transmission loss. Lapsley et al. improved the light coupling efficiency by incorporating two polymer lenses to collimate divergent beams on their chip [95]. Another method to enhance the sensitivity is to couple the light into a liquid core waveguide as the sensing arm [96–98]. The light and the sample fluid can transfer along the same channel to achieve a maximum light–matter interaction length.

3.2. Optofluidic Spectroscopy

Spectroscopy, as an important optical tool, can measure analytes at very low concentration level based on the absorption, fluorescence or Raman characteristics. The main limitation to integrate the spectroscopic function on a micro-chip is the short light–matter interaction length due to the inherently small size of the chip. However, the invention of the optofluidic waveguide enables an enhanced light–matter interaction to achieve maximum sensitivity by offering a way to transmit the light and analytes along the same guide (Figure 6b).

In the work of Yin et al. [99], an anti-resonant reflecting optical waveguide (ARROW) [100] was used to carry the analyte in the hollow liquid core. The excitation of fluorescence was implemented by coupling an external laser into the liquid core with the fluorescent molecules. The fluorescence signal can be carried by the propagation mode and transmitted through the ARROWs with enhanced interaction and collected by a photomultiplier tube. To improve the detection limit (DL), the same group proposed a configuration with the excitation waveguide perpendicularly aligned to fluorescence waveguide [101]. Single molecule detection can be achieved on a planar microfluidic network. Testa et al. proposed a configuration with the fluidic part fabricated with polymer which is stacked on an ARROW structure fabricated with silicone [102]. This configuration has the flexibility to replace the fluidic module with specific functions. The concept of using liquid core waveguide has also been demonstrated to measure the refractive index of the analyte [103] and the concentration of bio-molecules [104,105] with extremely high sensitivity by looking at the spectral shift of the transmitted light. Besides the waveguide, the optofluidic prism [35,36] and grating [106] can also find their applications in sensing the variation of RI of the analytes.

3.3. Optofluidic Cytometry

A traditional flow cytometer employs bulky optics to inspect the flowing particles of micro-size or biological cells aligned in a single line. The scattered light or fluorescence signal can be collected and analyzed to reveal the physical and chemical properties of the particles or cells. Since both delivery and collection of light involve external bulky lenses as well as the alignment of optics, the system can hardly be used for field test.

Recently, many efforts have been dedicated to downsize the cytometry system and make it portable. The microfluidic platform offers an efficient way to align the particles along a single streamline due to the laminar flow characteristic. The most common implementation is to utilize a flow carrying particles sandwiched by two sheath flows. The hydrodynamic focusing effect allows the particles following a line to pass through the inspection window [107] (Figure 6c). The inspection light can be simply delivered via external objective lens and the scattered light from the particles, which can be described by Mie theory, can be collected from either backscattering direction or forward-scattering direction. To further downsize the system, external optics needs to be integrated on the same microfluidic system. Optical fiber can be used for both light delivering and collecting [108,109]. However, the divergent beam at the output end of the fiber leads to inaccuracy of detection due to the enlarged inspection region. Several groups proposed to integrate a micro-lens by direct shaping of PDMS–air/liquid interface to focus the light beam delivered from an optical fiber [110–112]. The focused beam crossing the inspection window enables the detection of single particle and prevents the misinterpretation of the collected signal. Further improvement involves the use of an optofluidic lens configured by hydrodynamic spreading to focus the beam without any light scattering which might be caused by the lens with solid-air interface [78,113].

3.4. On-Chip Manipulation of Micro-Object

The manipulation of micro-object involves trapping, sorting, and separation, which can find their applications in the fields of chemistry, biology and colloidal science. The optical method, which is referred to as optical tweezers [114,115], has played an important role due to its accuracy and flexibility to handle single or multiple micro-objects.

Combining laminar flow characteristics in a microfluidic channel with optical trapping, sorting of particles or cells can also be achieved. The absence of turbulence in a microchannel gives the particle or cells a predictable drag force along flow direction. The optical trap will exert a gradient force along a defined direction, which combined with drag force results in a net force that can redirect particles from their original tracks into the designated fluidic branch [116,117]. Similar work was done by Applegate et al., who used laser diode bars and focused the beam in a microchannel to achieve particle sorting [118]. With such a configuration, bovine red blood cells can be captured by the optical trap with linear shape, which was placed at an angle to the flow direction, and the cells were conveyed along it before being released at its end, thereby sorting cells according their sizes.

Instead of using external optical instruments to manipulate the light, the evanescent field trapping is another type of optical method to manipulate particles which can employ on-chip waveguide to generate an intensity gradient field of light. Schmidt et al. investigated the forces exerted by the evanescent field, which can attract the particles to the surface of the waveguide and at the same time propel the particles along the propagation direction of light [93]. Particles with a diameter of 3 μm and velocity up to 28 μm/s were successfully sorted by their device. Besides the gradient force, the scattering force of light can also be used for on-chip particle manipulation. Both free space optics [119,120] and embedded fiber [121] have been demonstrated to deliver the scattering force on the flowing particles confined in a microfluidic environment and divert them to a new routing.

4. Commercialization of Optofluidics

Leveraging the superior reconfigurable properties, optofluidic systems can be dynamically tuned without any moving mechanical parts, which can be utilized to develop faster and more compact optical systems. Thus far, several concepts have been commercialized with their applications in a variety of fields. Due to the compatibility with conventional optical systems, the out-of-plane optofluidic configurations, especially lenses, were the forerunners during the development of optofluidic products. An acoustics-driven GRIN lens (TAG Optics Inc., Princeton, NJ, USA [122]) is available on market with its applications in machine-vision, biological microscopy and laser material processing. The technology relies on changing the refractive index of liquid confined in a cylindrical chamber by acoustic pressure waves. Since the wave travels at sound speed, the redistribution of RI gradient can respond within 10–30 μs. Electro-wetting based liquid lenses were commercialized by Varioptic [123]. Besides the imaging functions, the lens of this type possesses the capability of focusing and steering laser beam in *X*, *Y* and *Z* axis by selectively controlling the electrodes, which can find its application in ophthalmology. Other actuation methods, such as using deformable diaphragm and electro-active polymer, were also used in the production of tunable liquid lenses by Optotune [124].

For the lab-on-chip applications, the in-plane waveguides are commonly used to develop fluid sensor and particle tweezers due to the convenient fabrication and control. World Precision Instruments commercialized a device using liquid core waveguide for the measurement of light absorbance of analytes [125]. A NanoTweezer based on the evanescent-field trapping using integrated waveguide was developed by Optofluidics [126]. However the overall sizes of these devices are still bulky, since they require external optics, in particular lenses, to perform further analysis. Unlike the out-of-plane optofluidic lenses, the in-plane lenses are still in their early research stage. Several reasons impede the on-chip optofluidic in-plane lenses from their integration into commercial products. In terms of the hydrodynamically developed in-plane lenses, they consume liquid when at working state, and, to date, no technology has been demonstrated to recycle and reuse the functioning fluids for this type of lens. The printed polymer in-plane lens does not consume liquid in principle, but its surface quality depends on the fabrication process. Furthermore, the focal length of the printed lens is fixed after fabrication, on-chip optical alignment during the test is impossible. The capillary effect of liquid in micro-channels can be used to form lens interface without continuously consuming liquids, and the curvature of the interface can be tuned by balancing surface tension and other driving forces. Thus far, the fastest way to tune the liquid lens in micro-channels is by the control of pneumatic pressure [70]. However, the response is still slow when compared to sound or electro-wetting driven out-of-plane liquid lenses, and sensitive pressure controller might be needed to balance the surface tension of liquid in microscale, which will increase the entire cost of the system. Those might be the reasons that the currently demonstrated in-plane optofluidic lenses have not been widely utilized for commercial products.

Recently, several ways have been explored to manipulate the fluid interface at microscale with integrated control units in the field of droplet-based microfluidics. Thermal heating generated by integrated microheater has been demonstrated to effectively change the interfacial tension between two immiscible fluids [127,128]. Maxwell stress induced by integrated electrodes can also be used to change the interfacial tension [129,130]. The manipulation of interfacial tension can dynamically adjust the size and shape of micro-droplets which essentially can serve as tunable micro-lens. The concept of the out-of-plane acoustics-driven lens can also be leveraged to realize in-plane tunable GRIN lens by imposing surface acoustic wave (SAW) to generate gradient density field using affordable and self-aligned interdigitated transducers [131]. The authors believe all these methods can be leveraged to develop cost-effective in-plane optofluidic lenses with high tuning response and precise manipulation, which would benefit the further commercialization of on-chip optofluidic systems.

5. Discussions and Conclusions

This paper gives an overview of the developments of fundamental optical elements using fluid referred as optofluidic components. Driven by the demands of lab-on-chip applications, many efforts have been devoted to realize the integration of optofluidic components into microfluidic systems. With the advancements of MEMS technologies, a variety of optofluidic components can be miniaturized and printed on the same planar structure with other assay modules. Those optofluidic components can be categorized based on their working mechanisms: (1) liquid replacing; (2) RI gradient control; and (3) liquid/fluid interface deformation. Owing to the reconfigurable features, these optofluidic components possess a high level of tuneability to make the optical system highly adaptive. Adaptability as well as the miniaturization can significantly downsize the entire system and make the system portable for field test. Instead of using bulk optics in free space, the integrated optics can greatly reduce the cost of system and thus hold a promise for affordable point-of-care diagnosis even in those regions suffering from poverty.

Based on the developments of fundamental optofluidic components, many of the optical tools for biochemical analysis and treatment can be realized with miniaturized and simplified architectures. The invention of optofluidic waveguide enables an efficient way to print the optical arms of an interferometer on a planar structure with the sensing arm directly interacting with the analytes; the liquid core waveguide allows the transfer of sample fluid and light along the same guide. The enhanced interaction between light and fluid enables integration of spectroscopy into microfluidic platforms; the developments of optofluidic lenses enable enhanced detection of micro-particles or even single fluorescent molecule without using external bulky optics. It is envisioned that an all-optofluidic system can be practically realized with all optical functions integrated into a single chip in the near future.

To achieve this goal, several critical challenges need to be addressed carefully. Firstly, many of the optofluidic components consume functioning fluids even when working at steady state. An effective way needs to be figured out to recycle and reuse the fluid, which can further lower the operational cost of the system and make the system more practical. The stability of fluid pumping is another critical concern. Any discontinuity in pumping might introduce disturbance to the functioning of optofluidic components, which can result in inaccuracy of the analysis or malfunction of the optical treatment. Besides the discontinuity in pumping, the vibrational noise from the environment may also introduce perturbation to the optofluidic system. The environmental noise can be isolated for the free-space optics by mounting the optical elements on an anti-vibration table, but it is not practical to do the same for the optofluidic systems since they are born to be used for field test and point-of-diagnosis. Hence, cost-effective housing of on-chip optofluidic systems with effective anti-noise techniques has to be developed for the practical use and their further commercialization. Different from the manipulation of light in free space using glass-based optical elements, it is difficult to tightly focus a light beam with a high numerical aperture using optofluidic components, since most structural materials for microfluidic system have very high refractive index around 1.4 and the RI of liquids lies between 1.3 and 1.6 RIU. This drawback of optofluidic components may hamper the developments of optical tweezing with high strength and high-resolution imaging. Efforts need to be dedicated to exploit novel fabrication methods or materials for the light manipulation using on-chip elements.

Acknowledgments: This work was supported by the Fundamental Research Funds for the Central Universities, China University of Geosciences (Wuhan) (CUG170608).

Author Contributions: All authors contributed to the writing, reviewing, and editing of this paper.

Conflicts of Interest: The authors declare no conflict of interest.

References

1. Schmidt, H.; Hawkins, A.R. The photonic integration of non-solid media using optofluidics. *Nat. Photonics* **2011**, *5*, 598–604. [CrossRef]
2. Wood, R.W. The mercury paraboloid as a reflecting telescope. *Astrophys. J.* **1909**, *29*, 164–176. [CrossRef]
3. Monat, C.; Domachuk, P.; Eggleton, B.J. Integrated optofluidics: A new river of light. *Nat. Photonics* **2007**, *1*, 106–114. [CrossRef]
4. Petersen, K.E. Silicon torsional scanning mirror. *IBM J. Res. Dep.* **1980**, *24*, 631–637. [CrossRef]
5. Psaltis, D.; Quake, S.R.; Yang, C. Developing optofluidic technology through the fusion of microfluidics and optics. *Nature* **2006**, *442*, 381–386. [CrossRef] [PubMed]
6. Kim, E.; Baaske, M.D.; Vollmer, F. Towards next-generation label-free biosensors: Recent advances in whispering gallery mode sensors. *Lab Chip* **2017**, *17*, 1190–1205. [CrossRef] [PubMed]
7. Wang, Y.; Li, H.; Zhao, L.; Wu, B.; Liu, S.; Liu, Y.; Yang, J. A review of droplet resonators: Operation method and application. *Opt. Laser Technol.* **2016**, *86*, 61–68. [CrossRef]
8. Fan, X.; White, I.M.; Shopova, S.I.; Zhu, H.; Suter, J.D.; Sun, Y. Sensitive optical biosensors for unlabeled targets: A review. *Anal. Chim. Acta* **2008**, *620*, 8–26. [CrossRef] [PubMed]
9. Zhang, Y.; Watts, B.R.; Guo, T.; Zhang, Z.; Xu, C.; Fang, Q. Optofluidic device based microflow cytometers for particle/cell detection: A review. *Micromachines* **2016**, *7*, 70. [CrossRef]
10. Erickson, D.; Sinton, D.; Psaltis, D. Optofluidics for energy applications. *Nat. Photonics* **2011**, *5*, 583–590. [CrossRef]
11. Monat, C.; Domachuk, P.; Grillet, C.; Collins, M.; Eggleton, B.J.; Cronin-Golomb, M.; Mutzenich, S.; Mahmud, T.; Rosengarten, G.; Mitchell, A. Optofluidics: A novel generation of reconfigurable and adaptive compact architectures. *Microfluid. Nanofluid.* **2008**, *4*, 81–95. [CrossRef]
12. Huang, N.-T.; Zhang, H.; Chung, M.-T.; Seo, J.H.; Kurabayashi, K. Recent advancements in optofluidics-based single-cell analysis: Optical on-chip cellular manipulation, treatment, and property detection. *Lab Chip* **2014**, *14*, 1230–1245. [CrossRef] [PubMed]
13. Yang, T.; Bragheri, F.; Minzioni, P. A comprehensive review of optical stretcher for cell mechanical characterization at single-cell level. *Micromachines* **2016**, *7*, 90. [CrossRef]
14. Brennan, D.; Justice, J.; Corbett, B.; McCarthy, T.; Galvin, P. Emerging optofluidic technologies for point-of-care genetic analysis systems: A review. *Anal. Bioanal. Chem.* **2009**, *395*, 621–636. [CrossRef] [PubMed]
15. Mak, J.S.W.; Rutledge, S.A.; Abu-Ghazalah, R.M.; Eftekhari, F.; Irizar, J.; Tam, N.C.M.; Zheng, G.; Helmy, A.S. Recent developments in optofluidic-assisted raman spectroscopy. *Prog. Quantum Electron.* **2013**, *37*, 1–50. [CrossRef]
16. Schmidt, H.; Hawkins, A.R. Optofluidic waveguides: I. Concepts and implementations. *Microfluid. Nanofluid.* **2008**, *4*, 3–16. [CrossRef] [PubMed]
17. Testa, G.; Persichetti, G.; Bernini, R. Liquid core arrow waveguides: A promising photonic structure for integrated optofluidic microsensors. *Micromachines* **2016**, *7*, 47. [CrossRef]

18. Mishra, K.; van den Ende, D.; Mugele, F. Recent developments in optofluidic lens technology. *Micromachines* **2016**, *7*, 102. [CrossRef]

19. Nguyen, N.-T. Micro-optofluidic lenses: A review. *Biomicrofluidics* **2010**, *4*, 031501. [CrossRef] [PubMed]

20. Pan, M.; Kim, M.; Kuiper, S.; Tang, S.K.Y. Actuating fluid-fluid interfaces for the reconfiguration of light. *IEEE J. Sel. Top. Quantum Electron.* **2015**, *21*, 444–455. [CrossRef]

21. Domachuk, P.; Nguyen, H.C.; Eggleton, B.J. Transverse probed microfluidic switchable photonic crystal fiber devices. *IEEE Photonics Technol. Lett.* **2004**, *16*, 1900–1902. [CrossRef]

22. Intonti, F.; Vignolini, S.; Türck, V.; Colocci, M.; Bettotti, P.; Pavesi, L.; Schweizer, S.L.; Wehrspohn, R.; Wiersma, D. Rewritable photonic circuits. *Appl. Phys. Lett.* **2006**, *89*, 211117. [CrossRef]

23. Speijcken, N.W.L.; Dündar, M.A.; Bedoya, A.C.; Monat, C.; Grillet, C.; Domachuk, P.; Nötzel, R.; Eggleton, B.J.; van der Heijden, R.W. In situ optofluidic control of reconfigurable photonic crystal cavities. *Appl. Phys. Lett.* **2012**, *100*, 261107. [CrossRef]

24. Erickson, D.; Rockwood, T.; Emery, T.; Scherer, A.; Psaltis, D. Nanofluidic tuning of photonic crystal circuits. *Opt. Lett.* **2006**, *31*, 59–61. [CrossRef] [PubMed]

25. Arango, F.B.; Christiansen, M.B.; Gersborg-Hansen, M.; Kristensen, A. Optofluidic tuning of photonic crystal band edge lasers. *Appl. Phys. Lett.* **2007**, *91*, 223503. [CrossRef]

26. Shi, J.; Juluri, B.K.; Lin, S.C.S.; Lu, M.; Gao, T.; Huang, T.J. Photonic crystal composites-based wide-band optical collimator. *J. Appl. Phys.* **2010**, *108*, 043514. [CrossRef]

27. Mortensen, N.A.; Xiao, S. Slow-light enhancement of beer-lambert-bouguer absorption. *Appl. Phys. Lett.* **2007**, *90*, 141108. [CrossRef]

28. Smith, C.L.C.; Wu, D.K.C.; Lee, M.W.; Monat, C.; Tomljenovic-Hanic, S.; Grillet, C.; Eggleton, B.J.; Freeman, D.; Ruan, Y.; Madden, S.; et al. Microfluidic photonic crystal double heterostructures. *Appl. Phys. Lett.* **2007**, *91*, 121103. [CrossRef]

29. Domachuk, P.; Nguyen, H.C.; Eggleton, B.J.; Straub, M.; Gu, M. Microfluidic tunable photonic band-gap device. *Appl. Phys. Lett.* **2004**, *84*, 1838–1840. [CrossRef]

30. Campbell, K.; Groisman, A.; Levy, U.; Pang, L.; Mookherjea, S.; Psaltis, D.; Fainman, Y. A microfluidic 2*2 optical switch. *Appl. Phys. Lett.* **2004**, *85*, 6119–6121. [CrossRef]

31. Groisman, A.; Zamek, S.; Campbell, K.; Pang, L.; Levy, U.; Fainman, Y. Optofluidic 1*4 switch. *Opt. Express* **2008**, *16*, 13499–13508. [CrossRef] [PubMed]

32. Li, X.C.; Wu, J.; Liu, A.Q.; Li, Z.G.; Soew, Y.C.; Huang, H.J.; Xu, K.; Lin, J.T. A liquid waveguide based evanescent wave sensor integrated onto a microfluidic chip. *Appl. Phys. Lett.* **2008**, *93*, 193901. [CrossRef]

33. Schelle, B.; Dreb, P.; Franke, H.; Klein, K.F.; Slupek, J. Physical characterization of lightguide capillary cells. *J. Phys. D Appl. Phys.* **1999**, *32*, 3157–3163. [CrossRef]

34. Datta, A.; Eom, I.Y.; Dhar, A.; Kuban, P.; Manor, R.; Ahmad, I.; Gangopadhyay, S.; Dallas, T.; Holtz, M.; Temkin, H.; et al. Microfabrication and characterization of teflon af-coated liquid core waveguide channels in silicon. *IEEE Sens. J.* **2003**, *3*, 788–795. [CrossRef]

35. Llobera, A.; Wilke, R.; Buttgenbach, S. Poly(dimethylsiloxane) hollow abbe prism with microlenses for detection based on absorption and refractive index shift. *Lab Chip* **2004**, *4*, 24–27. [CrossRef] [PubMed]

36. Llobera, A.; Wilke, R.; Buttgenbach, S. Optimization of poly(dimethylsiloxane) hollow prisms for optical sensing. *Lab Chip* **2005**, *5*, 506–511. [CrossRef] [PubMed]

37. Erickson, D.; Heng, X.; Li, Z.; Rockwood, T.; Emery, T.; Zhang, Z.; Scherer, A.; Yang, C.; Psaltis, D. Optofluidics. *Proc. SPIE* **2005**, *5908*, 1–12.

38. Olshanky, R.; Keck, D.B. Pulse broadening in graded-index optical fibers. *Appl. Opt.* **1976**, *15*, 483–491. [CrossRef] [PubMed]

39. Koike, Y.; Ishigure, T.; Nihei, E. High-bandwidth graded-index polymer optical fiber. *J. Lightwave Technol.* **1995**, *13*, 1475–1489. [CrossRef]

40. Ishigure, T.; Koike, Y.; Fleming, J.W. Optimum index profile of the perfluorinated polymer-based GI polymer optical fiber and its dispersion properties. *J. Lightwave Technol.* **2000**, *18*, 178–184. [CrossRef]

41. Van Buren, M.; Riza, N.A. Foundations for low-loss fiber gradient-index lens pair coupling with the self-imaging mechanism. *Appl. Opt.* **2003**, *42*, 550–565. [CrossRef] [PubMed]

42. Pierscionek, B.K.; Chan, D.Y.C. Refractive index gradient of human lenses. *Optom. Vis. Sci.* **1989**, *66*, 822–829. [CrossRef] [PubMed]

43.   Ren, H.; Fan, Y.H.; Gauza, S.; Wu, S.T. Tunable-focus flat liquid crystal spherical lens. *Appl. Phys. Lett.* **2004**, *84*, 4789–4791. [CrossRef]

44.   Ren, H.; Wu, S.T. Tunable electronic lens using a gradient polymer network liquid crystal. *Appl. Phys. Lett.* **2003**, *82*, 22–24. [CrossRef]

45.   Jones, C.E.; Atchison, D.A.; Meder, R.; Pope, J.M. Refractive index distribution and optical properties of the isolated human lens measured using magnetic resonance imaging (MRI). *Vis. Res.* **2005**, *45*, 2352–2366. [CrossRef] [PubMed]

46.   Mao, X.; Lin, S.C.S.; Lapsley, M.I.; Shi, J.; Juluri, B.K.; Huang, T.J. Tunable liquid gradient refractive index (L-GRIN) lens with two degrees of freedom. *Lab Chip* **2009**, *9*, 2050–2058. [CrossRef] [PubMed]

47.   Huang, H.; Mao, X.; Lin, S.C.S.; Kiraly, B.; Huang, Y.; Huang, T.J. Tunable two-dimensional liquid gradient refractive index (L-GRIN) lens for variable light focusing. *Lab Chip* **2010**, *10*, 2387–2393. [CrossRef] [PubMed]

48.   Le, Z.; Sun, Y.; Du, Y. Liquid gradient refractive index microlens for dynamically adjusting the beam focusing. *Micromachines* **2015**, *6*, 1984–1995. [CrossRef]

49.   Zhao, H.T.; Yang, Y.; Chin, L.K.; Chen, H.F.; Zhu, W.M.; Zhang, J.B.; Yap, P.H.; Liedberg, B.; Wang, K.; Wang, G.; et al. Optofluidic lens with low spherical and low field curvature aberrations. *Lab Chip* **2016**, *16*, 1617–1624. [CrossRef] [PubMed]

50.   Tang, S.K.Y.; Mayers, B.T.; Vezenov, D.V.; Whitesides, G.M. Optical waveguiding using thermal gradients across homogeneous liquids in microfluidic channels. *Appl. Phys. Lett.* **2006**, *88*, 061112. [CrossRef]

51.   Chen, Q.; Jian, A.; Li, Z.; Zhang, X. Optofluidic tunable lenses using laser-induced thermal gradient. *Lab Chip* **2016**, *16*, 104–111. [CrossRef] [PubMed]

52.   Higginson, K.A.; Costolo, M.A.; Rietman, E.A. Adaptive geometric optics derived from nonlinear acoustic effects. *Appl. Phys. Lett.* **2004**, *84*, 843–845. [CrossRef]

53.   Higginson, K.A.; Costolo, M.A.; Rietman, E.A.; Ritter, J.M.; Lipkens, B. Tunable optics derived from nonlinear acoustic effects. *J. Appl. Phys.* **2004**, *95*, 5896–5904. [CrossRef]

54.   McLeod, E.; Arnold, C.B. Mechanics and refractive power optimization of tunable acoustic gradient lenses. *J. Appl. Phys.* **2007**, *102*, 033104. [CrossRef]

55.   McLeod, E.; Hopkins, A.B.; Arnold, C.B. Multiscale bessel beams generated by a tunable acoustic gradient index of refraction lens. *Opt. Lett.* **2006**, *31*, 3155–3157. [CrossRef] [PubMed]

56.   Mermillod-Blondin, A.; McLeod, E.; Arnold, C.B. High-speed varifocal imaging with a tunable acoustic gradient index of refraction lens. *Opt. Lett.* **2008**, *33*, 2146–2148. [CrossRef] [PubMed]

57.   Olivier, N.; Mermillod-Blondin, A.; Arnold, C.B.; Beaurepaire, E. Two-photon microscopy with simultaneous standard and extended depth of field using a tunable acoustic gradient-index lens. *Opt. Lett.* **2009**, *34*, 1684–1686. [CrossRef] [PubMed]

58.   Duocastella, M.; Sun, B.; Arnold, C.B. Simultaneous imaging of multiple focal planes for three-dimensional microscopy using ultra-high-speed adaptive optics. *J. Biomed. Opt.* **2012**, *17*, 050505. [CrossRef] [PubMed]

59.   Ahn, S.H.; Kim, Y.K. Proposal of human eye's crystalline lens-like variable focusing lens. *Sens. Actuators A Phys.* **1999**, *78*, 48–53. [CrossRef]

60.   Werber, A.; Zappe, H. Tunable microfluidic microlenses. *Appl. Opt.* **2005**, *44*, 3238–3245. [CrossRef] [PubMed]

61.   Zhang, D.Y.; Justis, N.; Lo, Y.H. Fluidic adaptive lens of transformable lens type. *Appl. Phys. Lett.* **2004**, *84*, 4194–4196. [CrossRef]

62.   Song, C.; Xi, L.; Jiang, H. Liquid acoustic lens for photoacoustic tomography. *Opt. Lett.* **2013**, *38*, 2930–2933. [CrossRef] [PubMed]

63.   López, C.A.; Hirsa, A.H. Fast focusing using a pinned-contact oscillating liquid lens. *Nat. Photonics* **2008**, *2*, 610–613. [CrossRef]

64.   Lee, S.W.; Lee, S.S. Focal tunable liquid lens integrated with an electromagnetic actuator. *Appl. Phys. Lett.* **2007**, *90*, 121129. [CrossRef]

65.   Malouin, B.A., Jr.; Vogel, M.J.; Olles, J.D.; Cheng, L.; Hirsa, A.H. Electromagnetic liquid pistons for capillarity-based pumping. *Lab Chip* **2010**, *11*, 393–397. [CrossRef] [PubMed]

66.   Gorman, C.B.; Biebuyck, H.A.; Whitesides, G.M. Control of the shape of liquid lenses on a modified gold surface using an applied electrical potential across a self-assembled monolayer. *Langmuir* **1995**, *11*, 2242–2246. [CrossRef]

67.   López, C.A.; Lee, C.C.; Hirsa, A.H. Electrochemically activated adaptive liquid lens. *Appl. Phys. Lett.* **2005**, *87*, 1–3. [CrossRef]

68. Mishra, K.; Murade, C.; Carreel, B.; Roghair, I.; Oh, J.M.; Manukyan, G.; van den Ende, D.; Mugele, F. Optofluidic lens with tunable focal length and asphericity. *Sci. Rep.* **2014**, *4*, 6378. [CrossRef] [PubMed]

69. Lien, V.; Berdichevsky, Y.; Lo, Y.H. Microspherical surfaces with predefined focal lengths fabricated using microfluidic capillaries. *Appl. Phys. Lett.* **2003**, *83*, 5563–5565. [CrossRef]

70. Shi, J.; Stratton, Z.; Lin, S.C.S.; Huang, H.; Huang, T.J. Tunable optofluidic microlens through active pressure control of an air-liquid interface. *Microfluid. Nanofluid.* **2010**, *9*, 313–318. [CrossRef]

71. Dong, L.; Jiang, H. Tunable and movable liquid microlens in situ fabricated within microfluidic channels. *Appl. Phys. Lett.* **2007**, *91*, 041109. [CrossRef]

72. Mao, X.; Stratton, Z.I.; Nawaz, A.A.; Lin, S.-C.S.; Huang, T.J. Optofluidic tunable microlens by manipulating the liquid meniscus using a flared microfluidic structure. *Biomicrofluidics* **2010**, *4*, 043007. [CrossRef] [PubMed]

73. Wolfe, D.B.; Conroy, R.S.; Garstecki, P.; Mayers, B.T.; Fischbach, M.A.; Paul, K.E.; Prentiss, M.; Whitesides, G.M. Dynamic control of liquid-core/liquid-cladding optical waveguides. *Proc. Natl. Acad. Sci. USA* **2004**, *101*, 12434–12438. [CrossRef] [PubMed]

74. Mao, X.; Waldeisen, J.R.; Juluri, B.K.; Huang, T.J. Hydrodynamically tunable optofluidic cylindrical microlens. *Lab Chip* **2007**, *7*, 1303–1308. [CrossRef] [PubMed]

75. Song, C.; Nguyen, N.T.; Asundi, A.K.; Tan, S.H. Tunable micro-optofluidic prism based on liquid-core liquid-cladding configuration. *Opt. Lett.* **2010**, *35*, 327–329. [CrossRef] [PubMed]

76. Xiong, S.; Liu, A.Q.; Chin, L.K.; Yang, Y. An optofluidic prism tuned by two laminar flows. *Lab Chip* **2011**, *11*, 1864–1869. [CrossRef] [PubMed]

77. Tang, S.K.Y.; Stan, C.A.; Whitesides, G.M. Dynamically reconfigurable liquid-core liquid-cladding lens in a microfluidic channel. *Lab Chip* **2008**, *8*, 395–401. [CrossRef] [PubMed]

78. Rosenauer, M.; Vellekoop, M.J. Characterization of a microflow cytometer with an integrated three-dimensional optofluidic lens system. *Biomicrofluidics* **2010**, *4*, 043005. [CrossRef] [PubMed]

79. Song, C.; Nguyen, N.-T.; Asundi, A.K.; Low, C.L.-N. Tunable optofluidic aperture configured by a liquid-core/liquid-cladding structure. *Opt. Lett.* **2011**, *36*, 1767–1769. [CrossRef] [PubMed]

80. Seow, Y.C.; Liu, A.Q.; Chin, L.K.; Li, X.C.; Huang, H.J.; Cheng, T.H.; Zhou, X.Q. Different curvatures of tunable liquid microlens via the control of laminar flow rate. *Appl. Phys. Lett.* **2008**, *93*, 084101. [CrossRef]

81. Song, C.; Nguyen, N.T.; Asundi, A.K.; Low, C.L.N. Biconcave micro-optofluidic lens with low-refractive-index liquids. *Opt. Lett.* **2009**, *34*, 3622–3624. [CrossRef] [PubMed]

82. Song, C.; Nguyen, N.T.; Tan, S.H.; Asundi, A.K. Modelling and optimization of micro optofluidic lenses. *Lab Chip* **2009**, *9*, 1178–1184. [CrossRef] [PubMed]

83. Song, C.; Nguyen, N.T.; Tan, S.H.; Asundi, A.K. A tuneable micro-optofluidic biconvex lens with mathematically predictable focal length. *Microfluid. Nanofluid.* **2010**, *9*, 889–896. [CrossRef]

84. Song, C.; Nguyen, N.T.; Tan, S.H.; Asundi, A.K. A micro optofluidic lens with short focal length. *J. Micromech. Microeng.* **2009**, *19*, 085012. [CrossRef]

85. Song, C.; Nguyen, N.T.; Yap, Y.F.; Luong, T.D.; Asundi, A.K. Multi-functional, optofluidic, in-plane, bi-concave lens: Tuning light beam from focused to divergent. *Microfluid. Nanofluid.* **2010**, *10*, 671–678. [CrossRef]

86. Kopp, D.; Lehmann, L.; Zappe, H. Optofluidic laser scanner based on a rotating liquid prism. *Appl. Opt.* **2016**, *55*, 2136–2142. [CrossRef] [PubMed]

87. Heideman, R.G.; Kooyman, R.P.H.; Greve, J. Development of an optical waveguide interferometric immunosensor. *Sens. Actuators B Chem.* **1991**, *4*, 297–299. [CrossRef]

88. Heideman, R.G.; Kooyman, R.P.H.; Greve, J. Performance of a highly sensitive optical waveguide mach-zehnder interferometer immunosensor. *Sens. Actuators B Chem.* **1993**, *10*, 209–217. [CrossRef]

89. Ingenhoff, J.; Drapp, B.; Gauglitz, G. Biosensors using integrated optical devices. *Fresenius J. Anal. Chem.* **1993**, *346*, 580–583. [CrossRef]

90. Gauglitz, G.; Ingenhoff, J. Design of new integrated optical substrates for immuno-analytical applications. *Fresenius J. Anal. Chem.* **1994**, *349*, 355–359. [CrossRef]

91. Schipper, E.F.; Brugman, A.M.; Dominguez, C.; Lechuga, L.M.; Kooyman, R.P.H.; Greve, J. The realization of an integrated mach-zehnder waveguide immunosensor in silicon technology. *Sens. Actuators B Chem.* **1997**, *40*, 147–153. [CrossRef]

92. Heideman, R.G.; Lambeck, P.V. Remote opto-chemical sensing with extreme sensitivity: Design, fabrication and performance of a pigtailed integrated optical phase-modulated Mach–Zehnder interferometer system. *Sens. Actuators B Chem.* **1999**, *61*, 100–127. [CrossRef]

93. Crespi, A.; Gu, Y.; Ngamsom, B.; Hoekstra, H.J.W.M.; Dongre, C.; Pollnau, M.; Ramponi, R.; van den Vlekkert, H.H.; Watts, P.; Cerullo, G.; et al. Three-dimensional mach-zehnder interferometer in a microfluidic chip for spatially-resolved label-free detection. *Lab Chip* **2010**, *10*, 1167–1173. [CrossRef] [PubMed]
94. Bedoya, A.C.; Monat, C.; Domachuk, P.; Grillet, C.; Eggleton, B.J. Measuring the dispersive properties of liquids using a microinterferometer. *Appl. Opt.* **2011**, *50*, 2408–2412. [CrossRef] [PubMed]
95. Lapsley, M.I.; Chiang, I.K.; Zheng, Y.B.; Ding, X.; Mao, X.; Huang, T.J. A single-layer, planar, optofluidic mach-zehnder interferometer for label-free detection. *Lab Chip* **2011**, *11*, 1795–1800. [CrossRef] [PubMed]
96. Dumais, P.; Callender, C.L.; Noad, J.P.; Ledderhof, C.J. Integrated optical sensor using a liquid-core waveguide in a mach-zehnder interferometer. *Opt. Express* **2008**, *16*, 18164–18172. [CrossRef] [PubMed]
97. Bernini, R.; Testa, G.; Zeni, L.; Sarro, P.M. Integrated optofluidic mach–zehnder interferometer based on liquid core waveguides. *Appl. Phys. Lett.* **2008**, *93*, 011106. [CrossRef]
98. Testa, G.; Huang, Y.; Sarro, P.M.; Zeni, L.; Bernini, R. High-visibility optofluidic Mach–Zehnder interferometer. *Opt. Lett.* **2010**, *35*, 1584–1586. [CrossRef] [PubMed]
99. Yin, D.; Barber, J.P.; Hawkins, A.R.; Schmidt, H. Highly efficient fluorescence detection in picoliter volume liquid-core waveguides. *Appl. Phys. Lett.* **2005**, *87*, 211111. [CrossRef]
100. Duguay, M.A.; Kokubun, Y.; Koch, T.L.; Pfeiffer, L. Antiresonant reflecting optical waveguides in SiO_2-Si multilayer structures. *Appl. Phys. Lett.* **1986**, *49*, 13–15. [CrossRef]
101. Yin, D.; Deamer, D.W.; Schmidt, H.; Barber, J.P.; Hawkins, A.R. Single-molecule detection sensitivity using planar integrated optics on a chip. *Opt. Lett.* **2006**, *31*, 2136–2138. [CrossRef] [PubMed]
102. Testa, G.; Persichetti, G.; Sarro, P.M.; Bernini, R. A hybrid silicon-PDMS optofluidic platform for sensing applications. *Biomed. Opt. Express* **2014**, *5*, 417–426. [CrossRef] [PubMed]
103. Campopiano, S.; Bernini, R.; Zeni, L.; Sarro, P.M. Microfluidic sensor based on integrated optical hollow waveguides. *Opt. Lett.* **2004**, *29*, 1894–1896. [CrossRef] [PubMed]
104. Dongre, C.; van Weerd, J.; Bellini, N.; Osellame, R.; Cerullo, G.; van Weeghel, R.; Hoekstra, H.J.W.M.; Pollnau, M. Dual-point dual-wavelength fluorescence monitoring of DNA separation in a lab on a chip. *Biomed. Opt. Express* **2010**, *1*, 729–735. [CrossRef] [PubMed]
105. Bernini, R.; De Nuccio, E.; Minardo, A.; Zeni, L.; Sarro, P.M. Integrated silicon optical sensors based on hollow core waveguide. *Proc. SPIE* **2007**, *6477*, 647714.
106. Yu, J.Q.; Yang, Y.; Liu, A.Q.; Chin, L.K.; Zhang, X.M. Microfluidic droplet grating for reconfigurable optical diffraction. *Opt. Lett.* **2010**, *35*, 1890–1892. [CrossRef] [PubMed]
107. Mao, X.; Lin, S.C.S.; Dong, C.; Huang, T.J. Single-layer planar on-chip flow cytometer using microfluidic drifting based three-dimensional (3D) hydrodynamic focusing. *Lab Chip* **2009**, *9*, 1583–1589. [CrossRef] [PubMed]
108. Tung, Y.C.; Zhang, M.; Lin, C.T.; Kurabayashi, K.; Skerlos, S.J. PDMS-based opto-fluidic micro flow cytometer with two-color, multi-angle fluorescence detection capability using pin photodiodes. *Sens. Actuators B Chem.* **2004**, *98*, 356–367. [CrossRef]
109. Pamme, N.; Koyama, R.; Manz, A. Counting and sizing of particles and particle agglomerates in a microfluidic device using laser light scattering: Application to a particle-enhanced immunoassay. *Lab Chip* **2003**, *3*, 187–192. [CrossRef] [PubMed]
110. Wang, Z.; El-Ali, J.; Engelund, M.; Gotsæd, T.; Perch-Nielsen, I.R.; Mogensen, K.B.; Snakenborg, D.; Kutter, J.P.; Wolff, A. Measurements of scattered light on a microchip flow cytometer with integrated polymer based optical elements. *Lab Chip* **2004**, *4*, 372–377. [CrossRef] [PubMed]
111. Godin, J.; Lien, V.; Lo, Y.-H. Demonstration of two-dimensional fluidic lens for integration into microfluidic flow cytometers. *Appl. Phys. Lett.* **2006**, *89*, 061106. [CrossRef]
112. Barat, D.; Benazzi, G.; Mowlem, M.C.; Ruano, J.M.; Morgan, H. Design, simulation and characterisation of integrated optics for a microfabricated flow cytometer. *Opt. Commun.* **2010**, *283*, 1987–1992. [CrossRef]
113. Song, C.; Luong, T.-D.; Kong, T.F.; Nguyen, N.-T.; Asundi, A.K. Disposable flow cytometer with high efficiency in particle counting and sizing using an optofluidic lens. *Opt. Lett.* **2011**, *36*, 657–659. [CrossRef] [PubMed]
114. Ashkin, A. Acceleration and trapping of particles by radiation pressure. *Phys. Rev. Lett.* **1970**, *24*, 156–159. [CrossRef]
115. Ashkin, A.; Dziedzic, J.M.; Bjorkholm, J.E.; Chu, S. Observation of a single-beam gradient force optical trap for dielectric particles. *Opt. Lett.* **1986**, *11*, 288–290. [CrossRef] [PubMed]

116. Lin, H.-C.; Hsu, L. Dynamic and Programmable Cell-Sorting by Using Microfluidics and Holographic Optical Tweezers. *Proc. SPIE* **2005**, *5930*, 59301–59308.

117. Wang, M.M.; Tu, E.; Raymond, D.E.; Yang, J.M.; Zhang, H.; Hagen, N.; Dees, B.; Mercer, E.M.; Forster, A.H.; Kariv, I.; et al. Microfluidic sorting of mammalian cells by optical force switching. *Nat. Biotechnol.* **2005**, *23*, 83–87. [CrossRef] [PubMed]

118. Applegate, R.W.; Squier, J.; Vestad, T.; Oakey, J.; Marr, D.W.M. Optical trapping, manipulation, and sorting of cells and colloids in microfluidic systems with diode laser bars. *Opt. Express* **2004**, *12*, 4390–4398. [CrossRef] [PubMed]

119. Lee, K.S.; Lee, K.H.; Kim, S.B.; Jung, J.H.; Ha, B.H.; Sung, H.J.; Kim, S.S. Refractive-index-based optofluidic particle manipulation. *Appl. Phys. Lett.* **2013**, *103*, 073701.

120. Lee, K.H.; Lee, K.S.; Jung, J.H.; Chang, C.B.; Sung, H.J. Optical mobility of blood cells for label-free cell separation applications. *Appl. Phys. Lett.* **2013**, *102*, 141911.

121. Kim, S.B.; Yoon, S.Y.; Sung, H.J.; Kim, S.S. Cross-type optical particle separation in a microchannel. *Anal. Chem.* **2008**, *80*, 2628–2630. [CrossRef] [PubMed]

122. Tag Optics. Available online: Http://www.tag-optics.com/TL2ProductLine.php (accessed on 1 April 2017).

123. Varioptic. Available online: Http://www.Varioptic.Com/products/variable-focus/ (accessed on 1 April 2017).

124. Optotune. Available online: Http://www.Optotune.Com/ (accessed on 1 April 2017).

125. World Precision Instruments. Available online: Https://www.Wpiinc.Com/ (accessed on 1 April 2017).

126. Nanotweezer Developed by Optofluidics. Available online: Http://opfluid.Com/products/nanotweezer/ (accessed on 1 April 2017).

127. Yit-Fatt, Y.; Say-Hwa, T.; Nam-Trung, N.; Murshed, S.M.S.; Teck-Neng, W.; Levent, Y. Thermally mediated control of liquid microdroplets at a bifurcation. *J. Phys. D Appl. Phys.* **2009**, *42*, 065503.

128. Say-Hwa, T.; Murshed, S.M.S.; Nam-Trung, N.; Teck Neng, W.; Levent, Y. Thermally controlled droplet formation in flow focusing geometry: Formation regimes and effect of nanoparticle suspension. *J. Phys. D Appl. Phys.* **2008**, *41*, 165501.

129. Xi, H.-D.; Guo, W.; Leniart, M.; Chong, Z.Z.; Tan, S.H. AC electric field induced droplet deformation in a microfluidic T-junction. *Lab Chip* **2016**, *16*, 2982–2986. [CrossRef] [PubMed]

130. Tan, S.H.; Semin, B.; Baret, J.-C. Microfluidic flow-focusing in ac electric fields. *Lab Chip* **2014**, *14*, 1099–1106. [CrossRef] [PubMed]

131. Ma, Z.; Teo, A.; Tan, S.; Ai, Y.; Nguyen, N.-T. Self-aligned interdigitated transducers for acoustofluidics. *Micromachines* **2016**, *7*, 216. [CrossRef]

micromachines

MDPI

Review

The Use of Microfluidics in Cytotoxicity and Nanotoxicity Experiments

Scott C. McCormick [1], **Frederik H. Kriel** [1], **Angela Ivask** [1,†], **Ziqiu Tong** [1,2], **Enzo Lombi** [1], **Nicolas H. Voelcker** [1,2] and **Craig Priest** [1,*]

[1] Future Industries Institute, University of South Australia, Mawson Lakes Blvd., Mawson Lakes, 5098 SA, Australia; scott.mccormick@mymail.unisa.edu.au (S.C.M.); Erik.Kriel@unisa.edu.au (F.H.K.); angela.ivask@kbfi.ee (A.I.); tommy.tong@monash.edu (Z.T.); Enzo.Lombi@unisa.edu.au (E.L.); nicolas.voelcker@monash.edu (N.H.V.)

[2] Monash Institute of Pharmaceutical Sciences, Monash University, Parkville, 3052 VIC, Australia

* Correspondence: Craig.Priest@unisa.edu.au; Tel.: +61-8-8302-5146

† Current address: National Institute of Chemical Physics and Biophysics, Akadeemia tee 23, Tallinn 12618, Estonia.

Academic Editors: Weihua Li, Hengdong Xi and Say Hwa Tan
Received: 28 February 2017; Accepted: 7 April 2017; Published: 12 April 2017

Abstract: Many unique chemical compounds and nanomaterials are being developed, and each one requires a considerable range of in vitro and/or in vivo toxicity screening in order to evaluate their safety. The current methodology of in vitro toxicological screening on cells is based on well-plate assays that require time-consuming manual handling or expensive automation to gather enough meaningful toxicology data. Cost reduction; access to faster, more comprehensive toxicity data; and a robust platform capable of quantitative testing, will be essential in evaluating the safety of new chemicals and nanomaterials, and, at the same time, in securing the confidence of regulators and end-users. Microfluidic chips offer an alternative platform for toxicity screening that has the potential to transform both the rates and efficiency of nanomaterial testing, as reviewed here. The inherent advantages of microfluidic technologies offer high-throughput screening with small volumes of analytes, parallel analyses, and low-cost fabrication.

Keywords: microfluidics; cytotoxicity; nanotoxicity; nanoparticles; screening

PACS: 87.17.UV Biotechnology of cell processes; 87.85.dh Cells on a chip

1. Introduction

Toxicity studies are important in biochemical and medical research, and essential prior to the commercial use of newly developed chemicals and nanomaterials. The health and safety of researchers, production workers, end-users, and bystanders who may come into contact with new or innovative products, and any secondary products that may arise from the degradation of such products, is of great concern to government regulatory bodies and society as a whole [1–3]. The benefits of the effective toxicity screening of chemicals and nanomaterials prior to their commercialization include better community health outlooks, reduced costs (healthcare and/or compensation payments), and faster paths to the market for new non-toxic products. Currently, biochemical and medical products are subjected to extensive testing before adoption or commercialization, but this is costly in terms of labour, time, and money. Many companies are specifically set up to assist with performing the biocompatibility (i.e., toxicology) screening for new formulations. Analysts predict that the in vitro toxicity testing market value will reach approximately 10 billion dollars in 2017 [4]. Thus, new technologies for toxicity screening are attractive for their perceived economic, social, and environmental benefits.

Cytotoxicity analysis at a cellular level is concerned with how a given toxic chemical affects a given cell's physical structure (e.g., membrane integrity) and its ability to viably replicate without damage to the daughter cell's genetic code or normal functionality [5]. Thorough cytotoxicity screening of a chemical requires the studied toxicant to be tested against different cell types, and is generally performed in static fluid in well plates. This approach requires laborious liquid handling, long hours of incubation, and large reagent volumes. The idea of an all-encompassing cytotoxicity test is a daunting task, as the adult human body contains trillions of eukaryotic cells with different phenotypes and functionality [6] and, according to Vickaryous [7], the number of unique cell types is 411, including 145 types of neuronal cells. The many different cell types in the human body makes the effective in vitro screening of potential toxic effects an enormous challenge, further complicated by the types of analysis required (e.g., viability, cell metabolism, and biochemistry). The growing number of novel chemicals and materials that are being suggested for various commercial applications multiplies the size of the challenge and calls for much faster and cheaper methods.

Nanotoxicity is an important subgroup of toxicity which considers the damaging effect of nanomaterials on cells. The first reports on the toxicity of nanomaterials on mammalian biology were reported in the 1990's by Jani et al. [8] and Penney et al. [9]. Now, an awareness of the potential toxicity of those materials has reached beyond the scientific community to include regulators and consumers, who, in many cases, are not equipped with enough information to guide their decisions. All the while, more nanomaterials are being created and incorporated into consumer products, from sunscreens [10,11] and cosmetics [2,12], to antibacterial and antifouling coatings [13–15]. A major apprehension pertaining to the exposure of the human body to nanoparticles is their physical size (typically 1–100 nm), which can allow them to enter cells via pathways that naturally transport biological and chemical species [5]. Many nanoparticles can form free radicals and reactive oxygen species (ROS) [16] from surrounding molecules, due to their increased reactivity and high surface area, which have the ability to cause damage to cellular membranes and proteins within cells, leading to inflammation and oxidative stress [17,18]. Nanomaterials can also disassociate into ionic species upon reacting with biological tissue and fluids such as gastric juices, which can lead to the release of reactive ions that can damage the cellular environment and cause toxic effects [19,20]. Genotoxicity can occur if a nanoparticle interferes with the delicate process of DNA transcription and replication, potentially knocking out one or more genes from the sequence and causing a range of negative effects, such as apoptosis (in which case the cells die off) [21–23] or mutation, which can lead to the cells becoming cancerous [24].

Meaningful nanotoxicity studies require high-throughput screening methods, as the toxic effects of a nanomaterial can be dependent on the core composition, size, shape, and surface modification that material possesses, suggesting that a high number of materials should be tested. The many different cell and tissue types of the human body can react in different ways to any given nanomaterial, meaning that an ideal toxicology screen should test every unique cell type. Thus, the desire to create a fast, stable screening process for the maximum number of combinations of nanoparticles and human cells possible is of great interest to the industry and health sectors. Where practical, these methods will enable the determination of safety exposure levels and maintain the health of both workers in the nanomaterial industries and the end-users of products containing nanomaterials. To this end, microfluidic approaches to toxicity screening have been investigated and are the topic of this review.

2. Nanomaterial Exposure Pathways in Biology

Nanomaterials can be taken up into a living body via the natural internalization pathways of ingestion, inhalation, or dermal uptake [25], or through direct injection if used in nanomedicine [26], and can pose a substantial risk to the viability of a cell, depending on the nature of their interaction with living cells. Once inside the body, nanomaterials can transmit from one tissue to another via the bloodstream and the surrounding tissues, potentially migrating into other organs such as the kidney or spleen. Once they enter the body, they come into contact with the body's cellular structure and potentially gain entrance into the cells themselves [27]. A diagram of nanoparticles and the bodily areas that they are able to access after uptake via inhalation is shown in Figure 1. The uptake of nanoparticles to the body and different organs is particle size dependent. As an example of size-dependency on the uptake of nanoparticles, it was shown by Jani et al. [8] that when a range of polystyrene particles ranging from 50 nm to 3 μm was introduced to a rat model via ingestion pathways, no particles above 100 nm reached the bone marrow and none larger than 300 nm were present in the bloodstream, whereas the 50 nm and 100 nm particles were absorbed at rates of 34% and 26%, respectively, into the liver, spleen, blood, and bone marrow tissues.

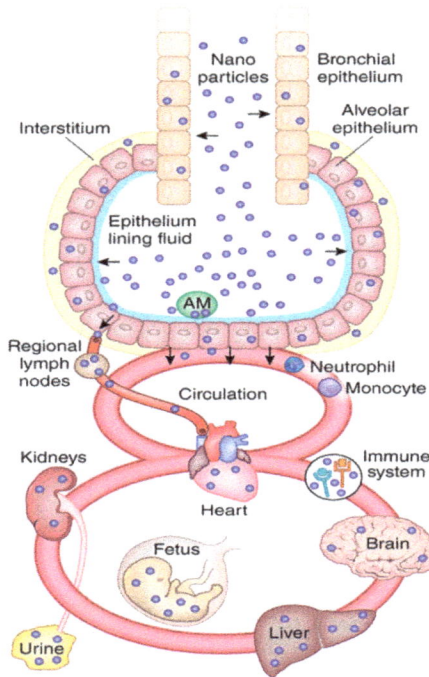

Figure 1. Schematic diagram of nanoparticle pathways through the human body after inhalation, Reprinted by permission from Macmillan Publishers Ltd: NATURE BIOTECHNOLOGY Kreyling, Hirn [28], copyright 2010.

Nanoparticles can pass through the cellular membrane via the passive transport mechanics of diffusion and osmosis, requiring no activation energy [29,30]. Alternatively, nanoparticles can be taken into the cell via active transport mechanisms, in which carrier proteins or ionic pumps within the cell membrane attach to the particle and use energy to move them across the cell membrane into the cytoplasm [27]. Inorganic nanoparticles that require this mechanism to cross the membrane are often blocked from entering cells, unless they are coated with a biomolecule (such as transferrin) that facilitates their uptake by the carrier proteins, as was shown in Yang et al. [31] in the case of transferrin-conjugated gold nanoparticles.

Nanoparticles that do not enter the cell via membrane diffusion or through membrane pores can still be transported into the cell via endocytosis [32]—that is, the cytoplasm of the cell extends around a particle and engulfs it, forming an endocytic vesicle that retains them in the inner cytoplasm of the cell. From here, the particles can either: escape the vesicle and remain in the cytoplasm; persist in the vesicle and be consumed by a lysosome (an organelle full of enzymes that serve to digest foreign bodies that exist in the cytoplasm); or combine with other vesicles to form a multi-vesicle endosome contained in a secondary membrane, which stabilizes and contains the individual vesicles [30].

Certain nanoparticles possess the ability to alter or bypass the membrane permeability, depending on their ionic potential or their shape. Nanoparticles that are shaped with sharp points or edges can mechanically damage the cell membrane, creating temporary nanochannels through which they can enter the cytoplasm [33]. This can be exploited to create drug delivery mechanisms by coating nanoparticles such as carbon nanotubes with biocompatible molecules which attach and enter cells, or manufacturing nanoneedles from materials such as silicon or polymers that can mechanically puncture cells to deliver drugs directly into the cytoplasm [34,35]; however, toxic nanomaterials could very easily enter cells by the same mechanism and induce cytotoxic effects.

The above pathways and cellular interactions are complex and very sensitive to the size, shape, chemistry, and surface charge of the nanomaterials, meaning that the importance of a high-throughput evaluation of nanotoxicity is growing commensurate with the rapid development of nanotechnology. In recent years, microfluidics technology has resulted in large impacts on the cytotoxicity screening of nanoparticles. This will be discussed in the following section, to underpin the later discussion of nanotoxicity screening using microfluidic chips.

3. Microfluidics in Cytotoxicity Screening

Microfluidics has garnered a great deal of interest in the field of in vitro cytotoxicity screening. Combining biological engineering with microfluidics is termed cell-laden microfluidics, in which living cells are affixed within a microfluidic channel, exposed to various chemical species in a flowing medium, and assayed to determine their post-exposure viability using viability assay dyes. Viability assay dyes are chemical compounds that are taken up into either living or dead cells, or bind to cell-death markers that are released outside of dead cells. They contain a fluorescent or coloured moiety that can be detected and quantified using optical detection methods [25,36]. These fluorescent methods only require a very small amount of dye per cell, and are therefore reasonable candidates for miniaturization into a microfluidic platform.

The use of microfluidic platforms in cytotoxicity screening is desirable due to a number of factors, i.e., small sample volumes [37], reduced costs [38,39], a controlled and reproducible laminar flow, and the ability to functionalize (e.g., antibodies) (36) or structure (e.g., compartments) [37] microchannels to produce a varied microarray of multiple cells. It also seeks to address the current paradigm of static well-plate testing that, as Cunha-Matos et al. stated, "these procedures provide averaged results, do not guarantee precise control over the delivery of nanoparticles to cells and cannot easily generate information about the dynamics of nanoparticle-cell interactions and/or nanoparticle-mediated compound delivery" [40]. There are multiple variations on cell-laden microfluidic methods and protocols, some of which are detailed in the following sections.

3.1. Cell Capture and Immobilisation

The concept of cell-laden microfluidics requires cells to be immobilized inside the channel in such a way that they can be analysed under precisely controlled (on-chip) conditions. A device that marries the concept of well plates and microfluidic flow for single-cell capture was designed by Hosokawa et al. [37]. The microfluidic device was assembled on top of a laser-perforated polymer microcavity array, which was created with conical cavities measuring 2 μm at the surface. This was assembled on top of a vacuum line which applied negative pressure to pull a cell suspension through the channel and produced a microarray of single cells as they settled in the perforations. Further, an on-chip chemical gradient generator was used to treat six channels with unique concentrations of a potential toxicant (potassium cyanide), which were subsequently stained with cell viability assay dyes (EthD-1, which stained exclusively the DNA of lysed cells), allowing the high-throughput screening of these different concentrations. The results of their experiments can be seen in Figure 2, where they showed increases in red fluorescence corresponding to greater numbers of lysed cells as the concentration of the toxicant increases.

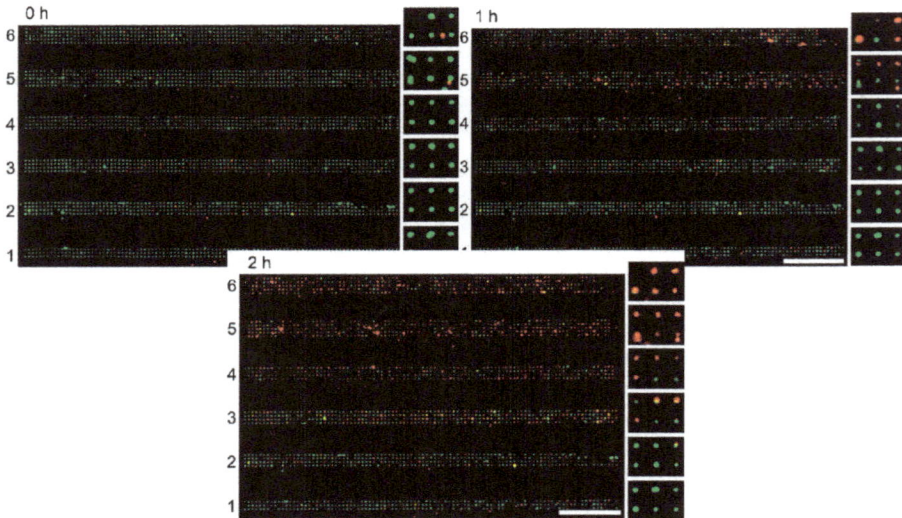

Figure 2. Six parallel channels containing single-cellular microwell arrays after exposure to gradients of KCN, green indicates viable cells and red indicates lysed cells, scale bar = 1 mm. Adapted with permission from Analytical Chemistry 83(10) Hosokawa, Hayashi [37]. Copyright 2011 American Chemical Society.

A device that utilized a larger-scale microwell environment was produced by the group of Weibull et al. [41]. In order to investigate the behaviour of cells to a concentration gradient without interference from intercellular paracrine signalling from neighbouring cells, they developed a system in which single cells could be analysed. They fabricated a microscope-slide-sized microwell plate with 672 (14 × 48) 500 nL wells, in which only single cells would fit by attaching a grid of etched silicon to a glass slide, allowing for high-resolution imaging. They then layered three PDMS channel designs over the top of this microwell plate, which produced a concentration gradient generator in which a reagent from the top layer mixes with a diluent from the middle layer, and flows down to the reaction chamber in the bottom layer, before exposing the mixture to the cells beneath. Their device was able to culture bovine aortic endothelial cells in the silicon microwells and expose the cells to differing concentrations of saponin to induce cell death. Live/dead staining could be performed by exposing dyes through the same ports.

Cell traps are useful for their ability to analyse non-adherent single cells, but often adherent cells are used in experiments that are usually found in large contiguous layers. To achieve larger areas of cell attachment, cellular microarrays have been produced by printing antibodies onto a substrate, before assembling the microfluidic device and exposing the printed area to a culture of cells. These methodologies are suitable for the robotic "spotting" of antibodies or proteins, and the subsequent binding of the target cells to these scaffolds [42]. By printing antibody spots that are selective for unique cell types, the cells bind to the surface and form a microarray, allowing non-selected cells to continue flowing and either bind to another spot, or flow out as waste. A two-dimensional array that binds a flat patch of cells can be produced on a substrate with a modified inkjet printer [43] or micro-contact printer [44].

Our group has investigated the potential of microfluidics for a high-throughput screening methodology. In a recent paper by Tong et al. [42], a device incorporating five parallel laminar streams crossed with five perpendicular streams was investigated, in order to bind different cell types in a single device, and subsequently delivered different chemical treatments to the cells to create an orthogonal microarray. Antibody-antigen binding or extracellular matrix (ECM) protein binding was utilized in a 5 × 5 array with a microcontact printer to anchor cells to a glass substrate under microfluidic flow. A model cytotoxicity assay was performed using various levels of osmotic stress in different laminar streams, and fluorescein diacetate/propidium iodide viability assay dyes were applied to achieve a fluorometric readout of cell viability.

In a reversal of the concept of micropatterning to create a binding surface for cells flowing in media, the group of Leclerc et al. [45] used a PDMS microfluidic channel with the microstamps on the top channel wall to press down and crush any bound cells within the area of the stamp. By first culturing a layer of cells on the bottom channel surface, when the crushed cells were washed away by perfusion, they allowed new cells to grow in the affected area. This technique could allow for the long-term culturing of the same cell sample, thus enabling many concurrent tests to be carried out on the same chip.

In microfluidic systems where larger cell binding areas are required for analysis, it is possible to flow a bio-functional binding agent through a microfluidic device and coat the entire channel surface. Pasirayi et al. [39] utilized a multilayer device that sandwiched a 10 μm PDMS membrane between two PDMS channels, held together by two rigid polymethylmethacrylate (PMMA) cover plates. One of the PDMS channel sides contained a concentration gradient generator, while the opposite side contained cell culture chambers separated by valve arrays to prevent cross-contamination. The valves could be opened and closed by the application of a mild vacuum, allowing for fine control over the exposure and conditions in the cell culture chambers. They coated the chambers with an extracellular matrix (ECM) protein, fibronectin, and attached model breast cancer and liver cells. The cells were kept viable by perfusing fresh media. Finally, the cell culture chambers were exposed to an antibiotic, pyocyanine, via the concentration gradient generator to achieve a range of 0–100 μM, and a fluorescent live-cell assay was performed with Calcein AM. A combination of drugs, paclitaxel and aspirin, was also tested to identify potential synergistic toxic effects, and assayed in the same manner as the pyocyanine.

In recent years, advances in lithographic techniques and cellular gels have enabled the production of cell arrays in microfluidic channels post-assembly. The production of natural hydrogels based on in vivo ECM proteins, i.e., collagen, or very similar synthetic products like Matrigel® (Corning, NY, USA) and alginate, has allowed for 3D scaffolds and structures that very closely mimic an in vivo cellular microenvironment [46]. Groups such as Toh et al. [47] have produced multiplexed 3D microfluidic cell culture systems in which primary hepatocytes could be cultured. Their chips utilized separating micropillars to divide the cell culture channels into a central cell culture area surrounded by two perfusion channels, and a linear concentration gradient generator delivered culture medium and drug solutions to the cells via the perfusion channels. They prepared a hydrogel of methylated collagen and terpolymer combined with hepatocytes that could be flowed through the microfluidic channels to settle in the cell culture area, which maintained function and produced albumin proteins. They tested five model hepatotoxic drugs using the concentration gradient generator and produced toxicity data by fluorescently staining the cells post-treatment.

Microfluidic chips can be used to study the ecotoxicity and cytotoxicity of prokaryotic cells. Yoo et al. [48] utilized a water-soluble photosensitive polymer to create patterns of bioluminescent bacteria inside a microfluidic chip. By flowing a mixture of the monomers of the photosensitive polymer and strains of genetically modified oxidative stress-induced bioluminescent *E. coli* through the channel and selectively exposing it to UV to cause gelling, they were able to bind the bacterial cells on the exposed areas and measure their luminescence intensities. A UV exposure of 10 mins was found to be short enough for bacterial cells to recover from the binding process, after which they could be used as a toxicity testing platform. Upon exposure to both hydrogen peroxide and phenol via the microfluidic flow, the bacterial cells underwent oxidative stress and presented a more intense luminescence in a dose-dependent manner, meaning that they could be utilized for chemical screening.

The emerging technology of organ-on-a-chip seeks to enable testing on human biomimetic environments, by producing three-dimensional cultures of human cells that possess a similar structure and shape as those in in vivo conditions. The group of Wagner et al. [49] utilized this beneficial technology by producing a microfluidic environment with multiple culture locations, connected by the flow pathway; in which they cultured both biopsied skin tissue and pre-grown liver microtissue aggregates. The chips were infused with just 300 µL of cell culture medium and sealed, with perfusion being provided by an on-chip micropump. The medium required only a 40% replacement at 12 h intervals for the first week of culture. To prove the usefulness of the devices in toxicity testing, they exposed the system to troglitazone, a drug with a known hepatotoxicity. A dose-dependent toxic response was detected by assaying the culture medium for glucose consumption and lactate production. There was a visible increase in the cytochrome concentration in the drug-exposed samples when the cells were immunostained after the device was disassembled. They also showed the potential of using the skin layer as an air-liquid interface for more realistic methods of applying topical drugs in future devices. Overall, this device shows the promise of multi-organ microfluidic devices for investigating specific uptake profiles and the run-on effects between different bodily organs.

Very few standardized microfluidic platforms are available for toxicology testing. One company called SynVivo [50] provides a standardized toxicity assay chip in which a ring of endothelial cells can be cultured around a choice of other tissue cells, i.e., cardiomyocytes and hepatocytes. Their platform enables optical and fluorescent imaging, as well as chemical assays such as an ROS assay, and has been shown to culture liver cells such that they successfully produce urea and responded to the toxicity of acetaminophen and doxorubicin. Platforms like these must become much more commonly produced if microfluidic cytotoxicity assays are to be accepted as standardized testing.

3.2. Channel Arrays and Laminar Flow

Microfluidics presents the ability to separate fluid streams from each other using physical barriers or the properties of laminar flow. One of the earlier examples of microfluidics' use in cytotoxicity experiments came from the group of Ma et al. in 2008 [51], who fabricated channels in a quartz chip and two additional channel-containing PDMS layers attached to opposite sides. The chip was designed to test both the cytotoxicity and cellular metabolism of drugs in human liver microsomes (HLMs), which catalyze drug metabolism. HLMs were applied to the devices' microwells in a homogenous sol-gel suspension, held in place by reversibly bonded PDMS. Liver carcinoma cells (hepG2) were cultured in chambers that were exposed to the metabolic products from the HLMs. The mixtures of liver-active drugs, acetaminophen, and phenytoin, in addition to the viability assay dyes, were introduced across the sol-gel columns via microfluidic flow. Viability was determined via fluorescence imaging, while drug metabolism was determined by UV absorbance spectroscopy, performed on the flowing media before it exited the device.

In order to show the reproducibility of microfluidic cytotoxicity experiments, multiple repeat experiments of a 64-chamber microfluidic chip were performed by Cooksey et al. [52]. Their cell culture chambers were arranged in an 8×8 pattern, and multiple different cell densities were seeded on fibronectin for an analysis of the expression of transfected destabilized green fluorescent protein (GFP) to show protein synthesis in healthy cells. Fluorescence data and time-constants could be analysed across each chamber to produce an 8×8 dataset for each individual chip, to compare different conditions. A toxic agent, cycloheximide, was applied across the device and the reduction in GFP activity could be quantitatively measured across an experimental duration of 60 h. Sub-lethal concentrations of cycloheximide would cause the GFP to decay, but could be recovered when perfused with fresh media. Different tubing was used across various tests to determine whether the leaching of gases through the plastic or pH changes would affect the results. When compared to tests that were run in static 96-well plates on tissue culture grade polystyrene and PDMS substrates, the datasets were found to be very similar in their decay time constants, but statistically superior in terms of their standard deviations and number of regions that could be analysed.

A three-dimensional flow cell microarray was produced by Eddings et al. [53], in which 48 individual flow cells were stacked in four rows of twelve, with individual inlets and outlets feeding each flow cell. Each individual microchannel was capable of being injected with a fluid, which could be spotted onto a sensor surface flush with the array head and electrochemically measured for the presence of certain analytes in solution, or used for the sequential patterning of ligands on the surface. The high-throughput nature of the three-dimensional flow cell array leads to the potential to use it in either "one on many" or "many on one" approaches with many different cell types, or with many different nanoparticles on a few cell types.

Wada et al. [54] demonstrated a cytotoxicity screening method using cells that were pre-transfected with a green fluorescent protein plasmid fused with a gene encoding for a heat-shock protein (HSP70B'), creating sensor cells that would express fluorescent protein biomarkers as the heat shock protein is expressed in the presence of cytotoxic compounds. They showed that these sensor cells could be bound and propagated inside a microfluidic channel to produce a near 100% coverage. Using laminar flow methods, the cells were exposed to an ionic cytotoxic compound ($CdCl_2$) for 1 h in multiple concentrations with an in-line negative control area of 0% concentration, and could be assayed for a relative fluorescent signal compared to the negative control. They were able to produce a gradient-flow chip that used in-line mixing of $CdCl_2$ and buffer solution to form eight unique concentrations in laminar streams that produced a fluorescence profile showing an increase in fluorescence with an increasing concentration of the cytotoxic compound. This technique shows that a pre-transfected cell line maybe very useful for high-throughput screening methods inside microchannels.

Cytotoxicity in anatomically relevant scenarios has been investigated by producing organ-on-a-chip devices where human cells are bound in a microfluidic channel or chamber. These are designed to mimic single or multiple human organs, and seek to provide information on specific diseases or cases of poor health. Gori et al. [55] produced a liver-on-a-chip device in which they were able to expose a 3D culture of healthy hepatic cells to solutions of free fatty acids, and observe the oxidative stress and resulting cytotoxicity from the overload of oils/fats. They proved that the liver cells could diffuse the nutrients and eliminate waste products due to being in the 3D culture environment, which leads to longer viability times and more accurate depictions of a true hepatic system compared to 2D analogous systems.

Some microfluidic devices do not rely on binding the cells to a specific area inside a microchannel, instead focusing on analyzing cells in suspension as they flow through the channel. The detection of biomarkers produced by ionizing radiation is often performed on expensive and complex flow cytometers, which reduces the amount of diagnoses available in remote or poorer areas. Also, in areas where radiation is particularly prevalent, such as power plants or space missions, tests should be performed to diagnose and prevent radiation sickness. In order to improve these issues, Wang et al. [56] produced a microfluidic device with a disposable chip that could perform the same fluorescence intensity readings for the most common radiation biomarker (γ-H2AX) in a hand-held format, shown in Figure 3, as well as provide information on the number of cells passing through the analysis area using a resistive pulse sensor (RPS) and thus obtain the ratio of damaged-to-undamaged cells. From this, they were able to analyze the extent of radiation damage from UV light on human lymphocytes, and obtained comparable results to a conventional flow cytometer. By designing the microfluidic channels with a detection spot or "sensing gate" of 15 μm and a channel height of only 30 μm, they were able to direct the cells to individually flow through the system under laminar flow for analysis. The device contained an on-board light emitting diode for exciting the fluorescent marker, and the software was able to acquire the data and produce readouts of the amount of radiation damage present in samples of lymphocyte cells sourced from anti-coagulant samples of human blood.

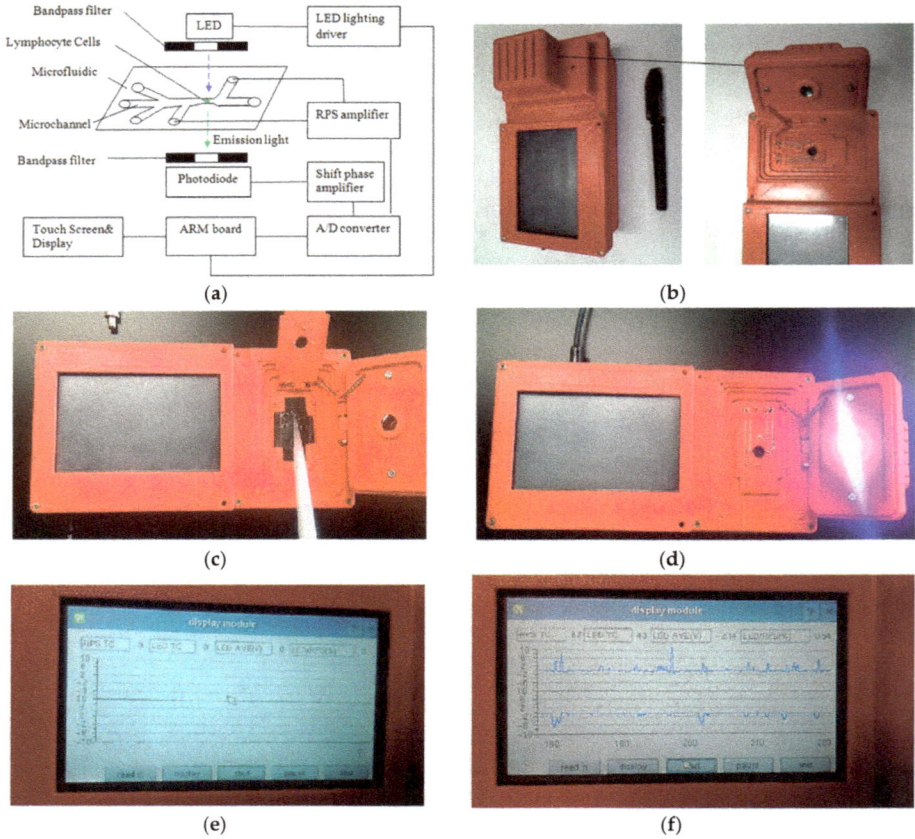

Figure 3. Handheld flow cytometer produced by Wang et al. (Scientific Reports, 6(2016) [56]), with miniaturized optical detectors, a disposable microfluidic analysis chip, and electronic display readouts. (**a**) schematic diagram; (**b**) device housing with space for microfluidic chip; (**c**) pipetting sample onto microfluidic chip; (**d**) UV LED active; (**e**) electronic display module; (**f**) electronic readout of sample electrophoretic data.

Recent advances in 3D printing technology have attempted to reduce the complexity of microfluidic fabrication and assembly. Devices with simple finger-tight joints have been created by groups such as Morgan et al. [57], who investigated not only the suitability of seeding encapsulated dental stem cells inside the printed microfluidics, but also optimized the transparency of the poly-lactic acid (PLA) polymer surface by controlling the printing parameters (layer thickness, print speed and fill patterning). This has been a major point of contention as to the usefulness of 3D printed structures for microfluidic devices, which require optically transparent flat surfaces for ideal imaging and analysis. Given that they were able to achieve this transparency, their testing was able to fluorescently visualize labelled cell aggregates and differentiate between live and dead cells. Thus, they are moving towards simplifying and standardizing the assembly of suitable microfluidics for toxicity testing, which will hopefully improve the speed at which they are considered, for more widespread use.

3.3. Droplet Microfluidics

Microfluidic droplet generation is a technology whereby immiscible phases are combined to produce droplets that are stabilized by a carrier medium. The group of Brouzes et al. [58] demonstrated that single mammalian cells could be encapsulated in aqueous droplets, stabilized by an oil medium. When placed in an incubator, the encapsulated cells were found to stay viable for up to four days inside the droplet. They then flowed these cell-containing droplets in sequence with droplets of viability assay dyes, which were then mixed inside a well affected by an AC electrical field, causing electrically-controlled droplet fusion. This fused droplet was held for 15 min incubation on-chip, and could then be driven towards an in-line laser excitation and detection area that would determine the cell viability from the assay dye's transmission wavelength. In this method, they were able to measure single-cell viabilities on a single human cell type after combining it with a library of optically labelled drug molecules. A histogram could then be produced, where each single-cell droplet showed a fluorescence signal corresponding to whether the encapsulated cell was alive or dead. These results show that this method could be used for the high-throughput screening of a large number of potential cytotoxic compounds, and could be scaled up to run many cell types in parallel. Figure 4 shows a schematic of their system, along with the microscope images of their on-chip procedure, and microscope image D shows that a large number of droplets can be produced in a single microfluidic environment, enabling a high-throughput of assay screening.

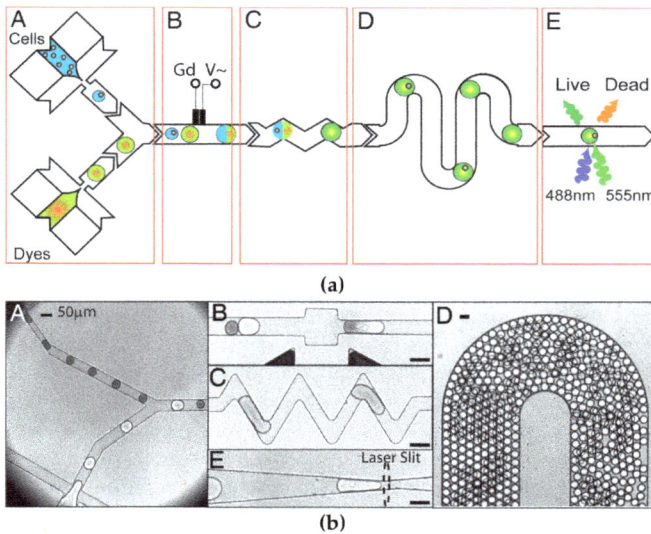

Figure 4. Schematic diagram and microscope images of viability assay by droplet generation reprinted with permission from Brouzes et al. Proc. Natl. Acad. Sci. USA., 106(34) (2009) Brouzes, Medkova [58], showing droplet fusion and reagent mixing, as well as high-throughput generation of cell-containing droplets. (**a-A**) Combination of cells and dyes, (**a-B**) in-line electrochemical sensor, (**a-C**) droplet fusion channels, (**a-D**) serpentine mixing channels (**a-E**) fluorescence assay. (**b-A**) micrograph of droplets entering channel, (**b-B**) droplet fusion, (**b-C**) droplet mixing, (**b-D**) high-throughput droplet generation in channel, (**b-E**) location of fluorescent excitation. Scale bars = 50 μm.

The group of Konry et al. [59] utilized droplet microfluidics to determine the cytotoxic effects of human immune system cells on cancer cells at the single cell level. By using the encapsulation method of flowing immiscible phases against each other, they were able to produce droplets with "distinct heterotypic cell pairs" and investigated the interactions of dendritic cells and T-lymphocytes. By introducing a cancer cell into the droplets, namely a multiple myeloma cell line, they were able to measure the speed at which CD8$^+$ T-lymphocytes could achieve cytolysis of the foreign body and what effect different levels of antigen activation had on the time taken for cell death. They determined that the presence of interferon gamma, secreted by the myeloma cells, reduced the reactivity of the T-lymphocytes, and subsequently showed that the addition of a "neutralizing antibody" could prevent this loss of reactivity and improve the immune cells' ability to kill off the cancer cells. As this technique is highly scalable, it could be used to test large variances in antigen concentration or different cell types in single runs.

4. Microfluidics for Nanotoxicity Screening

Microfluidic devices offer many advantages when it comes to cellular analysis with small sample volumes, reduced costs, controllability, and reproducibility. In addition, microfluidics offers the ability to introduce multiple biological conditions in a single device, and replicate in vivo conditions and dimensions. Thus, its usefulness in producing a high-throughput platform for toxicological experiments with nanomaterials is promising for screening applications. Small sample volumes can be very important when dealing with nanomaterial testing. Given that particulate matter may be produced from nanomaterials in extremely low concentrations, and that some nanomaterials are very expensive and produced in low quantities, there may be very limited amounts or diluted analytes to perform testing on. The behaviour of nanoparticles under flow is more difficult to quantify compared with that of macroscale particles. Therefore, the reproducible flow profiles and concentration gradients achieved in microfluidic nanotoxicity testing presents a distinct advantage.

Similar to cytotoxicity tests with other chemicals, the toxicity testing of nanomaterials is generally performed in bulk by seeding cells suspended in growth media into well plates using pipettes [17]. The cells are then exposed to nanomaterials in static conditions, which may cause the nanomaterials to adsorb or sediment onto the exposed surface of the cells under gravity [60]. Once these cells have been exposed to the nanomaterials for a certain time period, the cells can be assayed for their viability in a number of ways. Particular cell types may exhibit changed membrane permeability values and nanomaterial uptake properties when their morphology changes under flow, as compared to their sedentary morphologies, and thus, a static nanotoxicity test may provide inaccurate data.

A review by Mahto et al. [61] goes into exceptional detail on the subject of nanomaterials in microfluidic environments, and brings up a number of important details regarding nanotoxicity. First, it refers to a number of studies that show that nanoparticles, depending on their size and shape, are passively taken up into nearly all cell types via endocytotic pathways [27,62,63]. Secondly, it refers to the potential pitfalls of current in vitro nanotoxicity testing methods. Notably, nanoparticles often react with the organic dyes commonly used in cell-based assays, meaning that they cannot be properly assayed [64]. Cell exposure to nanomaterials is often improperly controlled, with aggregation and sedimentation leading to very different exposure profiles (as seen in Figure 5). Current nanotoxicity testing also utilizes immortalized cell lines which differ significantly from primary human tissue [65]. The review article mentions platforms that look to circumvent the current issues with nanotoxicity testing, such as the device produced by Richter et al. [66], which used non-invasive electrodes to electrochemically measure the amount of collagen production as a label-free marker of cell viability. This device could detect the nanotoxicity of silver nanoparticles after 2 h of exposure, as compared to the lack of nanotoxicity seen for gold nanoparticles over a period of 24 h. It could also detect reductions in collagen production, given a sub-lethal concentration of silver nanoparticles. For further discussion, the authors recommend Mahto's review paper to the reader.

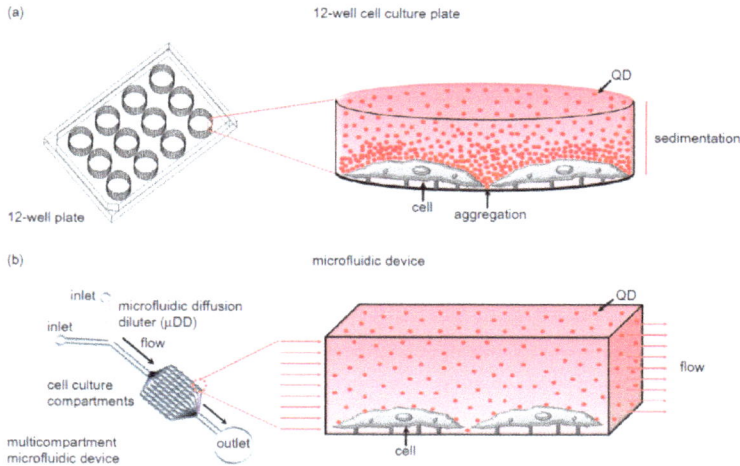

Figure 5. Diagram of nanoparticle behaviour inside well plates vs. microfluidic channel. (**a**) Well-plates without flow have particles aggregate under gravity, causing heterogeneous concentrations; (**b**) microfluidics can keep particles in homogenous suspension while under flow. Reprinted from Biomicrofluidics 4(2010) Mahto, Yoon [60], with the permission of AIP Publishing.

The well-plate methodology for nanotoxicity screening has been debated for its suitability in replicating in vivo conditions [27,67–70]. This is investigated in detail in a paper by Mahto et al. [60], where nanoparticles influenced by gravity in a static system formed a concentration gradient within a cell culture plate. The static conditions involved in well-plate analysis, i.e., pipetting nanomaterial solutions on top of a cell culture and allowing them to sediment on top of the cellular layer, are thought to have limitations in providing accurate nanotoxicology data for cells that are under shear stress from flowing biological fluids such as arterial, lymphatic, and renal cells in vivo. To compare the differences between static and flowing nanoparticles, they tested a sample of core/shell CdSe/ZnSe quantum dots in static tissue culture plates for 12 h at 8–80 pM to discover the optimal cytotoxic range, and then exposed the same quantum dots through a microfluidic concentration gradient generator in cell culture medium to murine embryonic fibroblast cells. When exposed to the quantum dots under flow, the cells exhibited apoptosis effects, namely detachment and dose-dependent morphological changes. However, the difference between the two exposure conditions at 40 pM was significant, in that the static conditions showed higher percentages of cell death and increased cell deformities, suggested to be due to the physicochemical stress of the sedimentation of quantum dots onto the cell membranes.

The effect of the shear-stress effect of flowing media over cells has been investigated by groups such as Kim et al. [71], utilizing bound endothelial cells in a single microchannel and exposing them to mesoporous silica nanoparticles. The shear-stress forces were tuned to mimic those expected in the arterial and capillary system of a healthy human (5–6 N/m^2), in order to observe any differences between these values and those of a static system. The nanoparticle concentration was also tuned to eliminate the effect of higher dosages during periods of a higher flow rate/higher shear-stress, so that the shear-stress forces were the primary variable. The unmodified silica nanoparticles used in this paper were found to increase in toxicity as the shear-stress increased, indicating an increase of cell membrane morphology and/or permeability under normal bodily shear-stress conditions, whereas the same dosages showed a reduced toxicity when applied in a static environment. This result suggests that static nanotoxicity tests may not be representative of the actual toxicity in a human body. If this is indeed the case, future toxicity tests would benefit from reproducing human vascular conditions in their flow and shear-stress properties.

In order to improve the cellular seeding and viability of hepatocytes in a microchannel, the group of Liu et al. [72] produced an electrospun biocompatible scaffold inside of a microfluidic device. By creating a 3D micro-environment of fibres for the liver cells, they were able to form a micro-perfusion environment which overcame the previous limitations of the lack of scaffold stiffness and the permeability to large molecules/cells. Using this platform, they were able to culture hepatocytes on the scaffold without microfluidic flow, then washed through and assayed for viability by measuring the albumin/urea secretion of the cells. The viability was determined to be higher under microfluidic perfusion than without perfusion flow or in static conditions. Upon the addition of silver nanoparticles, they could measure the amount of cell membrane damage with a commercial lactate dehydrogenase assay kit and found that the biomimetic 3D hepatocyte spheroids were more sensitive to silver nanoparticle damage than on a 2D tissue culture plate.

The use of cytometric methods can be integrated into microfluidic platforms to provide rapid and low cost nanotoxicity data, as was shown by Park et al. [73]. Their group cultured adherent cells (HeLa) directly into channels in a PDMS-glass microfluidic device and incubated the entire chip for 48 h to allow the cells to spread and grow over the analysis areas. They then introduced silver nanoparticles using a syringe pump via a concentration gradient generator. The silver nanoparticles induced both morphological changes in the cells and a colorimetric response to the MTT assay, which was also investigated in its conventional use in well-plates to compare against the microfluidic response. Optical brightfield images of the cell culture areas inside the channels were acquired post-exposure, to determine cell viability from the morphology and absorbance data. Dose-dependency was clearly observed for the toxicity of the silver nanoparticles and the half-lethal concentration (LC50) of the nanoparticles could be calculated. The LC50 from the microfluidic experiments was comparable to the estimated value from the conventional 96 well-plate method. The benefits of this technique using microfluidic approaches may include lower costs and the ease of use.

When single-cell responses to nanoparticle solutions are studied, using cell traps allows for very specific analyses to be performed on individual cells. Cunha-Matos et al. [67] formed cell traps designed to accept single cells inside a microfluidic channel using soft lithographic techniques, and then seeded the traps with functionalized gold nanorods, followed by Raman-active molecules and a coating of polyelectrolytes and proteins, that allowed them to bind primary bone marrow dendritic cells. The nanorods were then visible under surface-enhanced Raman spectroscopy (SERS), which allowed for the real-time visualization of nanoparticle concentration gradients as they were applied to the cells under flow. They used a live-cell incubator on a microscope stage to keep the cells in a biologically compatible environment for a duration of 24 h, over which they were able to assess each individual trapped cell for its response to nanoparticles. They were also able to add viability dyes to detect apoptosis and necrosis responses to the nanoparticles.

Adding electrodes to cell-trapping microwells via metallic deposition followed by chemical etching allows for electrophoretic measurements of cell viability. A thesis by Pratikkumar [74] detailed a microfluidic device that incorporates a combination of dielectrophoresis (DEP) and microwell methods of trapping single cells in wells aligned with gold microelectrodes, which allowed for the analysis of cells using electrochemical methods. The DEP forces could be switched on and off during the cell capture step, allowing the targeted capture of specific cells on the individual electrode/well features. Once the cells were captured, copper oxide nanoparticles were introduced into the microfluidic channel and flowed over the cell membranes to study the morphological response in each microwell. The author states that the impedance-based cell analysis was rapid, simple, label-free, and non-invasive. Measuring impedance vs. time revealed a significant drop in impedance after exposure to CuO nanoparticles. This correlated to a reduction in cell size and detachment from the electrode surface, which indicated a loss of viability due to toxic effects.

Similarly, a device containing impedance electrodes was also produced by the group of Rothbauer et al. [70], but with larger cell culture chambers instead of microwells. Human lung adenocarcinoma cells were cultured by on-chip perfusion in serum-containing media, until a confluent layer was formed across the electrode surface. Silica nanoparticles were administered under flow in serum-free media (in order to prevent contamination and a change in bioactivity of the nanoparticles). Serum-containing media was perfused again, in order to regenerate the tumour cells from their previous treatment. A metabolic assay was performed in parallel to the electrical impedance assay, and found that the AmSil30 silica nanoparticles caused a reduction in tumour regeneration and re-attachment to the electrode surface. Additionally, the presence of microfluidic flow in the device caused a reduction in regenerative capacity dependant on the flow velocity, indicating that the shear stress exerted on the cells played an important role in increasing the extent of nanoparticle uptake and thus the toxicity.

Organ-on-a-chip devices have only recently been used in nanotoxicity, and most often they utilize a single organ type. As their suitability for multi-organ toxicity assays becomes more fully realized, they will likely become more widely used. The group of Huh et al. [75] produced a lung-on-a-chip by seeding human alveolar epithelial cells and microvascular endothelial cells onto opposite sides of a porous PDMS membrane coated with an extracellular matrix protein. This membrane was sandwiched in between PDMS layers with large adjacent side channels, which were deformed by the application of a vacuum. This meant that the membrane could be subjected to mechanical stretching to simulate the action of breathing. The epithelial cells were exposed to air after their initial seeding was successful, while the endothelial cells remained exposed to culture medium with added blood-borne immune cells, providing an air-liquid interface to mimic the natural lung environment. In order to determine the device's response to nanomaterials, a solution of 12 nm silica nanoparticles in fluid was injected over the epithelial layer and aspirated to leave a thin layer, mimicking an aerosol uptake of the solution. The silica nanoparticles were found to promote the inflammation of the underlying endothelial layer, seen by an increase in the expression of Intercellular Adhesion Molecule 1 (ICAM-1) and the increased capture of neutrophils, a type of white blood cell. This device was considered to have increased the efficacy due to the mechanical breathing motions, as the motion only promoted ICAM-1 production upon exposure to the silica nanoparticles. This finding suggests that a lung-on-a-chip device with a breathing motion may give more accurate results on nanotoxicity than a static culture.

5. Outlook

Microfluidics nanotoxicity screening offers a range of potential advantages over traditional screening methods. The ability to integrate parallel streams on the same chip allows for high-throughput screening in a small form factor, as well as the reduced use of reagents/analytes and a reduction in the overall testing time. However, microfluidic screening is not yet widely employed in nanotoxicity testing. This may be due to the many parameters that require optimization if an agreed standard operating procedure is to be broadly accepted.

Indeed, the standardization of testing is the most critical roadblock against the adoption of microfluidic nanotoxicity screening. In order to achieve this task, many fundamental studies on the interactions between nanomaterials and channel-bound cells must be performed. Parameters that must be defined include: the effects of channel dimension and flow rate on the amount of nanomaterial exposure; differences in exposure along the length of the channel; and potential run-on effects of affected cells upon downstream cells. The viability assay methods must also be normalized, whereas groups are currently researching multiple variations on dye assays, microscopy, and flow cytometry.

Currently, microfluidics for nanotoxicity screening faces a significant challenge, in that the adoption of a standard device design is perhaps the first necessary step to produce an accepted industry method. However, most research groups involved in the study of this field have their own unique ideas and designs for their devices. It is rare for a research group to precisely follow the designs of another in their own experiments, and therefore, most fundamental studies are performed on different platforms. It would require a concerted effort between multiple groups to agree on a design and perform the required standardization testing to make it suitable for commercialization, which has not yet occurred.

The fact that many different methodologies are still being researched indicates that each method may have its own merits. While droplet microfluidic methods do offer a substantial increase in the number of tests able to be performed in a single device, they tend to be more suited to single-cell or single-cluster analyses. The latter is of interest to fundamental studies of particle or drug uptake in cells that are often found alone or in small clusters. However, the segregation of cellular analysis is less favourable for toxicity screening. Adherent cells, for example, will not adopt the same layer configuration found in normal biology and would therefore give less meaningful data. To address this, researchers are turning to multi-organ labs-on-a-chip platforms. These 3D cellular environments are likely to achieve more biologically relevant discoveries, including any run-on effects of toxic species between cell types. Coupling these 3D cellular environments with microfluidic flow will facilitate the collection of uptake profiles of toxicants under conditions that mimic arterial shear stresses on cell types, i.e., better models of toxicity in the complex environment of the human body.

To summarize, it is likely that microfluidic technologies that are taken-up by the nanotoxicity screening industry will provide data that is extremely difficult to obtain through current well-plate methods. The most likely candidate for this is the organ-on-a-chip style of device, due to the close mimicry of the human body, as discussed above. If these platforms prove to be successful (biologically relevant), fast, accurate, and inexpensive, microfluidics-enabled nanotoxicity screening may become a widely accepted testing platform for industry and regulators alike.

Acknowledgments: The author wishes to acknowledge the support of the University of South Australia and the Future Industries Institute, and the funding of the Australian Research Council's Discovery Projects grant DP150101774.

Author Contributions: All authors contributed to the writing, reviewing, and editing of this paper.

Conflicts of Interest: The authors declare no conflict of interest. The founding sponsors had no role in the design of the study; in the collection, analyses, or interpretation of data; in the writing of the manuscript, and in the decision to publish the results.

References

1. Monica, J.C.; Heintz, M.E.; Lewis, P.T. The perils of pre-emptive regulation. *Nat. Nano* **2007**, *2*, 68–70. [CrossRef] [PubMed]
2. Bowman, D.M.; van Calster, G.; Friedrichs, S. Nanomaterials and regulation of cosmetics. *Nat. Nano* **2010**, *5*, 92. [CrossRef] [PubMed]
3. Balas, F.; Arruebo, M.; Urrutia, J.; Santamaria, J. Reported nanosafety practices in research laboratories worldwide. *Nat. Nano* **2010**, *5*, 93–96. [CrossRef] [PubMed]
4. Hunter, R. *In Vitro Toxicity Testing: Technologies and Global Markets*; Global Markets: A BCC Research Report: Contract No.: PHM017E; BCC Research: Wellesley, MA, USA, 2014.
5. Alberts, B.; Johnson, A.; Lewis, J.; Raff, M.; Roberts, K.; Walter, P. *Molecular Biology of the Cell*, 4th ed.; Garland Science: New York, NY, USA, 2002.
6. Bianconi, E.; Piovesan, A.; Facchin, F.; Beraudi, A.; Casadei, R.; Frabetti, F.; Vitale, L.; Pelleri, M.C.; Tassani, S.; Piva, F.; et al. An estimation of the number of cells in the human body. *Ann. Hum. Biol.* **2013**, *40*, 463–471. [CrossRef] [PubMed]
7. Vickaryous, M.K.; Hall, B.K. Human cell type diversity, evolution, development, and classification with special reference to cells derived from the neural crest. *Biol. Rev. Camb. Philos. Soc.* **2006**, *81*, 425–455. [CrossRef] [PubMed]
8. Jani, P.; Halbert, G.W.; Langridge, J.; Florence, A.T. Nanoparticle uptake by the rat gastrointestinal mucosa: Quantitation and particle size dependency. *J. Pharm. Pharmacol.* **1990**, *42*, 821–826. [CrossRef] [PubMed]
9. Penney, D.P.; Ferin, J.; Oberdorster, G. Pulmonary retention of ultrafine and fine particles in rats. *Am. J. Respir. Cell Mol. Biol.* **1992**, *6*, 535–542.
10. Wakefield, G.; Green, M.; Lipscomb, S.; Flutter, B. Modified titania nanomaterials for sunscreen applications—Reducing free radical generation and DNA damage. *Mater. Sci. Technol.* **2004**, *20*, 985–988. [CrossRef]
11. Osmond, M.J.; McCall, M.J. Zinc oxide nanoparticles in modern sunscreens: An analysis of potential exposure and hazard. *Nanotoxicology* **2010**, *4*, 15–41. [CrossRef] [PubMed]

12. Mu, L.; Sprando, R.L. Application of nanotechnology in cosmetics. *Pharm. Res.* **2010**, *27*, 1746–1749. [CrossRef] [PubMed]

13. Hu, W.; Peng, C.; Luo, W.; Lv, M.; Li, X.; Li, D.; Huang, Q.; Fan, C. Graphene-Based Antibacterial Paper. *ACS Nano* **2010**, *4*, 4317–4323. [CrossRef] [PubMed]

14. Li, Q.; Mahendra, S.; Lyon, D.Y.; Brunet, L.; Liga, M.V.; Li, D.; Alvarez, P.J.J. Antimicrobial nanomaterials for water disinfection and microbial control: Potential applications and implications. *Water Res.* **2008**, *42*, 4591–4602. [CrossRef] [PubMed]

15. Marambio-Jones, C.; Hoek, E.M.V. A review of the antibacterial effects of silver nanomaterials and potential implications for human health and the environment. *J. Nanopart. Res.* **2010**, *12*, 1531–1551. [CrossRef]

16. Betteridge, D.J. What is oxidative stress? *Metabolism* **2000**, *49*, 3–8. [CrossRef]

17. Durán, N.; Guterres, S.S.; Alves, O.L. *Nanotoxicology: Materials, Methodologies, and Assessments*; Springer: Berlin, Germany, 2013.

18. Finkel, T.; Holbrook, N.J. Oxidants, oxidative stress and the biology of ageing. *Nature* **2000**, *408*, 239–247. [CrossRef] [PubMed]

19. Chen, Z.; Meng, H.; Xing, G.; Chen, C.; Zhao, Y.; Jia, G.; Wang, T.; Yuan, H.; Ye, C.; Zhao, F.; et al. Acute toxicological effects of copper nanoparticles in vivo. *Toxicol. Lett.* **2006**, *163*, 109–120. [CrossRef] [PubMed]

20. Meng, H.; Chen, Z.; Xing, G.; Yuan, H.; Chen, C.; Zhao, F.; Zhang, C.; Zhao, Y. Ultrahigh reactivity provokes nanotoxicity: Explanation of oral toxicity of nano-copper particles. *Toxicol. Lett.* **2007**, *175*, 102–110. [CrossRef] [PubMed]

21. Ye, Y.; Liu, J.; Xu, J.; Sun, L.; Chen, M.; Lan, M. Nano-SiO$_2$ induces apoptosis via activation of p53 and Bax mediated by oxidative stress in human hepatic cell line. *Toxicol. In Vitro* **2010**, *24*, 751–758. [CrossRef] [PubMed]

22. Ahamed, M. Silica nanoparticles-induced cytotoxicity, oxidative stress and apoptosis in cultured A431 and A549 cells. *Hum. Exp. Toxicol.* **2013**, *32*, 186–195. [CrossRef] [PubMed]

23. Park, E.-J.; Yi, J.; Chung, K.-H.; Ryu, D.-Y.; Choi, J.; Park, K. Oxidative stress and apoptosis induced by titanium dioxide nanoparticles in cultured BEAS-2B cells. *Toxicol. Lett.* **2008**, *180*, 222–229. [CrossRef] [PubMed]

24. Ko, K.S.; Kong, I.C. Toxic effects of nanoparticles on bioluminescence activity, seed germination, and gene mutation. *Appl. Microbiol. Biotechnol.* **2014**, *98*, 3295–3303. [CrossRef] [PubMed]

25. Moran, J.H.; Schnellmann, R.G. A rapid beta-NADH-linked fluorescence assay for lactate dehydrogenase in cellular death. *J. Pharmacol. Toxicol. Methods* **1996**, *36*, 41–44. [CrossRef]

26. Zhao, J.; Castranova, V. Toxicology of nanomaterials used in nanomedicine. *J. Toxicol. Environ. Health Part B Crit. Rev.* **2011**, *14*, 593–632. [CrossRef] [PubMed]

27. Wang, T.; Bai, J.; Jiang, X.; Nienhaus, G.U. Cellular uptake of nanoparticles by membrane penetration: A study combining confocal microscopy with FTIR spectroelectrochemistry. *ASC Nano* **2012**, *6*, 1251–1519. [CrossRef] [PubMed]

28. Kreyling, W.G.; Hirn, S.; Schleh, C. Nanoparticles in the lung. *Nat. Biotechnol.* **2010**, *28*, 1275. [CrossRef] [PubMed]

29. Guo, Y.; Terazzi, E.; Seemann, R.; Fleury, J.B.; Baulin, V.A. Direct proof of spontaneous translocation of lipid-covered hydrophobic nanoparticles through a phospholipid bilayer. *Sci. Adv.* **2016**, *2*, e1600261. [CrossRef] [PubMed]

30. Kettiger, H.; Schipanski, A.; Wick, P.; Huwyler, J. Engineered nanomaterial uptake and tissue distribution: From cell to organism. *Int. J. Nanomed.* **2013**, *8*, 3255–3269.

31. Yang, P.-H.; Sun, X.; Chiu, J.-F.; Sun, H.; He, Q.-Y. Transferrin-mediated gold nanoparticle cellular uptake. *Bioconjug. Chem.* **2005**, *16*, 494–496. [CrossRef] [PubMed]

32. Oh, N.; Park, J.-H. Endocytosis and exocytosis of nanoparticles in mammalian cells. *Int. J. Nanomed.* **2014**, *9*, 51–62.

33. Fischer, H.C.; Chan, W.C.W. Nanotoxicity: The growing need for in vivo study. *Curr. Opin.Biotechnol.* **2007**, *18*, 565–571. [CrossRef] [PubMed]

34. Chiappini, C.; Almeida, C. 8-Silicon nanoneedles for drug delivery. In *Semiconducting Silicon Nanowires for Biomedical Applications*; Imperial College Press: London, UK, 2014; pp. 144–167.

35. Kolhar, P.; Doshi, N.; Mitragotri, S. Polymer nanoneedle-mediated intracellular drug delivery. *Small* **2011**, *7*, 2094–2100. [CrossRef] [PubMed]

36. Carmona, H.; Valadez, H.; Yun, Y.; Sankar, J.; Estala, L.; Gomez, F.A. Development of microfluidic-based assays to estimate the binding between osteocalcin (BGLAP) and fluorescent antibodies. *Talanta* **2015**, *132*, 676–679. [CrossRef] [PubMed]

37. Hosokawa, M.; Hayashi, T.; Mori, T.; Yoshino, T.; Nakasono, S.; Matsunaga, T. Microfluidic device with chemical gradient for single-cell cytotoxicity assays. *Anal. Chem.* **2011**, *83*, 3648–3654. [CrossRef] [PubMed]

38. Ng, E.; Hoshino, K.; Zhang, X. Microfluidic immunodetection of cancer cells via site-specific microcontact printing of antibodies on nanoporous surface. *Methods* **2013**, *63*, 266–275. [CrossRef] [PubMed]

39. Pasirayi, G.; Scott, S.M.; Islam, M.; O'Hare, L.; Bateson, S.; Ali, Z. Low cost microfluidic cell culture array using normally closed valves for cytotoxicity assay. *Talanta* **2014**, *129*, 491–498. [CrossRef] [PubMed]

40. Cunha-Matos, C.A.; Millington, O.R.; Wark, A.W.; Zagnoni, M. Real-time assessment of nanoparticle-mediated antigen delivery and cell response. *Lab Chip* **2016**, *16*, 3374–3381. [CrossRef] [PubMed]

41. Weibull, E.; Matsui, S.; Sakai, M.; Andersson Svahn, H.; Ohashi, T. Microfluidic device for generating a stepwise concentration gradient on a microwell slide for cell analysis. *Biomicrofluidics* **2013**, *7*, 064115. [CrossRef] [PubMed]

42. Tong, Z.; Ivask, A.; Guo, K.; McCormick, S.; Lombi, E.; Priest, C.; Voelcker, N.H. Crossed flow microfluidics for high throughput screening of bioactive chemical-cell interactions. *Lab Chip* **2017**, *17*, 501–510. [CrossRef] [PubMed]

43. Roth, E.A.; Xu, T.; Das, M.; Gregory, C.; Hickman, J.J.; Boland, T. Inkjet printing for high-throughput cell patterning. *Biomaterials* **2004**, *25*, 3707–3715. [CrossRef] [PubMed]

44. Melamed, S.; Elad, T.; Belkin, S. Microbial sensor cell arrays. *Curr. Opin. Biotechnol.* **2012**, *23*, 2–8. [CrossRef] [PubMed]

45. Leclerc, E.; El Kirat, K.; Griscom, L. In situ micropatterning technique by cell crushing for co-cultures inside microfluidic biochips. *Biomed. Microdevices* **2008**, *10*, 169–177. [CrossRef] [PubMed]

46. Wu, J.; Chen, Q.; Liu, W.; He, Z.; Lin, J.-M. Recent advances in microfluidic 3D cellular scaffolds for drug assays. *TrAC Trends Anal. Chem.* **2017**, *87*, 19–31. [CrossRef]

47. Toh, Y.-C.; Lim, T.C.; Tai, D.; Xiao, G.; van Noort, D.; Yu, H. A microfluidic 3D hepatocyte chip for drug toxicity testing. *Lab Chip* **2009**, *9*, 2026–2035. [CrossRef] [PubMed]

48. Yoo, S.K.; Lee, J.H.; Yun, S.-S.; Gu, M.B.; Lee, J.H. Fabrication of a bio-MEMS based cell-chip for toxicity monitoring. *Biosens. Bioelectron.* **2007**, *22*, 1586–1592. [CrossRef] [PubMed]

49. Wagner, I.; Materne, E.M.; Brincker, S.; Süßbier, U.; Frädrich, C.; Busek, M.; Sonntag, F.; Sakharov, D.A.; Trushkin, E.V.; Tonevitsky, A.G.; et al. A dynamic multi-organ-chip for long-term cultivation and substance testing proven by 3D human liver and skin tissue co-culture. *Lab Chip* **2013**, *13*, 3538–3547. [CrossRef] [PubMed]

50. SynVivo. SynTox 3D Toxicology Model—Organ Specific Physiological Responses: SynVivo. Available online: http://www.synvivobio.com/syntox/ (accessed on 4 April 2017).

51. Ma, B.; Zhang, G.; Qin, J.; Lin, B. Characterization of drug metabolites and cytotoxicity assay simultaneously using an integrated microfluidic device. *Lab Chip* **2009**, *9*, 232–238. [CrossRef] [PubMed]

52. Cooksey, G.A.; Elliott, J.T.; Plant, A.L. Reproducibility and Robustness of a Real-Time Microfluidic Cell Toxicity Assay. *Anal. Chem.* **2011**, *83*, 3890–3896. [CrossRef] [PubMed]

53. Eddings, M.A.; Eckman, J.W.; Arana, C.A.; Papalia, G.A.; Connolly, J.E.; Gale, B.K.; Myszka, D.G. "Spot and hop": Internal referencing for surface plasmon resonance imaging using a three-dimensional microfluidic flow cell array. *Anal. Biochem.* **2009**, *385*, 309–313. [CrossRef] [PubMed]

54. Wada, K.-I.; Taniguchi, A.; Kobayashi, J.; Yamato, M.; Okano, T. Live Cells-Based Cytotoxic Sensorchip Fabricated in a Microfluidic System. *Biotechnol. Bioeng.* **2007**, *99*, 1513–1517. [CrossRef] [PubMed]

55. Gori, M.; Simonelli, M.C.; Giannitelli, S.M.; Businaro, L.; Trombetta, M.; Rainer, A. Investigating Nonalcoholic Fatty Liver Disease in a Liver-on-a-Chip Microfluidic Device. *PLoS ONE* **2016**, *11*, e0159729. [CrossRef] [PubMed]

56. Wang, J.; Fan, Z.; Zhao, Y.; Song, Y.; Chu, H.; Song, W.; Song, Y.; Pan, X.; Sun, Y.; Li, D. A new hand-held microfluidic cytometer for evaluating irradiation damage by analysis of the damaged cells distribution. *Sci. Rep.* **2016**, *6*, 23165. [CrossRef] [PubMed]

57. Morgan, A.J.L.; Hidalgo San Jose, L.; Jamieson, W.D.; Wymant, J.M.; Song, B.; Stephens, P.; Barrow, D.A.; Castell, O.K. Simple and versatile 3D printed microfluidics using fused filament fabrication. *PLoS ONE* **2016**, *11*, e0152023. [CrossRef] [PubMed]

58. Brouzes, E.; Medkova, M.; Savenelli, N.; Marran, D.; Twardowski, M.; Hutchinson, J.B.; Rothberg, J.M.; Link, D.R.; Perrimon, N.; Samuels, M.L. Droplet microfluidic technology for single-cell high-throughput screening. *Proc. Natl. Acad. Sci.* **2009**, *106*, 14195–14200. [CrossRef] [PubMed]

59. Konry, T.; Sarkar, S.; Sabhachandani, P.; Stroopinsky, D.; Palmer, K.; Cohen, N.; Rosenblatt, J.; Avigan, D.; Konry, T. Dynamic analysis of immune and cancer cell interactions at single cell level in microfluidic droplets. *Biomicrofluidics* **2016**, *10*, 054115.

60. Mahto, S.K.; Yoon, T.H.; Rhee, S.W. A new perspective on in vitro assessment method for evaluating quantum dot toxicity by using microfluidics technology. *Biomicrofluidics* **2010**, *4*, 034111. [CrossRef] [PubMed]

61. Mahto, S.K.; Charwat, V.; Ertl, P.; Rothen-Rutishauser, B.; Rhee, S.W.; Sznitman, J. Microfluidic platforms for advanced risk assessments of nanomaterials. *Nanotoxicology* **2015**, *9*, 381–395. [CrossRef] [PubMed]

62. Geiser, M.; Rothen-Rutishauser, B.; Kapp, N.; Schürch, S.; Kreyling, W.; Schulz, H.; Semmler, M.; Hof, V.I.; Heyder, J.; Gehr, P. Ultrafine particles cross cellular membranes by nonphagocytic mechanisms in lungs and in cultured cells. *Environ. Health Perspect.* **2005**, *113*, 1555–1560. [CrossRef] [PubMed]

63. Rothen-Rutishauser, B.M.; Schurch, S.; Haenni, B.; Kapp, N.; Gehr, P. Interaction of fine particles and nanoparticles with red blood cells visualized with advanced microscopic techniques. *Environ. Sci. Technol.* **2006**, *40*, 4353–4359. [CrossRef] [PubMed]

64. Balbus, J.M.; Maynard, A.D.; Colvin, V.L.; Castranova, V.; Daston, G.P.; Denison, R.A.; Dreher, K.L.; Goering, P.L.; Goldberg, A.M.; Kulinowski, K.M.; et al. Meeting report: Hazard assessment for nanoparticles–report from an interdisciplinary workshop. *Environ. Health Perspect.* **2007**, *115*, 1654–1659. [CrossRef] [PubMed]

65. Nel, A.E.; Madler, L.; Velegol, D.; Xia, T.; Hoek, E.M.V.; Somasundaran, P.; Klaessig, F.; Castranova, V.; Thompson, M. Understanding biophysicochemical interactions at the nano-bio interface. *Nat. Mater.* **2009**, *8*, 543–557. [CrossRef] [PubMed]

66. Richter, L.; Charwat, V.; Jungreuthmayer, C.; Bellutti, F.; Brueckl, H.; Ertl, P. Monitoring cellular stress responses to nanoparticles using a lab-on-a-chip. *Lab Chip* **2011**, *11*, 2551–2560. [CrossRef] [PubMed]

67. Cunha-Matos, C.A.; Millington, O.M.; Wark, A.W.; Zagnoni, M. (Eds.) Real-Time Multimodal Imaging of Nanoparticle-Cell Interactions in High-Throughput Microfluidics. In Proceedings of the 18th International Conference on Miniaturized Systems for Chemistry and Life Sciences, San Antonio, TX, USA, 26–30 October 2014.

68. Velve-Casquillas, G.; Le Berre, M.; Piel, M.; Tran, P.T. Microfluidic tools for cell biological research. *Nano Today* **2010**, *5*, 28–47. [CrossRef] [PubMed]

69. Shah, P.; Kaushik, A.; Zhu, X.; Zhang, C.; Li, C.-Z. Chip based single cell analysis for nanotoxicity assessment. *Analyst* **2014**, *139*, 2088–2098. [CrossRef] [PubMed]

70. Rothbauer, M.; Praisler, I.; Docter, D.; Stauber, R.H.; Ertl, P. Microfluidic impedimetric cell regeneration assay to monitor the enhanced cytotoxic effect of nanomaterial perfusion. *Biosensors* **2015**, *5*, 736–749. [CrossRef] [PubMed]

71. Kim, D.; Lin, Y.-S.; Haynes, C.L. On-chip evaluation of shear stress effect on cytotoxicity of mesoporous silica nanoparticles. *Anal. Chem.* **2012**, *83*, 8377–8382. [CrossRef] [PubMed]

72. Liu, Y.; Wang, S.; Wang, Y. Patterned fibers embedded microfluidic chips based on PLA and PDMS for Ag nanoparticle safety testing. *Polymers* **2016**, *8*, 402. [CrossRef]

73. Park, J.Y.; Yoon, T.H. Microfluidic image cytometry (µFIC) assessments of silver nanoparticle cytotoxicity. *Bull. Korean Chem. Soc.* **2012**, *33*, 4023–4027. [CrossRef]

74. Shah, P. Development of a Lab-on-a-Chip Device for Rapid Nanotoxicity Assessment In Vitro. Ph.D. Thesis, Florida International University, Miami, FL, USA, 2014.

75. Huh, D.; Matthews, B.D.; Mammoto, A.; Montoya-Zavala, M.; Hsin, H.Y.; Ingber, D.E. Reconstituting organ-level lung functions on a chip. *Science* **2010**, *328*, 1662–1668. [CrossRef] [PubMed]

micromachines

MDPI

Review

The Self-Propulsion of the Spherical Pt–SiO$_2$ Janus Micro-Motor

Jing Zhang [1], Xu Zheng [2,*], Haihang Cui [1] and Zhanhua Silber-Li [2]

[1] School of Environment and Municipal Engineering, Xi'an University of Architecture and Technology, Xi'an 710055, China; zhangjing102133@163.com (J.Z.); cuihaihang@xauat.edu.cn (H.C.)
[2] State Key Laboratory of Nonlinear Mechanics, Institute of Mechanics, Chinese Academy of Sciences, Beijing 100190, China; lili@imech.ac.cn
* Correspondence: zhengxu@lnm.imech.ac.cn; Tel.: +86-10-8254-3925

Academic Editor: Hengdong Xi
Received: 22 February 2017; Accepted: 5 April 2017; Published: 12 April 2017

Abstract: The double-faced Janus micro-motor, which utilizes the heterogeneity between its two hemispheres to generate self-propulsion, has shown great potential in water cleaning, drug delivery in micro/nanofluidics, and provision of power for a novel micro-robot. In this paper, we focus on the self-propulsion of a platinum–silica (Pt–SiO$_2$) spherical Janus micro-motor (JM), which is one of the simplest micro-motors, suspended in a hydrogen peroxide solution (H$_2$O$_2$). Due to the catalytic decomposition of H$_2$O$_2$ on the Pt side, the JM is propelled by the established concentration gradient known as diffusiophoretic motion. Furthermore, as the JM size increases to O (10 μm), oxygen molecules nucleate on the Pt surface, forming microbubbles. In this case, a fast bubble propulsion is realized by the microbubble cavitation-induced jet flow. We systematically review the results of the above two distinct mechanisms: self-diffusiophoresis and microbubble propulsion. Their typical behaviors are demonstrated, based mainly on experimental observations. The theoretical description and the numerical approach are also introduced. We show that this tiny motor, though it has a very simple structure, relies on sophisticated physical principles and can be used to fulfill many novel functions.

Keywords: Janus micromotor; self-diffusiophoresis; bubble propulsion

1. Introduction

Motors that are designed to convert chemical or electromagnetic energy into mechanical energy are ubiquitous in people's lives. In the last decade, micro/nano-motors have emerged, and they constitute a new field of technology attracting great interest from researchers. The pioneer work in micro/nano-motors involved a cylindrical micro/nano-motor using hydrogen peroxide (H$_2$O$_2$) as fuel, as proposed by Whitesides et al. [1] and the group of Sen and Mallouk [2]. Many other micro/nano-motors based on similar mechanisms were later fabricated [3–5]. These autonomous motors have two common features: (1) they have a metal-dielectric or bimetal double-faced structure, which is usually made of platinum–silica (Pt–SiO$_2$) or platinum–gold (Pt–Au); and (2) they operate on the catalytic decomposition of H$_2$O$_2$, which is the energy source. As a result, these motors were named after the Greek double-faced god, Janus. The Janus micro-motor (JM) benefits from the heterogeneous structure that can spontaneously establish a local gradient around the JM. One obvious advantage is that JMs are self-motile due to the local gradient, and no external energy is required [6–8]. Encouraged by this advantage, researchers have developed many novel functions for Janus micro/nano-motors in recent years, such as drug delivery or ion detection in micro/nanofluidic chips [9–15], and purification of polluted water [15–17]. The JM has also shown wide application prospects as the power unit of a micro-robot [18].

The concept of using the heterogeneous structure of a microparticle to build a local gradient field was proposed by the Nobel laureate P.G. de Gennes [19]. Basically, the JM's self-propulsion is in the low Reynolds number (*Re*) flow [8,20], in which viscous effect is dominant. In this regime, in order to achieve propulsion, a microswimmer has to break time-reversal of the Stokes flow by employing moving units or deforming, as demonstrated by the scallop theorem [20–22]. A geometrically symmetric JM has to make use of the adjacent heterogeneoty to overcome the constraint of the scallop theorem [20,23,24]. Different surfaces with distinct physical/chemical properties can be used to build a heterogeneous concentration, temperature, or electrical field. Correspondingly, the self-motile motion of the microparticle driven by each field gradient is called self-diffusiophoresis [5,25], self-thermophoresis [26,27], or self-electrophoresis [28,29], respectively. Taking the Pt-SiO$_2$ spherical micro-motor in H$_2$O$_2$ as an example, the concentration gradient is formed by the decomposition of H$_2$O$_2$ on the surface of the Pt hemisphere (Figure 1): $2H_2O_2 \rightarrow 2H_2O + O_2$. The higher molecular concentration on the Pt side provides the power to propel the JM in the other direction. Recalling the pioneer work of bubble propulsion based on microtubular engine developed by Mei and Solovev and their colleagues [30–34], the function of the JM could be significantly extended. It was recently found that O$_2$ molecules could nucleate to form microbubbles if the size of the JM was up to O (10 μm) [35]. The capability to generate microbubbles by a JM, inspired by the work of the microtubular engine, could shed light on a novel, high-efficiency, form of fast propulsion. The mechanisms that how bubbles are generated and how bubbles propel the micromotors still need to be clarified. It is of great interest to thoroughly understand the mechanisms, including fluid–particle interaction and bubble dynamics.

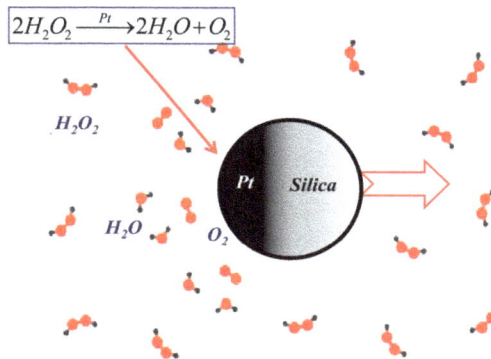

Figure 1. Schematic diagram of Pt–SiO$_2$ spherical Janus micro-motor (JM) self-propulsion in H$_2$O$_2$ solution.

Many experiments have been devoted to unveiling the physical mechanism behind a JM's self-propulsion. A common consensus is that the motion of a JM whose size is smaller than 5 μm is dominated by self-diffusiophoresis, while that of a JM larger than 10 μm is primarily due to bubble propulsion [24,35,36]. However, some key issues remain unclear. For slow self-diffusiophoresis, from a statistical point of view, what is its typical behavior? What is the contribution of the rotational motion? For fast bubble propulsion, what is the exact force pushing the JM forward? How do we understand the bubble dynamics? In this review, we first organize the major experimental results to exhibit the big picture regarding a JM's self-propulsion. By comparing with our results, we try to capture the typical behavior of the JM's motion and explore the physics. Furthermore, we introduce both theoretical and numerical approaches to describe the self-propulsion of a Pt–SiO$_2$ spherical JM. Finally, we discuss some important issues, such as improving the efficiency of JMs and manipulating them in real applications.

2. Self-Diffusiophoretic Motion

The motions of JMs with diameters of 1–5 μm have been measured in some experiments [5,36–38]. The fabrication of a Pt–SiO$_2$ JM usually gets help from e-beam evaporation to deposit a thin Pt layer on the hemisphere of a silica microsphere, for which details can be found in the literature [37–39] and the supporting information. As has been mentioned, these small JMs are propelled by the concentration gradient due to the catalytic reaction on the Pt side. This self-motile motion is called self-diffusiophoresis. The pioneer experiment of JM self-diffusiophoresis was performed by Howse et al. [5] using an optical microscope with the least observation time interval being about 10 ms. They showed the difference between self-diffusiophoresis and pure Brownian motion by comparing the trajectories of a JM in an H$_2$O$_2$ solution and pure water (Figure 2a) When the concentration of H$_2$O$_2$ was less than 1%, the trajectory of the JM was similar to that of Brownian random motion in pure water. With the increase of the H$_2$O$_2$ concentration, the self-diffusiophoretic trajectory exhibited the features of long-range directional movement, which was significantly different from the Brownian motion. Similar trajectories were reported in the literature [37–43]. The increased H$_2$O$_2$ concentration resulted in stronger catalytic reaction on the Pt surface. The chemical reaction kinetic was thus connected to the instantaneous speed of the JM movement. It was believed that the typical self-diffusiophoretic speed V_{DFP} is linearly proportional to the reaction rate per unit area k, i.e., $V_{DFP} \sim k$ [5,6]. In the experiment of Howse et al. [5], k was estimated to be about 5×10^{10} μm^{-2}·s^{-1}, which is consistent with our results.

The mean square displacement (MSD) of the JM's self-diffusiophoresis in H$_2$O$_2$ solution was often measured to quantitatively illustrate the swimming characteristics. A particle-tracking method [43,44] was usually employed to obtain the MSD $<L^2>$ based on the trajectories, where $<L^2>$ denotes an ensemble average. A special image-processing technique was used to assure high-precision determination of the JM's displacement. Figure 2b shows the typical MSD of JMs with a diameter d of 1.6 μm measured by Howse et al. [5]. Figure 2c shows the MSD of JMs with diameters of 1 or 2 μm measured by our group [37]. Different from the MSD of simple Brownian motion in water that increases linearly with time, the MSD of a JM's self-diffusiophoresis in H$_2$O$_2$ solution shows nonlinear behavior at shorter times, and it turns linear at longer times. The parabolic-like MSD suggests a typical "ballistic motion" [45] driven by the concentration gradient. The transition time of the MSD is affected by the JM's size. As shown in Figure 2c, the transition time of a 2-μm JM is approximately 5–6 s, while that of a 1-μm JM is only about 0.5 s.

We note that this transition time is close to the rotation characteristic time $\tau_R = \pi \mu d^3 / k_B T$ of a spherical JM (μ is viscosity, k_B is Boltzmann constant, and T is temperature) [46]. Thus, the MSD results can be non-dimensionalized as shown in Figure 2d. The vertical axis is $<L^2>/d^2$, and the horizontal axis is $\tau = t/\tau_R$. A green line with a constant slope of 1 is drawn to show the linear Brownian behavior. The dimensionless results show a three-stage behavior of the JM's self-diffusiophoresis [37]: (1) At very short times $\tau < 10^{-2}$, the curve of the dimensionless MSD is similar to that of linear Brownian motion in pure water. This indicates that Brownian motion still dominates the JM's motion because the concentration gradient has not yet been established. This stage is usually less than 10 ms, which was too short to be noticed in most previous experiments; (2) At intermediate times $\tau = 10^{-2}$–1, the curves of dimensionless MSDs all exhibit a slope of about 2. This slope is a typical signal of the ballistic motion driven by a concentration gradient, which is also called super-diffusive [5,46]; (3) The long-time stage begins at $\tau = 1$, which is also $t = \tau_R$. It is interesting to see that the slope of the MSD returns to 1, which gives the name "Brownian-like motion" to this stage. However, it is obvious that the motion of the JM in the third stage is different from Brownian motion. It is the rotational motion that varies the directional propulsion in the second stage and decreases the slope back to 1. The above results reveal that the three stages of motion of the JM are dominated by different scales of physical effects.

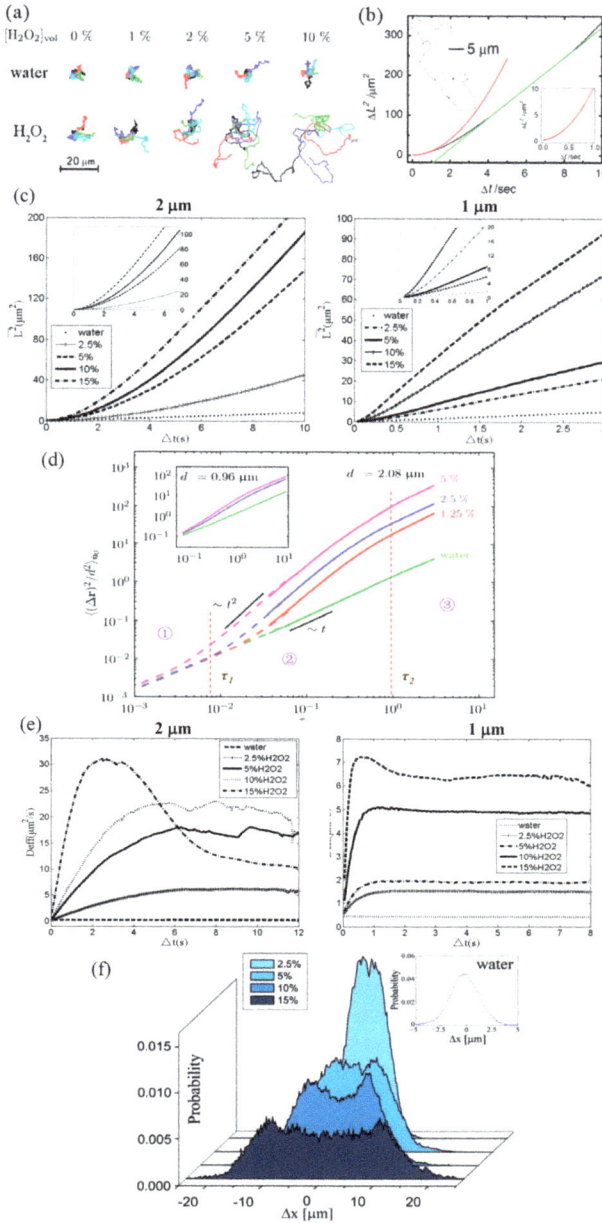

Figure 2. Characteristics of a JM's translational diffusiophoresis. (**a**) The typical trajectory and (**b**) the time-varied mean square displacement (MSD) of JMs with a diameter of 1.6 μm [5]; (**c**) The typical MSDs of JMs with diameters of 2 μm and 1 μm, measured by our group; (**d**) The three-stage dimensionless MSD [37]; (**e**) The effective diffusion coefficient of JMs; (**f**) The double-peaked displacement probability distribution (DPD) [37]. Figure 2a,b is reproduced with permission from Howse et al. [5]; Published by APS, 2007. Figure 2d,f is reproduced with permission from Zheng et al. [37]; Published by APS, 2013.

The effective diffusion coefficient D_{eff} is introduced to describe the self-diffusiophoresis of the JM. We can calculate D_{eff} based on $D_{eff} = <L^2>/4t$, and the results are shown in Figure 2e. In pure water, the D_{eff} is approximately constant. The D_{eff} of a 2-μm JM is 0.20 μm^2/s, and that of a 1-μm JM is 0.43 μm^2/s; these are consistent with the results calculated by the Stokes–Einstein equation $D = k_B T/6\pi\mu d$. In H$_2$O$_2$ solutions with different concentrations, D_{eff} first increases rapidly and linearly with time, and then reaches an approximately constant value. The plateau values of D_{eff} of 2-μm and 1-μm JMs are about 10–30 μm^2/s and 1.5–7.2 μm^2/s, respectively in 2.5%–15% H$_2$O$_2$ solutions. These values are increased by 1–2 orders of magnitude compared to that of simple Brownian motion in pure water. This shows the great potential of using super-diffusive microparticles in industry. From the Langevin equation, it has been deduced that a JM's D_{eff} can be written as [5,25]:

$$D_{eff} = \frac{4R^2}{3\tau_R} + \frac{1}{4}V_{DFP}{}^2\tau_R + \frac{V^2\tau_R{}^2}{8t}(e^{-2t/\tau_R} - 1), \tag{1}$$

where V_{DFP} is the typical self-diffusiophoretic velocity of the JM. On the right side of Equation (1), the first term is the contribution of pure Brownian motion described by the Stokes–Einstein equation, the second term represents the contribution of the translational ballistic motion of self-diffusiophoresis, and the third term is the contribution of the rotation. It shows that the rotational time τ_R is an important time scale in the dimensionless law of Figure 2d.

It is of great interest to investigate the statistical behavior of a JM's self-diffusiophoresis [37,45]. The nonlinear behavior mentioned above should result in non-Gaussian behavior that is distinct from linear Brownian motion. The Displacement probability distribution (DPD) is measured in different H$_2$O$_2$ solutions to demonstrate the non-Gaussian behavior (Figure 2f). For Brownian motion in pure water, the DPD is consistent with the Gaussian distribution (the subplot in the top-right corner of Figure 2f). However, in different H$_2$O$_2$ solutions, every measured DPD exhibits a non-Gaussian double-peaked distribution. This double-peaked structure becomes more obvious with increasing concentrations of H$_2$O$_2$. The double peaks represent the most probable displacements of the JM's self-diffusiophoresis motion, so they are of significance in controlling a JM's motion. More interestingly, the non-Gaussian double-peaked structures are still evident even in the long-time stage ($t > \tau_R$), although the MSD has reverted to linearity. Further quantitative analysis of the non-Gaussian behavior relies on kurtosis, which is defined based on the fourth moment of the displacement. The kurtosis is negative for self-diffusiophoresis, as shown in our previous result [37]. More details of the statistical characteristics of the JM's self-propulsion can be found in the literature [41,42,47,48].

The origin of the JM's rotational motion is a topic that requires clarification. Figure 3a shows the rotational angle probability distribution of a JM in water and in 10% H$_2$O$_2$ solution, respectively. For Brownian motion in water, when $t = 0.05$ s, the rotational angle of the JM is mainly between $-40°$ and $40°$, and the peak of the probability distribution is 0.34. Then, the distribution turns flat. At $t = 15$ s, the probability distribution is almost uniform in all directions. The results in different H$_2$O$_2$ solutions are approximately the same as those in water. This indicates that the concentration gradient does not produce torque to affect the JM's rotation. The only effect dominating the JM's rotation is Brownian torque. The theory of Brownian rotation has been well established, with a typical time scale τ_R. To control the rotation of the JM, ideas based on designing special geometrical shapes have been proposed. For example, "L-type" or boomerang-like micro-motors were fabricated to produce spiral motion [49].

The interaction between a wall and a JM is also interesting. The orientation and rotation of the Janus microsphere were observed to be influenced by the wall in our previous experiment [43]. Figure 3b illustrates the difference between a 3-D rotation in low-concentration H$_2$O$_2$ solution (<2%) and a 2-D rotation in the horizontal plane in high-concentration H$_2$O$_2$ solution (10%). This phenomenon has also been observed in a recent experiment by Das et al. [50]. The wall confinement was used to control the orientation and trajectory of the JM, as reported by Simmchen et al. [51]. The change of the

JM's direction near the wall is influenced by the symmetry-breaking of both the concentration and velocity fields. The confined self-diffusiophoresis of JMs has been a topic of great interest recently [52,53].

In addition, it should be noted that background flow has seldom been involved in previous experiments. In practical application, the self-diffusiophoresis of JMs should be coupled with surrounding fluid flow. Because the flow near the JM's surface, the concentration gradient, and even the rotation of the JM is influenced by the background convective flow, the behavior of the JM may be changed significantly [54–56]. For example, Zottl and Stark reported that helical motion of a JM occurs in a microchannel with a square cross-section [54]. Due to the complexity of an analytic solution, there is still no consensus on this issue, and experimental research is urgently needed.

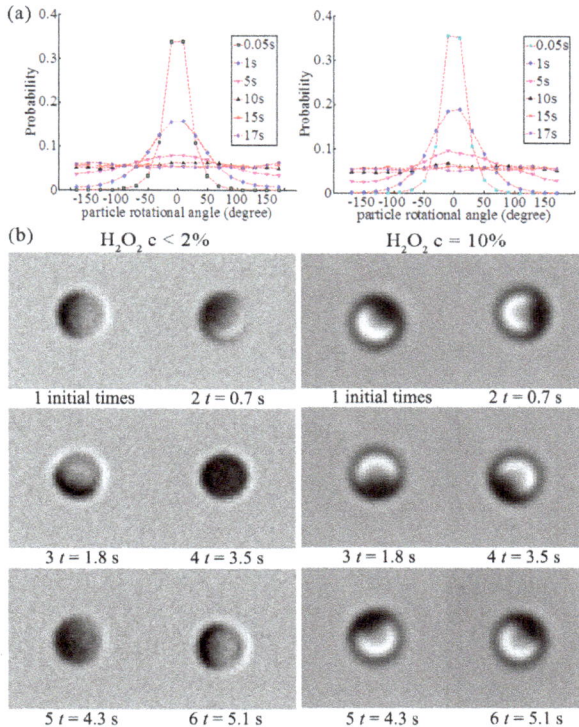

Figure 3. Rotational characteristics of the JM [43]. (**a**) A comparison between the rotational angle probability distribution of a JM with a diameter of 2 μm in different H_2O_2 solutions; (**b**) The JM exhibits 3-D rotation in an H_2O_2 solution with low concentration, while it exhibits 2-D behavior in an H_2O_2 solution with high concentration. Figure 3 is reproduced with permission from Zheng et al. [43]; Published by Springer, 2015.

3. Fast Microbubble Propulsion

Following the microtubular engine, spherical JMs can also achieve fast microbubble propulsion. When the diameter of a spherical JM is larger than 10 μm, it is observed that the oxygen molecules generated by the decomposition of H_2O_2 can nucleate to form microbubbles. During this time, the self-propelled motion of the Janus microsphere is driven by the microbubbles. The mechanism of bubble nucleation remains unclear. It is believed that the curvature of the microsphere surface must be small enough [57]. In this section, we focus on the microbubble propulsion of a JM whose diameter ranges from 20 to 50 μm. These larger JMs move much faster than the smaller JMs described in the last section. This breaks the constraint that the phoretic speed of a JM is proportional to $1/d$ [58], which sheds light on the development of a faster JM. Obviously, the dynamics of the microbubble play an important role in the JM's propulsion, although it is still under debate [24,59].

We will first show the characteristics of the JM's motion propelled by the microbubble. Figure 4a shows the image sequences of a JM with a diameter of 33.2 μm in a microbubble growth-collapse period. Just one bubble is generated throughout the period, and this is connected to the Pt surface of the JM by point-contact. The maximum bubble diameter is about 1.4 times larger than the JM. The complete cycle of one bubble growth–collapse period lasts about 83.0 ms. There is no visible microbubble at the Pt side of the JM until $t > 9.0$ ms. At $t \approx 9.5$ ms, a bubble appears at the Pt side and grows gradually. At $t = 45.0$ ms, the diameter of the bubble becomes as large as that of the JM. The bubble's diameter reaches its maximum value at $t = 81.6$ ms. The bubble suddenly collapses at about $t = 81.8$ ms. The collapse only takes about 10 μs according to our observation using a high-speed camera. At $t = 83.0$ ms, the JM's propulsion stops due to the viscous effect, and the cycle is over. It is interesting to note that the H_2O_2 concentration does not influence the bubble growth significantly. The dominant factor of bubble growth is believed to be surface tension rather than chemical reaction rate. An early model suggested that the mechanism of bubble propulsion is the impulse produced by bubble disengagement [59], which was extended by Li et al. to bubble-propelled micro-motors with different shapes [60]. The model could explain the propulsion mechanism of microtubular motors studied by Mei and Solovev et al. [30–33,61]. However, it is not suitable for explaining the observation in our experiment that the bubble maintains direct contact with the JM directly.

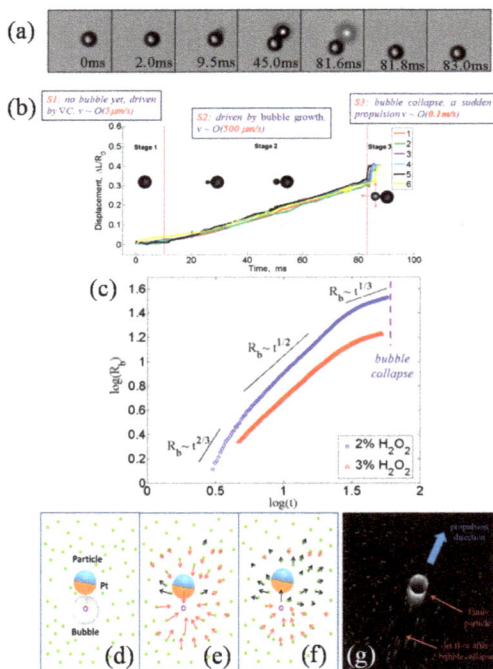

Figure 4. The JM's motion propelled by a microbubble. (**a**) Image series of a JM's motion in one bubble period which lasts about 83 ms; (**b**) Three-stage behavior of microbubble propulsion; (**c**) Different scaling laws during microbubble growth; (**d–f**) Variation of the flow field after microbubble collapse; (**g**) The microjet and wake vortices behind the JM shown by tracers. R_b: microbubble radius.

Figure 4b shows the displacement of the same JM in six consecutive periods. Three stages in each period can be observed: (1) In the first stage (S1), no visible bubble is generated, and the JM's typical speed is less than 10 μm/s. In S1, the motion of the JM is mainly affected by the concentration gradient, resulting in a low diffusiophoretic speed; (2) In the second stage (S2), the JM's displacement increases almost linearly accompanying the bubble growth. The motion of the JM is propelled by the bubble growth force F_{bubble}, and the typical speed is about 500 μm/s; (3) In the third stage (S3), the bubble collapses and the JM is pushed forward instantaneously. The speed of the JM is up to about 0.1 m/s in S3, resulting in a high Re number ($Re \sim 10$) in micro flow. However, we must emphasize that there is a back-pull phenomenon between S2 and S3. Before the strong forward motion, the JM sometimes withdraws after the bubble collapses. Manjare et al. [35] proposed that the drag from the bubble caused this back-pull, which is known as quasi-oscillatory motion. Below we will provide an explanation based on fluid mechanics.

We further analyze the bubble growth process. Figure 4c shows the growth of the microbubble radius R_b with time t in 2% and 3% H_2O_2 solutions, respectively. One can see that there are different scaling laws during bubble-growth: at short times $R_b \sim t^{2/3}$, at intermediate times $R_b \sim t^{1/2}$, and at the end near bubble collapse $R_b \sim t^{1/3}$. The bubble-growth process is generally described by the Rayleigh–Plesset (R–P) equation [35,62]:

$$P_b - P_\infty = \frac{2\sigma}{R_b} + \frac{4\mu}{R_b}\dot{R}_b + \rho(R_b\ddot{R}_b + \frac{3}{2}\dot{R}_b^2), \tag{2}$$

where P_b and P_∞ are the bubble internal pressure and fluid pressure very far from the bubble, respectively. It is assumed that the bubble radius increases exponentially with time, $R_b \sim t^m$. Based on the ideal gas law $PV = nR_gT$ (the bubble volume is $V = 4\pi R_b^3/3$, the oxygen molar number is $n = 2\pi R^2kt$, and k is a constant reaction rate), and neglecting high-order terms, the following three scaling laws are derived:

1. At short times, the microbubble radius R_b is very small and the bubble pressure is dominated by the viscous term, $P_b \sim 4\mu\dot{R}_b/R_b$. Thus $R_b \sim t^{2/3}$ is derived.
2. At intermediate times, the bubble radius R_b is about 10 μm and the bubble pressure is dominated by the surface tension, $P_b \sim 2\sigma/R_b$. Then, $R_b \sim t^{1/2}$ is obtained.
3. At the end, the bubble radius R_b is quite large and the bubble pressure is close to the adjacent fluid pressure, $P_b \sim P_\infty$. Then, $R_b \sim t^{1/3}$ is derived.

Finally, bubble collapse occurs when the bubble reaches its maximum size. A balance $P_b \sim P_\infty$ is reached, however, it will not be sustained because the oxygen supply becomes insufficient for the bubble with maximum size. The bubble suddenly shrinks due to $P_b < P_\infty$, and the high order term in Equation (2) becomes negative. The theory of bubble cavitation is usually employed to describe this process, as mentioned in [35,62,63].

The above three scaling laws are in good agreement with experimental results, indicating there are three physical mechanisms dominating each stage of bubble growth. A previous study [36] used only a single power law, similar to $R_b \sim t^{1/2}$, to describe the bubble growth. Obviously, some physical mechanisms are missing in that description. In addition, although the R–P equation is derived based on the assumption of infinite unbounded fluids and uniform mass transfer through the bubble interface, we find that the R–P equation could still be used to approximately describe bubble dynamics near a JM.

Another key issue needing clarification is the origin of fast, instantaneous propulsion after bubble collapse. We used polystyrene (PS) micro-tracers suspended in the fluid to visualize the flow field. Figure 4d–f shows a series of schematic diagrams based on the real experimental video (see Supplementary Materials, Section 3) to illustrate the motions of the JM and the tracers. Figure 4d is the situation just before bubble collapse; Figure 4e,f is two successive frames after bubble collapse. The point "o" is the bubble center and the arrows represent the velocity vectors of the JM and the

PS tracers. In the last moment before bubble collapse (Figure 4d), the bubble reaches its maximum size and maintains direct contact with the Pt side of the JM. In the first frame after bubble collapse (Figure 4e), the JM and the PS tracers around the bubble all move toward "o". This can be described by the Stokes sink in fluid mechanics, around which the fluid velocity measured by the tracers is approximately $U(r) \sim 1/r^2$ (r is the distance to the sink center "o"). This is why the JM withdraws at the beginning of S3, as mentioned above. In the second frame after bubble collapse (Figure 4f), the JM and the tracers in front of the microbubble are pushed forward and away from point "o", while the tracers behind the bubble move toward "o". This is due to the fact that the pressure near point "o" is much lower than that of the adjacent fluids (as described by the R–P equation). However, the existence of the JM whose mobility is much lower than that of the fluid molecules hinders the flow from the JM's side. As a result, a jet appears flowing through point "o" and pointing to the JM. In fluid dynamics, this is known as the cavitation-induced-jet arising from the asymmetry of the medium around the bubble [63,64], as shown by Figure 4g. Considering the fact that the bubble and the JM usually do not locate in the same horizontal plane due to the difference of their density, the jet flow is not strictly in the horizontal plane. It is the horizontal component of the jet propelling the JM forward, as observed in Figure 4g. The instantaneous speed of the fluid could reach 1 m/s, about one order of magnitude larger than the maximum speed of the JM. At the same time, a pair of vortices appears behind the JM.

The above results show that the main source of the fast bubble propulsion is the horizontal component of the microjet produced by the bubble collapse rather than the impulse when the bubble leaves the surface of the JM. The impulse model proposes that the JM is propelled by the impulse of bubbles detaching from the JM's surface. The experimental results shown in Figure 4 do not favor impulse mode, as the bubble never detaches from the JM. The microjet can focus the energy on propelling the JM, and therefore it can significantly improve the propulsion efficiency. This bubble cavitation-induced microjet provides a novel and important physical mechanism for propelling objects in micro-scale [35,65–67].

In addition, when the local concentration of micromotors is high, a large bubble emerges whose growth will significantly influence the collective motion of a group of JMs nearby [68]. The presence of the large bubble and the non-uniform temperature distribution along the liquid–gas interface will introduce a Marangoni flow near the bubble. This flow is stronger than the diffusiophoresis of the JM, and could be comparable to the bubble propulsion in some cases. The study of the collective motion of JMs is lacking, which could be applied in establishing self-assemble structures [69]. This new phenomenon provides an approach for manipulating collective motion of JMs based on bubbles.

4. Theoretical Description and Numerical Approach

Unlike the extensive experimental studies, theoretical or numerical approaches are not frequently seen in the literature. Nonetheless, they are helpful for their applications. The surface catalytic reaction, mass transfer process, Brownian motion, and even bubble dynamics are involved in the self-propelled motions mentioned above. The dimensionless numbers of self-propulsion reflect the basic physical characteristics. For the mass-transfer process, the Peclet number $Pe = V_p d_p / D_{O_2} \sim 10$ is obtained for propulsion of larger JM bubbles, while $Pe \sim 0.01$ for smaller JM's self-diffusiophoresis (V_p is the self-propelled velocity, d_p is the diameter of the JM, and D_{O_2} is the diffusion coefficient of dissolved O_2). This means that self-diffusiophoresis of smaller JMs is dominated by diffusion, while for the larger JMs, neither convection nor diffusion can be neglected. The Reynolds number Re is between 10^{-6} and 10^{-3}, so self-propulsion of the JM is a low-Re flow. In this section, we will first introduce the theoretical description coupling the velocity and concentration fields, based on continuum equations. The chemical reaction flux and slip velocity on a JM's surface are critical boundary conditions. As an alternative way, the Langevin equation can be used to describe the motion of the JM. The key issue in Langevin simulation is establishing the expression of the driving force due to the concentration gradient. Finally, for microbubble propulsion, the volume of fluid (VOF) method should be used to solve a JM's motion and bubble dynamics, where the bubble is described by the phase-transition equation (see Supplementary Materials, Section 4).

4.1. Theoretical Description Based on Continuum Mechanics

The simplest theoretical description taking both velocity and concentration distribution into account was given by Anderson [8]. The key dynamic process was considered to occur in a very thin layer close to the liquid–solid surface, where an effective slip on the surface was proposed. The concentration equilibrium is reached as described by the Boltzmann distribution: $C = C_s \exp(-\Phi/k_B T)$, where C_s is the concentration at the surface and Φ is the potential energy. The momentum equations normal to the surface and along the surface are, respectively [8]:

$$\frac{\partial p}{\partial y} + C\frac{d\Phi}{dy} = 0 \tag{3}$$

$$\mu\frac{\partial^2 u_x}{\partial y^2} - \frac{\partial p}{\partial x} = 0 \tag{4}$$

The typical velocity solved from the equations gives $u_x \sim k_B T \nabla C/\mu$. Correspondingly, the phoretic speed of the JM is approximately $u \sim \nabla C$, which can be also written as $u \sim \nabla F$ and is extended to other phoretic motions driven by a field F [6,8]. A detailed analysis about the scaling law of phoretic speed was given by Ebbens et al. [58]; this considered the JM size and the catalytic reaction activity.

For a complete numerical simulation, the low Re Stokes equation and the convection–diffusion equation should be coupled:

$$\nabla \cdot U = 0 \tag{5}$$

$$\mu\nabla^2 U - \nabla p = 0 \tag{6}$$

$$U\nabla C - D\nabla^2 C = R_{dec} \tag{7}$$

The source term R_{dec} in (7) is due to the decomposition of H_2O_2.

When the JM is far from the solid–liquid or gas–liquid interface, the simulation can be simplified to a 2-D symmetric problem. Relative coordinates are usually used: the JM is fixed and the fluid with a far-field velocity V_p flows past the JM. The flow boundary condition on the surface of the JM is the slip boundary condition, $U_{slip} = C_{slip}\nabla C_t$ [6]. The flux boundary condition on the Pt surface is determined by the consumption flux of H_2O_2 that follows $f_{H_2O_2} = -k_r C_{H_2O_2}$, where the reaction rate k_r is about 2.5×10^{-3} m/s [40]. Thus, the production flux of O_2 is given by $f_{dec} = 0.5k_r C_{H_2O_2}$. On the SiO_2 side, a zero-flux condition is applied. To get the solution of the above equations, the viscous drag should be balanced by the diffusiophoretic force due to the concentration gradient.

The wall effect is crucial when the JM moves close to the substrate wall. The near-wall effect will slow the motion of the JM and greatly increase the drag on the one hand, and will break the symmetry of both the flow and concentration fields around the JM on the other hand. The asymmetric fields will produce a torque that changes the JM's orientation. Thus, to fully demonstrate the JM's motion near a wall, 3-D numerical simulation is required, in which the slip coefficient σ, the equilibrium position δ, and the orientation angle ϕ are three parameters to be solved for in the simulation. Accordingly, three equations about force or torque equilibrium are needed. For the force equilibrium, in the horizontal direction the diffusiophoretic force is balanced by the Stokes drag force: $F_{Stokes-X} + F_{DFP-X} = 0$, while in the vertical direction gravity should be involved: $F_{Stokes-Y} + F_{DFP-Y} = G$. Due to the symmetry of geometric and flow conditions, the torques in the X-axis and Z-axis are naturally balanced. For the torque equilibrium, as the flow field and concentration field are not symmetric, the torque generated by Stokes drag T_{Stokes} should be balanced by the torque generated by diffusiophoretic force T_{DFP}: $T_{Stokes} + T_{DFP} = 0$. It is worthwhile to note that all the forces and torques listed above are dependent on the three parameters σ, δ, and ϕ. Solutions based on this approach have been obtained [40], which could help us to understand the wall effects mentioned in previous literature.

4.2. Kinetic Motion Solved by Langevin Equation

When a JM's size is small enough, the random thermal disturbance $R(t)$ will become important enough to result in Brownian motion. In 1908, Langevin introduced $R(t)$ as a random force into the Newton equation and established the Langevin equation of single particle motion. The effect of external fields can also be directly introduced into the Langevin equation to solve the motion of a JM under multiple physical fields. In the present case of self-diffusiophoresis due to concentration gradient, the Langevin equation is established as:

$$m\frac{d^2x}{dt^2} = F_{Stokes} + F_{Brownian} + F_{DFP},$$

(8)

where $F_{Brownian}$ is a random force whose time-average value is zero, $F_{Stokes} = 6\pi R_p \mu V_p$ is the Stokes drag, R_p is the radius of the JM, V_p is the velocity of the JM, and F_{DFP} is the self-diffusiophoretic force. $F_{Brownian}$ is produced by the impact of adjacent fluid molecules, and could be considered an equivalent force acting on the center of the JM:

$$F_{Brownian} = \xi_1 \sqrt{\frac{12\pi k_B T_0 \mu R_p}{\Delta t}},$$

(9)

where ξ_1 is a random number, k_B is Boltzmann's constant, T_0 is the thermodynamic temperature, and Δt is the observation time interval. Obviously, $F_{Brownian}$ decreases with an increasing time interval, and approaches zero for a long time interval. F_{DFP} cannot be directly defined. However, it has been reported that F_{DFP} is proportional to the drift velocity (V_{DFP}) of a JM under pure self-diffusiophoresis [5,35]. Thus, $F_{DFP} = 6\pi\mu R_p V_{DFP} \vec{\theta}$ is proposed, where $\vec{\theta}$ is the rotational angle of the JM pointed from the Pt side to the SiO$_2$ side.

At any moment, $\vec{\theta}$ is still unknown due to the Brownian random torque. To solve the Langevin equation, the change of $\vec{\theta}$ due to Brownian torque must be introduced. The rotational angular velocity Ω is used to describe the change of the rotational angle:

$$\Omega = \frac{d\vec{\theta}}{dt} = \frac{\Gamma_\theta}{f_r} = \frac{F_{Brownian} R_p}{8\pi\mu R_p^3} = \xi_1 \sqrt{\frac{3k_B T_0}{16\pi\mu R_p^3 \Delta t}}$$

(10)

where Γ_θ is the torque and f_r is the rotational friction coefficient of the viscous fluid. Combining the Langevin equation of translational motion and the rotational equation of Ω, the kinetic motion of a small JM can be fully solved. This approach can reveal the competition between Brownian motion and pure diffusiophoretic motion, which is hardly achieved by the methods based on continuum mechanics except for fluctuating hydrodynamics.

5. Discussion

5.1. Propulsion Efficiency

Improving the efficiency of the JM is crucial in application. Unfortunately, the energy transfer efficiencies η of existing JMs as reported in the literature are unsatisfactory [70,71]. The efficiency η is usually too low to be applied (10^{-10}) [70], mainly because the kinetic energy of the JM will rapidly dissipate in a low *Re* viscous flow. We estimate the energy transfer efficiency of fast microbubble propulsion based on the energy variation of the system before and after bubble collapse. At the onset of bubble collapse, the bubble surface energy, defined by $E_b = 4\pi R_b^2 \sigma \sim 10^{-9}$ J, reaches its maximum value. Based on the observed speed V_p of the JM after bubble collapse, the JM's kinetic energy is estimated as $E_k = 2\pi\rho R_{JC}^3 V_p^2/3 \sim 10^{-11}$ J. Therefore, the energy efficiency $\eta \sim E_k/E_b$ is about 1%. This efficiency is 7–8 orders of magnitude larger than that estimated by Wang et al. for phoretic JMs [70]. This high efficiency occurs because the bubble cavitation-induced microjet focuses the energy on propelling the JM forward rather than being transferred to the adjacent liquid.

5.2. Microfluidic Applications

Many studies have been devoted to realizing more and better functions of the JM [9–12]. A challenge of manipulating a JM in microfluidic application is to control its direction. For small JMs, since Brownian motion significantly influences both translational and rotational motion of the JM, new techniques should be developed to overcome the randomness. In this section, a microshuttle technique based on dielectrophoresis is introduced. The Pt–SiO_2 type JM under high-frequency AC voltage always exhibits a positive dielectrophoresis (pDEP) response [72], and the JM is attracted to the electrode border, keeping its Pt side outward and its SiO_2 side toward the electrode. In the experiment, JMs with a diameter of 2 μm are immersed in 5% H_2O_2 solution [73,74]. Then, a pulsed AC field (voltage 2 V, frequency 10 MHz, switching frequency 0.2 Hz, see Figure 5a) is applied to the stripe-like ITO electrodes (width 20 μm). The dielectrophoretic force will suppress the self-diffusiophoretic motion of the JM and trap the JM at the border of the electrode when the AC voltage is on (Figure 5b). The pDEP response of the JM disappears immediately after the AC voltage is turned off, and the self-propulsion becomes dominated. Because the orientation of the JM is toward the electrode, the JM will move along its orientation as long as it has not rotated too much in a short time. Since one "on–off" period is 5 s in the experiment, the self-propelled time of the JM is 2.5 s. Based on the typical self-diffusiophoretic speed (5–10 μm/s) mentioned in Section 2, the JM can move to the position near the opposite border of the electrode. When the AC voltage is on again, the JM will soon be trapped at the opposite border by the pDEP force and change its orientation (Figure 5b). When the AC voltage is off again, the JM will start to return to its original position. Thus, the JM's motion in a microfluidic system can be manipulated like a microshuttle under an AC voltage with a suitable frequency [73,74]. By controlling the frequency of the AC voltage, the JM can also move back and forth between different electrodes.

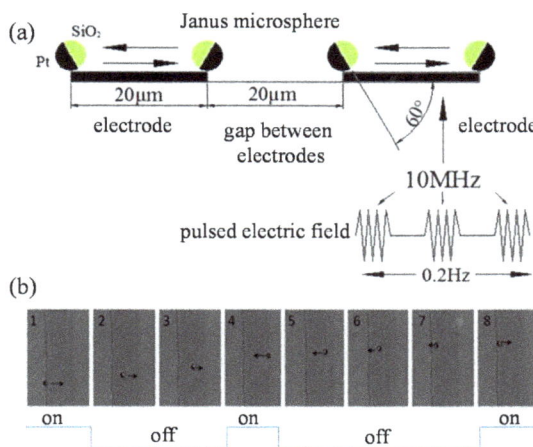

Figure 5. The back-and-forth motion of a "microshuttle" JM between different electrodes based on positive dielectrophoresis (pDEP) [73,74]. (**a**) The schematic diagram; and (**b**) the image series of a JM's reciprocating microshuttle motion in one electrode. Figure 5 is reproduced with permission from Chen et al. [73]; Published by AIP, 2014.

Micromachines **2017**, *8*, 123

6. Conclusions

In this paper, we introduce the mechanisms of two distinct self-propulsion methods of the spherical Pt–SiO$_2$ JMs: self-diffusiophoresis and microbubble propulsion. Major experimental results and theoretical/numerical approaches are reviewed. The former motion occurs for JMs whose diameter is roughly smaller than 5–10 µm, and it results from the concentration gradient established by decomposition of H$_2$O$_2$ on the Pt surface. The MSD of the self-diffusiophoresis exhibits a three-stage behavior: simple Brownian motion at short times, ballistic motion at intermediate times, and Brownian-like motion at long times. The self-diffusiophoresis can be seen as a superposition of a translational ballistic motion and a Brownian rotation. The typical time scale is the rotational time of the microsphere τ_R. The statistical characteristics of self-diffusiophoresis consist of the double-peaked structure of DPD and negative kurtosis. The latter motion occurs when the JM's diameter is larger than 10 µm. In this case, the JM moves with a speed about 500 µm/s during microbubble growth, and surprisingly, its speed can reach 0.1 m/s during bubble collapse. The strong instantaneous propulsion originates from the bubble cavitation-induced microjet. This microjet can significantly improve the energy efficiency, and thus provides a novel power source for future micro-motors. It is found that the R–P equation can be approximately used to describe the bubble dynamics.

To develop novel JMs in the future, we still need to pay attention to several issues. First, the materials and functions of micro/nano-motors are rapidly developed. There are many new types of micro/nano-motors rather than the spherical Janus ones. Second, the commonly used JMs or other micromotors work in chemical solutions that could have toxicity. It is important to develop micro/nano-motors that can simply work in water or nontoxic solutions [11,75,76]. Third, the bio-compatibility of the materials of micro/nano-motors is also important for bio-medicine applications [77–82]. Fourth, developing multi-function and high-efficiency micro/nano-motors requires the involvement of other physical fields. JMs and other types of micro/nano-motors driven or controlled by ultrasound, electromagnetic field, and light pressure are now under extensive investigation [83–88]. A thorough understanding of the physics is very helpful for designing powerful micro/nano-motors.

Supplementary Materials: The following are available online at www.mdpi.com/2072-666X/8/4/123/s1, Figure S1: Three successive images during bubble collapse.

Acknowledgments: The authors are grateful for financial support from the National Natural Science Foundation of China (Grants Nos. 11572335, 11272322 and 11602187), the CAS Strategic Priority Research Program (Grant No. XDB22040403), and Major Science and Technology Program for Water Pollution Control and Treatment (Grant No. 2014ZX07305-002-01). We thank LetPub for its linguistic assistance during the preparation of this manuscript.

Author Contributions: X.Z. and H.C. conceived and designed the experiments; J.Z. and X.Z. performed the experiments; J.Z. and H.C. performed theoretical and numerical study; Z.S. analyzed the data; X.Z. and J.Z. wrote the paper.

Conflicts of Interest: The authors declare no conflict of interest.

References

1. Ismagilov, R.; Schwartz, A.; Bowden, N.; Whitesides, G. Autonomous movement and self-assembly. *Angew. Chem. Int. Ed.* **2002**, *41*, 652–654. [CrossRef]
2. Paxton, W.F.; Kistler, K.C.; Olmeda, C.C.; Sen, A.; St. Angelo, S.K.; Cao, Y.; Mallouk, T.E.; Lammert, P.E.; Crespi, V.H. Catalytic nanomotors: Autonomous movement of striped nanorods. *J. Am. Chem. Soc.* **2004**, *126*, 13424. [CrossRef] [PubMed]
3. Wang, W.; Duan, W.; Ahmed, S.; Mallouk, T.E.; Sen, A. Small power: Autonomous nano- and micromotors propelled by self-generated gradients. *Nano Today* **2013**, *8*, 531–554. [CrossRef]
4. Laocharoensuk, R.; Burdick, J.; Wang, J. Carbon-nanotube-induced acceleration of catalytic nanomotors. *ACS Nano* **2008**, *2*, 1069–1075. [CrossRef] [PubMed]

5. Howse, J.; Jones, R.; Ryan, A.; Gough, T.; Vafabakhsh, R.; Golestanian, R. Self-Motile colloidal particles: From directed propulsion to random walk. *Phys. Rev. Lett.* **2007**, *99*, 048102. [CrossRef] [PubMed]

6. Golestanian, R.; Liverpool, T.B.; Ajdari, A. Designing phoretic micro- and nano-swimmers. *New J. Phys.* **2007**, *9*, 126. [CrossRef]

7. Singh, V.V.; Soto, F.; Kaufmann, K.; Wang, J. Micromotor-based energy generation. *Angew. Chem. Int. Ed.* **2015**, *54*, 6896–6899. [CrossRef] [PubMed]

8. Anderson, J.L. Colloidal transport by interfacial forces. *Annu. Rev. Fluid Mech.* **1989**, *21*, 61–99. [CrossRef]

9. Duan, W.; Wang, W.; Das, S.; Yadav, V.; Mallouk, T.; Sen, A. Synthetic nano- and micromachines in analytical chemistry: Sensing, migration, capture, delivery and separation. *Annu. Rev. Anal. Chem.* **2015**, *8*, 311–333. [CrossRef] [PubMed]

10. Wong, F.; Dey, K.K.; Sen, A. Synthetic micro/nanomotors and pumps: Fabrication and applications. *Annu. Rev. Mater. Res.* **2016**, *46*, 407–432. [CrossRef]

11. Wu, Z.G.; Li, J.X.; de Avila, B.E.F.; Li, T.L.; Gao, W.W.; He, Q.; Zhang, L.F.; Wang, J. Water-powered cell-mimicking Janus micromotor. *Adv. Funct. Mater.* **2015**, *25*, 7497–7501. [CrossRef]

12. Wang, J.; Gao, W. Nano/Microscale motors: Biomedical opportunities and challenges. *ACS Nano* **2012**, *6*, 5745–5751. [CrossRef] [PubMed]

13. Baraban, L.; Makarov, D.; Streubel, R.; Monch, I.; Grimm, D.; Sanchez, S.; Schmidt, O. Catalytic Janus Motors on Microfluidic Chip: Deterministic Motion for Targeted Cargo Delivery. *ACS Nano* **2012**, *6*, 3383–3389. [CrossRef] [PubMed]

14. Baraban, L.; Tasinkevych, M.; Popescu, M.N.; Sanchez, S.; Dietrich, S.; Schmidt, O.G. Transport of cargo by catalytic Janus micro-motors. *Soft Matter* **2012**, *8*, 48–52. [CrossRef]

15. Parmar, J.; Jang, S.; Soler, L.; Kim, D.P.; Sanchez, S. Nano-photocatalysts in microfluidics, energy conversion and environmental applications. *Lab Chip* **2015**, *15*, 2352–2356. [CrossRef] [PubMed]

16. Jurado-Sanchez, B.; Sattayasamisathit, S.; Gao, W.; Santos, L.; Fedorak, Y.; Singh, V.; Orozco, J.; Galarnyk, M.; Wang, J. Self-Propelled activated carbon Janus micromotors for efficient water purification. *Small* **2015**, *11*, 499–506. [CrossRef] [PubMed]

17. Soler, L.; Sanchez, S. Catalytic nanomotors for environmental monitoring and water remediation. *Nanoscale* **2014**, *6*, 7175–7182. [CrossRef] [PubMed]

18. Sanchez, S.; Solovev, A.; Harazim, S.M.; Schmidt, O. Microbots swimming in the flowing streams of microfluidic channels. *J. Am. Chem. Soc.* **2011**, *133*, 701–703. [CrossRef] [PubMed]

19. De Gennes, P.G. Soft matter (Nobel lecture). *Angew. Chem. Int. Ed.* **1992**, *31*, 842–845. [CrossRef]

20. Purcell, E.M. Life at low Reynolds-number. *Am. J. Phys.* **1977**, *45*, 3–11. [CrossRef]

21. Lauga, E. Life around the scallop theorem. *Soft Matter* **2010**, *7*, 3060–3065. [CrossRef]

22. Moran, J.L.; Posner, J.D. Phoretic Self-Propulsion. *Annu. Rev. Fluid Mech.* **2017**, *49*, 511–540. [CrossRef]

23. Sakes, A.; van der Wiel, M.; Henselmans, P.W.; van Leeuwen, J.L.; Dodou, D.; Breedveld, P. Shooting Mechanisms in Nature: A Systematic Review. *PLoS ONE* **2016**, *11*, e0158277. [CrossRef] [PubMed]

24. Wang, L.; Chen, L.; Zhang, J.; Duan, J.M.; Silber-Li, Z.-H.; Zheng, X.; Cui, H.H. Propulsion and dynamic hovering of microswimmer by a cyclical bubble-cavitation catapult. *Small* **2017**. submitted.

25. Palacci, J.; Cottin-Bizonne, C.; Ybert, C.; Bocquet, L. Sedimentation and effective temperature of active colloidal suspensions. *Phys. Rev. Lett.* **2010**, *105*, 088304. [CrossRef] [PubMed]

26. Jiang, H.R.; Yoshinaga, N.; Sano, M. Active motion of a Janus particle by self-thermophoresis in a defocused laser beam. *Phys. Rev. Lett.* **2010**, *105*, 268302. [CrossRef] [PubMed]

27. Wu, Y.J.; Si, T.Y.; Shao, J.X.; Wu, Z.G.; He, Q. Near-infrared light-driven Janus capsule motors: Fabrication, propulsion, and simulation. *Nano Res.* **2016**, *9*, 3747–3756. [CrossRef]

28. Liu, R.; Sen, A. Autonomous nanomotor based on copper-platinum segmented nanobattery. *J. Am. Chem. Soc.* **2011**, *133*, 20064–20067. [CrossRef] [PubMed]

29. De Buyl, P.; Kapral, R. Phoretic self-propulsion: A mesoscopic description of reaction dynamics that powers motion. *Nanoscale* **2013**, *5*, 1337–1344. [CrossRef] [PubMed]

30. Mei, Y.F.; Huang, G.S.; Solovev, A.; Urena, E.B.; Monch, I.; Ding, F.; Reindl, T.; Fu, R.; Chu, P.; Schimidt, O. Versatile approach for integrative and functionalized tubes by strain engineering of nanomembranes on polymers. *Adv. Mater.* **2008**, *20*, 4085–4090. [CrossRef]

31. Solovev, A.; Mei, Y.F.; Urena, E.B.; Huang, G.S.; Schmidt, O. Catalytic microtubular jet engines self-propelled by accumulated gas bubbles. *Small* **2009**, *5*, 1688–1692. [CrossRef] [PubMed]

32. Solovev, A.; Xi, W.; Gracias, D.; Harazim, S.; Deneke, C.; Sanchez, S.; Schmidt, O.G. Self-propelled Nanotools. *ACS Nano* **2012**, *6*, 1751–1756. [CrossRef] [PubMed]

33. Sanchez, S.; Ananth, A.; Fomin, V.; Viehrig, M.; Schmidt, O.G. Superfast motion of catalytic microjet engines at physiological temperature. *J. Am. Chem. Soc.* **2011**, *133*, 14860–14863. [CrossRef] [PubMed]

34. Soler, L.; Magdanz, V.; Fomin, V.M.; Sanchez, S.; Schmidt, O.G. Self-propelled micromotors for cleaning polluted water. *ACS Nano* **2013**, *7*, 9611–9620. [CrossRef] [PubMed]

35. Manjare, M.; Yang, B.; Zhao, Y.P. Bubble driven quasioscillatory translational motion of catalytic micromotors. *Phys. Rev. Lett.* **2012**, *109*, 128305. [CrossRef] [PubMed]

36. Wang, S.J.; Wu, N. Selecting the Swimming mechanisms of colloidal particles: Bubble propulsion versus self-diffusiophoresis. *Langmuir* **2014**, *30*, 3477–3486. [CrossRef] [PubMed]

37. Zheng, X.; ten Hagen, B.; Kaiser, A.; Wu, M.; Cui, H.H.; Silber-Li, Z.H.; Lowen, H. Non-Gaussian statistics for the motion of self-propelled Janus particles: Experiment versus theory. *Phys. Rev. E* **2013**, *88*, 032304. [CrossRef] [PubMed]

38. Ebbens, S.; Howse, J.R. Direct observation of the direction of motion for spherical catalytic swimmers. *Langmuir* **2011**, *27*, 12293–12296. [CrossRef] [PubMed]

39. Ke, H.; Ye, S.R.; Carroll, R.L.; Showalter, K. Motion analysis of self-propelled Pt-Silica particles in hydrogen peroxide solutions. *J. Phys. Chem. A* **2010**, *114*, 5462–5467. [CrossRef] [PubMed]

40. Cui, H.H.; Tan, X.J.; Zhang, H.Y.; Chen, Li. Experiment and numerical study on the characteristics of self-propellant Janus microspheres near the wall. *Acta Phys. Sin.* **2015**, *64*, 134705.

41. Elgeti, J.; Winkler, R.G.; Gompper, G. Physics of microswimmers-single particle motion and collective behavior: Review. *Rep. Prog. Phys.* **2015**, *78*, 056601. [CrossRef] [PubMed]

42. Menzel, A.M. Tuned, driven, and active soft matter. *Phys. Rep. Rev. Sect. Phys. Lett.* **2015**, *554*, 1–45. [CrossRef]

43. Zheng, X.; Wu, M.L.; Kong, F.D.; Cui, H.H.; Silber-Li, Z.H. Visualization and measurement of the self-propelled and rotational motion of the Janus microparticles. *J. Vis.* **2015**, *18*, 425–435. [CrossRef]

44. Zheng, X.; Kong, G.P.; Silber-Li, Z.H. The influence of nano-particle tracers on the slip length measurements by microPTV. *Acta Mech. Sin.* **2013**, *29*, 411–419. [CrossRef]

45. Ten Hagen, B.; van Teeffelen, S.; Lowen, H. Non-Gaussian behaviour of a self-propelled particle on a substrate. *J. Phys. Condens. Matter* **2011**, *12*, 725–738. [CrossRef]

46. Probstein, R.F. *Physicochemical Hydrodynamics*; Butterworth Publishers: Stoneham, MA, USA, 1989.

47. Maggi, C.; Marconi, U.M.B.; Gnan, N.; Di Leonardo, R. Multidimensional stationary probability distribution for interacting active particles. *Sci. Rep.* **2015**, *5*, 10742. [CrossRef] [PubMed]

48. Guerin, T.; Levernier, N.; Benichou, O.; Voituriez, R. Mean first-passage times of non-Markovian random walkers in confinement. *Nature* **2016**, *534*, 356–359. [CrossRef] [PubMed]

49. Kummel, F.; ten Hagen, B.; Wittkowski, R.; Buttinoni, I.; Eichhorn, R.; Volpe, G.; Lowen, H.; Bechinger, C. Circular motion of asymmetric self-propelling particles. *Phys. Rev. Lett.* **2013**, *110*, 198302. [CrossRef] [PubMed]

50. Das, S.; Garg, A.; Campbell, A.; Howse, J.; Sen, A.; Velogol, D.; Golestanian, R.; Ebbens, S. Boundaries can steer active Janus spheres. *Nat. Commun.* **2015**, *6*, 8999. [CrossRef] [PubMed]

51. Simmchen, J.; Katuri, J.; Uspal, W.; Popescu, M.; Tasinkevych, M.; Sanchez, S. Topographical pathways guide chemical microswimmers. *Nat. Commun.* **2016**, *7*, 10598. [CrossRef] [PubMed]

52. Gomez-Solano, J.R.; Blokhuis, A.; Bechinger, C. Dynamics of self-propelled Janus particles in viscoelastic fluids. *Phys. Rev. Lett.* **2016**, *116*, 138301. [CrossRef] [PubMed]

53. Yang, F.C.; Qian, S.Z.; Zhao, Y.P.; Qiao, R. Self-diffusiophoresis of Janus catalytic micromotors in confined geometries. *Langmuir* **2016**, *32*, 5580–5592. [CrossRef] [PubMed]

54. Zottl, A.; Stark, H. Nonlinear dynacmis of a microswimmer in Poiseuille flow. *Phys. Rev. Lett.* **2012**, *108*, 218104. [CrossRef] [PubMed]

55. Ten Hagen, B.; Wittkowski, R.; Lowen, H. Brownian dynamics of a self-propelled particle in shear flow. *Phys. Rev. E* **2011**, *84*, 031105. [CrossRef] [PubMed]

56. Zhu, L.; Lauga, E.; Brandt, L. Low-Reynolds-number swimming in a capillary tube. *J. Fluid Mech.* **2013**, *726*, 285–311. [CrossRef]

57. Fletcher, N.H. Size effect in heterogeneous nucleation. *J. Chem. Phys.* **1958**, *29*, 572–576. [CrossRef]

58. Ebbens, S.; Tu, M.H.; Howse, J.R.; Golestanian, R. Size dependence of the propulsion velocity for catalytic Janus-sphere swimmers. *Phys. Rev. E* **2012**, *85*, 020401. [CrossRef] [PubMed]

59. Manjare, M.; Yang, B.; Zhao, Y.P. Bubble-propelled microjets: Model and experiment. *J. Phys. Chem. C* **2013**, *117*, 4657–4665. [CrossRef]
60. Li, L.Q.; Wang, J.Y.; Li, T.L.; Song, W.P.; Zhang, G.Y. A unified model of drag force for bubble-propelled catalytic micro/nano-motors with different geometries in low Reynolds number flows. *J. Appl. Phys.* **2015**, *117*, 104308. [CrossRef]
61. Mei, Y.F.; Solovev, A.; Sanchez, S.; Schmidt, O.G. Rolled-up Nanotech on Polymers: From Basic Perception to Self-Propelled Catalytic Microengines. *Chem. Soc. Rev.* **2011**, *40*, 2109. [CrossRef] [PubMed]
62. Brennen, C. *Cavitation and Bubble Dynamics*; Oxford University Press: Oxford, UK, 1995.
63. Zwaan, E.; Le Gac, S.; Tsuji, K.; Ohl, C.-D. Controlled Cavitation in Microfluidic systems. *Phys. Rev. Lett.* **2007**, *98*, 254501. [CrossRef] [PubMed]
64. Blake, J.; Leppinen, D.; Wang, Q. Cavitation and bubble dynamics: The Kelvin impulse and its applications. *Interface Focus* **2016**, *5*, 20150017. [CrossRef] [PubMed]
65. Poulain, S.; Guenoun, G.; Gart, S.; Crowe, W.; Jung, S. Particle motion induced by bubble cavitation. *Phys. Rev. Lett.* **2015**, *114*, 214501. [CrossRef] [PubMed]
66. Versluis, M.; Schmits, B.; von der Veydt, A.; Lohse, D. How snapping shrimp snap: Through cavitating bubbles. *Science* **2000**, *289*, 2114–2117. [CrossRef] [PubMed]
67. Yang, F.; Manjare, M.; Zhao, Y.; Qiao, R. On the peculiar bubble formation, growth, and collapse behaviors in catalytic micro-motor systems. *Microfluid. Nanofluid.* **2017**, *21*, 1–11. [CrossRef]
68. Manjare, M.; Yang, F.; Qiao, R.; Zhao, Y.P. Marangoni flow induced collective motion of catalytic micromotros. *J. Phys. Chem. C* **2015**, *119*, 28361–28367. [CrossRef]
69. Gao, W.; Pei, A.; Feng, X.M.; Hennessy, C.; Wang, J. Organized self-assembly of Janus micromotors with hydrophobic hemispheres. *J. Am. Chem. Soc.* **2013**, *135*, 998–1001. [CrossRef] [PubMed]
70. Wang, W.; Chiang, T.Y.; Velegol, D.; Mallouk, T.E. Understanding the efficiency of autonomous nano-and microscale motors. *J. Am. Chem. Soc.* **2013**, *135*, 10557–10565. [CrossRef] [PubMed]
71. Kreissl, P.; Holm, C.; de Graaf, J. The efficiency of self-phoretic propulsion mechanisms with surface reaction heterogeneity. *J. Chem. Phys.* **2016**, *144*, 204902. [CrossRef] [PubMed]
72. Zhang, L.; Zhu, Y. Dielectrophoresis of Janus particles under high frequency ac-electric fields. *Appl. Phys. Lett.* **2010**, *96*, 141902. [CrossRef]
73. Chen, J.L.; Zhang, H.Y.; Zheng, X.; Cui, H.H. Janus particle microshuttle: 1D directional self-propulsion modulated by AC electrical field. *AIP Adv.* **2014**, *4*, 031325. [CrossRef]
74. Wu, M.L.; Zhang, H.Y.; Zheng, X.; Cui, H.H. Simulation of diffusiophoresis force and the confinement effect of Janus particles with the continuum method. *AIP Adv.* **2014**, *4*, 031326. [CrossRef]
75. Chen, C.; Karshalev, E.; Li, J.X.; Soto, F.; Castillo, R.; Campos, I.; Mou, F.Z.; Guan, J.G.; Wang, J. Transient Micromotors That Disappear When No Longer Needed. *ACS Nano* **2016**, *10*, 10389–10396. [CrossRef] [PubMed]
76. Dong, R.-F.; Li, J.X.; Rozen, I.; Ezhilan, B.; Xu, T.L.; Christianson, C.; Gao, W.; Saintillan, D.; Ren, B.Y.; Wang, J. Vapor-Driven Propulsion of Catalytic Micromotors. *Sci. Rep.* **2015**, *5*, 13226. [CrossRef] [PubMed]
77. Wu, Z.G.; Lin, X.K.; Si, T.Y.; He, Q. Recent progress on bioinspired self-propelled micro/nanomotors via controlled molecular self-assembly. *Small* **2016**, *12*, 3080–3093. [CrossRef] [PubMed]
78. Wang, H.; Gu, X.Y.; Wang, C.Y. Self-propelling hydrogel/emulsion-hydrogel soft motors for water purification. *ACS Appl. Mater. Interfaces* **2016**, *8*, 9413–9422. [CrossRef] [PubMed]
79. Sanchez, S.; Soler, L.; Katuri, J. Chemically powered micro- and nanomotors. *Angew. Chem. Int. Ed.* **2015**, *54*, 1414–1444. [CrossRef] [PubMed]
80. Ge, Y.; Liu, M.; Liu, L.; Sun, Y.Y.; Zhang, H.; Dong, B. Dual-Fuel-Driven Bactericidal Micromotor. *Nano-Micro Lett.* **2016**, *8*, 157–164. [CrossRef]
81. Ma, X.; Wang, X.; Hahn, K. Motion Control of Urea-Powered Biocompatible Hollow Microcapsules. *ACS Nano* **2016**, *10*, 3597–3605. [CrossRef] [PubMed]
82. Guix, M.; Meyer, A.K.; Koch, B.; Schmidt, O.G. Carbonate-based Janus micromotors moving in ultra-light acidic environment generated by HeLa cells in situ. *Sci. Rep.* **2016**, *6*, 21701. [CrossRef] [PubMed]
83. Tao, Y.G.; Kapral, R. Dynamics of chemically powered nanodimer motors subject to an external force. *J. Chem. Phys.* **2009**, *131*, 024113. [CrossRef] [PubMed]
84. Dong, R.F.; Zhang, Q.L.; Gao, W.; Pei, A.; Ren, B.Y. Highly efficient light-driven TiO2-Au Janus micromotors. *ACS Nano* **2016**, *10*, 839–844. [CrossRef] [PubMed]

85. Li, J.X.; Rozen, I.; Wang, J. Rocket science at the nanoscale. *ACS Nano* **2016**, *10*, 5619–5634. [CrossRef] [PubMed]

86. Simoncelli, S.; Summer, J.; Nedev, S.; Kuhler, P.; Feldmann, J. Combined optical and chemical control of a microsized photofueled Janus particle. *Small* **2016**, *12*, 2854–2858. [CrossRef] [PubMed]

87. Wang, W.; Castro, L.A.; Hoyos, M.; Mallouk, T. Autonomous motion of metallic microrods propelled by ultrasound. *ACS Nano* **2012**, *6*, 6122–6132. [CrossRef] [PubMed]

88. Torrori, S.; Zhang, L.; Qiu, F.; Krawczyk, K.K.; Franco-Obregon, A.; Nelson, B.J. Magnetic helical micromachines: Fabrication controlled swimming and cargo transport. *Adv. Mater.* **2012**, *24*, 811–816.

micromachines

MDPI

Review

Advances in Single Cell Impedance Cytometry for Biomedical Applications

Chayakorn Petchakup [1], King Ho Holden Li [1,*] and Han Wei Hou [2]

[1]   Mechanical and Aerospace Engineering, Nanyang Technological University, Singapore 639798, Singapore; chayakor001@e.ntu.edu.sg

[2]   Lee Kong Chian School of Medicine, Nanyang Technological University, Singapore 636921, Singapore; hwhou@ntu.edu.sg

*   Correspondence: holdenli@ntu.edu.sg; Tel.: +65-6790-6398

Academic Editors: Weihua Li, Hengdong Xi and Say Hwa Tan
Received: 31 January 2017; Accepted: 7 March 2017; Published: 12 March 2017

Abstract: Microfluidics impedance cytometry is an emerging research tool for high throughput analysis of dielectric properties of cells and internal cellular components. This label-free method can be used in different biological assays including particle sizing and enumeration, cell phenotyping and disease diagnostics. Herein, we review recent developments in single cell impedance cytometer platforms, their biomedical and clinical applications, and discuss the future directions and challenges in this field.

Keywords: impedance spectroscopy; flow cytometry; single cell analysis

---

1. Introduction

Single cell analysis has gained considerable attention for biological assays and system biology in the past decade due to the increasing importance of studying cell populations that are highly heterogeneous, as well as sampling of complex biofluids such as blood. Bulk measurement can only reflect the average value, leading to a loss of valuable information about rare sub-populations (diseased cells or abnormal cells) present in the sample [1,2]. In bio-related studies, coulter counter and fluorescence-activated cell sorting (FACS) are widely used as high throughput cell counting and classification methods. Coulter counter detects a change in direct current (DC) or low frequency alternating current (AC) impedance signal caused by particle or cell passing through the detection region which can provide information about particle size [3,4]. FACS is a more powerful technique and requires fluorescent cell labelling to enable counting, characterization and sorting based on optical characteristics. However, several drawbacks including laborious sample preparation, and expensive equipment and reagents (antibodies) significantly limit its use for point-of-care (POC) testing.

Characterization of electrical impedance at different frequencies provides important information about biological cells and the suspending medium, making it attractive tool for single cell analysis [5–7]. In the kHz range, there is α-dispersion which originates from the displacement of counterions around the charged shell of the cell. This is hard to measure as it is hindered by the effects of electrode polarization which leads to high impedance below 1 MHz [8]. β-dispersion occurring in MHz range arises from the interfacial polarization of cellular components such as the cell membrane. The polarization of protein and other organic molecules also contributes to different parts of β-dispersion [9]. In addition to cells and cellular components information, dynamic studies of red blood cell (RBC) aggregation [10] and blood coagulation [11] can also be obtained from this range. γ-dispersion above 1 GHz is due to dipolar relaxation of water bound molecules in the cytoplasm and the external medium [12].

Advances in microfluidics and biomedical microelectromechanical systems (BioMEMS) are important in the development of impedance cytometers as it enables manipulation of small fluid volumes, and highly-sensitive measurement with the close proximity of microfabricated electrodes to single cells in microchannels [13,14]. Besides reducing reagent and sample volume consumption and expensive equipment, a key advantage over the traditional methods is that sample preparation and impedance detection modules can be readily integrated in a single device, commonly known as lab-on-a-chip (LOC), for POC testing.

In this review, we summarise recent developments of impedance based microfluidic cytometry for biomedical research. We will first provide a brief overview of the working principles and designs of impedance based microfluidic cytometry. For more detailed information, the readers can refer to other excellent reviews by Morgan et al. [15,16] and Chen et al. [17]. Next, we will focus on the diagnostics and phenotyping capabilities of impedance cytometry in different biomedical applications, and present reported work according to cell types including blood cells (leukocytes and RBCs), cancer cells, microbes and stem cells. Lastly, we will highlight future directions and challenges in this field based on our findings.

2. Design Principles

2.1. Theory

Electrical impedance is defined as the ratio between excitation voltage and response current of cell in suspension,

$$Z^* = \frac{V^*}{I^*} \tag{1}$$

where Z^* is electrical impedance (Ohm), V^* is excitation voltage (Volt) and I^* is current response (Amp) and superscript * denotes complex number.

Various approaches have been utilized to simulate or to interpret impedance results of particle in suspension such as finite element method (FEM), Maxwell's mixture theory (MMT) and equivalent circuit model (ECM). A comparison of three abovementioned approaches has been reported elsewhere [16,18–22].

Maxwell's mixture theory describes the dielectric property of particle in suspension [23]. The complex permittivity of the mixture can be determined by three key parameters, which are the complex permittivity of the cell, complex permittivity of its suspending medium and volume fraction, which is the ratio of volume of the cell to the volume of the channel.

$$\varepsilon^*_{mix} = \varepsilon^*_{med} \frac{2(1 - \varphi) + (1 + 2\varphi)\frac{\varepsilon^*_{cell}}{\varepsilon^*_{med}}}{(2 + \varphi) + (1 - \varphi)\frac{\varepsilon^*_{cell}}{\varepsilon^*_{med}}} \tag{2}$$

where $\varepsilon^* = \varepsilon - j\frac{\sigma}{\omega}$ denotes complex permittivity, $j^2 = -1$, ω is the angular frequency, and φ is the volume fraction which is the ratio of volume of cell to volume of medium inside the detection channel. The subscript "mix", "cell" and "med" represent mixture, cell and medium, respectively.

The complex permittivity of cell can also be determined in the same manner as complex permittivity of the mixture above. To determine complex permittivity of cell properly, several models have been proposed to describe cell or particle based on its internal complexity such as particle, single shelled model (cell consisting of cytoplasm and cell membrane) and double shelled model (cell consisting of cytoplasm, cell membrane and nucleus or vacuole) [24–26].

The impedance of the mixture containing cell (Z_{mix}) can be calculated from the following equation.

$$Z_{mix} = \frac{1}{jw\varepsilon^*_{mix}lG} \tag{3}$$

where G is a cell constant to correct the effect of non-uniform electric field and fringing field. l is width of the channel. The calculation of cell constant of different electrode configurations has been shown in previous literature [21,22,27–29].

2.2. Electrode Designs

In this section, we describe three common configurations used in impedance based microfluidic cytometry: coplanar electrodes, parallel electrodes, and constriction channel. Figure 1A–C shows microfluidics impedance cytometers using coplanar electrodes design (Figure 1A), parallel electrodes design (Figure 1B), and constriction channel design (Figure 1C). Each design is based on a similar detection principle, with excitation electrode and sensing electrodes embedded inside microfluidic channel to establish electrical measurement. As a cell flows between a pair of electrodes (A and C), the electric field between these two electrodes is disrupted, resulting in a current change that can be measured at point A. The current measured at this position corresponds to the impedance of cell and its suspending medium. To determine impedance of medium, the current at point B is also acquired simultaneously and the impedance of cell can be acquired from the difference between current at point A and at point C (Figure 1A). Typically, the setup consists of pre-amplifier, lock-in amplifier and data acquisition system (Figure 1D). The excitation signal is supplied to excitation electrode by function generator or lock-in amplifier, and sensing electrodes are connected to bridge circuit or trans-impedance amplifiers to measure current response of system. The amplifiers' output is connected to lock-in amplifier to demodulate current signal at excitation frequency. The data are sent to data acquisition system for post processing.

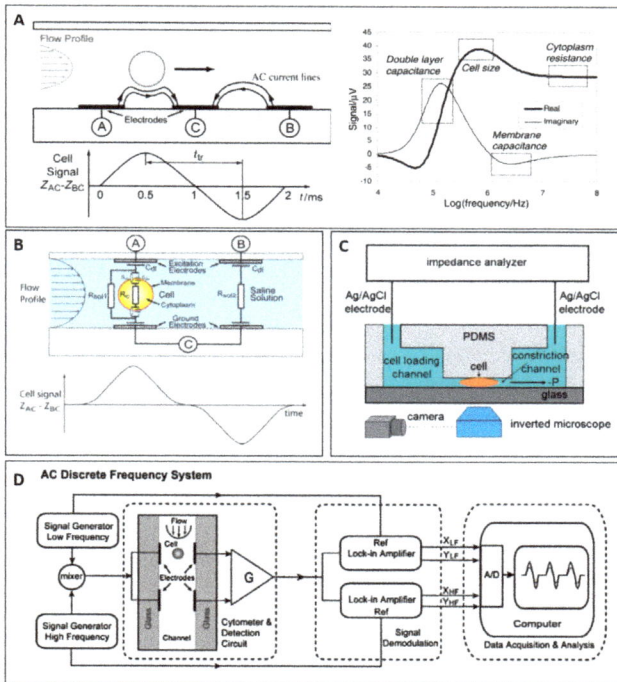

Figure 1. (**A**) (**Left**) Illustration of coplanar electrodes design and impedance signal response when a cell flows through the detection region. (**Right**) Impedance response at different frequencies carries different information regard the cell. Reproduced with permission from [30], copyright 2001, Royal Society of Chemistry. (**B**) Illustration of parallel electrodes design. Reproduced with permission from [31], copyright 2005, John Wiley and Sons. (**C**) Illustration of constriction channel design. Reproduced with permission from [32], copyright 2011, Royal Society of Chemistry. (**D**) Diagram shows the measurement setup. Reproduced with permission from [33], copyright 2008, Springer;

2.2.1. Coplanar Electrode Design

Coplanar electrode configuration was first proposed by Gawad et al. [30]. In this design, coplanar metal electrodes were integrated in microchannel and non-homogeneous electric field was generated. The authors carried out the simulation of cell impedance from equivalent circuit model and their simulation result showed that different parts of impedance spectra contain different information of cell components as presented in Figure 1A. Furthermore, they showed that opacity or a ratio of high frequency impedance magnitude to low frequency impedance magnitude does not depend on position of cell in the channel. Since then, opacity is widely used as characterization parameter in impedance cytometry.

The fabrication process of coplanar electrodes design starts with the patterning of electrode layer on glass substrate. The channel layer is then fabricated or bonded on glass substrate, creating a microfluidic device with integrated electrodes. The whole process can be easily fabricated since only a single alignment is needed to guide electrodes to the desirable position inside the channel.

Due to non-uniform electric field created by coplanar electrode configuration, the impedance measurement relied on the vertical position of cell in the detection region considerably. To reduce the effect of vertical position of cell on impedance, another coplanar electrode configuration called liquid electrodes was used [34–36]. In this case, the electrodes were placed at bottom of lateral channels perpendicular to main channel, as shown in Figure 2A. As a result, homogeneous electrical field over the channel height was generated, mitigating the height dependence. However, this design had several drawbacks. Firstly, the sensitivity is poorer than traditional coplanar electrode design due to the increase in detection volume as the distance between the electrode pair needs to be placed far enough in order to generate homogenous electrical field across main channel. Secondly, the effect of lateral position rises due to fringing effect at edges of electrodes. In this work, they used dielectrophoresis (DEP) force generated by liquid electrodes to focus the cell at the centre of channel. Shaker et al. used the combination of conventional and liquid coplanar electrodes configuration shown in Figure 2B [37]. Longitudinal measurement and transverse measurement provide different characteristics that can be exploited to detect a shape of particle.

Besides down-scaling channel dimension to achieve higher sensitivity of coplanar electrodes, several techniques have been demonstrated to focus the cell to channel centre or control the vertical position of the cell in the channel using DEP [31,34,37–40] and hydrodynamic focusing [41–46].

For hydrodynamic focusing, there are two approaches: 1D hydrodynamic focusing [41–43] and 2D hydrodynamic focusing [44–46]. For these devices, the channels are typically larger and low conductivity sheath fluids such as deionized water [40,41,45] or oil [40,42] are used to achieve particle focusing. A three-inlet device was designed, in which two additional focalisation lateral inlets were used to provide focusing stream for pinching sample stream. Not only does it allow single particles to flow through detection region, but the detection volume between electrodes can also be adjustable to fit a wide range of particles or cells sizes. Moreover, the use of a large channel greatly reduces the chance of channel blockage. Besides utilization of 1D hydrodynamic focusing, 2D hydrodynamic focusing was adapted in several devices [44–46], aiming to control vertical position of cell leading to better sensitivity than 1D hydrodynamic focusing.

To align particle to the centre of channel using DEP, several designs such as top and bottom taper shaped electrodes [39], coplanar deflecting electrodes [31] and liquid electrodes [34,37] have been reported. Noteworthy, the utilization of DEP focusing only provides the control of particle position in the channel, whereas hydrodynamic focusing can control both the particle position and detection volume.

Besides abovementioned particle focusing techniques, the effect of particle position on impedance can be also be corrected by multiple electrodes design and signal processing as demonstrated recently by De Ninno et al. [47]. Additional electrodes affect the measured signal profile which conveys information on particle position as well. Hence, the measured characteristic signal can be exploited to correct the signal of off-centre particle leading to accurate particle sizing. However, introduction of additional electrodes covers a larger region in the channel and a higher particle coincidence (two or more particles measured simultaneously) can occur if particle concentration is too high.

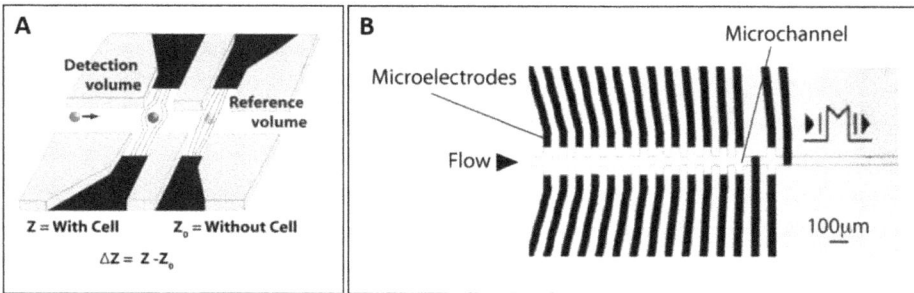

Figure 2. (**A**) Illustration of liquid electrode design. Reproduced with permission from [36], copyright 2010, Royal Society of Chemistry; and (**B**) illustration of combination approach of conventional coplanar electrode design and liquid electrode design. Reproduced with permission from [37], copyright 2014, Royal Society of Chemistry;

2.2.2. Parallel Electrode Design

Parallel electrodes configuration was first developed by Gawad et al. [48]. In this configuration, electrodes were placed at top and bottom or at sidewall of microchannel. Similar to previous design, two pairs of electrodes were used to measure impedance of cell passing between electrodes and impedance of the medium as depicted in Figure 1B. With parallel electrodes, electric field distribution was less divergent, leading to better sensitivity as compared to coplanar electrode design. However, this design also suffers from the measured signal dependence on cell position inside detection volume [49].

The fabrication process is more complex as compared to coplanar electrodes design. For top and bottom electrodes configuration, two alignment steps are needed for aligning channel to electrode pattern and aligning two chips with electrodes together. Precise alignment is needed to make the measurement reproducible. For the sidewall electrodes configuration, sidewall electrodes were fabricated by electroplating followed by SU-8 channel fabrication on top of the electrodes. This can be done with single alignment. However, there is always a vertical gap between the sidewall electrodes and microchannel, resulting in an inhomogeneity of the electric field [50]. This can possibly lead to a slightly poorer performance as compared to top-bottom configuration.

Several studies utilized hydrodynamic focusing [42] and DEP to control particle position inside microchannel. Additionally, multi-electrodes design used to correct the signal of off-centre particles was proposed by Spencer et al. [51]. In this work, they used five pairs of parallel electrodes. However, unlike multiple coplanar electrodes design approach, this requires four transimpedance amplifiers to get parameters for correction, resulting in a more complicated setup.

2.2.3. Constriction Channel Design

Lack of direct contact between electrodes and cell can introduce current leakage issues, in which current tends to pass though high conductivity fluid surrounding the cell. In order to solve this problem, the constriction channel design was introduced by Chen et al. [32]. In this design, the detection region was designed to be smaller than cell (Channel: 6 μm × 6 μm) as shown in Figure 3A. The Ag/AgCl electrodes placed at inlet and outlet were used instead of thin film electrodes on substrate. When cell was aspirated into the channel, the electric field across two electrodes was altered leading to the change in impedance which can be implied as impedance of cell. Moreover, mechanical properties such as cell deformability can be measured when the cell squeezes through the smaller channel, enabling multi-parametric mechanical and electrical cell characterization. Based on equivalent circuit model shown in Figure 3B, multi-frequencies measurement (at 1 kHz and 100 kHz) are used to determine size-independent electrical properties such as specific membrane capacitance ($C_{specific\ membrane}$) and cytoplasm conductivity ($\sigma_{cytoplasm}$). The drawbacks of this design are that it is prone to clogging and has lower throughput as compared to other designs.

The fabrication process of constriction channel design is simple, as only single alignment is needed in channel fabrication process.

The comparison of each design is shown in Table 1.

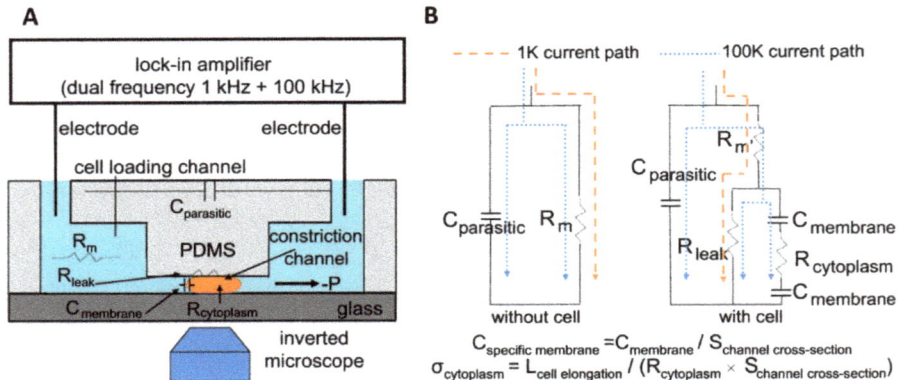

Figure 3. (**A**) Illustration of constriction channel design with dual frequencies measurement; and (**B**) equivalent circuit model and current paths at low and high frequency when constriction channel with cell and without cell. Reproduced with permission from [52], copyright 2013, Elsevier;

Table 1. Comparisons between different impedance cytometers design.

Design		Advantages	Disadvantages
Coplanar electrodes design	Coplanar electrodes	Simple fabrication High throughput	Vertical position dependence Low sensitivity
	Liquid electrodes	Simple fabrication High throughput	Lateral position dependence Low sensitivity [1]
Parallel electrodes design	Top-Bottom configuration	High sensitivity High throughput	Vertical position dependence Complex fabrication
	Sidewalls configuration	High sensitivity High throughput	Lateral position dependence Complex fabrication
Constriction channel		Simple fabrication High sensitivity Size independent electrical parameters Mechanical property characterization	Prone to clogging Low throughput

[1] Sensitivity comparison of each design was presented in several studies [30,34].

3. Biomedical Applications

In this section, we highlight the biomedical applications of impedance cytometers based on cell types (Table 2).

Table 2. List of developed impedance cytometry applications based on cell type. WBCs: white blood cells; RBCs: red blood cells; TRAP: thrombin receptor activating peptide; PDMS: polydimethylsiloxane.

Category	Author	Summary	Characterization parameters	Ref.
	Watkins et al. (2009)	CD4 T-cells counting using impedance cytometer with 2D hydrodynamic focusing	Impedance at 50 kHz	[44]
	Holmes et al. (2009)	Discrimination of leukocyte subpopulation	Impedance signal at 1.7 MHz and 503 kHz	[18]
	Holmes et al. (2010)	Discrimination of T-cells and CD4 T-cells conjugated with CD4 beads	Impedance signal at 10 MHz and 503 kHz	[53]
	Watkins et al. (2011)	Differential count of CD4 T-cells by reverse-flow technique with integrated cell capture chamber	Impedance signal at 1.1 MHz	[54]
	van Berkel et al. (2011)	Differential counting of blood cells using impedance cytometer with off-chip sample treatment	Impedance signal at 1.7 MHz and 500 kHz	[55]
White blood cells	Han et al. (2011)	Evaluation of RBC lysis chip for differential counting of WBCs	Impedance signal at 1.7 MHz and 444 kHz	[56]
	Watkins et al. (2013)	Integrated sample treatment and cell capture chamber for differential CD-4 and CD-8 T-cell counting	Impedance at 303 kHz and 1.7 MHz	[57]
	Spencer et al. (2014)	Integrated optical detection coupling with compound air lens for differential counting of blood cells	Impedance signal at 2 MHz and 500 kHz Fluorescence signal	[58]
	Frankowski et al. (2015)	Evaluation of parallel electrode designs for leukocyte sub-population counting	Impedance signal at 4 MHz and 500 kHz and fluorescence signal	[50]
	Hassan et al. (2016)	Integrated sample treatment and cell capture chamber for differential CD-4 and CD-8 T-cell counting	Impedance at 303 kHz and 1.7 MHz	[59]
	Gawad et al. (2001)	Discrimination of beads, RBCs and ghost RBCs	Impedance signal at 15 MHz and 1.72 MHz	[30]
	Cheung et al. (2005)	Discrimination of normal RBCs, ghost RBCs and glutaraldehyde fixed RBCs	Impedance signal at 10 MHz and 602 kHz	[31]
	Sun et al. (2007)	Utilization of maximum length sequences (MLS) for characterization of RBCs	Impedance spectrum up to 500 kHz	[60]
Red blood cells	Kuttel et al. (2007)	Discrimination of RBCs, B. Bovis infected RBCs and ghost RBCs	Impedance signal at 8.7 MHz	[61]
	Zheng et al. (2012)	Characterization of adult RBCs and neonatal RBCs	Transit time, amplitude ratio and phase shift at 100 kHz	[62]
	Du et al. (2013)	Discrimination of different states of Plasmodium falciparum infected RBCs	Combination of phase shift and magnitude shift in impedance at 2 MHz	[63]
	Evander et al. (2013)	Discrimination of RBCs, platelets and TRAP treated platelets.	Impedance signal at 284 kHz, 1.20 MHz, 2.39 MHz and 4.02 MHz	[40]

Table 2. *Cont.*

Category	Author	Summary	Characterization parameters	Ref.
	Benazzi et al. (2007)	Discrimination of three different types of phytoplankton	Impedance signal at 6 MHz and 327 kHz and fluorescence signal	[64]
	Rodriguez-Trujillo et al. (2007)	Discrimination of 20 μm polystyrene beads and 5 μm yeast cells	Impedance at 120 kHz	[45]
	Bernabini et al. (2010)	Discrimination of 1 μm, 2 μm beads and *Escherichia coli*	Impedance signal at 503 kHz	[42]
	Mernier et al. (2012)	Characterization of yeast cells before and after electrical lysis or thermal lysis	Impedance signal at 10 kHz	[65]
Microbes	Shaker et al. (2014)	Single cell morphology discrimination of budding yeasts' division stage by using liquid electrodes	Impedance signal at 427 kHz and 533 kHz	[37]
	Haandbaek et al. (2014)	Discrimination of wild-type yeasts and mutant yeasts	Impedance signal at 100 MHz and 0.5 MHz	[66]
	Haandbaek et al. (2014)	Discrimination of bacteria (*E. coli* and *B. subtilis*) and 2 μm beads by using resonator circuit	Signal polarity at 87.2 MHz and 89.2 MHz	[67]
	Haandbaek et al. (2016)	Discrimination of single and budding yeast cells	Impedance signal at 20 MHz, 9 MHz, 1 MHz and 0.55 MHz	[68]
	Schade-Kampmann et al. (2008)	Discrimination of mouse fibroblast, adipocytes, human monocytes, dendritic cells and macrophages	Impedance signal at 2 MHz, 5 MHz and 14 MHz	[69]
	Nikolic-Jaric et al. (2009)	Discrimination of different sized polystyrene beads, yeast cells and Chinese hamster ovary (CHO) cells	Capacitance at 1.5 GHz	[70]
	Gou et al. (2011)	Discrimination of liver tumour cells at normal, apoptotic and necrotic status and leukaemia cells	Resistance and capacitance change at 100 kHz	[71]
Tumors	Chen et al. (2011)	Characterization of osteoblasts and osteocytes/EMT6 cells and EMT6/AR1.0 cells	Cell elongation, transit time and impedance amplitude ratio at 100 kHz	[32]
	Mernier et al. (2012)	Utilization of lateral liquid electrodes for focusing and for discrimination of live and dead CHO cells	Impedance signal at 500 kHz and 15 MHz	[34]
	Zheng et al. (2012)	Characterization of 3249 AML-2 cells and 3398 HL-60 cells	Membrane capacitance and cytoplasm conductivity	[72]
	Zhao et al. (2013)	Characterization of kidney tumour cells (786-O) and vascular smooth muscle cells (T2)	Membrane capacitance and cytoplasm conductivity	[52]
	Zhao et al. (2013)	Characterization of lung cancer cell lines (CRL-5803 cells and CCL-185)	Membrane capacitance and cytoplasm conductivity	[73]

Table 2. *Cont.*

Category	Author	Summary	Characterization parameters	Ref.
	Zhao et al. (2014)	Characterization of various kinds of tumour such as 95C and 95D/549 and A549 CypA-KD	Membrane capacitance and cytoplasm conductivity	[74]
	Kirkegaard et al. (2014)	Characterization of HeLa cells and Paclitaxel treated HeLa cells	Impedance signal at 1.57 MHz and 82 kHz	[75]
	Spencer et al. (2014)	Detection of MCF7 cells spiked in whole blood	Impedance signal at 4 MHz and 500 kHz	[76]
	Bürgel et al. (2015)	Inversion of flow direction enabling impedance measurement of HeLa and CHO-K1 cells before and after electroporation	Impedance spectra from 20 kHz to 20 MHz with 8 steps	[77]
Tumors	Zhao et al. (2015)	Characterization of mouse tumour cell lines (A549 and H1299)	Membrane capacitance and cytoplasm conductivity	[78]
	Huang et al. (2015)	Characterization of normal PC-3 cells and PC-3 cells with membrane staining and/or fixation (4 conditions)	Membrane capacitance and cytoplasm conductivity	[79]
	Yuan et al. (2016)	Utilization of Ag PDMS as sidewall electrodes for discrimination of AML-2 and HL-60	Impedance signal ranging from 11 kHz–6 MHz	[80]
	Babahosseini et al. (2016)	Study the effect of different drug delivery approaches on electrical properties of MDA-MB-231	Impedance at 1 kHz, 10 kHz, 100 kHz and 1 MHz	[81]
	Xie et al. (2017)	Discrimination of apoptotic, necrotic and live HeLa cells	Conductance and susceptance at 1 MHz	[82]
	Song et al. (2013)	Characterization of mouse embryonic carcinoma cell (P19) differentiation	Impedance signal at 50 kHz, 250 kHz, 500 kHz and 1 MHz	[83]
Stem cells	Zhao et al. (2016)	Characterization of neural stem cell in differentiation	Membrane capacitance and cytoplasm conductivity	[84]
	Song et al. (2016)	Characterization of human mesenchymal stem cells and osteoblasts	Opacity at 500 kHz and relative angle at 3 MHz	[85]

3.1. Blood Cells

Previous works have used microfluidic impedance cytometry to study dielectric properties of various blood cells, including red blood cells [30,31,61,63,86] and white blood cells [18,38,44,50,53,55–59].

3.1.1. White Blood Cells

Holmes et al. proposed impedance labelling technique for counting of CD4+ T-cells [53]. Anti-CD4 antibody coated beads (1.8–2.4 μm) were mixed with lysed whole blood and bound to the monocyte and CD4 expressing (CD4+) T-cells. As a result, the population of CD4+ cells were larger due to the beads bounded on their surfaces, resulting in an increase in opacity (10 MHz and 0.5 MHz) and impedance signal at 0.5 MHz (Figure 4A). This method of using impedance labelling enables enumeration of sub-population. Spencer et al. also described a novel sheathless microfluidic cytometer with on-chip waveguide (Figure 4B (left)) that can measure four parameters: fluorescence signal, large angle side scatter and impedance at two different frequencies (0.5 MHz and 2 MHz) (Figure 4B (right)).

In another application for whole blood enumeration, van Berkel et al. developed an integrated microfluidic platform with sample pretreatment module for a three-part differential leukocyte counting together with red blood cells and platelets counting [55]. Figure 4C shows sample pretreatment design. Blood sample supplied to the sample pretreatment module was divided into two branches: (1) dilution process of subsequent RBC and platelets counting based on impedance signal at 0.5 MHz; (2) RBC lysis and quenching followed by white blood cells (WBCs) discrimination based on opacity (1.7 MHz and 0.5 MHz) and impedance signal at 0.5 MHz.

Recently, Hassan et al. reported a microfluidic impedance cytometer for simultaneous CD4+ and CD8+ T-cells counting [57,59,87]. In this design (Figure 4D), they included on-chip sample preparation and capture chamber specifically designed for capturing CD4+ or CD8+ T-cells. Cell population was electrically characterized before and after capture chamber, providing the number of cells captured in the capture chamber. The device was clinically validated in a cohort of healthy subjects and HIV+ patients, and they showed that the microfluidic measurements were strongly correlated to flow cytometry analysis.

3.1.2. Red Blood Cells and Platelets

Gawad et al. demonstrated the discrimination of normal red blood cells and ghost red blood cells (their cytoplasm replaced by hypotonic solution) based on impedance signal at 15 MHz, indicating the differences was due to cytoplasm conductivity between both RBCs populations [30].

Cheung et al. used parallel electrodes design to distinguish three kinds of RBCs, healthy, ghost and glutaraldehyde-fixed, at different concentrations [31]. In this study, they showed the identification of RBCs and glutaraldehyde-fixed RBCs based on impedance signal at 10 MHz and 602 kHz (Figure 5A). The impedance signal at 10 MHz of fixed RBCs was higher than that of normal RBCs, indicating a decrease in cytoplasm conductivity or increase in opacity.

Parasite invasion of RBCs can alter dielectric properties of RBCs in malaria [61,63,88]. Kuttel et al. used coplanar electrodes design impedance cytometer to detect *Babesia bovis* infected red blood cells [61]. Figure 5B shows the difference in impedance signal at 8.7 MHz of normal red blood cells and *Babesia bovis* infected red blood cells.

For platelet analysis, Evander et al. demonstrated the detection of red blood cells and platelets as shown in Figure 5C (top) [40]. Moreover, the group also successfully classified non-activated platelets from thrombin receptor activating peptide (TRAP) activated platelets based on discriminant analysis from impedance signal at four frequencies (284 kHz, 1.20 MHz, 2.39 MHz and 4.02 MHz), which will be useful to study thrombosis or platelet dysfunctions (Figure 5C (bottom)).

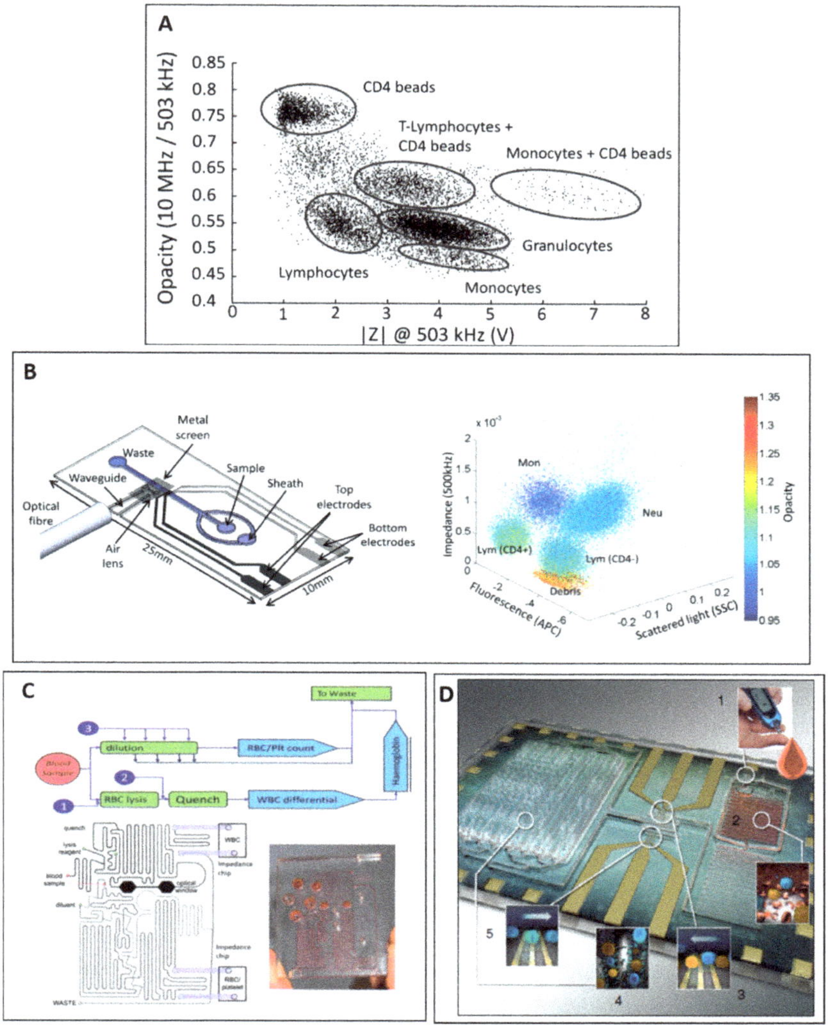

Figure 4. (A) Plot of opacity (10 MHz/503 kHz) versus impedance signal at 503 kHz for white blood cells population after addition of CD4 beads. Reproduced with permission from [53], copyright 2010, American Chemical Society. **(B)** **(Left)** Schematic of impedance cytometer with on-chip waveguide; and **(Right)** 3-D scatter plot of side scatter impedance and fluorescence for CD4 labelled white blood cells population. Colour represents opacity magnitude. Reproduced with permission from [58], copyright 2014, Royal Society of Chemistry. **(C)** Schematic of sample pretreatment module proposed by van Berkel et al. [55] for whole blood processing prior impedance detection. The design includes two pathways: (1) sample dilution followed by red blood cells and platelets counting; and (2) red blood cells lysis and quenching followed by white blood cells differential counting. Reproduced with permission from [55], copyright 2011, Royal Society of Chemistry. **(D)** Schematic of on-chip sample pretreatment with capture chamber for differential counting of CD4 or CD8: (1) inlets for loading of whole blood and reagent solutions (lysing buffer and quenching buffer); (2) lysing followed by quenching; (3) entrance counting; (4) CD4 or CD8 capture chamber; and (5) exit counter. Reproduced with permission from [59], copyright 2016, Nature Publishing Group;

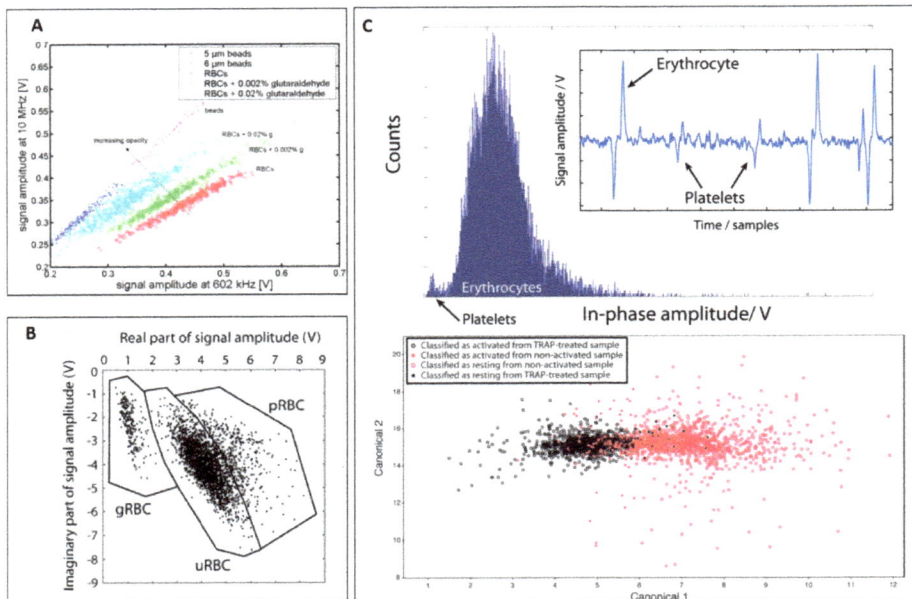

Figure 5. (**A**) Scatter plot of signal amplitude at 10 MHz versus 602 kHz for 5.6-μm beads, red blood cells and glutaraldehyde fixed red blood cells at different concentrations. Reproduced with permission from [31], copyright 2005, John Wiley and Sons. (**B**) Real part of signal amplitude versus imaginary part of signal magnitude of different kinds of red blood cells (ghost, normal, and parasite infected). Reproduced with permission from [61], copyright 2007, Elsevier. (**C**) (**Top**) Histogram shows the distribution of in-phase amplitude from platelets and red blood cells. (**Bottom**) Scatter plot of discriminant analysis of non-activated platelets and TRAP activated platelets. Reproduced with permission from [40], copyright 2013, Royal Society of Chemistry;

3.2. Cancer Cells

Several impedance cytometers were developed to characterize various types of tumours and cancers [32,52,72–74,76,78,79].

Previous studies reported distinct differences in dielectric properties of white blood cells and tumour cells, which generally have larger membrane capacitance and size [89,90]. Due to these differences, Spencer et al. demonstrated the use of parallel design microfluidic cytometer to distinguish breast tumour cells (MCF-7) from leukocytes when spiked in whole blood (Figure 6A) [76].

Zhao et al. characterized H1299 and A549 cells using a constriction channel design [78]. Specific membrane capacitance and cytoplasm conductivity of each population were acquired, enabling rapid discrimination of two tumour types (Figure 6B).

3.3. Microbes

Besides the discrimination of blood cells and tumours, impedance cytometry has also been utilized to characterize various kinds of samples such as yeast [37,66,68], bacteria [67] and plankton [64].

3.3.1. Yeasts

Haandbæk et al. demonstrated the use of high frequency impedance (>50 MHz) to characterize wild-type yeast from a mutant based on impedance at 0.5 MHz and 100 MHz to reflect size and vacuole property [66]. They found that the distribution of mutant shifted toward a higher opacity magnitude

(100 MHz to 0.5 MHz), indicating the difference in vacuole property or vacuole size. Interestingly, the difference in electrical volume profile at 0.5 MHz corresponded to the yeasts' sub-populations (large mother cells and small daughter cells). In a follow-up study, they further investigated different yeast phenotypes based on impedance at four different frequencies (0.55–9.08 MHz), particle velocity and fluorescence signal (Figure 7A) [68].

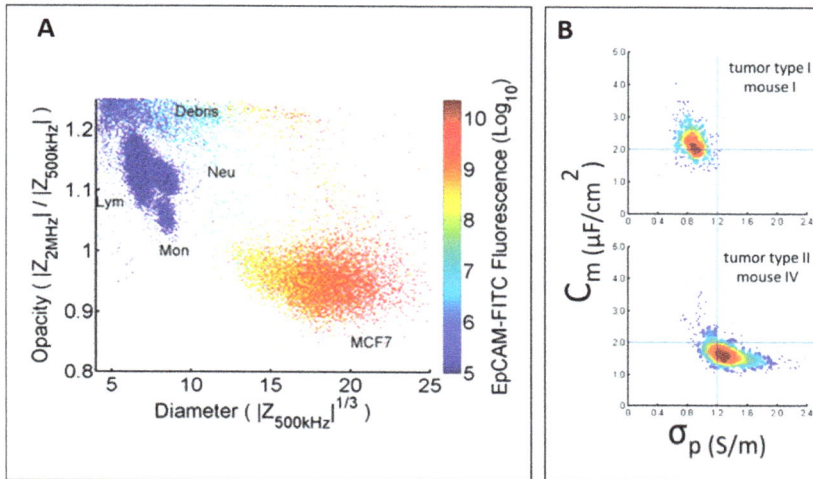

Figure 6. (**A**) Scatter plot of opacity (2 MHz/500 kHz) and diameter (impedance signal at 500 kHz) for leukocytes and MCF7 cells. Colour represents fluorescence signal. Reproduced with permission from [76], copyright 2014, AIP Publishing LLC. (**B**) Scatter plot of $C_{specific\ membrane}$ versus $\sigma_{cytoplasm}$ for A549 (mouse I) and H1299 (mouse IV). Reproduced with permission from [78], copyright 2016, Nature Publishing Group;

3.3.2. Bacteria

For bacteria detection, Haandbæk et al. reported a novel resonator enhanced impedance based cytometer for the detection of sub-micrometre beads and bacteria [67]. By adding a series resonator circuit at excitation part, the sensitivity at high frequency can be improved. Instead of using impedance magnitude, they used the phase shift (at 89.2 MHz) as the characterization parameters to distinguish bacteria and 2-μm beads. Interestingly, bacteria and 2-μm beads can be discriminated by using phase polarity at 87.2 MHz and 89.2 MHz due to difference in dielectric properties of their internal structure. However, the proposed technique cannot be used to distinguish different types of bacteria, as their dielectric properties of cytoplasm would be too similar (Figure 7B).

3.4. Stem Cells

Previous studies also showed the feasibility of using impedance cytometers to characterize stem cells differentiation [83–85].

Zhao et al. studied the differentiation of neural stem cells using constriction channel based impedance cytometer [84]. In this study, murine neural stem cells were cultured and sampled for several days. Figure 8A shows the distribution of specific cell membrane capacitance and cytoplasm conductivity. Initially, the population had wide distribution of cytoplasm conductivity which corresponded to nature of collected neurospheres. Over time, the distribution of specific cell membrane capacitance changed continuously which indicates active changes in cell membrane of the population. These data suggest the potential of using electrical measurements to monitor cell differentiation process.

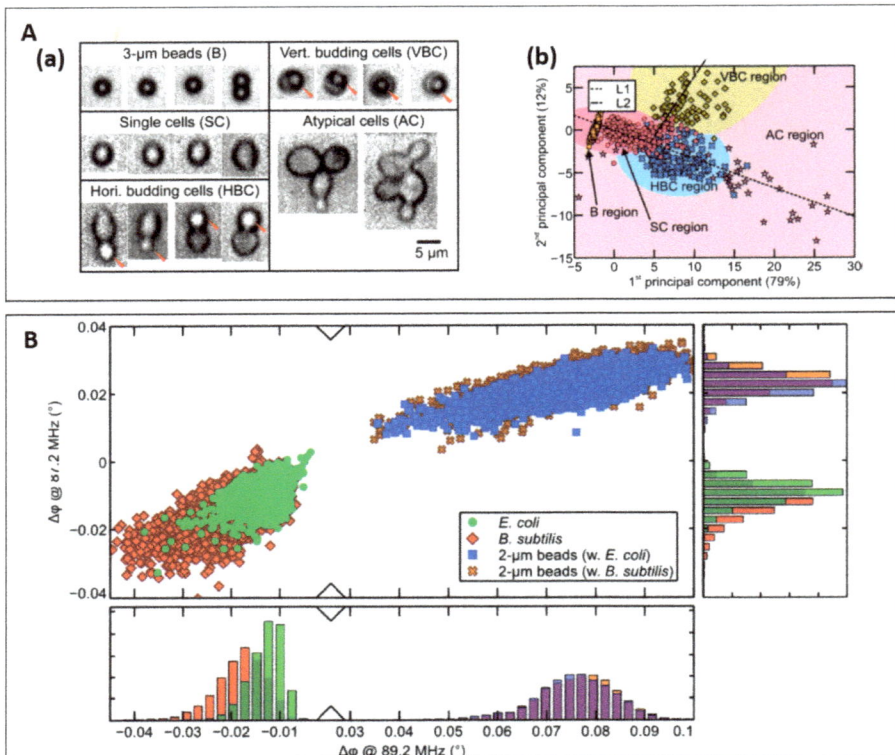

Figure 7. (**A**) (**a**) Different phenotypes of yeast cells used in Haandbæk's studies [68]. (**b**) Scatter plot of the first two principal components as result from principal components analysis of the impedance information (impedance at four frequencies and velocity). Reproduced with permission from [68], copyright 2016, American Chemical Society. (**B**) The result from another Haandbæk's studies. Scatter plot of phase shift at 87.2 MHz versus phase shift at 89.2 MHz for two experiments: (1) *E. coli* (green) spiked with 2-µm beads (blue); and (2) *B. subtilis* (red) spiked with 2-µm beads (yellow). Reproduced with permission from [67], copyright 2014, Royal Society of Chemistry;

Recently, Song et al. studied the differentiation states of mesenchymal stem cells [85]. In this study, human mesenchymal stem cells (hMSC) were induced to differentiate into osteoblasts and impedance was measured on Days 7 and 14 (Figure 8B). The classification model was trained by using relative angle at 3 MHz and opacity at 500 kHz of control hMSC and osteoblast population to determine osteoblast differentiation.

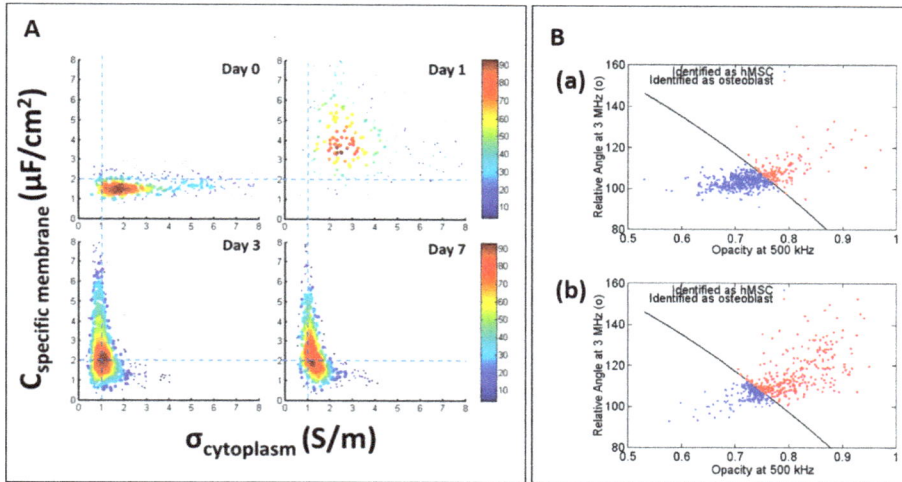

Figure 8. (**A**) Scatter plot of $C_{\text{specific membrane}}$ versus $\sigma_{\text{cytoplasm}}$ of cultured neural stem cells for different days. Reproduced with permission from [84], copyright 2016, Public Library of Science. (**B**) Classification results of human mesenchymal stem cells (hMSC) and osteoblast measured at: seven days (**a**); and 14 days (**b**). Reproduced with permission from [85], copyright 2016, Royal Society of Chemistry;

4. Conclusions and Future Directions

In this review, we present the developments of single cell impedance cytometry using microfluidics in the past decade. There are three developed designs: coplanar electrode design, parallel electrode design and constriction channel. Impedance cytometer can be utilized in a wide range of applications, from differential blood cell counting for disease diagnostics to monitoring cell phenotypic changes and microbial studies.

In terms of throughput, coplanar electrode design and parallel electrode design are much higher (~1000 cell/s) than constriction channel (~100 cells/s). Developing high-throughput cytometer to achieve traditional flow cytometry level remains a key challenge, as there are trade-offs between throughput and signal quality. Nevertheless, throughput can be increased using data acquisition with high sampling rate, or having multiple detection channels to further improve detection sensitivity and speed.

While most impedance measurements are based on cell size and membrane dielectric properties, impedance characterization of intracellular vacuole or nucleus is still at its infancy. Two possible reasons are the high frequency requirement (above 100 MHz) and difficulties in quantifying intracellular organelles position. In future studies, we envision that high frequency measurement of cells will be important as it can provide interesting insights about intracellular nucleus and organelles, which will be useful for developmental biology or genomics studies.

To facilitate user operations for biomedical applications, significant research efforts are focused on integrating important functionalities to microfluidic impedance cytometer such as optical detection, sample processing, and cell sorting. Optical detection such fluorescence labelling allows simultaneous characterization of cell phenotype with impedance measurement to study their associations, and further assess the potential of impedance-based biomarkers in clinical testing. Sample processing is another crucial feature for POC testing as most biofluids are complex and it is necessary to isolate the target cells prior analysis. Post measurement sorting is also attractive as it helps to further reduce the gap between microfluidic cytometer and conventional flow cytometry. Noteworthy, sorting based

Micromachines **2017**, *8*, 87

on impedance signature can be a label-free analytical tool which enables the separation of rare or abnormal cells without known good markers.

In summary, there is a great potential of using impedance cytometer for biomedical applications and clinical diagnostics. In addition to technological improvements, large scale clinical validation will be necessary to determine feasibility of single cell impedance as novel biomarkers for disease diagnosis.

Acknowledgments: This research is supported by the Singapore Ministry of Education Academic Research Fund Tier 1-RG37/15 awarded to King Ho Holden Li and Han Wei Hou would like to acknowledge support from the Lee Kong Chian School of Medicine (LKCMedicine) Postdoctoral Fellowship, as well as Singapore National Research Foundation under CBRG NIG, and administered by the Singapore Ministry of Health's National Medical Research Council (NMRC-08/2015-BNIG).

Author Contributions: All authors contributed equally to this work.

Conflicts of Interest: The authors declare no conflict of interest.

References

1. Altschuler, S.J.; Wu, L.F. Cellular heterogeneity: Do differences make a difference? *Cell* **2010**, *141*, 559–563. [CrossRef] [PubMed]
2. Sims, C.E.; Allbritton, N.L. Analysis of single mammalian cells on-chip. *Lab Chip* **2007**, *7*, 423–440. [CrossRef] [PubMed]
3. Coulter, W.H. High speed automatic blood cell counter and cell size analyzer. In Proceedings of the National Electronics Conference, Chicago, IL, USA, 1–3 October 1956; pp. 1034–1040.
4. Hoffman, R.; Johnson, T.; Britt, W. Flow cytometric electronic direct current volume and radiofrequency impedance measurements of single cells and particles. *Cytometry Part A* **1981**, *1*, 377–384. [CrossRef] [PubMed]
5. Schwan, H.P. Electrical properties of tissue and cell suspensions. *Adv. Biol. Med. Phys.* **1956**, *5*, 147–209.
6. Schwan, H. Electrical properties of tissues and cell suspensions: Mechanisms and models. In Proceedings of the 16th Annual International Conference of the IEEE, Chicago, IL, USA, 26–30 August 1994; Volume 71, pp. A70–A71.
7. Schwan, H.P. Determination of biological impedances. *Phys. Tech. Biol. Res.* **1963**, *6*, 323–407. [CrossRef]
8. Schwan, H. Electrode polarization impedance and measurements in biological material. *Ann. N. Y. Acad. Sci.* **1968**, *148*, 191–209. [CrossRef] [PubMed]
9. Gabriel, C.; Gabriel, S.; Corthout, E. The dielectric properties of biological tissues: I. Literature survey. *Phys. Med. Biol.* **1996**, *41*, 2231. [CrossRef] [PubMed]
10. Asami, K.; Hanai, T. Dielectric monitoring of biological cell sedimentation. *Colloid Polym. Sci.* **1992**, *270*, 78–84. [CrossRef]
11. Hayashi, Y.; Katsumoto, Y.; Omori, S.; Yasuda, A.; Asami, K.; Kaibara, M.; Uchimura, I. Dielectric coagulometry: A new approach to estimate venous thrombosis risk. *Anal. Chem.* **2010**, *82*, 9769–9774. [CrossRef] [PubMed]
12. Kirkwood, J.G. The dielectric polarization of polar liquids. *J. Chem. Phys.* **1939**, *7*, 911–919. [CrossRef]
13. Whitesides, G.M. The origins and the future of microfluidics. *Nature* **2006**, *442*, 368–373. [CrossRef] [PubMed]
14. Yager, P.; Edwards, T.; Fu, E.; Helton, K.; Nelson, K.; Tam, M.R.; Weigl, B.H. Microfluidic diagnostic technologies for global public health. *Nature* **2006**, *442*, 412–418. [CrossRef] [PubMed]
15. Morgan, H.; Spencer, D. Microfluidic impedance cytometry for blood cell analysis. *Microfluidics Med. Appl.* **2014**, *36*, 213.
16. Sun, T.; Morgan, H. Single-cell microfluidic impedance cytometry: A review. *Microfluidics Nanofluidics* **2010**, *8*, 423–443. [CrossRef]
17. Chen, J.; Xue, C.; Zhao, Y.; Chen, D.; Wu, M.-H.; Wang, J. Microfluidic impedance flow cytometry enabling high-throughput single-cell electrical property characterization. *Int. J. Mol. Sci.* **2015**, *16*, 9804. [CrossRef] [PubMed]
18. Holmes, D.; Pettigrew, D.; Reccius, C.H.; Gwyer, J.D.; van Berkel, C.; Holloway, J.; Davies, D.E.; Morgan, H. Leukocyte analysis and differentiation using high speed microfluidic single cell impedance cytometry. *Lab Chip* **2009**, *9*, 2881–2889. [CrossRef] [PubMed]

19. Hywel, M.; Tao, S.; David, H.; Shady, G.; Nicolas, G.G. Single cell dielectric spectroscopy. *J. Phys. D Appl. Phys.* **2007**, *40*, 61.

20. Tao, S.; Shady, G.; Nicolas, G.G.; Hywel, M. Dielectric spectroscopy of single cells: Time domain analysis using maxwell's mixture equation. *J. Phys. D Appl. Phys.* **2007**, *40*, 1.

21. Sun, T.; Green, N.G.; Morgan, H. Analytical and numerical modeling methods for impedance analysis of single cells on-chip. *Nano* **2008**, *3*, 55–63. [CrossRef]

22. Sun, T.; Green, N.G.; Gawad, S.; Morgan, H. Analytical electric field and sensitivity analysis for two microfluidic impedance cytometer designs. *IET Nanobiotechnol.* **2007**, *1*, 69–79. [CrossRef] [PubMed]

23. Maxwell, J.C. *A Treatise on Electricity and Magnetism*; Clarendon Press: Oxford, UK, 1881; Volume 1.

24. Irimajiri, A.; Hanai, T.; Inouye, A. A dielectric theory of "multi-stratified shell" model with its application to a lymphoma cell. *J. Theoretical Biol.* **1979**, *78*, 251–269. [CrossRef]

25. Willis, M.R. Dielectric and electronic properties of biological materials by R Pethig. pp 376. John Wiley & Sons, Chichester and New York. 1979. £15. *Biochem. Educ.* **1980**, *8*, 31.

26. Hanai, T.; Asami, K.; Koizumi, N. Dielectric theory of concentrated suspensions of shell. *Bull. Inst. Chem. Res. Kyoto Univ.* **1979**, *57*, 297–305.

27. Hong, J.; Yoon, D.S.; Kim, S.K.; Kim, T.S.; Kim, S.; Pak, E.Y.; No, K. AC frequency characteristics of coplanar impedance sensors as design parameters. *Lab Chip* **2005**, *5*, 270–279. [CrossRef] [PubMed]

28. Linderholm, P.; Renaud, P. Comment on "AC frequency characteristics of coplanar impedance sensors as design parameters" by Jongin Hong, Dae Sung Yoon, Sung Kwan Kim, Tae Song Kim, Sanghyo Kim, Eugene Y. Pak and Kwangsoo No. Lab Chip 2005, 5, 270. *Lab Chip* **2005**, *5*, 1416–1417, author reply 1418.

29. Sun, T.; Morgan, H.; Green, N.G. Analytical solutions of AC electrokinetics in interdigitated electrode arrays: Electric field, dielectrophoretic and traveling-wave dielectrophoretic forces. *Phys. Rev. E* **2007**, *76*, 046610. [CrossRef] [PubMed]

30. Gawad, S.; Schild, L.; Renaud, P. Micromachined impedance spectroscopy flow cytometer for cell analysis and particle sizing. *Lab Chip* **2001**, *1*, 76–82. [CrossRef] [PubMed]

31. Cheung, K.; Gawad, S.; Renaud, P. Impedance spectroscopy flow cytometry: On-chip label-free cell differentiation. *Cytometry Part A* **2005**, *65A*, 124–132. [CrossRef] [PubMed]

32. Chen, J.; Zheng, Y.; Tan, Q.; Shojaei-Baghini, E.; Zhang, Y.L.; Li, J.; Prasad, P.; You, L.; Wu, X.Y.; Sun, Y. Classification of cell types using a microfluidic device for mechanical and electrical measurement on single cells. *Lab Chip* **2011**, *11*, 3174–3181. [CrossRef] [PubMed]

33. Sun, T.; van Berkel, C.; Green, N.G.; Morgan, H. Digital signal processing methods for impedance microfluidic cytometry. *Microfluidics Nanofluidics* **2009**, *6*, 179–187. [CrossRef]

34. Mernier, G.; Duqi, E.; Renaud, P. Characterization of a novel impedance cytometer design and its integration with lateral focusing by dielectrophoresis. *Lab Chip* **2012**, *12*, 4344–4349. [CrossRef] [PubMed]

35. Demierre, N.; Braschler, T.; Linderholm, P.; Seger, U.; van Lintel, H.; Renaud, P. Characterization and optimization of liquid electrodes for lateral dielectrophoresis. *Lab Chip* **2007**, *7*, 355–365. [CrossRef] [PubMed]

36. Valero, A.; Braschler, T.; Renaud, P. A unified approach to dielectric single cell analysis: Impedance and dielectrophoretic force spectroscopy. *Lab Chip* **2010**, *10*, 2216–2225. [CrossRef] [PubMed]

37. Shaker, M.; Colella, L.; Caselli, F.; Bisegna, P.; Renaud, P. An impedance-based flow microcytometer for single cell morphology discrimination. *Lab Chip* **2014**, *14*, 2548–2555. [CrossRef] [PubMed]

38. Holmes, D.; She, J.K.; Roach, P.L.; Morgan, H. Bead-based immunoassays using a micro-chip flow cytometer. *Lab Chip* **2007**, *7*, 1048–1056. [CrossRef] [PubMed]

39. Morgan, H.; Holmes, D.; Green, N.G. High speed simultaneous single particle impedance and fluorescence analysis on a chip. *Curr. Appl. Phys.* **2006**, *6*, 367–370. [CrossRef]

40. Evander, M.; Ricco, A.J.; Morser, J.; Kovacs, G.T.A.; Leung, L.L.K.; Giovangrandi, L. Microfluidic impedance cytometer for platelet analysis. *Lab Chip* **2013**, *13*, 722–729. [CrossRef] [PubMed]

41. Rodriguez-Trujillo, R.; Mills, C.A.; Samitier, J.; Gomila, G. Low cost micro-coulter counter with hydrodynamic focusing. *Microfluidics Nanofluidics* **2007**, *3*, 171–176. [CrossRef]

42. Bernabini, C.; Holmes, D.; Morgan, H. Micro-impedance cytometry for detection and analysis of micron-sized particles and bacteria. *Lab Chip* **2011**, *11*, 407–412. [CrossRef] [PubMed]

43. Barat, D.; Spencer, D.; Benazzi, G.; Mowlem, M.C.; Morgan, H. Simultaneous high speed optical and impedance analysis of single particles with a microfluidic cytometer. *Lab Chip* **2012**, *12*, 118–126. [CrossRef] [PubMed]

44. Watkins, N.; Venkatesan, B.M.; Toner, M.; Rodriguez, W.; Bashir, R. A robust electrical microcytometer with 3-dimensional hydrofocusing. *Lab Chip* **2009**, *9*, 3177–3184. [CrossRef] [PubMed]
45. Rodriguez-Trujillo, R.; Castillo-Fernandez, O.; Garrido, M.; Arundell, M.; Valencia, A.; Gomila, G. High-speed particle detection in a micro-coulter counter with two-dimensional adjustable aperture. *Biosens. Bioelectron.* **2008**, *24*, 290–296. [CrossRef] [PubMed]
46. Scott, R.; Sethu, P.; Harnett, C. Three-dimensional hydrodynamic focusing in a microfluidic coulter counter. *Rev. Sci. Instrum.* **2008**, *79*, 046104. [CrossRef] [PubMed]
47. De Ninno, A.; Errico, V.; Bertani, F.R.; Businaro, L.; Bisegna, P.; Caselli, F. Coplanar electrode microfluidic chip enabling accurate sheathless impedance cytometry. *Lab Chip* **2017**. [CrossRef] [PubMed]
48. Gawad, S.; Cheung, K.; Seger, U.; Bertsch, A.; Renaud, P. Dielectric spectroscopy in a micromachined flow cytometer: Theoretical and practical considerations. *Lab Chip* **2004**, *4*, 241–251. [CrossRef] [PubMed]
49. Spencer, D.; Morgan, H. Positional dependence of particles in microfludic impedance cytometry. *Lab Chip* **2011**, *11*, 1234–1239. [CrossRef] [PubMed]
50. Frankowski, M.; Simon, P.; Bock, N.; El-Hasni, A.; Schnakenberg, U.; Neukammer, J. Simultaneous optical and impedance analysis of single cells: A comparison of two microfluidic sensors with sheath flow focusing. *Eng. Life Sci.* **2015**, *15*, 286–296. [CrossRef]
51. Spencer, D.; Caselli, F.; Bisegna, P.; Morgan, H. High accuracy particle analysis using sheathless microfluidic impedance cytometry. *Lab Chip* **2016**, *16*, 2467–2473. [CrossRef] [PubMed]
52. Zhao, Y.; Chen, D.; Li, H.; Luo, Y.; Deng, B.; Huang, S.-B.; Chiu, T.-K.; Wu, M.-H.; Long, R.; Hu, H.; et al. A microfluidic system enabling continuous characterization of specific membrane capacitance and cytoplasm conductivity of single cells in suspension. *Biosens. Bioelectronics* **2013**, *43*, 304–307. [CrossRef] [PubMed]
53. Holmes, D.; Morgan, H. Single cell impedance cytometry for identification and counting of CD4 T-cells in human blood using impedance labels. *Anal. Chem.* **2010**, *82*, 1455–1461. [CrossRef] [PubMed]
54. Watkins, N.N.; Sridhar, S.; Cheng, X.; Chen, G.D.; Toner, M.; Rodriguez, W.; Bashir, R. A microfabricated electrical differential counter for the selective enumeration of CD4+ T lymphocytes. *Lab Chip* **2011**, *11*, 1437–1447. [CrossRef] [PubMed]
55. van Berkel, C.; Gwyer, J.D.; Deane, S.; Green, N.; Holloway, J.; Hollis, V.; Morgan, H. Integrated systems for rapid point of care (PoC) blood cell analysis. *Lab Chip* **2011**, *11*, 1249–1255. [CrossRef] [PubMed]
56. Han, X.; van Berkel, C.; Gwyer, J.; Capretto, L.; Morgan, H. Microfluidic lysis of human blood for leukocyte analysis using single cell impedance cytometry. *Anal. Chem.* **2012**, *84*, 1070–1075. [CrossRef] [PubMed]
57. Watkins, N.N.; Hassan, U.; Damhorst, G.; Ni, H.; Vaid, A.; Rodriguez, W.; Bashir, R. Microfluidic CD4+ and CD8+ T lymphocyte counters for point-of-care HIV diagnostics using whole blood. *Sci. Transl. Med.* **2013**, *5*, 214ra170. [CrossRef] [PubMed]
58. Spencer, D.; Elliott, G.; Morgan, H. A sheath-less combined optical and impedance micro-cytometer. *Lab Chip* **2014**, *14*, 3064–3073. [CrossRef] [PubMed]
59. Hassan, U.; Watkins, N.N.; Reddy B., Jr.; Damhorst, G.; Bashir, R. Microfluidic differential immunocapture biochip for specific leukocyte counting. *Nat. Protocols* **2016**, *11*, 714–726. [CrossRef] [PubMed]
60. Tao, S.; Shady, G.; Catia, B.; Nicolas, G.G.; Hywel, M. Broadband single cell impedance spectroscopy using maximum length sequences: Theoretical analysis and practical considerations. *Meas. Sci. Technol.* **2007**, *18*, 2859.
61. Küttel, C.; Nascimento, E.; Demierre, N.; Silva, T.; Braschler, T.; Renaud, P.; Oliva, A.G. Label-free detection of babesia bovis infected red blood cells using impedance spectroscopy on a microfabricated flow cytometer. *Acta Tropica* **2007**, *102*, 63–68. [CrossRef] [PubMed]
62. Zheng, Y.; Shojaei-Baghini, E.; Azad, A.; Wang, C.; Sun, Y. High-throughput biophysical measurement of human red blood cells. *Lab Chip* **2012**, *12*, 2560–2567. [CrossRef] [PubMed]
63. Du, E.; Ha, S.; Diez-Silva, M.; Dao, M.; Suresh, S.; Chandrakasan, A.P. Electric impedance microflow cytometry for characterization of cell disease states. *Lab Chip* **2013**, *13*, 3903–3909. [CrossRef] [PubMed]
64. Benazzi, G.; Holmes, D.; Sun, T.; Mowlem, M.C.; Morgan, H. Discrimination and analysis of phytoplankton using a microfluidic cytometer. *IET Nanobiotechnol.* **2007**, *1*, 94–101. [CrossRef] [PubMed]
65. Mernier, G.; Hasenkamp, W.; Piacentini, N.; Renaud, P. Multiple-frequency impedance measurements in continuous flow for automated evaluation of yeast cell lysis. *Sens. Actuators B Chem.* **2012**, *170*, 2–6. [CrossRef]

66. Haandbaek, N.; Burgel, S.C.; Heer, F.; Hierlemann, A. Characterization of subcellular morphology of single yeast cells using high frequency microfluidic impedance cytometer. *Lab Chip* **2014**, *14*, 369–377. [CrossRef] [PubMed]

67. Haandbaek, N.; With, O.; Burgel, S.C.; Heer, F.; Hierlemann, A. Resonance-enhanced microfluidic impedance cytometer for detection of single bacteria. *Lab Chip* **2014**, *14*, 3313–3324. [CrossRef] [PubMed]

68. Haandbæk, N.; Bürgel, S.C.; Rudolf, F.; Heer, F.; Hierlemann, A. Characterization of single yeast cell phenotypes using microfluidic impedance cytometry and optical imaging. *ACS Sens.* **2016**, *1*, 1020–1027. [CrossRef]

69. Schade-Kampmann, G.; Huwiler, A.; Hebeisen, M.; Hessler, T.; Di Berardino, M. On-chip non-invasive and label-free cell discrimination by impedance spectroscopy. *Cell Prolif.* **2008**, *41*, 830–840. [CrossRef] [PubMed]

70. Nikolic-Jaric, M.; Romanuik, S.F.; Ferrier, G.A.; Bridges, G.E.; Butler, M.; Sunley, K.; Thomson, D.J.; Freeman, M.R. Microwave frequency sensor for detection of biological cells in microfluidic channels. *Biomicrofluidics* **2009**, *3*, 034103. [CrossRef] [PubMed]

71. Gou, H.-L.; Zhang, X.-B.; Bao, N.; Xu, J.-J.; Xia, X.-H.; Chen, H.-Y. Label-free electrical discrimination of cells at normal, apoptotic and necrotic status with a microfluidic device. *J. Chromatogr. A* **2011**, *1218*, 5725–5729. [CrossRef] [PubMed]

72. Zheng, Y.; Shojaei-Baghini, E.; Wang, C.; Sun, Y. Microfluidic characterization of specific membrane capacitance and cytoplasm conductivity of singlecells. *Biosens. Bioelectron.* **2013**, *42*, 496–502. [CrossRef] [PubMed]

73. Zhao, Y.; Chen, D.; Luo, Y.; Li, H.; Deng, B.; Huang, S.-B.; Chiu, T.-K.; Wu, M.-H.; Long, R.; Hu, H.; et al. A microfluidic system for cell type classification based on cellular size-independent electrical properties. *Lab Chip* **2013**, *13*, 2272–2277. [CrossRef] [PubMed]

74. Zhao, Y.; Zhao, X.T.; Chen, D.Y.; Luo, Y.N.; Jiang, M.; Wei, C.; Long, R.; Yue, W.T.; Wang, J.B.; Chen, J. Tumor cell characterization and classification based on cellular specific membrane capacitance and cytoplasm conductivity. *Biosens. Bioelectron.* **2014**, *57*, 245–253. [CrossRef] [PubMed]

75. Kirkegaard, J.; Clausen, C.; Rodriguez-Trujillo, R.; Svendsen, W. Study of paclitaxel-treated HeLa cells by differential electrical impedance flow cytometry. *Biosensors* **2014**, *4*, 257. [CrossRef] [PubMed]

76. Spencer, D.; Hollis, V.; Morgan, H. Microfluidic impedance cytometry of tumour cells in blood. *Biomicrofluidics* **2014**, *8*, 064124. [CrossRef] [PubMed]

77. Bürgel, S.C.; Escobedo, C.; Haandbæk, N.; Hierlemann, A. On-chip electroporation and impedance spectroscopy of single-cells. *Sens. Actuators B Chem.* **2015**, *210*, 82–90. [CrossRef]

78. Zhao, Y.; Jiang, M.; Chen, D.; Zhao, X.; Xue, C.; Hao, R.; Yue, W.; Wang, J.; Chen, J. Single-cell electrical phenotyping enabling the classification of mouse tumor samples. *Sci. Rep.* **2016**, *6*, 19487. [CrossRef] [PubMed]

79. Huang, S.-B.; Zhao, Y.; Chen, D.; Liu, S.-L.; Luo, Y.; Chiu, T.-K.; Wang, J.; Chen, J.; Wu, M.-H. Classification of cells with membrane staining and/or fixation based on cellular specific membrane capacitance and cytoplasm conductivity. *Micromachines* **2015**, *6*, 163. [CrossRef]

80. Yuan, W.; Zhensong, X.; Mark, A.C.; John, N.; Yi, Z.; Chen, W.; Yu, S. Embedded silver PDMS electrodes for single cell electrical impedance spectroscopy. *J. Micromech. Microeng.* **2016**, *26*, 095006.

81. Babahosseini, H.; Srinivasaraghavan, V.; Zhao, Z.; Gillam, F.; Childress, E.; Strobl, J.S.; Santos, W.L.; Zhang, C.; Agah, M. The impact of sphingosine kinase inhibitor-loaded nanoparticles on bioelectrical and biomechanical properties of cancer cells. *Lab Chip* **2016**, *16*, 188–198. [CrossRef] [PubMed]

82. Xie, X.; Cheng, Z.; Xu, Y.; Liu, R.; Li, Q.; Cheng, J. A sheath-less electric impedance micro-flow cytometry device for rapid label-free cell classification and viability testing. *Anal. Methods* **2017**, *9*, 1201–1212. [CrossRef]

83. Song, H.; Wang, Y.; Rosano, J.M.; Prabhakarpandian, B.; Garson, C.; Pant, K.; Lai, E. A microfluidic impedance flow cytometer for identification of differentiation state of stem cells. *Lab Chip* **2013**, *13*, 2300–2310. [CrossRef] [PubMed]

84. Zhao, Y.; Liu, Q.; Sun, H.; Chen, D.; Li, Z.; Fan, B.; George, J.; Xue, C.; Cui, Z.; Wang, J.; et al. Electrical property characterization of neural stem cells in differentiation. *PLoS ONE* **2016**, *11*, e0158044. [CrossRef] [PubMed]

85. Song, H.; Rosano, J.M.; Wang, Y.; Garson, C.J.; Prabhakarpandian, B.; Pant, K.; Klarmann, G.J.; Perantoni, A.; Alvarez, L.M.; Lai, E. Identification of mesenchymal stem cell differentiation state using dual-micropore microfluidic impedance flow cytometry. *Anal. Methods* **2016**, *8*, 7437–7444. [CrossRef]

86. Sun, T.; Holmes, D.; Gawad, S.; Green, N.G.; Morgan, H. High speed multi-frequency impedance analysis of single particles in a microfluidic cytometer using maximum length sequences. *Lab Chip* **2007**, *7*, 1034–1040. [CrossRef] [PubMed]

87. Hassan, U.; Reddy, B.; Damhorst, G.; Sonoiki, O.; Ghonge, T.; Yang, C.; Bashir, R. A microfluidic biochip for complete blood cell counts at the point-of-care. *Technology* **2015**, *3*, 201–213. [CrossRef] [PubMed]

88. Gascoyne, P.; Pethig, R.; Satayavivad, J.; Becker, F.F.; Ruchirawat, M. Dielectrophoretic detection of changes in erythrocyte membranes following malarial infection. *Biochim. Biophys. Acta Biomembr.* **1997**, *1323*, 240–252. [CrossRef]

89. Gascoyne, P.R.C.; Shim, S.; Noshari, J.; Becker, F.F.; Stemke-Hale, K. Correlations between the dielectric properties and exterior morphology of cells revealed by dielectrophoretic field-flow fractionation. *Electrophoresis* **2013**, *34*, 1042–1050. [CrossRef] [PubMed]

90. Becker, F.F.; Wang, X.B.; Huang, Y.; Pethig, R.; Vykoukal, J.; Gascoyne, P.R. Separation of human breast cancer cells from blood by differential dielectric affinity. *Proc. Natl. Acad. Sci. USA* **1995**, *92*, 860–864. [CrossRef] [PubMed]

micromachines

MDPI

Review

Droplet Microfluidics for the Production of Microparticles and Nanoparticles

Jianmei Wang [1,2], Yan Li [2], Xueying Wang [2], Jianchun Wang [2], Hanmei Tian [2], Pei Zhao [2], Ye Tian [3], Yeming Gu [4], Liqiu Wang [2,3,*] and Chengyang Wang [1,*]

[1] School of Chemical Engineering and Technology, Tianjin University, Tianjin 300072, China; wangjm@sderi.cn
[2] Energy Research Institute, Shandong Academy of Sciences, Jinan 250014, China; liyan@sderi.cn (Y.L.); wangxy@sderi.cn (X.W.); wangjc@sderi.cn (J.W.); tianhm@sderi.cn (H.T.); zhaop@sderi.cn (P.Z.)
[3] Department of Mechanical Engineering, The University of Hong Kong, Hong Kong, China; tianye@hku.hk
[4] Shandong Shengli Co., Ltd., Jinan 250101, China; perfectgu@126.com
* Correspondence: lqwang@hku.hk (L.W.); cywang@tju.edu.cn (C.W.);
   Tel.: +86-531-8872-8326 (L.W.); +86-22-2789-0481 (C.W.)

Academic Editor: Say Hwa Tan
Received: 17 October 2016; Accepted: 6 January 2017; Published: 14 January 2017

Abstract: Droplet microfluidics technology is recently a highly interesting platform in material fabrication. Droplets can precisely monitor and control entire material fabrication processes and are superior to conventional bulk techniques. Droplet production is controlled by regulating the channel geometry and flow rates of each fluid. The micro-scale size of droplets results in rapid heat and mass-transfer rates. When used as templates, droplets can be used to develop reproducible and scalable microparticles with tailored sizes, shapes and morphologies, which are difficult to obtain using traditional bulk methods. This technology can revolutionize material processing and application platforms. Generally, microparticle preparation methods involve three steps: (1) the formation of micro-droplets using a microfluidics generator; (2) shaping the droplets in micro-channels; and (3) solidifying the droplets to form microparticles. This review discusses the production of microparticles produced by droplet microfluidics according to their morphological categories, which generally determine their physicochemical properties and applications.

Keywords: microfluidics; microparticles; nanoparticles; monodisperse; emulsion; droplet microfluidics

1. Introduction

Monodisperse microparticles and nanoparticles with uniform sizes and morphologies are used in bio-pharmaceuticals [1,2], drug delivery applications [3,4], electro/optic devices [5,6] and catalysis [7–9] because of their unique properties. Many efforts are made to produce uniform microparticles with custom sizes, shapes and morphologies using traditional methods, such as precipitation polymerization [10,11], emulsion polymerization [12,13], dispersion polymerization [14,15], SPG (Shirasu Porous Glass) membrane emulsification [16,17], and layer-by-layer assemblies [18]. However, these conventional emulsion droplet methods with bulk shearing forces are uncontrollable, and the resulting droplets, especially the non-spherical particles, are disparate in size and shape. Because of the interfacial tensions between the two phases, the emulsion droplets automatically shrink into spheres, making it difficult for traditional methods to prepare quality custom shaped particles. Moreover, traditional methods are complex, inflexible and expensive. Therefore, better methods are needed to produce monodisperse microparticles with tailored sizes, shapes and morphologies.

Droplet microfluidics techniques, including active and passive method are promising methods to generate monodisperse emulsions. The main difference between active and passive methods is based on the external forces. Active droplet generators are usually achieved by incorporating additional forces into microfluidic systems, such as electrical [19], magnetic [20], pneumatic [21], acoustic [22] and thermal [23]. Co-flow [24,25], flow-focusing [26–28], T-junction [29,30], step emulsification [31] and microchannel terraces [32,33] are basic passive droplet generators. The active control offers more flexibility in manipulating droplet than passive droplet generation. However, the active methods suffer from difficulties in fabrication and miniaturization. In this review, we will mainly focus on the passive methods. These approaches operate in the laminar flow region and generate one drop at a time. The conditions are identical for each droplet as it breaks off [34,35], and the produced emulsions are uniform in size, structure and composition [36]. In addition, the microfluidic devices provide much more flexibility because only the device structures need to be changed to produce complex structure droplets, such as single-emulsions, double-emulsions and multi-emulsions [37,38].

The superior properties of droplet microfluidics are advantageous for precise microparticle manufacturing, especially when used as templates to prepare microparticles and nanoparticles with various morphologies. In general, highly monodisperse emulsified droplets form in a microfluidic device and are simultaneously used as a template, and then are solidified to form microparticles and nanoparticles by chemical, photochemical or physical methods [39]. This review discusses recent advances in microparticle and nanosphere fabrication with droplet microfluidics. Because a single emulsion can be a template for solid spheres, and double or multi-emulsions can be templates for core shells, Janus or other morphology spheres, we classified the emulsion droplets according to their structures, and focused on how the emulsion droplets evolved into various structures and morphologies, as shown in Figure 1. This review provides a general background to those new to droplet microfluidics.

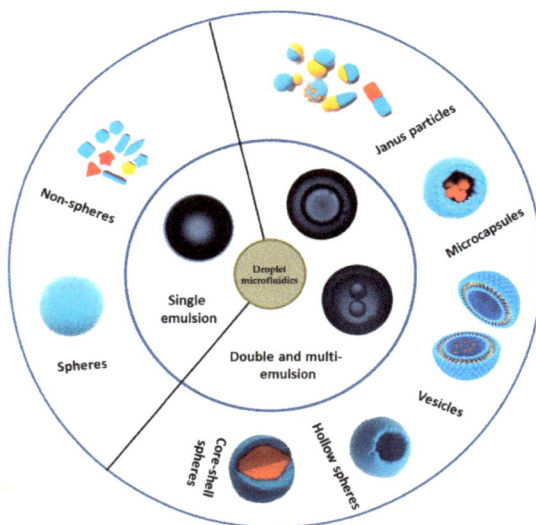

Figure 1. Summary of applications for microparticles and nanoparticles.

2. Microparticles and Nanoparticles with Single Emulsion Template

Single emulsions are droplets of one phase fluid dispersed in another immiscible phase fluid. The key step in forming monodisperse microparticles is to form monodisperse droplets with microfluidic devices. The most frequently used systems to generate monodisperse droplets are co-flow, cross-flow, and flow focusing, and the coefficient of variation (CV, defined as the ratio of standard deviation to the mean of the droplet radius) of droplets is usually less than 5% [40]. The size of the droplets generated from co-flow and cross-flow is often related to the dispersed channel dimension, whereas flow-focusing structure is different from the above two types. In flow-focusing, the inner fluid is hydrodynamically flow focused by the outer fluid through the orifice, and it allows generating droplets with smaller size than that of the orifice. Based on this feature, we can use a larger orifice to make droplets for those fluids with suspended particles, to minimize the probability of clogging the orifice.

Chong et al. [41] and Zhu et al. [42] gave a very detailed summary on the active and passive droplet generation methods with microfluidic devices, in which they overviewed the different droplet generators and the characteristics and mechanisms of breakup modes of droplet generation. The five breaking modes in passive generation, which are squeezing, dripping, jetting, tip-streaming and tip-multi-breaking have their own unique characteristics, and can be applied to various fields, for example, to perform chemical and biochemical reactions where droplets used as microreactors [43,44] and to synthesis microparticles with droplets as templates [45–48]. In material science, this is a superior tool for engineering micromaterials and nanomaterials, we can change the component and the structure of the droplets to produce polymer particles, inorganic nanoparticles and metal particles [49]. In this section, we will discuss how single emulsions are used as templates for generating solid particles, including spherical and non-spherical particles with different materials.

2.1. Spherical Particles

2.1.1. Polymer Microspheres

Polymer microspheres are commonly used in pharmaceutical and medical applications. Polyvinyl alcohol (PVA), poly(lactic-co-glycolic acid) (PLGA), sodium alginate, polyethylene glycol (PEG) and gelatin microspheres have been successfully used as drug carriers [26,39,50–56]. Several methods, such as spray-drying, coacervation and emulsification, are used to prepare polymer microspheres. However, these conventional bulk procedures cannot be precisely controlled and result in polydispersed and irregular shapes, which limits their practical applications [57].

Xu et al. used flow-focusing geometry to generate PLGA droplets in PDMS devices [58]. They produced monodisperse particles with defined sizes ranging from 10 to 50 μm by simply tuning the flow rates of the continuous phase and the disperse phase. After loading bupivacaine (an amphiphilic drug), they found that the drug release kinetics of the monodisperse particles were different from those of the polydisperse particles produced by conventional methods. The monodisperse PLGA microparticles had significantly reduced burst releases and slower overall release rates than those of the polydisperse particles under the similar conditions.

Chu et al. used a glass capillary-based single emulsion device as shown in Figure 2 to make monodisperse Poly(Nisopropylacrylamide) (PNIPAm) microgels [46]. The spherical voids were introduced in a controlled manner into the microgels. They found that the microgels with voids would swell and shrink in response to temperature changes faster than those with voidless microgels. In addition, the response rates were finely tuned by changing the size and number of spherical voids inside the microgels.

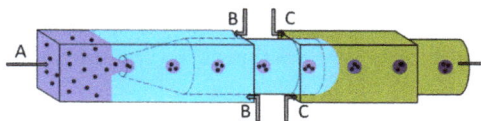

Figure 2. Schematic illustration of a capillary-based single emulsion device for fabricating monodisperse PNIPAm microgels. This device was used by Chu et al. [46]. **A** containing the monomer (N-isopropylacrylamide), a crosslinker (N,N′-methylene-bis-acrylamide), and a reaction initiator (ammonium persulfate) and solid polystyrene particles; **B** containing the kerosene and a surfactant; and **C** amphipathic reaction accelerator (N,N,N′,N′-tetramethyethylenediamine) dissolved in kerosene.

Seo et al. studied the emulsification of four kinds of monomer acrylates to synthesize polymer particles in a microfluidic flow-focusing device [59]. They detailed the effects of hydrodynamic conditions, channel geometry and micro-device materials on the size of the droplets. The selection of an appropriate material for the device was a vital stage in the generation of the droplets, and phase inversion occurred because of the higher affinity of the droplet phase for the material of the microfluidic device, such as glass, silicon, polydimethylsiloxane (PDMS) and polyurethane (PU). Though the channel surface could be modified by surfactants, the modified effect disappeared after several hours of emulsification. In addition, the rate of monomer polymerization with UV-light affected the quality of the sphere. Rapid polymerization caused enough heat to induce an explosion and a vacuum for low monomer-to-polymer conversion caused the sphere to collapse. Serra et al. produced polymer particles without surfactant or pretreatment in a co-flow device, and studied the effect of the viscosity of the continuous phase on the particle size [60]. They found high viscosity of the continuous phase could prevent the droplets from phase inversion and generate smaller particles for a given fluid flow rate. It could be of particular interest for the synthesis of particles with a functionalized surface.

Controlling the composition of the particles allows for a variety of functional properties. When dyes, semiconductor quantum dots, magnetic nanoparticles or liquid crystals are added to the particles, they have an optical, magnetic and actuation performance [61–65]. Carrying chemo-therapeutic or radio-therapeutic agents in the embolism microspheres greatly improves the treatment effects [66,67]. Magnetically guided drug carriers for medical imaging and therapeutic applications have been studied for decades and are currently ready for clinical trials [68,69]. These applications are available by adding another functional composition in the disperse phase to form stable droplet, and then by delivering the droplet into the microsphere.

2.1.2. Inorganic Microspheres

Monodisperse inorganic microspheres, including those composed of silica, carbon, and titanium, have received considerable attention for their potential applications in biomolecules, drug delivery [70,71], sensors [6] and catalysts [9]. The classical Stöber method is a general approach for the synthesis of silica spheres based on sol-gel chemistry. The synthesis involves the hydrolysis and condensation of silicon alkoxides in alcohol solvents with ammonia as the catalyst, and produces monodisperse silica spheres of predetermined sizes in the range 0.05–2 μm [72]. However, the particle sizes are not precisely reproducible with this method. Liu et al. extended this method to prepare monodisperse resorcinol-formaldehyde resin polymers (RFs) and carbon spheres, and the particle sizes of the RFs and carbon spheres were tuned from 200 nm to 1000 nm by varying the concentration of the reactant [73]. However, this method is not generally used to prepare other materials, especially when preparations require precise control over the particle sizes across wider ranges and shapes.

Lee et al. [47], Carroll and another researchers [74,75] reported a one-step method to manipulate ordered mesoporous silica (OMS) in a microfluidic device. This method combined a microfluidic emulsification technique and a rapid solvent diffusion induced self-assembly (DISA) technique. Monodisperse droplets were generated at the flow-focusing orifice and assembled into mesostructured silica/surfactant composite spheres within the microchannel, as shown in Figure 3. The sizes and the surface morphologies were easily controlled by changing the synthesis parameters, such as the geometry of the microfluidic channels, the flow rate of the precursor solution and oil, and the type of oil.

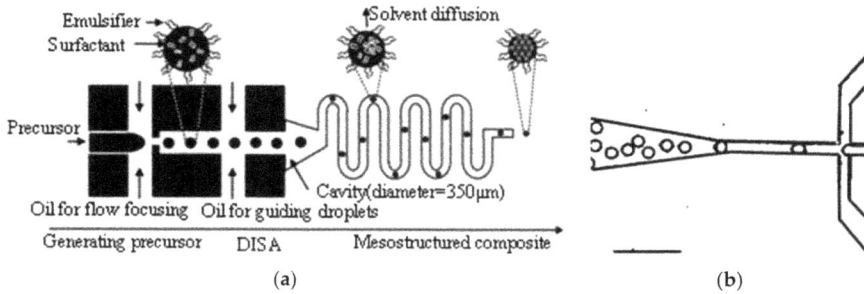

Figure 3. (a) Schematic illustration of the synthesis of OMS particles using microfluidic DISA used by Lee et al. [47]. Reproduced with permission from Lee, I., et al., *Advanced Functional Materials*; published by John Wiley and Sons, 2008. (b) Optical microscopy image of droplets of the silica precursor solution emulsified in a flow focusing microfluidic device in hexadecane from Carroll et al. [74]. The channel dimensions of the orifice are 25 μm in width and 30 μm in length. The scale bar is 100 μm. Reproduced with permission from Carrol, N.J., et al., *Langmuir*; published by American Chemical Society, 2008.

2.1.3. Noble Metal Nanospheres

Noble metal nanoparticles, such as gold, silver and platinum, are interesting materials because of their size and shape dependent properties [49,76–84]. However, the individual nanoparticles tend to coagulate and precipitate to lower the surface free energy. Therefore, these materials are difficult to use to obtain a desired size and size distribution.

The control of the crystal structure of the nanoparticles is another key issue in nanoparticle synthesis. Heat and mass control is very important to crystal growth, and the droplet microfluidics device provides a unique platform to precisely control heat and mass, which results from the fast heat and mass-transfer rates in the microchannel due to the short diffusion pathways induced by the small characteristic lengths of microfluidic device.

Gold nanoparticles have received considerable attention because of their broad range of applications. Recently, gold nanoparticles were synthesized from spherical gold nanoparticle seeds <4 nm in size in the microfluidic device [85]. Various shapes such as spheres, spheroids, rods and extended sharp-edged structures were obtained by tuning the concentrations of reagents and feed rates of the individual aqueous streams.

2.1.4. Semiconductor Nanospheres

Droplet microfluidics are commonly used in synthesizing nanoparticles at room temperature [86], but the pyrolytic synthesis of high quality semiconductor nanocrystals, such as CdSe, requires higher reaction temperatures of 200–350 °C, which renders this method impractical. The droplets and carrier fluids should be stable, non-interacting, non-volatile and immiscible from ambient to reaction temperatures, and the microfluidic reactor must have thermal and chemical stability [87].

Chan et al. used octadecene (ODE) as the solvent, long-chained perfluorinated polyethers (PFPEs) with high-boiling points as the continuous fluids and glass as the microreactor material. All fluids and device materials were stable at the reaction temperature [87]. However, this system had a low interfacial tension (γ) (5–25 mN/m) and a high viscosity (μ) (>100 mP·s) for the high-boiling PFPEs, which induced a high value of C_a ($C_a = \mu v/\gamma$) and a low value of viscosity ratios (λ where $\lambda = \mu_{disperse}/\mu_{continuous}$). This was undesirable for droplet formation because the interfacial velocity (γ/v) was not fast enough relative to v (m·s^{-1}) to relax the strained interface into forming droplets, in addition, it need large values of shear rate to rupture the interface at low viscosity ratios, which can generate high pressures with viscous PFPEs as carrier fluids [27,88,89].To solve this problem, Chan et al. designed a microdevice with a stepped microstructure increasing in channel height, as shown in Figure 4. With this device, ODE droplets in Fomblin Y 06/6 PFPEs were generated at a flow-focusing orifice, and CdSe nanocrystals were produced when the droplets going through the glass microreactor had temperatures of 240–300 °C, although C_a was 0.81 and λ was 0.035.

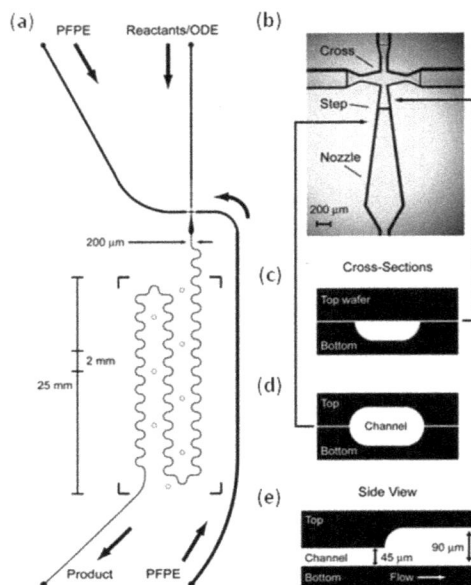

Figure 4. The microreactor channel design with droplet jet injector used by Chan et al. [87]: (**a**) Channel schematic showing dimensions, inlets (●), thermocouple wells (○), and boundaries of Kapton heater [87] (**b**) Optical micrograph of droplet injection cross. ODE is injected into the top channel, while the PFPE is injected in the side channels (**c**) Lateral "D"-shaped cross section of channel etched on the bottom wafer only (**d**) Cross-section of ellipsoidal channel etched on both top and bottom wafers. (**e**) Axial cross-section showing the 45 μm stepped up in channel height. Reproduced with permission from Chan, E.M., et al., *Journal of the American Chemical Society*; published by American Chemical Society, 2005.

2.2. Non-Spherical Particles

Non-spherical particles offer unique properties compared to those of spherical particles [90–93]. For example, in optics, rod-shaped particles often have superior optical properties due to the optical antenna effect. Prior studies suggested that anisotropically shaped nanoparticles can avoid bio-elimination more effectively than spherical particles under the same conditions [94]. These findings are promising and will lead to additional studies on irregular shapes and corresponding applications [49,85,95–98]. Many strategies were developed to fabricate non-spherical particles, including template molding [99], seeded emulsion polymerization [100] and self-assembly [101]. However, these methods are still difficult for producing high quality, monodisperse, non-spherical particles with tailored geometries and shapes.

Recent advances in droplet microfluidic technologies offer new approaches for the fabrication of non-spherical particles. One approach is to confine the droplets in microfluidic channels with different sizes and shapes. If the volume of the droplet is larger than that of the largest sphere which could be accommodated in the channel, the droplet will be deformed into a disk, ellipsoid or a rod, and the non-spherical particles can be generated after they are solidified in the confined channel [49,102]. Another approach is through the combination of photo-chemistry and photomasking; the photomask with desired patterns is used as a template for the final particles. When the droplet periodically flows through the mask, photo-initiated polymerization will occur. With this approach, polymer particles with complicated shapes can be generated [91,103].

Xu et al. produced particles with different sizes and shapes using the first method [49]. They produced monodisperse droplets in a flow-focusing device, shaped the droplets in the confined channel, and solidified these droplets in situ. This method is applicable to a variety of materials, such as gels, metals and polymers. Some products are shown in Figure 5a,b. Dendukuri et al. easily produced complex and multifunctional particles using the second approach [103]. They synthesized various shapes, such as polygonal shapes, non-symmetric or curved objects, and high-aspect-ratio objects, as shown in Figure 5c,d. The shape and the size of particle is only controlled by the mask, and the morphology and the chemistry of the particles can be independently chosen to form a large number of unique particles for applications in coding, drug delivery and biosensors.

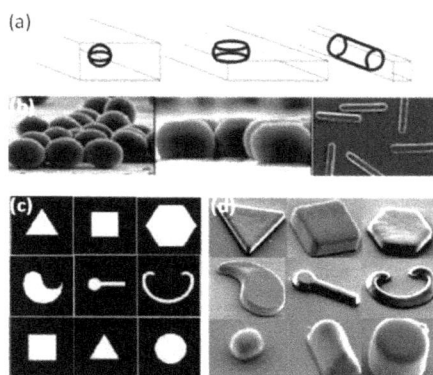

Figure 5. Non-spherical particles formed using the two approaches. (**a**) The spherical droplets deformed by the confinement of the outlet channel; (**b**) The optical microscopy images of particles: ellipsoids, disks, rods with (**a**) approaches [49]. Reproduced with permission from Xu, S., et al., *Angewandte Chemie*; published by John Wiley and Sons, 2005; (**c**) The transparency mask used to make nonspherical particles; (**d**) The SEM images of corresponding particles with (**c**) approaches [103]. Reproduced with permission from Dendukuri, D., et al., *Nature Materials*; published by Nature Publishing Group, 2006.

3. Microparticles with Double or Multi-Emulsions as the Template

Double or multi-emulsions are droplets with smaller droplets encapsulated in larger drops. Core shell microparticles are typically made using double emulsion droplets as templates. Theoretically, micro-devices combined with co-flowing and/or flow-focusing geometries can easily produce monodisperse double or multiple droplets, most of which are capillary microfluidic devices, as shown in Figure 6. There are three fluids flowing in different capillaries, including the inner fluid in the injection capillary, the middle and the outer fluid in square capillary. The inner fluid is sheared by the middle fluid to form single droplets, and the middle fluid containing one or more single droplets is pinched off by the outer fluid to form double or multiple droplets. This technique eliminates the difficulties of precisely controlling the shell thickness, secondary nucleation and aggregation, and non-uniform in the traditional processes [104–106]. However, the precise size and morphology control in this technology is very important and still a challenge to the preparation of microparticles.

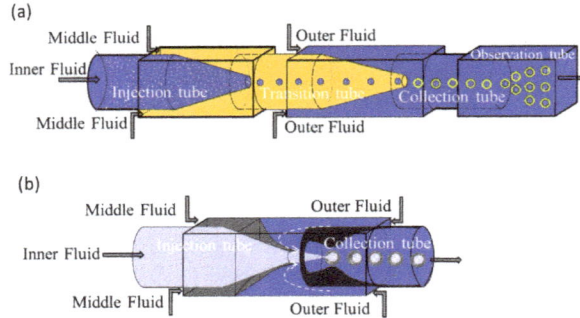

Figure 6. Fabricate of double emulsions in microfluidic devices. (**a**) Schematic of a capillary microfluidic device that combines double co-flowing geometry adapted from Figure 1 in [107]; (**b**) Schematic of a capillary microfluidic device that combines co-flow and flow-focusing geometry adapted from Figure 1 in [37].

3.1. Size Control of Core Shell Microparticles

Accurate control of the particle size is essential and in turn affects the release of inner active materials [108]. Dripping and jetting are two droplet formation regimes for each inner and middle fluid, as shown in Figure 7 [109]. The dripping regime may be the best choice to form a controllable core shell structure. The inner and outer drops should all be formed in the dripping regime to produce uniform microparticles, and the thickness of the shell can be precisely controlled.

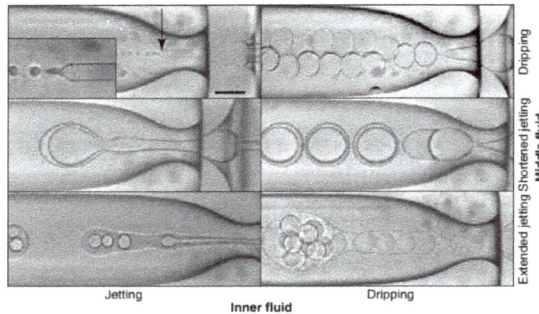

Figure 7. Different morphologies of double emulsions produced by a microcapillary device [109]. Reproduced with permission from Utada, A.S., et al., *MRS Bulletin*; published by Cambridge University Press, 2007.

Kim et al. used a microfluidic device to illustrate the predicted model of the radius droplet, as shown in Figure 8b [110]. At lower flow speeds, droplets will form in the dripping regime where droplets are very close to the orifice, and the mass flux is related to cross-sectional area. The size is controlled by the ratio of the flow rates of the sum of inner and middle fluids to the outer fluid (Q_{sum}/Q_{OF}). Equation (1) gives the relationship of the flow rate (Q) and each radius (R), where R_{thread} is the radius of the fluid thread that breaks into drops, and $R_{orifice}$ is the radius of the exit orifice. The values of $R_{thread}/R_{orifice}$ are predicted from Equation (1), with no adjustable parameters, and are consistent with the measured values in Figure 8a (open symbols). A comparison of the measured radii of the drops and the thread shows that $R_{drop} = 1.82R_{thread}$ [111]. Based on the above research, R_{drop} can be predicted, which provides important guidance in creating double emulsions of a desired size.

$$\frac{Q_{sum}}{Q_{OF}} = \frac{\pi R_{thread}^2}{\pi R_{orifice}^2 - \pi R_{thread}^2} \tag{1}$$

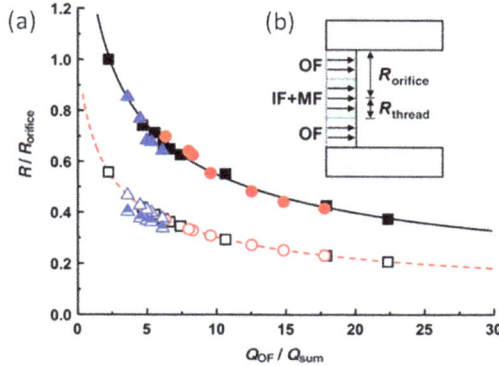

Figure 8. (**a**) Dependence of $R_{thread}/R_{orifice}$ on the scaled flow rate Q_{OF}/Q_{sum} [110]. The open symbols represent the R_{thread} for different liquids and double emulsions consisting of a single silicon drop surrounded by a liquid shell (3 $Q_{IF} = Q_{MF}$, triangle). The dashed line represents the predicted R_{thread}. R_{drop} values are represented with solid identical symbols. Half-filled triangles correspond to the radius of the internal droplets of the double emulsions. The solid line represents the predicted R_{drop}; (**b**) The flat velocity profile of the flow as it enters the capillary tube [110]. Reproduced with permission from Kim, J.W., et al., *Angewandte Chemie*; published by John Wiley and Sons, 2007.

Chang et al. designed a two co-axial capillaries microfluidic device to produce double droplets, from experiments they extracted an empirical law to predict core and shell sizes [112]. Since the core and the overall core shell drop has the same formation time, the core and the shell size could be predicted using the following equations, shown as Equations (2) and (3) (Q_I and Q_M are the inner and middle fluid flow rate respectively):

$$d_{core} = \sqrt[3]{\frac{Q_I}{Q_I + Q_M}} d_{drop} \tag{2}$$

$$d_{shell} = \frac{1}{2}(1 - \sqrt[3]{\frac{Q_I}{Q_I + Q_M}}) d_{drop} \tag{3}$$

3.2. Morphology Control of Core Shell Microparticles

As found by Dowding and Shum, when the polymer was formed by phase separation, the emulsion stabilizer had to be adjusted to control the particle morphology [113,114]. For combinations of a range of liquids, the final equilibrium morphology can be a core shell, or "acorn"-shaped, as shown in Figure 9. The transitions between different topologies of the double droplet can be described as the interfacial tensions model [115]. In the interfacial tension model, the balance of forces acting on the three-phase contact line is expressed in the form of relations (Neumann triangle Law, Figure 9a) between the contact angles and the interfacial tensions:

$$\gamma_{AB}\cos\theta_B + \gamma_B + \gamma_A(\theta_A + \theta_B) = 0 \tag{4}$$

$$\gamma_{AB}\cos\theta_A + \gamma_A + \gamma_B(\theta_A + \theta_B) = 0 \qquad (5)$$

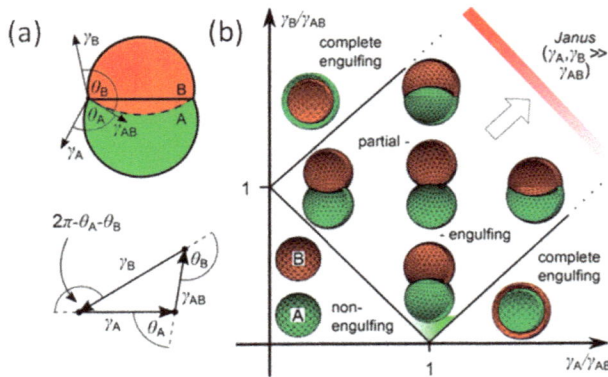

Figure 9. (**a**) Schematic of a double droplet with indicated contact angles θ_A and θ_B and the Neumann's triangle [115]; (**b**) Stability diagram representing the possible morphologies of a double droplet of phase A and B in the case $\kappa = V_B/V_A = 1$, κ is the ratio of the liquid volumes [115]. Reproduced with permission from Guzowski, J., et al., *Soft Matter*; published by Royal Society of Chemistry, 2012.

The existence and type of solution of Equations (4) and (5) depends on the values of the interfacial tensions. Guzowski and Korczyk used interfacial tensions to describe the transitions between different topologies of the droplets, marked by solid lines in Figure 9b [115]. They found three possible equilibrium topologies:

1. Complete-wetting: $\gamma_B > \gamma_{AB} + \gamma_A$, where a droplet of phase A is entirely encapsulated by phase B; vice versa ($\gamma_A > \gamma_{AB} + \gamma_B$), typical core shell structure;
2. Non-wetting: ($\gamma_{AB} > \gamma_A + \gamma_A$), droplets of phase A and B are separated by the outer phase;
3. Partial-wetting: droplets of phase A and B have a common interface and are both exposed to the external phase, corresponding to acorn-shaped or Janus.

This model has been generally applied to rationalize the particle morphologies observed when the polymer was caused to phase separate within the emulsified droplets, and in the presence of various core oils and aqueous emulsifier combinations [116]. By controlling the composition of the organic middle phase, the evolvement process will be conveniently transformed from initial core shell to the desired acorn-like configuration. Each of these morphologies has different potential applications. For example, the complete wetting state can be used for the encapsulation of active compounds, and partial wetting can be used to synthesize asymmetric and non-spherical functional particles [117].

3.3. Janus Particles

Janus particles are a class of anisotropic colloids, two sides of which have different compositions, polarities or surface modifications [117,118]. These particles have a wide range of potential applications in emulsion stabilization and dual-functionalized optical, electronic, sensor devices and incompatible drug delivery [119–123]. The "Janus" term has been also used for describing asymmetric dendritic macromolecules or unimolecular micelles based on block copolymers in solutions. Nevertheless, we will focus on hard and permanent Janus structures in this review, including biocompartmental, dumbbell-like, snowman-like, acorn-like and half-raspberry-like particles, as shown as Figure 10. In terms of materials, the reported Janus includes hydrogels and amphiphilic Janus.

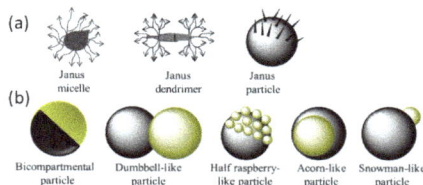

Figure 10. Schematic representation of Janus-like morphologies. (**a**) Extended Janus morphologies; (**b**) Solid Janus particles. (Note: spheres symbolize particles, diamonds and triangles symbolize chemical functions).

3.3.1. Hydrogel Janus

Hydrogel Janus is composed of two hydrogel phases produced from two completely immiscible hydrophilic monomer fluids. Unlike the principle for fabricating homogeneous particles, here, two separate streams are co-flowing through the same channel of the microfluidic device. However, the two fluids must remain parallel and the interface between them must be stable at different temporal and spatial scales, and any perturbations can lead to the formation of particles with mixed internal morphologies instead of particles with two distinct sides [24].

Shepherd and Conrad reported a scalable microfluidic assembly route for creating monodisperse silica colloid-filled hydrogel Janus [124]. Drops were formed by shearing a concentrated silica colloid-acrylamide aqueous suspension in a continuous oil phase using a sheath-flow device, as shown in Figure 11. Then, they immobilized the colloids within each drop by photo polymerizing the acrylamide to form a hydrogel. Seiffert et al. demonstrated a microfluidic technique to produce functional Janus microgels from prefabricated, cross-linkable precursor polymers [125]. This approach separated the particle formation from the synthesis of the polymer material, which allowed the droplet templating and functionalization of the matrix polymer to be controlled independently. Therefore, Janus particles were created with very specific, well-defined modifications of the two sides, even on a molecular level. In addition, they fabricated hollow microcapsules with two different sides (Janus shells), using the method as shown in Figure 12, and the size of the resultant droplets were controlled by adjusting the fluid flow rates and channel geometry.

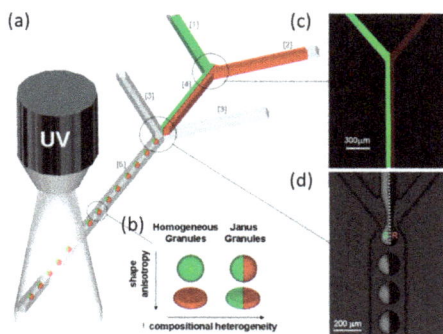

Figure 11. (**a**) Schematic representation of sheath-flow microfluidic device used to produce monodisperse colloid-filled hydrogel granules; (**b**) Schematic view of granule shapes and compositions explored; (**c**) Fluorescent image of Y-junction formed by inlets [1,2] for the production of Janus spheres; (**d**) Backlit fluorescence image (green excitation) illustration that the fluorescein isothiocyanate (FITC)-silica microspheres remain sequestered in the left hemisphere of each granule generated. Pictures are from [124]. Reproduced with permission from Shepherd, R.F., et al., *Langmuir*; published by American Chemical Society, 2006.

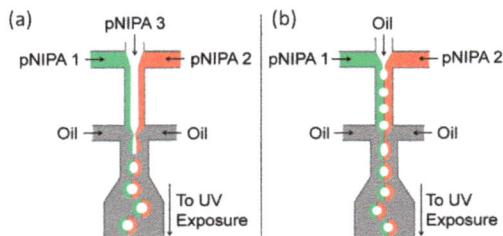

Figure 12. Formation of Janus microgels and microshells [125]. (**a**) Schematic of a microfluidic device forming aqueous droplets from three independent semidilute pNIPPAm solutions. The center phase (white) is assembled in the core of the droplets, whereas the right- and left-flowing phases (red and green) form the shell (**b**) operating the device in a modified way yields oil-water-oil double emulsions with Janus-shaped middle phases. Reproduced with permission from Seiffert, S., et al., *Langmuir*; published by American Chemical Society, 2010.

Although the mixing between liquids in a laminar flow was weak, it was enhanced by the hydrodynamic focusing of liquid threads before the break-up into droplets, which led to a gradual change in composition along the direction normal to the interface [126]. This limits the Janus application since a sharp interface between the phases is required.

3.3.2. Amphiphilic Janus

Amphiphilic Janus is formed from different immiscible fluids using double-emulsion droplets as templates. This method eliminates the problem of the mixing in the interface, however, the interfacial tension γ or spreading coefficients S_i must be within a certain range to assure the stability of double droplets within the partial-wetting area (in Figure 9). Thus, the portfolio of chemicals from a broad range of immiscible fluids is a key problem in the preparation of Janus particles.

Chen et al. fabricated acorn-like particles using W/O/W double emulsions as templates in PDMS microfluidic devices [80]. Moreover, the inner and the outer droplet size were adjusted by varying each fluid flow rate, and the swelling of the particles was controlled by varying the cross-linker concentration. Dendukuri et al. reported the synthesis and self-assembly of wedge-shaped particles bearing segregated hydrophilic and hydrophobic sections using continuous flow lithography technology (CFL) [127]. Monodisperse dumbbell-like hybrid Janus microspheres with organic and inorganic parts were prepared using fused perfluoroplyethers (the organic phase) and hydrolytic allylhydridopolycarbosilane (the inorganic phase) droplets as the template in a cross-flowing microfluidic device [128]. The particles had distinctive surface properties. The hydrophobic hemisphere had a smooth surface and the hydrophilic region had a rough, porous surface.

3.4. Microcapsules

Microcapsules are commonly used in pharmaceuticals, foods, cosmetics and absorbent [85,129–131]. The solid polymer shells provide effective encapsulation, protect the encapsulated drug or the active materials from hazardous environmental conditions, and give a release profile for a desired period. Though solid microspheres can also be used as drug delivery supporters, the drug distribution is largely dependent on the microsphere size [132,133], and the active-release mechanism is the diffusion-degradation [57], which limits improvement for medication compliance.

Microcapsules provide another way to control the drug-release rate and the release mechanism [2,90]. First, when a drug is localized in the core matrices, the shell prolongs the diffusion path of water-in and drug-out, and hence lowers the initial burst release [134]. Second, altering the properties of the shell, such as the shell thickness, may change the active transport kinetics [90]. For example, when increasing the polylactic acid (PLA) shell thickness to 10µm, the release profile will shift from a biphasic shape for pure PLGA microspheres to a zero-order piroxicam release [135]. However, it is still difficult to produce a microcapsule with a predicted size and morphology.

Equations (1)–(3) show a quantitative analysis of the relationship between the flow rate and the diameters of the inner and outer drops, and predicts the number of inner droplets [62,112]. Chu and Utada fabricated highly monodisperse multiple emulsions with controlled sizes and inner structures using capillary microfluidics, and obtained a liner relationship between the relative flow rate with drop diameters for both the inner and outer drops, and with the number of encapsulated droplets in the double emulsions [136]. By simply incorporating alternate emulsification schemes, more complicated multiple emulsions and microcapsules can be fabricated, as shown in Figure 13.

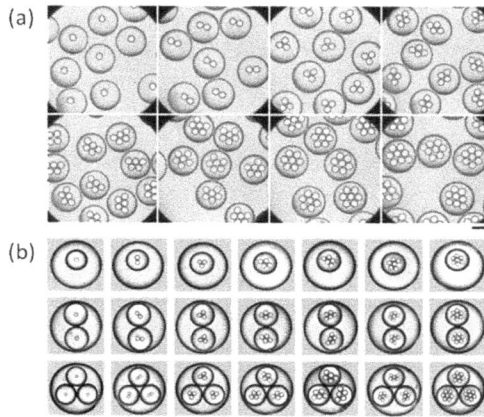

Figure 13. (**a**) Optical micrographs of double emulsions that contain a controlled number of inner droplets; (**b**) Optical micrographs of triple emulsions that contain a controlled number of inner and middle droplets. The scale bar in all images is 200 μm. Pictures are from [136]. Reproduced with permission from Chu, L.-Y., et al., *Angewandte Chemie International Edition*; published by John Wiley and Sons, 2007.

Carbon microspheres are of great interest due to their potential applications as cellular delivery vehicles, drug delivery carriers and absorbents [137–140]. Zhang et al. prepared monodisperse poly(furfuryl alcohol) (PFA) hollow microspheres by the microfluidic technique in a T-junction device through interfacial polymerization of FA in H_2SO_4 solution droplets. After pyrolysis, they produced carbon microspheres with mean particle sizes of 0.7–1.2 μm [141,142]. If some solid nanobeads or microbeads are added in the inner fluid, this method can be adapted to make microcapsules with controlled holes [107]. Other functional nanoparticles such as SiO_2, Au, Co, Fe_3O_4 and fluorescence nanoparticles have been introduced into the inner fluid to fabricate functional microcapsules [117,143,144].

3.5. Vesicles

An ideal encapsulating structure should not only have high encapsulation efficiency, but also should be easily triggered to release the actives. Vesicles are a good choice for an encapsulating structure. Vesicles are a compartment of one fluid enclosed by a bilayer of amphiphilic molecules, such as phospholipids, polypeptides and diblock copolymers, which correspond to liposomes and polymersomes. The bilayer membrane can encapsulate hydrophilic and hydrophobic molecules simultaneously. Thus multiple drugs with different properties stored in a single carrier can be released at the same time. The bilayer membrane also induces semi-permeability of small molecules, such as water, leading to inflated or deflated responses to osmotic pressure differences between the aqueous core and surrounding environment.

Liposomes are biocompatible vesicles with phospholipid bilayers. They have attracted much attention because phospholipids constitute the majority of biological membranes found in nature, such as plasma membranes. Thus, they have great potential for encapsulation and targeted drug delivery considering their biocompatible properties. The disadvantage is that they are more fragile than polymersomes, and their preparation is more delicate.

Lorenceau and Utada [37] used water in oil in water (W/O/W) double emulsions as templates to form monodisperse polymersomes by using a diblock copolymer poly(normal-butyl acrylate)-poly(acrylic acid) (PBA-PAA) in a capillary microfluidic device. The amphiphilic PBA-PAA in the middle fluid stabilized the two oil-water interfaces, and self-assembled on the interface during the dewetting oil phase upon solvent evaporation. The concentration of the amphiphilic molecules was a key control variable in the fabrication of polymersomes in this fabrication process [38]. If the concentration was lower than the amount required to fully cover the interfaces, the polymersomes were unstable. However, too much excess created a depletion interaction.

By integrating the advantages of the diblock copolymer and the microdroplet based-on microfluidic technology, the polymersome can be easily fabricated with excellent encapsulation efficiency, high levels of loading and tunable wall properties. For example, since the three fluids can be controlled individually, the encapsulate can contain the same amount of active material in each polymersomes and guarantee that the encapsulation efficiency will reach 100% [38]. In addition, the character of each block in the diblock copolymer can be tuned to fit the desired application. For example, tuning the molecular weight ratio of the hydrophilic and the hydrophobic blocks, the wetting angle of the polymer-containing solvent phase on the polyersomes will change in the emulsion-to-vesicles transition [114], as shown in Figure 14. The polymerization degree of the individual diblock molecules will affect the membrane thickness, whereas the elasticity and permeability of the membrane can be adjusted by changing the glass transition temperature of the hydrophobic block [145–147].

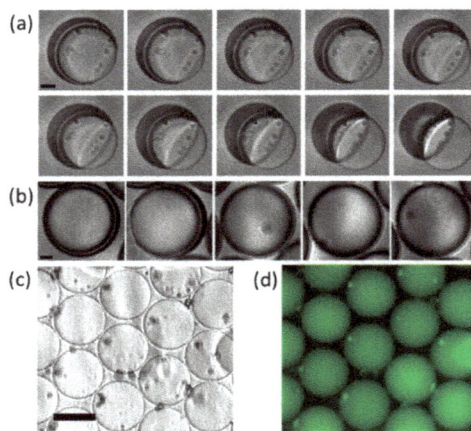

Figure 14. (a) Bright-field microscope images of a PEG(5000)-b-PLA(5000) polymersome undergoing dewetting transition; (b) Bright-field microscope images of a PEG(1000)-b-PLA(5000) polymersome undergoing the evaporation of the organic solvent shell; (c,d) Bright-field and fluorescence microscope images of a dried capsule formed from the PEG(1000)-b-PLA(5000) diblock copolymer. Pictures are from [114]. Reproduced with permission from Shum, H.C., et al., *Journal of the American Chemical Society*; published by American Chemical Society, 2008.

Core shell structured fibers are also excellent delivery vehicles for medicines. Wang et al. fabricated fibers with core shell structures by emulsion electrospinning [148]. The water in oil (W/O) emulsions were composed of deionized water or a phosphate buffer saline, and a poly(lactic-co-glycolic acid) (PLGA)/chloroform solution that contained a surfactant was used as a module to produce core shell structured fibers. This study demonstrated the evolution of the core shell structured fibers by investigating the water phase morphology in jets or fibers at different locations of the jet or fiber trajectory during the electrospinning process, as shown in Figure 15. They found that the water phase in emulsion jets experienced multi-level stretching and broke up. The Rayleigh capillary instability and the solvent evaporation rate significantly affected the breakup of water droplets.

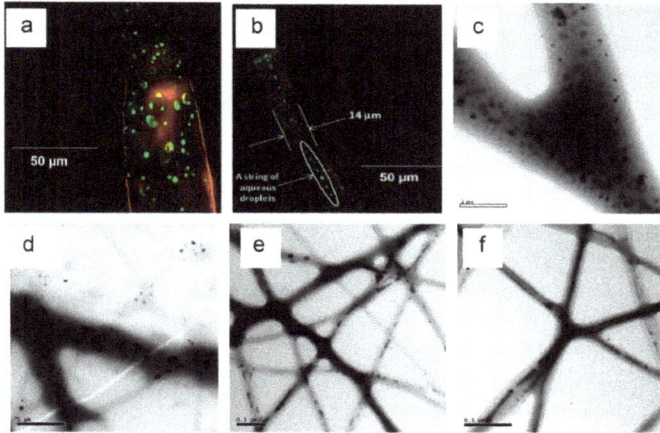

Figure 15. Morphological evaluation of the water phase in electrospun fibers collected at different locations of the emulsion jet path: (**a**,**b**) fluorescence microscopy images of fibers; (**c**–**f**) TEM micrographs at different magnifications. Pictures are from [148]. Reproduced with permission from Wang, C., et al., *Materials Letters*; published by Elsevier, 2014.

Since the mechanical properties of core shell microparticles made from materials with dramatically different elastic properties are also important factors in medical applications, the mechanical properties of these microparticles were measured and predicted by a microfluidic approach [149]. By forcing the particles through a tapered capillary and analyzing their deformation, the shear and compressive moduli were easily measured in one single experiment. The results showed that the moduli of these core shell structures were determined both by the material composition of the core shell microparticles and by their microstructures.

4. Conclusions

Droplet microfluidics can provide environments with properties of exceptional control and good stability, which can be used for performing microparticles synthesis with unique properties. We reviewed the different microfluidic systems for controlling droplet formation, the influencing factors for regulating droplet size and structure, and droplets solidification methods according to different materials. To make the applications of droplet microfluidics in preparation of micro- and nanoparticles more clear, we introduced the preparation methods of different materials, different structures and different functional microspheres in detail. Because these particles are characterized by particle size uniformity, structure controllability and component controllability, they are more widely applicable than those prepared by traditional methods, especially in new medicine (embolization treatment of tumor, drug controlled-release, multi-drug loading microspheres), adsorption separation, dual-functionalized optical, electrical and magnetic devices, etc. This technique has become a powerful platform in material fabrication, and biological and medical research. However, one obstacle in the

Micromachines **2017**, *8*, 22

practical application in microparticles preparation is their low-throughput, which limits the production efficiency. Recent advances have partially overcome this barrier by parallelizing droplet generation and enhanced the production rate by a factor of 100, but these parallel experiments were used primarily for the generation of spherical single droplets and didnot report on the generation of non-spherical, double or multi-emulsion, and nor did perform on droplets in situ curing parallelization. More efforts are needed to resolve these technical issues. Given the distinctive properties of this technique and its tremendous demand in applications, droplet microfluidics will fundamentally modify the future of micro- and nano-manufacturing and drug delivery.

Acknowledgments: The financial support from the National Natural Science Foundation of China (No. 21605094), Shandong Province Natural Science Foundation (No. ZR2015YL006, No. ZR2016BB15), The Youth Science fund of Shandong Academy of Sciences (No. 2016QN006), The Pilot Project Scheme and Basic Research Grant (2015.4-2017.4) of Shandong Academy of Sciences, The Research Grants Council of Hong Kong (GRF 17237316, 17211115, 17207914 and 717613E) and the University of Hong Kong (URC 201511159108, 201411159074 and 201311159187) are gratefully acknowledged.

Author Contributions: Jianmei Wang was mainly involved in writing the manuscript; Chengyang Wang and Liqiu Wang conceived and reviewed the manuscipt; Yan Li contributed in discussion and helped modify the manuscript; Xueying Wang and Jianchun Wang helped drawing the figures; Hanmei Tian, Pei Zhao, Ye Tian and Yeming Gu helped searching the literatures.

Conflicts of Interest: The authors declare no conflict of interest.

References

1. Cazado, C.P.S.; Pinho, S.C.D. Effect of different stress conditions on the stability of quercetin-loaded lipid microparticles produced with babacu (*Orbignya speciosa*) oil: Evaluation of their potential use in food applications. *Food Sci. Technol.* **2016**. ahead. [CrossRef]
2. Wu, J.; Kong, T.; Yeung, K.W.K.; Shum, H.C.; Cheung, K.M.C.; Wang, L.; To, M.K.T. Fabrication and characterization of monodisperse PLGA–alginate core–shell microspheres with monodisperse size and homogeneous shells for controlled drug release. *Acta Biomater.* **2013**, *9*, 7410–7419. [CrossRef] [PubMed]
3. Nidhi, R.M.; Kaur, V.; Hallan, S.S.; Sharma, S.; Mishra, N. Microparticles as controlled drug delivery carrier for the treatment of ulcerative colitis: A brief review. *Saudi Pharm. J.* **2016**, *24*, 458–472. [CrossRef] [PubMed]
4. Elsherbiny, I.M.; Abbas, Y. Janus Nano- and Microparticles as Smart Drug Delivery Systems. *Curr. Pharm. Biotechnol.* **2016**, *17*, 673–682. [CrossRef]
5. Shen, Y.; Zhao, Q.; Li, X.; Zhang, D. Monodisperse $Ca_{0.15}Fe_{2.85}O_4$ microspheres: Facile preparation, characterization, and optical properties. *J. Mater. Sci.* **2012**, *47*, 3320–3326. [CrossRef]
6. Yamada, H.; Nakamura, T.; Yamada, Y.; Yano, K. Colloidal-Crystal Laser Using Monodispersed Mesoporous Silica Spheres. *Adv. Mater.* **2009**, *21*, 4134–4138. [CrossRef]
7. Horák, D.; Kučerová, J.; Korecká, L.; Jankovi, B.; Mikulášek, P.; Bílková, Z. New Monodisperse Magnetic Polymer Microspheres Biofunctionalized for Enzyme Catalysis and Bioaffinity Separations. *Macromol. Biosci.* **2012**, *12*, 647–655. [CrossRef] [PubMed]
8. Fang, Q.; Cheng, Q.; Xu, H.; Xuan, S. Monodisperse magnetic core/shell microspheres with Pd nanoparticles-incorporated-carbon shells. *Dalton Trans.* **2013**, *43*, 2588–2595. [CrossRef] [PubMed]
9. Deng, Y.; Cai, Y.; Sun, Z.; Liu, J.; Liu, C.; Wei, J.; Li, W.; Liu, C.; Wang, Y.; Zhao, D. Multifunctional mesoporous composite microspheres with well-designed nanostructure: A highly integrated catalyst system. *J. Am. Chem. Soc.* **2010**, *132*, 8466–8473. [CrossRef] [PubMed]
10. Sang, E.S.; Yang, S.; Choi, H.H.; Choe, S. Fully crosslinked poly(styrene-co-divinylbenzene) microspheres by precipitation polymerization and their superior thermal properties. *J. Polym. Sci. Part A Polym. Chem.* **2004**, *42*, 835–845.
11. Goh, E.C.C.; Stöver, H.D.H. Cross-Linked Poly(methacrylic acid-co-poly(ethylene oxide) methyl ether methacrylate) Microspheres and Microgels Prepared by Precipitation Polymerization: A Morphology Study. *Macromolecules* **2002**, *35*, 9983–9989. [CrossRef]
12. Chu, Y.; Zhang, P.; Hu, J.; Yang, W.; Wang, C. Synthesis of Monodispersed Co(Fe)/Carbon Nanocomposite Microspheres with Very High Saturation Magnetization. *J. Phys. Chem C* **2009**, *113*, 4047–4052. [CrossRef]

13. Li, Y.; Chen, J.; Xu, Q.; He, L.; Chen, Z. Controllable Route to Solid and Hollow Monodisperse Carbon Nanospheres. *J. Phys. Chem. C* **2009**, *113*, 10085–10089. [CrossRef]

14. Choi, J.; Kwak, S.Y.; Kang, S.; Lee, S.S.; Park, M.; Lim, S.; Kim, J.; Choe, C.R.; Hong, S.I. Synthesis of highly crosslinked monodisperse polymer particles: Effect of reaction parameters on the size and size distribution. *J. Polym. Sci. Part A Polym. Chem.* **2002**, *40*, 4368–4377. [CrossRef]

15. Lee, J.; Ha, J.U.; Choe, S.; Lee, C.S.; Shim, S.E. Synthesis of highly monodisperse polystyrene microspheres via dispersion polymerization using an amphoteric initiator. *J. Colloid Interface Sci.* **2006**, *298*, 663–671. [CrossRef] [PubMed]

16. Akamatsu, K.; Chen, W.; Suzuki, Y.; Ito, T.; Nakao, A.; Sugawara, T.; Kikuchi, R.; Nakao, S. Preparation of monodisperse chitosan microcapsules with hollow structures using the SPG membrane emulsification technique. *Langmuir ACS J. Surf. Colloids* **2010**, *26*, 14854–14860. [CrossRef] [PubMed]

17. Feng, Q.; Wu, J.; Yang, T.; Ma, G.; Su, Z. Mechanistic studies for monodisperse exenatide-loaded PLGA microspheres prepared by different methods based on SPG membrane emulsification. *Acta Biomater.* **2014**, *10*, 4247–4256.

18. Caruso, F.; Spasova, M.; Susha, A.; Giersig, M.; Caruso, R.A. Magnetic Nanocomposite Particles and Hollow Spheres Constructed by a Sequential Layering Approach. *Chem. Mater.* **2001**, *13*, 109–116. [CrossRef]

19. Tan, S.H.; Maes, F.; Semin, B.; Vrignon, J.; Baret, J.C. The Microfluidic Jukebox. *Sci. Rep.* **2014**, *4*, 597–600. [CrossRef] [PubMed]

20. Tan, S.H.; Nguyen, N.T. Generation and manipulation of monodispersed ferrofluid emulsions: The effect of a uniform magnetic field in flow-focusing and T-junction configurations. *Phys. Rev. E* **2011**, *84*, 2299–2304. [CrossRef] [PubMed]

21. Zeng, S.; Li, B.; Su, X.; Qin, J.; Lin, B. Microvalve-actuated precise control of individual droplets in microfluidic devices. *Lab Chip* **2009**, *9*, 1340–1343. [CrossRef] [PubMed]

22. Ma, Z.; Teo, A.; Tan, S.; Ai, Y.; Nguyen, N.T. Self-Aligned Interdigitated Transducers for Acoustofluidics. *Micromachines* **2016**, *7*, 216. [CrossRef]

23. Tan, S.-H.; Sohel Murched, S.M.; Nguyen, N.-T.; Neng, T.N.; Wong, T.; Yobas, L. Thermally controlled droplet formation in flow focusing geometry: Formation regimes and effect of nanoparticle suspension. *J. Phys. D Appl. Phys.* **2008**, *41*, 165501. [CrossRef]

24. Nisisako, T.; Torii, T.; Takahashi, T.; Takizawa, Y. Synthesis of Monodisperse Bicolored Janus Particles with Electrical Anisotropy Using a Microfluidic Co-Flow System. *Adv. Mater.* **2006**, *18*, 1152–1156. [CrossRef]

25. Othman, R.; Vladisavljević, G.T.; Bandulasena, H.C.H.; Nagy, Z.K. Production of polymeric nanoparticles by micromixing in a co-flow microfluidic glass capillary device. *Chem. Eng. J.* **2015**, *280*, 316–329. [CrossRef]

26. Dang, T.D.; Kim, Y.H.; Kim, H.G.; Kim, G.M. Preparation of monodisperse PEG hydrogel microparticles using a microfluidic flow-focusing device. *J. Ind. Eng. Chem.* **2012**, *18*, 1308–1313. [CrossRef]

27. Anna, S.L.; Bontoux, N.; Stone, H.A. Formation of dispersions using "flow focusing" in microchannels. *Appl. Phys. Lett.* **2003**, *82*, 364–366. [CrossRef]

28. Cohen, C.; Giles, R.; Sergeyeva, V.; Mittal, N.; Tabeling, P.; Zerrouki, D.; Baudry, J.; Bibette, J.; Bremond, N. Parallelised production of fine and calibrated emulsions by coupling flow-focusing technique and partial wetting phenomenon. *Microfluid. Nanofluid.* **2016**, *17*, 959–966. [CrossRef]

29. Garstecki, P.; Fuerstman, M.J.; Stone, H.A.; Whitesides, G.M. Formation of droplets and bubbles in a microfluidic T-junction-scaling and mechanism of break-up. *Lab Chip* **2006**, *6*, 437–446. [CrossRef] [PubMed]

30. Wu, Y.; Fu, T.; Ma, Y.; Li, H.Z. Active control of ferrofluid droplet breakup dynamics in a microfluidic T-junction. *Microfluid. Nanofluid.* **2015**, *18*, 19–27. [CrossRef]

31. Chokkalingam, V.; Herminghaus, S.; Seemann, R. Self-synchronizing pairwise production of monodisperse droplets by microfluidic step emulsification. *Appl. Phys. Lett.* **2008**, *93*. [CrossRef]

32. Kobayashi, I.; Nakajima, M.; Nabetani, H.; Kikuchi, Y.; Shohno, A.; Satoh, K. Preparation of micron-scale monodisperse oil-in-water microspheres by microchannel emulsification. *J. Am. Oil Chem. Soc.* **2001**, *78*, 797–802. [CrossRef]

33. Khalid, N.; Kobayashi, I.; Neves, M.A.; Uemura, K.; Nakajima, M.; Nabetani, H. Monodisperse W/O/W emulsions encapsulating L-ascorbic acid: Insights on their formulation using microchannel emulsification and stability studies. *Colloids Surf. A Physicochem. Eng. Asp.* **2014**, *458*, 69–77. [CrossRef]

34. Kang, D.-K.; Monsur Ali, M.; Zhang, K.; Pone, E.J.; Zhao, W. Droplet microfluidics for single-molecule and single-cell analysis for cancer research, diagnosis and therapy. *TrAC Trends Anal. Chem.* **2014**, *58*, 145–153. [CrossRef]

35. Streets, A.M.; Huang, Y. Microfluidics for biological measurements with single-molecule resolution. *Curr. Opin. Biotechnol.* **2014**, *25*, 69–77. [CrossRef] [PubMed]

36. Thorsen, T.; Roberts, R.W.; Arnold, F.H.; Quake, S.R. Dynamic Pattern Formation in a Vesicle-Generating Microfluidic Device. *Phys. Rev. Lett.* **2001**, *86*, 4163–4166. [CrossRef] [PubMed]

37. Utada, A.; Lorenceau, E.; Link, D.; Kaplan, P.; Stone, H.; Weitz, D. Monodisperse double emulsions generated from a microcapillary device. *Science* **2005**, *308*, 537–541. [CrossRef] [PubMed]

38. Shah, R.K.; Shum, H.C.; Rowat, A.C.; Lee, D.; Agresti, J.J.; Utada, A.S.; Chu, L.-Y.; Kim, J.-W.; Fernandez-Nieves, A.; Martinez, C.J.; Weitz, D.A. Designer emulsions using microfluidics. *Mater. Today* **2008**, *11*, 18–27. [CrossRef]

39. Zhang, H.; Tumarkin, E.; Sullan, R.M.A.; Walker, G.C.; Kumacheva, E. Exploring Microfluidic Routes to Microgels of Biological Polymers. *Macromol. Rapid Commun.* **2007**, *28*, 527–538. [CrossRef]

40. Zhao, C.X.; Middelberg, A.P.J. Two-phase microfluidic flows. *Chem. Eng. Sci.* **2011**, *66*, 1394–1411. [CrossRef]

41. Chong, Z.Z.; Tan, S.H.; Gañán-Calvo, A.M.; Tor, S.B.; Loh, N.H.; Nguyen, N.T. Active droplet generation in microfluidics. *Lab Chip* **2016**, *16*, 35. [CrossRef] [PubMed]

42. Zhu, P.; Wang, L. Passive and active droplet generation with microfluidics: A review. *Lab Chip* **2017**, *17*, 34–75. [CrossRef] [PubMed]

43. Zeng, Y.; Shin, M.; Wang, T. Programmable active droplet generation enabled by integrated pneumatic micropumps. *Lab Chip* **2012**, *13*, 267–273. [CrossRef] [PubMed]

44. Beer, N.R.; Wheeler, E.K.; Leehoughton, L.; Watkins, N.; Nasarabadi, S.; Hebert, N.; Leung, P.; Arnold, D.W.; Bailey, C.G.; Colston, B.W. On-chip single-copy real-time reverse-transcription PCR in isolated picoliter droplets. *Anal. Chem.* **2007**, *80*, 1854–1858. [CrossRef] [PubMed]

45. Willingale, J.; Manzarpour, A.; Mantle, P.G. Continuous synthesis of gold nanoparticles in a microreactor. *Nano Lett.* **2005**, *5*, 685–691.

46. Chu, L.Y.; Kim, J.W.; Shah, R.K.; Weitz, D.A. Monodisperse Thermoresponsive Microgels with Tunable Volume-Phase Transition Kinetics. *Adv. Funct. Mater.* **2007**, *17*, 3499–3504. [CrossRef]

47. Lee, I.; Yoo, Y.; Cheng, Z.; Jeong, H.K. Generation of monodisperse mesoporous silica microspheres with controllable size and surface morphology in a microfluidic device. *Adv. Funct. Mater.* **2008**, *18*, 4014–4021. [CrossRef]

48. Datta, S.S.; Abbaspourrad, A.; Amstad, E.; Fan, J.; Kim, S.H.; Romanowsky, M.; Shum, H.C.; Sun, B.; Utada, A.S.; Windbergs, M. 25th anniversary article: Double emulsion templated solid microcapsules: Mechanics and controlled release. *Adv. Mater.* **2014**, *26*, 2205–2218. [CrossRef] [PubMed]

49. Xu, S.; Nie, Z.; Seo, M.; Lewis, P.; Kumacheva, E.; Stone, H.A.; Garstecki, P.; Weibel, D.B.; Gitlin, I.; Whitesides, G.M. Generation of monodisperse particles by using microfluidics: Control over size, shape, and composition. *Angew. Chem.* **2005**, *117*, 734–738. [CrossRef]

50. Pelage, J.-P.; Laurent, A.; Wassef, M.; Bonneau, M.; Germain, D.; Rymer, R.; Flaud, P.; Martal, J.; Merland, J.J. Uterine Artery Embolization in Sheep: Comparison of Acute Effects with Polyvinyl Alcohol Particles and Calibrated Microspheres. *Radiology* **2002**, *224*, 436–445. [CrossRef] [PubMed]

51. Gomes, A.S.; Rosove, M.H.; Rosen, P.J.; Amado, R.G.; Sayre, J.W.; Monteleone, P.A.; Busuttil, RW. Triple-drug transcatheter arterial chemoembolization in unresectable hepatocellular carcinoma: Assessment of survival in 124 consecutive patients. *Am. J. Roentgenol.* **2009**, *193*, 1665–1671. [CrossRef] [PubMed]

52. Beaujeux, R.; Laurent, A.; Wassef, M.; Casasco, A.; Gobin, Y.-P.; Aymard, A.; Rüfenacht, D.; Merland, J.J. Trisacryl gelatin microspheres for therapeutic embolization, II: Preliminary clinical evaluation in tumors and arteriovenous malformations. *Am. J. Neuroradiol.* **1996**, *17*, 541–548. [PubMed]

53. Carugo, D.; Capretto, L.; Willis, S.; Lewis, A.L.; Grey, D.; Hill, M.; Zhang, X. A microfluidic device for the characterisation of embolisation with polyvinyl alcohol beads through biomimetic bifurcations. *Biomed. Microdevice* **2012**, *14*, 153–163. [CrossRef] [PubMed]

54. Kong, T.; Wu, J.; To, M.; Wai Kwok Yeung, K.; Cheung Shum, H.; Wang, L. Droplet based microfluidic fabrication of designer microparticles for encapsulation applications. *Biomicrofluidics* **2012**, *6*, 34104. [CrossRef] [PubMed]

55. Lai, B.; Wei, Q.; Sun, C.-B.; Ma, G.-H.; Su, Z.-G. Preparation of Uniform-sized PELA Microspheres Containing Lysozyme by Membrane Emulsification and Double Emulsion-Solvent Removal Method. *Chin. J. Process Eng.* **2008**, *8*, 327–332.

56. Vladisavljević, G.T.; Schubert, H. Influence of process parameters on droplet size distribution in SPG membrane emulsification and stability of prepared emulsion droplets. *J. Membr. Sci.* **2003**, *225*, 15–23. [CrossRef]

57. Berkland, C.; Kipper, M.J.; Narasimhan, B.; Kim, K.K.; Pack, D.W. Microsphere size, precipitation kinetics and drug distribution control drug release from biodegradable polyanhydride microspheres. *J. Control. Release* **2004**, *94*, 129–141. [CrossRef] [PubMed]

58. Xu, Q.; Hashimoto, M.; Dang, T.T.; Hoare, T.; Kohane, D.S.; Whitesides, G.M.; Langer, R.; Anderson, D.G. Preparation of Monodisperse Biodegradable Polymer Microparticles Using a Microfluidic Flow-Focusing Device for Controlled Drug Delivery. *Small* **2009**, *5*, 1575–1581. [CrossRef] [PubMed]

59. Seo, M.; Nie, Z.; Xu, S.; Mok, M.; Lewis, P.C.; Graham, R.; Kumacheva, E. Continuous Microfluidic Reactors for Polymer Particles. *Langmuir* **2005**, *21*, 11614–11622. [CrossRef] [PubMed]

60. Serra, C.; Berton, N.; Bouquey, M.; Prat, L.; Hadziioannou, G. A Predictive Approach of the Influence of the Operating Parameters on the Size of Polymer Particles Synthesized in a Simplified Microfluidic System. *Langmuir* **2007**, *23*, 7745–7750. [CrossRef] [PubMed]

61. Theberge, A.B.; Courtois, F.; Schaerli, Y.; Fischlechner, M.; Abell, C.; Hollfelder, F.; Huck, W.S.T. Microdroplets in Microfluidics: An Evolving Platform for Discoveries in Chemistry and Biology. *Angew. Chem. Int. Ed.* **2010**, *49*, 5846–5868. [CrossRef] [PubMed]

62. Abate, A.R.; Seiffert, S.; Utada, A.S.; Shum, A.; Shah, R.; Thiele, J.; Duncanson, W.J.; Abbaspourad, A.; Lee, M.H.; Akartuna, I.; et al. Microfluidic Techniques for Synthesizing Particles. Available online: http://weitzlab.seas.harvard.edu/publications/Bookchapter_Microfluidic_techniques.pdf (accessed on 13 January 2017).

63. Pankhurst, Q.A.; Connolly, J.; Jones, S.K.; Dobson, J. Applications of magnetic nanoparticles in biomedicine. *J. Phys. D Appl. Phys.* **2003**, *36*, R167. [CrossRef]

64. Fleischmann, E.K.; Ohm, C.; Serra, C.; Zentel, R. Preparation of Soft Microactuators in a Continuous Flow Synthesis Using a Liquid-Crystalline Polymer Crosslinker. *Macromol. Chem. Phys.* **2012**, *213*, 1871–1878. [CrossRef]

65. Ohm, C.; Fleischmann, E.K.; Kraus, I.; Serra, C.; Zentel, R. Control of the Properties of Micrometer-Sized Actuators from Liquid Crystalline Elastomers Prepared in a Microfluidic Setup. *Adv. Funct. Mater.* **2010**, *20*, 4314–4322. [CrossRef]

66. Muvaffak, A.; Gurhan, I.; Gunduz, U.; Hasirci, N. Preparation and characterization of a biodegradable drug targeting system for anticancer drug delivery: Microsphere-antibody conjugate. *J. Drug Target.* **2005**, *13*, 151–159. [CrossRef] [PubMed]

67. Amesur, N.; Zajko, A.; Carr, B. Chemo-embolization for Unresectable Hepatocellular Carcinoma with Different Sizes of Embolization Particles. *Dig. Dis. Sci.* **2008**, *53*, 1400–1404. [CrossRef] [PubMed]

68. Lee, K.-H.; Liapi, E.; Vossen, J.A.; Buijs, M.; Ventura, V.P.; Georgiades, C.; Hong, K.; Kamel, I.; Torbenson, M.S.; Geschwind, J.F. Distribution of Iron Oxide–containing Embosphere Particles after Transcatheter Arterial Embolization in an Animal Model of Liver Cancer: Evaluation with MR Imaging and Implication for Therapy. *J. Vasc. Interv. Radiol.* **2008**, *19*, 1490–1496. [CrossRef] [PubMed]

69. Wang, Z.-Y.; Song, J.; Zhang, D.-S. Nanosized As_2O_3/Fe_2O_3 complexes combined with magnetic fluid hyperthermia selectively target liver cancer cells. *World J. Gastroenterol.* **2009**, *15*, 2995–3002. [CrossRef] [PubMed]

70. Slowing, I.I.; Vivero-Escoto, J.L.; Wu, C.W.; Lin, V.S. Mesoporous silica nanoparticles as controlled release drug delivery and gene transfection carriers. *Adv. Drug Deliv. Rev.* **2008**, *60*, 1278–1288. [CrossRef] [PubMed]

71. Gu, J.; Su, S.; Li, Y.; He, Q.; Shi, J. Hydrophilic mesoporous carbon nanoparticles as carriers for sustained release of hydrophobic anti-cancer drugs. *Chem. Commun.* **2011**, *47*, 2101–2103. [CrossRef] [PubMed]

72. Stöber, W.; Fink, A.; Bohn, E. Controlled growth of monodisperse silica spheres in the micron size range. *J. Colloid Interface Sci.* **1968**, *26*, 62–69. [CrossRef]

73. Liu, J.; Qiao, S.Z.; Liu, H.; Chen, J.; Orpe, A.; Zhao, D.; Qing, G. Extension of the Stöber Method to the preparation of monodisperse resorcinol–formaldehyde resin polymer and carbon spheres. *Angew. Chem. Int. Ed.* **2011**, *50*, 5947–5951. [CrossRef] [PubMed]

74. Carroll, N.J.; Rathod, S.B.; Derbins, E.; Mendez, S.; Weitz, D.A.; Petsev, D.N. Droplet-based microfluidics for emulsion and solvent evaporation synthesis of monodisperse mesoporous silica microspheres. *Langmuir* **2008**, *24*, 658–661. [CrossRef] [PubMed]

75. Rao, G.V.R.; López, G.P.; Bravo, J.; Pham, H.; Datye, A.K.; Xu, H.F.; Ward, T.L. Monodisperse Mesoporous Silica Microspheres Formed by Evaporation-Induced Self Assembly of Surfactant Templates in Aerosols. *Adv. Mater.* **2002**, *14*, 1301–1304.

76. Song, H.; Tice, J.D.; Ismagilov, R.F. A Microfluidic System for Controlling Reaction Networks in Time. *Angew. Chem. Int. Ed.* **2003**, *42*, 768–772. [CrossRef] [PubMed]

77. Habault, D.; Dery, A.; Leng, J.; Lecommandoux, S.; Le Meins, J.F.; Sandre, O. Droplet Microfluidics to Prepare Magnetic Polymer Vesicles and to Confine the Heat in Magnetic Hyperthermia. *IEEE Trans. Magn.* **2013**, *49*, 182–190. [CrossRef]

78. Zhang, J.; Coulston, R.J.; Jones, S.T.; Geng, J.; Scherman, O.A.; Abell, C. One-Step Fabrication of Supramolecular Microcapsules from Microfluidic Droplets. *Science* **2012**, *335*, 690–694. [CrossRef] [PubMed]

79. Zhao, C.-X.; He, L.; Qiao, S.Z.; Middelberg, A.P. Nanoparticle synthesis in microreactors. *Chem. Eng. Sci.* **2011**, *66*, 1463–1479. [CrossRef]

80. Chen, C.-H.; Shah, R.K.; Abate, A.R.; Weitz, D.A. Janus particles templated from double emulsion droplets generated using microfluidics. *Langmuir* **2009**, *25*, 4320–4323. [CrossRef] [PubMed]

81. Song, Y.; Modrow, H.; Henry, L.L.; Saw, C.K.; Doomes, E.; Palshin, V.; Hormes, J.; Kumar, C.S.S.R. Microfluidic synthesis of cobalt nanoparticles. *Chem. Mater.* **2006**, *18*, 2817–2827. [CrossRef]

82. Wagner, J.; Köhler, J.M. Continuous Synthesis of Gold Nanoparticles in a Microreactor. *Nano Lett.* **2005**, *5*, 685–691. [CrossRef] [PubMed]

83. Hler, J.M.; Romanus, H.; Bner, U.; Wagner, J. Formation of Star-Like and Core-Shell AuAg Nanoparticles during Two- and Three-Step Preparation in Batch and in Microfluidic Systems. *J. Nanomater.* **2007**, *2007*, 98134.

84. Köhler, J.M.; Held, M.; Hübner, U.; Wagner, J. Formation of Au/Ag Nanoparticles in a Two Step Micro Flow-Through Process. *Chem. Eng. Technol.* **2007**, *30*, 347–354. [CrossRef]

85. Duraiswamy, S.; Khan, S.A. Droplet-Based Microfluidic Synthesis of Anisotropic Metal Nanocrystals. *Small* **2009**, *5*, 2828–2834. [CrossRef] [PubMed]

86. Wei, X.; Kong, T.; Wang, L. Copper Nanoparticles and Nanofluids-based W/O Emulsions Synthesized with Droplet Microreactors. *Curr. Nanosci.* **2012**, *8*, 117–119. [CrossRef]

87. Chan, E.M.; Alivisatos, A.P.; Mathies, R.A. High-temperature microfluidic synthesis of CdSe nanocrystals in nanoliter droplets. *J. Am. Chem. Soc.* **2005**, *127*, 13854–13861. [CrossRef] [PubMed]

88. Tice, J.D.; Lyon, A.D.; Ismagilov, R.F. Effects of viscosity on droplet formation and mixing in microfluidic channels. *Anal. Chim. Acta* **2004**, *507*, 73–77. [CrossRef]

89. Bentley, B.J.; Leal, L.G. An experimental investigation of drop deformation and breakup in steady, two-dimensional linear flows. *J. Fluid Mech.* **1986**, *167*, 241–283. [CrossRef]

90. Kong, T.; Liu, Z.; Song, Y.; Wang, L.; Shum, H.C. Engineering polymeric composite particles by emulsion-templating: Thermodynamics versus kinetics. *Soft Matter.* **2013**, *9*, 9780–9784. [CrossRef]

91. Lu, Y.; Yin, Y.; Xia, Y. Three-Dimensional Photonic Crystals with Non-spherical Colloids as Building Blocks. *Adv. Mater.* **2001**, *13*, 415–420. [CrossRef]

92. Langer, R.; Tirrell, D.A. Designing materials for biology and medicine. *Nature* **2004**, *428*, 487–492. [CrossRef] [PubMed]

93. Bonderer, L.J.; Studart, A.R.; Gauckler, L.J. Bioinspired design and assembly of platelet reinforced polymer films. *Science* **2008**, *319*, 1069–1073. [CrossRef] [PubMed]

94. Liu, Z.; Cai, W.; He, L.; Nakayama, N.; Chen, K.; Sun, X.; Chen, X.; Dai, H. In vivo biodistribution and highly efficient tumour targeting of carbon nanotubes in mice. *Nat. Nanotechnol.* **2007**, *2*, 47–52. [CrossRef] [PubMed]

95. Shum, H.C.; Abate, A.R.; Lee, D.; Studart, A.R.; Wang, B.; Chen, C.H.; Thiele, J.; Shah, R.K.; Krummel, A.; Weitz, D.A. Droplet Microfluidics for Fabrication of Non-Spherical Particles. *Macromol. Rapid Commun.* **2010**, *31*, 108–118. [CrossRef] [PubMed]

96. Haghgooie, R.; Toner, M.; Doyle, P.S. Squishy Non-Spherical Hydrogel Microparticles. *Macromol. Rapid Commun.* **2010**, *31*, 128–134. [CrossRef] [PubMed]

97. Subramaniam, A.B.; Abkarian, M.; Mahadevan, L.; Stone, H.A. Colloid science: Non-spherical bubbles. *Nature* **2005**, *438*, 930. [CrossRef] [PubMed]

98. Nie, Z.; Xu, S.; Seo, M.; Lewis, P.C.; Kumacheva, E. Polymer particles with various shapes and morphologies produced in continuous microfluidic reactors. *J. Am. Chem. Soc.* **2005**, *127*, 8058–8063. [CrossRef] [PubMed]

99. Rolland, J.P.; Maynor, B.W.; Euliss, L.E.; Exner, A.E.; Denison, G.M.; DeSimone, J.M. Direct fabrication and harvesting of monodisperse, shape-specific nanobiomaterials. *J. Am. Chem. Soc.* **2005**, *127*, 10096–10100. [CrossRef] [PubMed]

100. Kim, J.W.; Larsen, R.J.; Weitz, D.A. Uniform nonspherical colloidal particles with tunable shapes. *Adv. Mater.* **2007**, *19*, 2005–2009. [CrossRef]

101. Sacanna, S.; Pine, D.J. Shape-anisotropic colloids: Building blocks for complex assemblies. *Curr. Opin. Colloid Interface Sci.* **2011**, *16*, 96–105. [CrossRef]

102. Dendukuri, D.; Tsoi, K.; Hatton, T.A.; Doyle, P.S. Controlled Synthesis of Nonspherical Microparticles Using Microfluidics. *Langmuir* **2005**, *21*, 2113–2116. [CrossRef] [PubMed]

103. Dendukuri, D.; Pregibon, D.C.; Collins, J.; Hatton, T.A.; Doyle, P.S. Continuous-flow lithography for high-throughput microparticle synthesis. *Nat. Mater.* **2006**, *5*, 365–369. [CrossRef] [PubMed]

104. Yang, J.; Lind, J.U.; Trogler, W.C. Synthesis of hollow silica and titania nanospheres. *Chem. Mater.* **2008**, *20*, 2875–2877. [CrossRef]

105. Wu, D.; Ge, X.; Zhang, Z.; Wang, M.; Zhang, S. Novel one-step route for synthesizing CdS/polystyrene nanocomposite hollow spheres. *Langmuir* **2004**, *20*, 5192–5195. [CrossRef] [PubMed]

106. Wang, X.; Feng, J.; Bai, Y.; Zhang, Q.; Yin, Y. Synthesis, Properties, and Applications of Hollow Micro-/Nanostructures. *Chem. Rev.* **2016**, *116*, 10983–11060. [CrossRef] [PubMed]

107. Wang, W.; Zhang, M.J.; Xie, R.; Ju, X.J.; Yang, C.; Mou, C.L.; Weitz, D.A.; Chu, L.Y. Hole–Shell Microparticles from Controllably Evolved Double Emulsions. *Angew. Chem. Int. Ed.* **2013**, *52*, 8084–8087. [CrossRef] [PubMed]

108. Khan, I.U.; Serra, C.A.; Anton, N.; Vandamme, T. Continuous-flow encapsulation of ketoprofen in copolymer microbeads via co-axial microfluidic device: Influence of operating and material parameters on drug carrier properties. *Int. J. Pharm.* **2013**, *441*, 809–817. [CrossRef] [PubMed]

109. Utada, A.; Chu, L.-Y.; Fernandez-Nieves, A.; Link, D.; Holtze, C.; Weitz, D. Dripping, jetting, drops, and wetting: The magic of microfluidics. *MRS Bull.* **2007**, *32*, 702–708. [CrossRef]

110. Kim, J.W.; Utada, A.S.; Hu, Z.; Weitz, D.A. Fabrication of monodisperse gel shells and functional microgels in microfluidic devices. *Angew. Chem. Int. Ed.* **2007**, *46*, 1819–1822. [CrossRef] [PubMed]

111. Tomotika, S. Breaking up of a drop of viscous liquid immersed in another viscous fluid which is extending at a uniform rate. *Proc. R. Soc. Lond. Ser. A Math. Phys. Sci.* **1936**, *153*, 302–318. [CrossRef]

112. Chang, Z.; Serra, C.A.; Bouquey, M.; Prat, L.; Hadziioannou, G. Co-axial capillaries microfluidic device for synthesizing size- and morphology-controlled polymer core-polymer shell particles. *Lab Chip* **2009**, *9*, 3007–3011. [CrossRef] [PubMed]

113. Dowding, P.J.; Atkin, R.; Vincent, B.; Bouillot, P. Oil core-polymer shell microcapsules prepared by internal phase separation from emulsion droplets. I. Characterization and release rates for microcapsules with polystyrene shells. *Langmuir* **2004**, *20*, 11374–11379. [CrossRef] [PubMed]

114. Shum, H.C.; Kim, J.-W.; Weitz, D.A. Microfluidic fabrication of monodisperse biocompatible and biodegradable polymersomes with controlled permeability. *J. Am. Chem. Soc.* **2008**, *130*, 9543–9549. [CrossRef] [PubMed]

115. Guzowski, J.; Korczyk, P.M.; Jakiela, S.; Garstecki, P. The structure and stability of multiple micro-droplets. *Soft Matter* **2012**, *8*, 7269–7278. [CrossRef]

116. Loxley, A.; Vincent, B. Preparation of poly (methylmethacrylate) microcapsules with liquid cores. *J. Colloid Interface Sci.* **1998**, *208*, 49–62. [CrossRef] [PubMed]

117. Kim, J.H.; Jeon, T.Y.; Choi, T.M.; Shim, T.S.; Kim, S.-H.; Yang, S.-M. Droplet Microfluidics for Producing Functional Microparticles. *Langmuir* **2014**, *30*, 1473–1488. [CrossRef] [PubMed]

118. Perro, A.; Reculusa, S.; Ravaine, S.; Bourgeat-Lami, E.; Duguet, E. Design and synthesis of Janus micro-and nanoparticles. *J. Mater. Chem.* **2005**, *15*, 3745–3760. [CrossRef]

119. Schick, I.; Lorenz, S.; Gehrig, D.; Schilmann, A.-M.; Bauer, H.; Panthoefer, M.; Fischer, K.; Strand, D.; Laquai, F.; Tremel, W. Multifunctional Two-Photon Active Silica-Coated Au@ MnO Janus Particles for Selective Dual Functionalization and Imaging. *J. Am. Chem. Soc.* **2014**, *136*, 2473–2483. [CrossRef] [PubMed]

120. Walther, A.; Hoffmann, M.; Müller, A.H. Emulsion polymerization using Janus particles as stabilizers. *Angew. Chem.* **2008**, *120*, 723–726. [CrossRef]

121. Binks, B.; Fletcher, P. Particles adsorbed at the oil-water interface: A theoretical comparison between spheres of uniform wettability and "Janus" particles. *Langmuir* **2001**, *17*, 4708–4710. [CrossRef]

122. Gangwal, S.; Cayre, O.J.; Velev, O.D. Dielectrophoretic assembly of metallodielectric Janus particles in AC electric fields. *Langmuir* **2008**, *24*, 13312–13320. [CrossRef] [PubMed]

123. Khan, I.U.; Serra, C.A.; Anton, N.; Xiang, L.; Akasov, R.; Messaddeq, N.; Kraus, I.; Vandamme, T.F. Microfluidic conceived drug loaded Janus particles in side-by-side capillaries device. *Int. J. Pharm.* **2014**, *473*, 239–249. [CrossRef] [PubMed]

124. Shepherd, R.F.; Conrad, J.C.; Rhodes, S.K.; Link, D.R.; Marquez, M.; Weitz, D.A.; Lewis, J.A. Microfluidic assembly of homogeneous and janus colloid-filled hydrogel granules. *Langmuir* **2006**, *22*, 8618–8622. [CrossRef] [PubMed]

125. Seiffert, S.; Romanowsky, M.B.; Weitz, D.A. Janus microgels produced from functional precursor polymers. *Langmuir* **2010**, *26*, 14842–14847. [CrossRef] [PubMed]

126. Nie, Z.; Li, W.; Seo, M.; Xu, S.; Kumacheva, E. Janus and ternary particles generated by microfluidic synthesis: Design, synthesis, and self-assembly. *J. Am. Chem. Soc.* **2006**, *128*, 9408–9412. [CrossRef] [PubMed]

127. Dendukuri, D.; Hatton, T.A.; Doyle, P.S. Synthesis and self-assembly of amphiphilic polymeric microparticles. *Langmuir* **2007**, *23*, 4669–4674. [CrossRef] [PubMed]

128. Prasad, N.; Perumal, J.; Choi, C.H.; Lee, C.S.; Kim, D.P. Generation of monodisperse inorganic–organic Janus microspheres in a microfluidic device. *Adv. Funct. Mater.* **2009**, *19*, 1656–1662. [CrossRef]

129. Windbergs, M.; Zhao, Y.; Heyman, J.A.; Weitz, D.A. Biodegradable core-shell carriers for simultaneous encapsulation of synergistic actives. *J. Am. Chem. Soc.* **2013**. [CrossRef] [PubMed]

130. Sekhar, A.S.; Meera, C.; Ziyad, K.; Gopinath, C.S.; Vinod, C. Synthesis and catalytic activity of monodisperse gold–mesoporous silica core–shell nanocatalysts. *Catal. Sci. Technol.* **2013**, *3*, 1190–1193. [CrossRef]

131. DiLauro, A.M.; Abbaspourrad, A.; Weitz, D.A.; Phillips, S.T. Stimuli-Responsive Core–Shell Microcapsules with Tunable Rates of Release by Using a Depolymerizable Poly(phthalaldehyde) Membrane. *Macromolecules* **2013**, *46*, 3309–3313. [CrossRef]

132. Dawes, G.; Fratila-Apachitei, L.; Mulia, K.; Apachitei, I.; Witkamp, G.-J.; Duszczyk, J. Size effect of PLGA spheres on drug loading efficiency and release profiles. *J. Mater. Sci. Mater. Med.* **2009**, *20*, 1089–1094. [CrossRef] [PubMed]

133. Siepmann, J.; Faisant, N.; Akiki, J.; Richard, J.; Benoit, J. Effect of the size of biodegradable microparticles on drug release: Experiment and theory. *J. Control. Release* **2004**, *96*, 123–134. [CrossRef] [PubMed]

134. Lee, T.H.; Wang, J.; Wang, C.-H. Double-walled microspheres for the sustained release of a highly water soluble drug: Characterization and irradiation studies. *J. Control. Release* **2002**, *83*, 437–452. [CrossRef]

135. Berkland, C.; Cox, A.; Kim, K.K.; Pack, D.W. Three-month, zero-order piroxicam release from monodispersed double-walled microspheres of controlled shell thickness. *J. Biomed. Mater. Res. Part A* **2004**, *70*, 576–584. [CrossRef] [PubMed]

136. Chu, L.-Y.; Utada, A.S.; Shah, R.K.; Kim, J.-W.; Weitz, D.A. Controllable monodisperse multiple emulsions. *Angew. Chem. Int. Ed.* **2007**, *46*, 8970–8974. [CrossRef] [PubMed]

137. Sun, X.; Li, Y. Colloidal Carbon Spheres and Their Core/Shell Structures with Noble-Metal Nanoparticles. *Angew. Chem. Int. Ed.* **2004**, *43*, 597–601. [CrossRef] [PubMed]

138. Guo, S.R.; Gong, J.Y.; Jiang, P.; Wu, M.; Lu, Y.; Yu, S.H. Biocompatible, Luminescent Silver@ Phenol Formaldehyde Resin Core/Shell Nanospheres: Large-Scale Synthesis and Application for In Vivo Bioimaging. *Adv. Funct. Mater.* **2008**, *18*, 872–879. [CrossRef]

139. Fang, Y.; Gu, D.; Zou, Y.; Wu, Z.; Li, F.; Che, R.; Deng, Y.; Tu, B.; Zhao, D. A Low-Concentration Hydrothermal Synthesis of Biocompatible Ordered Mesoporous Carbon Nanospheres with Tunable and Uniform Size. *Angew. Chem. Int. Ed.* **2010**, *49*, 7987–7991. [CrossRef] [PubMed]

140. Guo, L.; Zhang, L.; Zhang, J.; Zhou, J.; He, Q.; Zeng, S.; Cui, X.; Shi, J. Hollow mesoporous carbon spheres—An excellent bilirubin adsorbent. *Chem. Commun.* **2009**, *40*, 6071–6073. [CrossRef] [PubMed]

141. Ju, M.; Zeng, C.; Wang, C.; Zhang, L. Preparation of ultrafine carbon spheres by controlled polymerization of furfuryl alcohol in microdroplets. *Ind. Eng. Chem. Res.* **2014**, *53*, 3084–3090. [CrossRef]

142. Pan, Y.; Ju, M.; Wang, C.; Zhang, L.; Xu, N. Versatile preparation of monodisperse poly (furfuryl alcohol) and carbon hollow spheres in a simple microfluidic device. *Chem. Commun.* **2010**, *46*, 3732–3734. [CrossRef] [PubMed]

143. Zhu, T.; Cheng, R.; Sheppard, G.R.; Locklin, J.; Mao, L. Magnetic-Field-Assisted Fabrication and Manipulation of Nonspherical Polymer Particles in Ferrofluid-Based Droplet Microfluidics. *Langmuir* **2015**, *31*, 8531–8534. [CrossRef] [PubMed]

144. Sánchez-Ferrer, A.; Carney, R.P.; Stellacci, F.; Mezzenga, R.; Isa, L. Isolation and Characterization of Monodisperse Core–Shell Nanoparticle Fractions. *Langmuir* **2015**, *31*, 11179–11185. [CrossRef] [PubMed]

145. Bermudez, H.; Brannan, A.K.; Hammer, D.A.; Bates, F.S.; Discher, D.E. Molecular weight dependence of polymersome membrane structure, elasticity, and stability. *Macromolecules* **2002**, *35*, 8203–8208. [CrossRef]

146. Lee, J.C.M.; Bermudez, H.; Discher, B.M.; Sheehan, M.A.; Won, Y.Y.; Bates, F.S.; Discher, D.E. Preparation, stability, and in vitro performance of vesicles made with diblock copolymers. *Biotechnol. Bioeng.* **2001**, *73*, 135–145. [CrossRef] [PubMed]

147. Aranda-Espinoza, H.; Bermudez, H.; Bates, F.S.; Discher, D.E. Electromechanical limits of polymersomes. *Phys. Rev. Lett.* **2001**, *87*, 208301. [CrossRef] [PubMed]

148. Wang, C.; Wang, L.; Wang, M. Evolution of core–shell structure: From emulsions to ultrafine emulsion electrospun fibers. *Mater. Lett.* **2014**, *124*, 192–196. [CrossRef]

149. Kong, T.; Wang, L.; Wyss, H.M.; Shum, H.C. Capillary micromechanics for core-shell particles. *Soft Matter* **2014**, *10*, 3271–3276. [CrossRef] [PubMed]

micromachines

MDPI

Review

Polymer Microfluidics: Simple, Low-Cost Fabrication Process Bridging Academic Lab Research to Commercialized Production

Chia-Wen Tsao

Department of Mechanical Engineering, National Central University, Taoyuan 32001, Taiwan;
cwtsao@ncu.edu.tw; Tel.: +886-3-426-7343

Academic Editors: Weihua Li, Say Hwa Tan and Nam-Trung Nguyen
Received: 26 September 2016; Accepted: 7 December 2016; Published: 10 December 2016

Abstract: Using polymer materials to fabricate microfluidic devices provides simple, cost effective, and disposal advantages for both lab-on-a-chip (LOC) devices and micro total analysis systems (µTAS). Polydimethylsiloxane (PDMS) elastomer and thermoplastics are the two major polymer materials used in microfluidics. The fabrication of PDMS and thermoplastic microfluidic device can be categorized as front-end polymer microchannel fabrication and post-end microfluidic bonding procedures, respectively. PDMS and thermoplastic materials each have unique advantages and their use is indispensable in polymer microfluidics. Therefore, the proper selection of polymer microfabrication is necessary for the successful application of microfluidics. In this paper, we give a short overview of polymer microfabrication methods for microfluidics and discuss current challenges and future opportunities for research in polymer microfluidics fabrication. We summarize standard approaches, as well as state-of-art polymer microfluidic fabrication methods. Currently, the polymer microfluidic device is at the stage of technology transition from research labs to commercial production. Thus, critical consideration is also required with respect to the commercialization aspects of fabricating polymer microfluidics. This article provides easy-to-understand illustrations and targets to assist the research community in selecting proper polymer microfabrication strategies in microfluidics.

Keywords: polymer microfluidics; polymer microfabrication; thermoplastics; polydimethylsiloxane

1. Introduction

With the introduction of microfluidics, micro total analysis system (µTAS), and lab-on-a-chip (LOC) devices in the 1900s, the use of microfluidic devices has increased tremendously due to the great potential in biomedical, point-of-care testing, and healthcare applications. The early development of microfluidic devices commonly involved silicon and glass materials as basic substrates. However, with the concept of using polymer materials in microfluidics been proposed in the late 1990s [1], the use of silicon and glass materials has shifted to polymers, primarily due to their simple and low-cost advantages. Compared to silicon and glass, polymers are inexpensive materials and feature a wide variety of material properties for meeting the various application requirements of disposable biomedical microfluidics devices, as well as many promising applications [2–4].

Fabrication of polymer microfluidic devices is relatively simple and no hazardous etching reagent is required to create the polymer microstructures. The fabrication tools for making polymer devices are also much cheaper than those for making semiconductor infrastructures, such as wet benches or reactive-ion etching facilities. These factors make it possible for polymer microfluidics devices to be easily fabricated in average research labs, a fact which has driven the development of polymer microfluidics academically, and further toward industrial applications. After years of polymer microfluidics investigations, various polymer microfabrication technologies have been developed using simple and low-cost formats. However, polymer microfabrication is not a straightforward process and, as yet, there is no one-fits-for-all fabrication technique for creating polymer microfluidic devices. Proper determination of polymer microfabrication strategies is critical for successful polymer microfluidic device functionality. In this paper, we examine polymer microfabrication with respect to the raw materials, facility costs, and general and state-of-art fabrication processes, as well as commercialization considerations.

2. Selection of Polymer Material and Microfabrication Processes Selection

In the polymer microfabrication process, the first step is to identify its application and requirements. Once the microfluidic chip application is identified, the microchannel/chamber layouts can be designed. Next, one selects an appropriate polymer material and determines the fabrication strategy to create a polymer microfluidic device that will meet the specific microfluidic application requirements. The polymer materials typically used in microfluidic applications can be divided into two major categories: polydimethylsiloxane (PDMS) and thermoplastics. Figure 1 shows the polymer microfluidics fabrication procedures and selection strategies associated with PDMS (blue line) and thermoplastics (red line). PDMS is one of the major materials used in polymer microfluidics because of material elasticity, gas permittivity, and other several unique advantages. PDMS is an elastomer material that can be deformed under the application of force or air pressure. The PDMS valve was invented to control microchannel fluidic transportation, which also enables very large scale integration in high-throughput applications [5–7]. Both PDMS and thermoplastics materials have shown high biocompatibility for many biomolecules and cells [8,9]. Due to its high gas permittivity property and high optical transmissivity, PDMS is the main material choice for cell-based microfluidic devices [7,10,11]. Although PDMS has advantages, it also has several limitations in microfluidic applications. Problems, such as channel deformation, low solvent and acid/base resistivity, evaporation, sample absorption, leaching, and hydrophobic recovery, are the fundamental challenges associated with PDMS in microfluidic devices [12,13]. Thermoplastics are synthetic polymers that have various surface properties for microfluidic application. Thermoplastics such as poly(methyl methacrylate) (PMMA), polycarbonate (PC), polystyrene (PS), polyvinyl chloride (PVC), polyimide (PI), and the family of cyclic olefin polymers (i.e., cyclic olefin copolymer (COC), cyclic olefin polymer (COP), and cyclic block copolymer (CBC)) have been widely used in microfluidics [14–16]. Thermoplastics are rigid polymer materials that have good mechanical stability, a low water-absorption percentage, and organic-solvent, and acid/base resistivity, which are critical factors in many bioanalytical microfluidic applications, such as high-pressure liquid chromatography (HPLC) microfluidic applications [17], that involve a high-pressure solvent injection procedure. PDMS may suffer from solvent swelling and channel deformation issues, which makes thermoplastics (like COC) an ideal choice for the polymer material. Table 1 summarizes the typical mechanical, optical, chemical (solvent and acid/base resistance), and material costs for PDMS and thermoplastics commonly used in microfluidics.

Table 1. Summary of physical properties and suppliers for common polymer microfluidic materials.

Polymer		PDMS	Thermoplastics			
			PC	PMMA	PS	COC/COP/CBC
Mechanical property		Elastomer	Rigid	Rigid	Rigid	Rigid
Thermal property [1]		~80 °C	140~150 °C	100~125 °C	90~100 °C	70~155 °C
Solvent resistance		Poor	Good	Good	Poor	Excellent
Acid/base resistance		Poor	Good	Good	Good	Good
Optical transmissivity	Visible range	Excellent	Excellent	Excellent	Excellent	Excellent
	UV range	Good	Poor	Good	Poor	Excellent
Biocompatibility		Good	Good	Good	Good	Good
Material cost [2]		~150 $/Kit (1 Kg) [3]	<3 $/Kg [3]	2~4 $/Kg	<3 $/Kg	20~25 $/Kg [4]

[1] Thermal property is determined based on the PDMS curing temperature and thermoplastic glass transition (T_g) temperature; [2] The cost information is provided by a local supplier. Cost may be different in different regions. Thermoplastic material is in pellets; [3] Suppliers: Dow Corning, Midland, MI, USA; [4] Suppliers: JSR ARTON (Tokyo, Japan), ZEON Chemicals (Louisville, KY, USA), TOPAS Advanced Polymers (Florence, KY, USA), USI Corporation (Taipei, Taiwan).

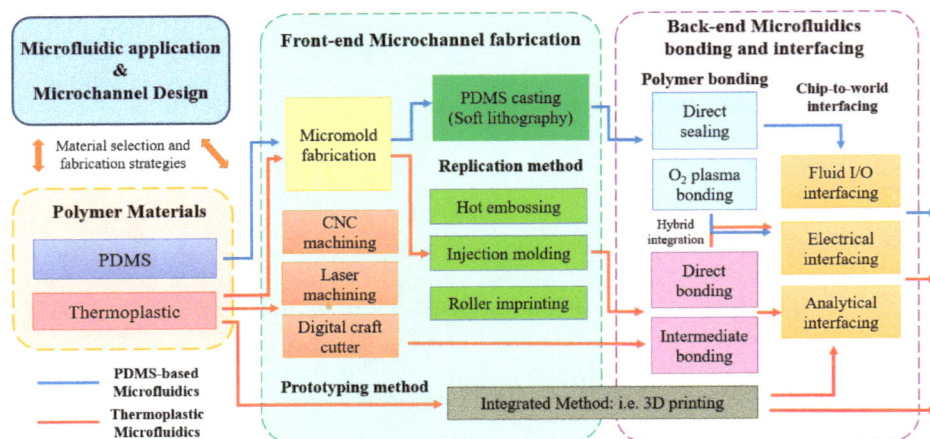

Figure 1. Polymer microfluidics fabrication process chart. The blue line indicates the PDMS-based microfluidics fabrication procedure, and the red line indicates the thermoplastic microfluidics fabrication procedure.

3. Polymer Microfluidics Fabrication Procedure

3.1. PDMS and Thermoplastic-Based Polymer Microfluidics

The fabrication process of PDMS microfluidic chips is relatively straightforward [18,19]. As shown in the blue process lines in Figure 1, the PDMS microchannel is mainly fabricated by a simple soft lithography process in which the PDMS reagent is directly cast onto a master micromold [20], followed by a bonding process [21]. The typical PDMS casting procedure is performed by mixing a PDMS base with a curing reagent in a 10:1 ratio, followed by curing at 80 °C for 1–2 h. The PDMS layer is then released from the micromold to complete the casting procedure. Since the casting process is such a simple process and the layer is easily released from the micromold, PDMS casting is a reliable and high yield procedure. SU-8 resin and standard photoresist (PR) can be used as micromolds in the PDMS procedure [22]. Sealing of the PDMS microstructure to enclose a microfluidic channel or chamber also involves a simple and reliable procedure. A PDMS layer can be directly sealed/stuck to another PDMS or glass substrate via van der Waals forces without the need for further fabrication procedures. To meet high bonding strength requirements, the PDMS bond strength can be enhanced by tuning the process parameter [21] or, more commonly, using oxygen plasma treatment to form an O–Si–O covalent bond at the PDMS interface [23,24].

In the thermoplastic microfluidic chip fabrication procedure (Figure 1, red lines), there are various fabrication options for making thermoplastic microchannels. Thermoplastic microchannels can be created either by rapid prototyping or replication methods. Rapid prototyping methods, such as computer numerical controlled (CNC) milling [25–27], and laser ablation [28,29] are available for generating microchannels on the thermoplastic substrate. Recently, a low-cost rapid prototyping method by a digital craft cutter was proposed to create microchannels on a thin transparent thermoplastic film [30–32]. Although CNC, laser ablation, or digital craft cutter methods have limits with respect to microchannel resolution and surface roughness, they are important procedures in thermoplastic microfluidics fabrication because rapid prototyping is a simple process for researchers to establish proof-of-concept without the need for micromold fabrication. For mass production, thermoplastic microchannels can be fabricated by replication processes, such as hot embossing/imprinting [33–35], roller imprinting [36,37], and injection molding [38,39], which are common polymer replication methods for massively reproducing thermoplastic microchips. In the thermoplastic fabrication process, bonding is a critical last step that determines the bonding strength, geometry stability, optical transmissivity, and surface chemistry of the produced microfluidic device. In some bonding processes, there can be bottlenecks in the mass production of thermoplastic microfluidic devices. Issues associated with bonding throughput are detailed in Section 4. A comprehensive review of thermoplastic bonding methods have been reported by Tsao and DeVoe [40]. Generally, thermoplastic bonding is achieved either by direct bonding or an intermediate bonding approach. Direct bonding is a bonding process that uses no intermediate material at the bonding interface. Methods such as thermal fusion bonding [41,42], ultrasonic welding [43], surface modification [44–46], and solvent bonding [47,48] are categorized as direct bonding methods. Indirect bonding is defined as bonding that involves the use of an additional material or chemical reagents to assist in the bonding, such as epoxy, adhesive tape, or chemical reagents. Indirect thermoplastic bonding methods, such as adhesive bonding [49–51] or microwave bonding [52], use an intermediate layer, such as metal or a chemical reagent.

After completing the microchannel fabrication and bonding process to seal the microchannels, the last step is to connect the microfluidic device for chip-to-world interface. Surface modification procedures are sometimes applied in the polymer microfluidics to meet specific application requirements [53–59]. Microfluidics interfacing issues remain a challenge and have been given less emphasis in the microfluidic community. A good interface is a critical aspect that determines the success of practical applications and commercialization potential. A recent review by Temiz et al. summarized methods on how to "plug" chips for post-end fluidics, electronics, and analytical interfaces [60]. Both PDMS and thermoplastic microfluidics chips commonly use a standard Luer lock/cone, or peek connector [61] for the fluidic interface. Solutions, such as surgical needles [62] or customer-designed connectors [63] (Figure 2a), for the fluidic inlet/outlet have also been proposed. In particular, due to the thermoplastic substrate's rigidity, needles can achieve a tight-fit insertion into the thermoplastic in high-pressure fluidic connections, which makes thermoplastics an appealing material for high-pressure applications. For microfluidic devices to provide control and detection functions, based on electrical principles (i.e., electrophoresis, electrowetting, electrochemical sensing), electrode pads for power and electrical connections are required. The use of stainless surgical needles [62] in fluidic connection has demonstrated good electrical contact or power connection for electrophoresis or isoelectric focusing applications [64]. For integrated on-chip electrode pads, electrical contacts can be deposited on the polymer surface by thermal or electron beam [64,65], screen printing [66], or 3D ion implantation electrode [67]. Many microfluidic devices are analyzed by optical detection. Since all polymer materials are optically transparent (Table 1), optical detection can be directly performed on a microscope without any additional interface setup on the chip. However, for other detection methods, an analytical interface is required. For example, when interfacing with mass spectrometry analysis, an electrospray or droplet deposition orifice must be incorporated into the microfluidic device [68].

3.2. Advances of Polymer Microfluidics Fabrication

Today, polymer microfluidics continues to be an intriguing research topic. Various polymer materials have been demonstrated in the microfluidic applications with better performance [69]. For example, thermoset polyester (TPE) was proposed in microfluidics as an alternative material to PDMS providing better chemical and solvent compatibility. The TPE fabrication process is also compatible with standard replica molding, as well as advanced rapid high-pressure injection procedures [70–72]. In polymer microfluidics fabrication, there is no one-fit-for-all polymer fabrication technique and research is ongoing to identify techniques that are more reliable, simple, versatile, and robust. In polymer replication, Beebe et al. recently reported a thermoplastic bonding method that combines hot embossing and milling for faster replication, based on the hot embossing method [73]. Several novel micromold technologies have been developed to realize better polymer replication performance. For example, the thermoplastic building blocks technique (Figure 2c) offers micromold design flexibility for producing PDMS microfluidic devices for diverse geometries and functionalities [74]. Liquid metal alloys (bulk metallic glass) can also be integrated into microfluidic molding technology to achieve more robust and versatile polymer fabrication [75,76]. Recently, 3D printing technologies [77,78] (Figure 2b) have become a popular prototyping method for fabricating the polymer microfluidic devices.

With respect to the post-end microfluidic bonding advances, reversible bonding, based on re-melting the wax [79] or a magnetic force [80] enables the production of dismountable and reusable microfluidic devices. Bonding a heterogeneous material to make a "hybrid" device is also an important method for making advanced integrated microfluidic devices. Bonding PDMS with thermoplastic material enables a wider range of microfluidic applications. Tan et al. introduced a PMMA–PDMS pneumatic micropump as a hybrid microfluidic device using optically-clear adhesive film [81]. Li et al. used a selective stamp bonding technique to transfer epoxy to bond a PDMS–polystyrene (PS)/poly(ethylene terephthalate) (PET) microfluidic device for human lung epithelial cells analysis [82]. A doubly cross-linked nano-adhesive method has also been reported for sealing PDMS with polyimide (PI) or polyethylene terephthalate (PET) [83]. Bonding polymer with paper can integrate thermoplastic material with novel microfluidic paper-based analytical devices (µPADs) [84].

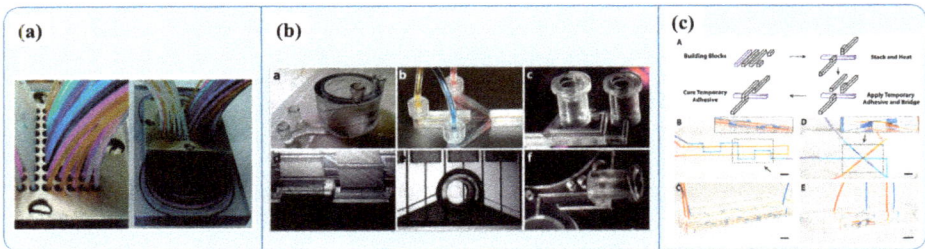

Figure 2. (**a**) Custom-designed chip-to-world multichannel interfacing. Reproduced from [63] with permission of The Royal Society of Chemistry; (**b**) polymer microfluidics device fabricated by 3D printing process. Reproduced from [78] with permission of The Royal Society of Chemistry; and (**c**) thermoplastic building blocks for versatile PDMS microfluidics. Reproduced from [74] with permission of The Royal Society of Chemistry.

4. Commercialization Considerations for Polymer Microfluidics Fabrication

Since publication of the first polymer microfluidic paper [85], the idea of using polymer material in microfluidics has become increasingly popular in the research community. With almost 20 years of development, polymer microfluidic technology has become the major material choice for microfluidics due to its advantages of low cost and disposability, and many effective bioanalytical applications have been demonstrated. Microfluidic devices are currently in the technology transfer stage from the research lab to commercial production. PDMS and thermoplastics each have their own advantages for microfluidics applications, which are also indispensable factors in choosing materials for commercialized products. For example, Fluidigm Inc.'s (South San Francisco, CA, USA) integrated fluidic circuits are generated using a PDMS soft lithography process and the HPLC chips from Agilent Technologies (Santa Clara, CA, USA) are based on a PMMA thermoplastic substrate. Several microfabrication foundries, such as MiniFAB or Micralyne, have provided a fabrication-services business model for the mass fabrication of polymer microfluidic devices. Many emerging microfluidic devices are also currently being transferred from research prototypes into products. In general, a low-volume (<200 pieces per month) polymer production rate is appropriate for academic or research labs developing prototypes for proof-of-concepts. For commercialized microfluidic devices, medium volume (200–2000 pieces per month) or preferably high-volume (>2000 piece month), mass production strategies should be considered. In addition to the material properties and performance of polymer microfluidic devices, fabrication throughput is a particularly important consideration for the commercialization of polymer microfluidic devices.

Figure 3 shows the key polymer microfluidic fabrication procedures (microchannel fabrication and chip bonding) in terms of their facility cost and fabrication throughput. From the fabrication perspective, the PDMS casting process is time-consuming, normally taking 0.5–1 h to complete a casting cycle, and can thus provide only 150–300 devices per month at a standard research-lab scale. This may potentially constrain the production of high-volume quantities of PDMS microfluidic devices. With respect to PDMS bonding, because PDMS can be directly sealed to the glass or PDMS layer, bonding can be achieved by a simple attachment procedure without the need for any bonding facility. Even for high bond strength O_2 plasma bonding, PDMS bonding can be achieved within 10 min using O_2 plasma activation. By combining the PDMS casting and sealing procedures, PDMS microfluidic devices can achieve medium-volume fabrication. Additoinally, PDMS chip fabrication facilities can be developed in low-budget conditions (i.e., hot plate/vacuum oven and plasma cleaner) while achieving good microfluidic throughput for research investigations. As such, PDMS has sometimes been a more popular microfluidic chip choice than thermoplastics in academic research labs.

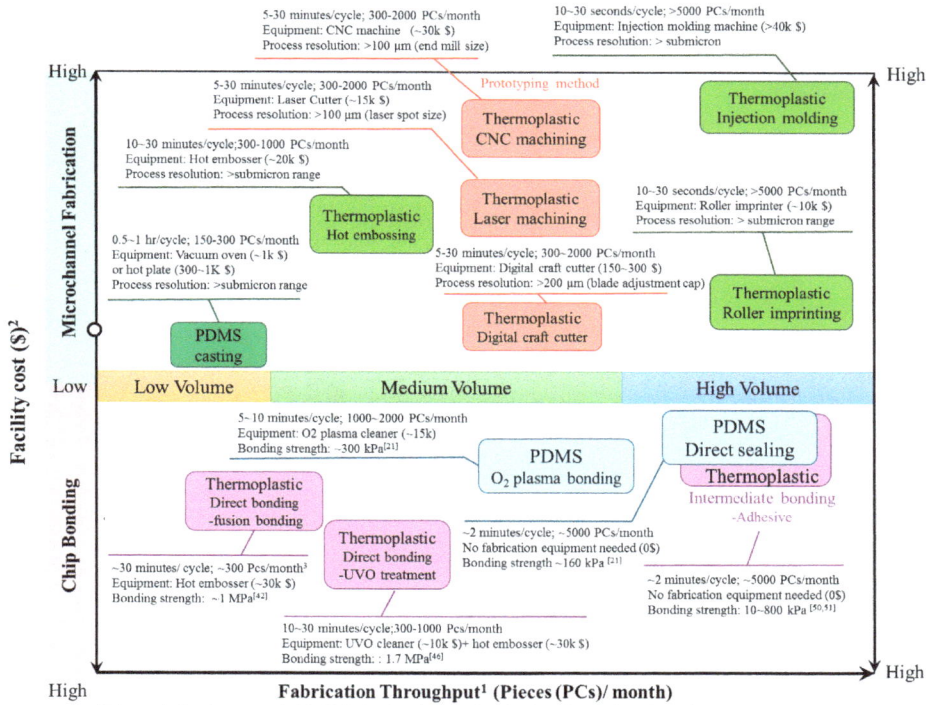

Figure 3. Estimation of fabrication throughput (*x*-axis, PCs/month) and facility cost (*y*-axis, in US dollars) of critical polymer microfabrication procedures.

With respect to thermoplastics, because there is more variety of choice in the fabrication process, thermoplastic microdevices can be generated either by low–medium throughput prototyping/replication or by high-throughput replication methods. For the commonly used hot embossing process, depending on the heating/cooling conditions, microchannels can be replicated in a medium-volume production range at a rate of 10–30 min/cycle. In particular, methods such as injection molding and the continuous reel-to-reel roller imprinting method can be complete a replication cycle within seconds, which is ideal for producing large numbers of replicas per day, as required for commercial manufacture. Regarding the thermoplastic bonding process, a wide variety of thermoplastic bonding methods have been reviewed previously [40]. For comparison with other fabrication methods, in Figure 3, we show three commonly used bonding methods: thermal fusion bonding, surface treatment (UV/ozone, UVO), and adhesive bonding. In the direct fusion bonding method, because it requires thermoplastic heating above T_g to "fuse" the bonding pairs, a longer time of around 30 min/cycle is required to bond a chip. A surface treatment bonding method has been proposed to effectively reduce the processing temperature below T_g or even to room temperature. Depending on the bonding temperature, the process cycle time can be reduced to ~10 min/cycle for the evaluated production volume. Using adhesive bonding, the thermoplastic can be bonded at room temperature, so chips can be bonded rapidly within 2 min to achieve high-volume polymer fabrication. We note that the facilities costs and fabrication throughputs in Figure 3 are estimated values, and the price and processing times may vary depending on the tool brand and the fabrication resolution. Nevertheless, Figure 3 provides a useful comparison of the fabrication throughput and cost aspects. With the selection of the more appropriate fabrication method, both PDMS and thermoplastic materials can reach medium- to high-volume fabrication throughput to meet mass commercial production requirements.

5. Conclusions

PDMS and thermoplastics are two important substrate materials in polymer microfluidics and polymer microfabrication techniques that have been well developed for both to begin transferring microfluidics prototypes from academic research labs to commercialized production. The proper selection of polymer material and polymer microfabrication strategy are critical to ensure success in polymer microfluidics research and commercialization. Future development of polymer microfabrication techniques should further explore the microfabrication performance (i.e., minimum channel resolution, bonding strength, etc.), but also consider commercial aspects (i.e., fabrication throughput and cost) to bridge polymer microfluidic devices from research prototypes into commercialized products.

Acknowledgments: The author thank the Ministry of Science and Technology, Taiwan, for financially supporting this project under Grant No. MOST 105-2221-E-008-061.

Conflicts of Interest: The authors declare no conflict of interest.

References

1. Manz, A.; Graber, N.; Widmer, H.M. Miniaturized total chemical-analysis systems—A novel concept for chemical sensing. *Sens. Actuators B Chem.* **1990**, *1*, 244–248. [CrossRef]
2. Becker, H.; Locascio, L.E. Polymer microfluidic devices. *Talanta* **2002**, *56*, 267–287. [CrossRef]
3. Tan, S.H.; Maes, F.; Semin, B.; Vrignon, J.; Baret, J.C. The microfluidic jukebox. *Sci. Rep.* **2014**, *4*, 4787. [CrossRef] [PubMed]
4. Chong, Z.Z.; Tan, S.H.; Ganan-Calvo, A.M.; Tor, S.B.; Loh, N.H.; Nguyen, N.T. Active droplet generation in microfluidics. *Lab Chip* **2016**, *16*, 35–58. [CrossRef] [PubMed]
5. Araci, I.E.; Quake, S.R. Microfluidic very large scale integration (mvlsi) with integrated micromechanical valves. *Lab Chip* **2012**, *12*, 2803–2806. [CrossRef] [PubMed]
6. McDonald, J.C.; Whitesides, G.M. Poly(dimethylsiloxane) as a material for fabricating microfluidic devices. *Acc. Chem. Res.* **2002**, *35*, 491–499. [CrossRef] [PubMed]
7. Sia, S.K.; Whitesides, G.M. Microfluidic devices fabricated in poly(dimethylsiloxane) for biological studies. *Electrophoresis* **2003**, *24*, 3563–3576. [CrossRef] [PubMed]
8. Alrifaiy, A.; Lindahl, O.A.; Ramser, K. Polymer-based microfluidic devices for pharmacy, biology and tissue engineering. *Polymers* **2012**, *4*, 1349–1398. [CrossRef]

9. Van Midwoud, P.M.; Janse, A.; Merema, M.T.; Groothuis, G.M.M.; Verpoorte, E. Comparison of biocompatibility and adsorption properties of different plastics for advanced microfluidic cell and tissue culture models. *Anal. Chem.* **2012**, *84*, 3938–3944. [CrossRef] [PubMed]
10. Mehling, M.; Tay, S. Microfluidic cell culture. *Curr. Opin. Biotech.* **2014**, *25*, 95–102. [CrossRef] [PubMed]
11. Wu, M.H.; Huang, S.B.; Lee, G.B. Microfluidic cell culture systems for drug research. *Lab Chip* **2010**, *10*, 939–956. [CrossRef] [PubMed]
12. Berthier, E.; Young, E.W.K.; Beebe, D. Engineers are from pdms-land, biologists are from polystyrenia. *Lab Chip* **2012**, *12*, 1224–1237. [CrossRef] [PubMed]
13. Halldorsson, S.; Lucumi, E.; Gómez-Sjöberg, R.; Fleming, R.M.T. Advantages and challenges of microfluidic cell culture in polydimethylsiloxane devices. *Biosens. Bioelectron.* **2015**, *63*, 218–231. [CrossRef] [PubMed]
14. Liu, K.; Fan, Z.H. Thermoplastic microfluidic devices and their applications in protein and DNA analysis. *Analyst* **2011**, *136*, 1288–1297. [CrossRef] [PubMed]
15. Becker, H.; Nevitt, M.; Gray, B.L. Selecting and designing with the right thermoplastic polymer for your microfluidic chip: A close look into cyclo-olefin polymer. *Proc. SPIE* **2013**, *8615*, 86150F.
16. Bhattacharyya, A.; Klapperich, C.M. Thermoplastic microfluidic device for on-chip purification of nucleic acids for disposable diagnostics. *Anal. Chem.* **2006**, *78*, 788–792. [CrossRef] [PubMed]
17. Liu, J.K.; Chen, C.F.; Tsao, C.W.; Chang, C.C.; Chu, C.C.; Devoe, D.L. Polymer microchips integrating solid-phase extraction and high-performance liquid chromatography using reversed-phase polymethacrylate monoliths. *Anal. Chem.* **2009**, *81*, 2545–2554. [CrossRef] [PubMed]
18. Friend, J.; Yeo, L. Fabrication of microfluidic devices using polydimethylsiloxane. *Biomicrofluidics* **2010**, *4*, 026502. [CrossRef] [PubMed]
19. Xi, H.D.; Guo, W.; Leniart, M.; Chong, Z.Z.; Tan, S.H. AC electric field induced droplet deformation in a microfluidic T-junction. *Lab Chip* **2016**, *16*, 2982–2986. [CrossRef] [PubMed]
20. Xia, Y.; Whitesides, G.M. Soft lithography. *Angew. Chem. Int. Ed. Engl.* **1998**, *37*, 550–575. [CrossRef]
21. Eddings, M.A.; Johnson, M.A.; Gale, B.K. Determining the optimal PDMS-PDMS bonding technique for microfluidic devices. *J. Micromech. Microeng.* **2008**, *18*, 067001. [CrossRef]
22. Unger, M.A.; Chou, H.P.; Thorsen, T.; Scherer, A.; Quake, S.R. Monolithic microfabricated valves and pumps by multilayer soft lithography. *Science* **2000**, *288*, 113–116. [CrossRef] [PubMed]
23. Chong, Z.Z.; Tor, S.B.; Loh, N.H.; Wong, T.N.; Ganan-Calvo, A.M.; Tan, S.H.; Nguyen, N.T. Acoustofluidic control of bubble size in microfluidic flow-focusing configuration. *Lab Chip* **2015**, *15*, 996–999. [CrossRef] [PubMed]
24. Tan, S.H.; Nguyen, N.T.; Chua, Y.C.; Kang, T.G. Oxygen plasma treatment for reducing hydrophobicity of a sealed polydimethylsiloxane microchannel. *Biomicrofluidics* **2010**, *4*, 032204. [CrossRef] [PubMed]
25. Guckenberger, D.J.; de Groot, T.E.; Wan, A.M.D.; Beebe, D.J.; Young, E.W.K. Micromilling: A method for ultra-rapid prototyping of plastic microfluidic devices. *Lab Chip* **2015**, *15*, 2364–2378. [CrossRef] [PubMed]
26. Rahmanian, O.; DeVoe, D.L. Pen microfluidics: Rapid desktop manufacturing of sealed thermoplastic microchannels. *Lab Chip* **2013**, *13*, 1102–1108. [CrossRef] [PubMed]
27. Okagbare, P.I.; Emory, J.M.; Datta, P.; Goettert, J.; Soper, S.A. Fabrication of a cyclic olefin copolymer planar waveguide embedded in a multi-channel poly(methyl methacrylate) fluidic chip for evanescence excitation. *Lab Chip* **2010**, *10*, 66–73. [CrossRef] [PubMed]
28. Suriano, R.; Kuznetsov, A.; Eaton, S.M.; Kiyan, R.; Cerullo, G.; Osellame, R.; Chichkov, B.N.; Levi, M.; Turri, S. Femtosecond laser ablation of polymeric substrates for the fabrication of microfluidic channels. *Appl. Surf. Sci.* **2011**, *257*, 6243–6250. [CrossRef]
29. Liu, K.; Xiang, J.; Ai, Z.; Zhang, S.; Fang, Y.; Chen, T.; Zhou, Q.; Li, S.; Wang, S.; Zhang, N. PMMA microfluidic chip fabrication using laser ablation and low temperature bonding with OCA film and LOCA. *Microsyst. Technol.* **2016**, 1–6. [CrossRef]
30. Yuen, P.K.; Goral, V.N. Low-cost rapid prototyping of flexible microfluidic devices using a desktop digital craft cutter. *Lab Chip* **2010**, *10*, 384–387. [CrossRef] [PubMed]
31. Cassano, C.L.; Simon, A.J.; Liu, W.; Fredrickson, C.; Hugh Fan, Z. Use of vacuum bagging for fabricating thermoplastic microfluidic devices. *Lab Chip* **2015**, *15*, 62–66. [CrossRef] [PubMed]
32. Islam, M.; Natu, R.; Martinez-Duarte, R. A study on the limits and advantages of using a desktop cutter plotter to fabricate microfluidic networks. *Microfluid. Nanofluid.* **2015**, *19*, 973–985. [CrossRef]

33. Abgrall, P.; Low, L.N.; Nguyen, N.T. Fabrication of planar nanofluidic channels in a thermoplastic by hot-embossing and thermal bonding. *Lab Chip* **2007**, *7*, 520–522. [CrossRef] [PubMed]

34. Peng, L.; Deng, Y.; Yi, P.; Lai, X. Micro hot embossing of thermoplastic polymers: A review. *J. Micromech. Microeng.* **2014**, *24*, 013001. [CrossRef]

35. Yang, S.; DeVoe, D.L. Microfluidic device fabrication by thermoplastic hot-embossing. In *Microfluidic Diagnostics: Methods and Protocols*; Jenkins, G., Mansfield, C.D., Eds.; Humana Press: Totowa, NJ, USA, 2013; pp. 115–123.

36. Focke, M.; Kosse, D.; Muller, C.; Reinecke, H.; Zengerle, R.; von Stetten, F. Lab-on-a-foil: Microfluidics on thin and flexible films. *Lab Chip* **2010**, *10*, 1365–1386. [CrossRef] [PubMed]

37. Velten, T.; Schuck, H.; Richter, M.; Klink, G.; Bock, K.; Khan Malek, C.; Roth, S.; Schoo, H.; Bolt, P.J. Microfluidics on foil: State of the art and new developments. *Proc. Inst. Mech. Eng. B J. Eng. Manuf.* **2008**, *222*, 107–116. [CrossRef]

38. Mair, D.A.; Geiger, E.; Pisano, A.P.; Frechet, J.M.J.; Svec, F. Injection molded microfluidic chips featuring integrated interconnects. *Lab Chip* **2006**, *6*, 1346–1354. [CrossRef] [PubMed]

39. Attia, U.M.; Marson, S.; Alcock, J.R. Micro-injection moulding of polymer microfluidic devices. *Microfluid. Nanofluid.* **2009**, *7*, 1–28. [CrossRef]

40. Tsao, C.W.; DeVoe, D.L. Bonding of thermoplastic polymer microfluidics. *Microfluid. Nanofluid.* **2009**, *6*, 1–16. [CrossRef]

41. Roy, S.; Yue, C.Y.; Wang, Z.Y.; Ananda, L. Thermal bonding of microfluidic devices: Factors that affect interfacial strength of similar and dissimilar cyclic olefin copolymers. *Sens. Actuators B Chem.* **2012**, *161*, 1067–1073. [CrossRef]

42. Sun, Y.; Kwok, Y.C.; Nguyen, N.T. Low-pressure, high-temperature thermal bonding of polymeric microfluidic devices and their applications for electrophoretic separation. *J. Micromech. Microeng.* **2006**, *16*, 1681–1688. [CrossRef]

43. Yu, H.; Tor, S.B.; Loh, N.H. Rapid bonding enhancement by auxiliary ultrasonic actuation for the fabrication of cyclic olefin copolymer (COC) microfluidic devices. *J. Micromech. Microeng.* **2014**, *24*, 115020. [CrossRef]

44. Yu, H.; Chong, Z.Z.; Tor, S.B.; Liu, E.; Loh, N.H. Low temperature and deformation-free bonding of pmma microfluidic devices with stable hydrophilicity via oxygen plasma treatment and PVA coating. *RSC Adv.* **2015**, *5*, 8377–8388. [CrossRef]

45. Tsao, C.W.; Hromada, L.; Liu, J.; Kumar, P.; DeVoe, D.L. Low temperature bonding of PMMA and COC microfluidic substrates using UV/ozone surface treatment. *Lab Chip* **2007**, *7*, 499–505. [CrossRef] [PubMed]

46. Shinohara, H.; Mizuno, J.; Shoji, S. Studies on low-temperature direct bonding of VUV, VUV/O$_3$ and O$_2$ plasma pretreated cyclo-olefin polymer. *Sens. Actuators A Phys.* **2011**, *165*, 124–131. [CrossRef]

47. Keller, N.; Nargang, T.M.; Runck, M.; Kotz, F.; Striegel, A.; Sachsenheimer, K.; Klemm, D.; Lange, K.; Worgull, M.; Richter, C.; et al. Tacky cyclic olefin copolymer: A biocompatible bonding technique for the fabrication of microfluidic channels in COC. *Lab Chip* **2016**, *16*, 1561–1564. [CrossRef] [PubMed]

48. Wan, A.M.D.; Sadri, A.; Young, E.W.K. Liquid phase solvent bonding of plastic microfluidic devices assisted by retention grooves. *Lab Chip* **2015**, *15*, 3785–3792. [CrossRef] [PubMed]

49. Salvo, P.; Verplancke, R.; Bossuyt, F.; Latta, D.; Vandecasteele, B.; Liu, C.; Vanfleteren, J. Adhesive bonding by SU-8 transfer for assembling microfluidic devices. *Microfluid. Nanofluid.* **2012**, *13*, 987–991. [CrossRef]

50. Lu, C.M.; Lee, L.J.; Juang, Y.J. Packaging of microfluidic chips via interstitial bonding. *Electrophoresis* **2008**, *29*, 1407–1414. [CrossRef] [PubMed]

51. Lai, S.; Cao, X.; Lee, L.J. A packaging technique for polymer microfluidic platforms. *Anal. Chem.* **2004**, *76*, 1175–1183. [CrossRef] [PubMed]

52. Toossi, A.; Moghadas, H.; Daneshmand, M.; Sameoto, D. Bonding pmma microfluidics using commercial microwave ovens. *J. Micromech. Microeng.* **2015**, *25*, 085008. [CrossRef]

53. Soper, S.A.; Henry, A.C.; Vaidya, B.; Galloway, M.; Wabuyele, M.; McCarley, R.L. Surface modification of polymer-based microfluidic devices. *Anal. Chim. Acta* **2002**, *470*, 87–99. [CrossRef]

54. Subramanian, B.; Kim, N.; Lee, W.; Spivak, D.A.; Nikitopoulos, D.E.; McCarley, R.L.; Soper, S.A. Surface modification of droplet polymeric microfluidic devices for the stable and continuous generation of aqueous droplets. *Langmuir* **2011**, *27*, 7949–7957. [CrossRef] [PubMed]

55. Vourdas, N.; Tserepi, A.; Boudouvis, A.G.; Gogolides, E. Plasma processing for polymeric microfluidics fabrication and surface modification: Effect of super-hydrophobic walls on electroosmotic flow. *Microelectron. Eng.* **2008**, *85*, 1124–1127. [CrossRef]

56. Hu, S.; Ren, X.; Bachman, M.; Sims, C.E.; Li, G.P.; Allbritton, N. Surface modification of poly(dimethylsiloxane) microfluidic devices by ultraviolet polymer grafting. *Anal. Chem.* **2002**, *74*, 4117–4123. [CrossRef] [PubMed]

57. Zhou, J.W.; Khodakov, D.A.; Ellis, A.V.; Voelcker, N.H. Surface modification for pdms-based microfluidic devices. *Electrophoresis* **2012**, *33*, 89–104. [CrossRef] [PubMed]

58. Kitsara, M.; Ducree, J. Integration of functional materials and surface modification for polymeric microfluidic systems. *J. Micromech. Microeng.* **2013**, *23*, 033001. [CrossRef]

59. Zilio, C.; Sola, L.; Damin, F.; Faggioni, L.; Chiari, M. Universal hydrophilic coating of thermoplastic polymers currently used in microfluidics. *Biomed. Microdevices* **2014**, *16*, 107–114. [CrossRef] [PubMed]

60. Temiz, Y.; Lovchik, R.D.; Kaigala, G.V.; Delamarche, E. Lab-on-a-chip devices: How to close and plug the lab? *Microelectron. Eng.* **2015**, *132*, 156–175. [CrossRef]

61. Van Heeren, H. Standards for connecting microfluidic devices? *Lab Chip* **2012**, *12*, 1022–1025. [CrossRef] [PubMed]

62. Chen, C.F.; Liu, J.; Hromada, L.P.; Tsao, C.W.; Chang, C.C.; DeVoe, D.L. High-pressure needle interface for thermoplastic microfluidics. *Lab Chip* **2009**, *9*, 50–55. [CrossRef] [PubMed]

63. Wilhelm, E.; Neumann, C.; Duttenhofer, T.; Pires, L.; Rapp, B.E. Connecting microfluidic chips using a chemically inert, reversible, multichannel chip-to-world-interface. *Lab Chip* **2013**, *13*, 4343–4351. [CrossRef] [PubMed]

64. Zou, Z.W.; Kai, J.H.; Rust, M.J.; Han, J.; Ahn, C.H. Functionalized nano interdigitated electrodes arrays on polymer with integrated microfluidics for direct bio-affinity sensing using impedimetric measurement. *Sens. Actuators A Phys.* **2007**, *136*, 518–526. [CrossRef]

65. Gärtner, C.; Kirsch, S.; Anton, B.; Becker, H. In Hybrid microfluidic systems: Combining a polymer microfluidic toolbox with biosensors. *Proc. SPIE* **2007**, *6465*, 64650F.

66. Godino, N.; Gorkin, R.; Bourke, K.; Ducree, J. Fabricating electrodes for amperometric detection in hybrid paper/polymer lab-on-a-chip devices. *Lab Chip* **2012**, *12*, 3281–3284. [CrossRef] [PubMed]

67. Choi, J.W.; Rosset, S.; Niklaus, M.; Adleman, J.R.; Shea, H.; Psaltis, D. 3-dimensional electrode patterning within a microfluidic channel using metal ion implantation. *Lab Chip* **2010**, *10*, 783–788. [CrossRef] [PubMed]

68. Oedit, A.; Vulto, P.; Ramautar, R.; Lindenburg, P.W.; Hankemeier, T. Lab-on-a-chip hyphenation with mass spectrometry: Strategies for bioanalytical applications. *Curr. Opin. Biotech.* **2015**, *31*, 79–85. [CrossRef] [PubMed]

69. Nge, P.N.; Rogers, C.I.; Woolley, A.T. Advances in microfluidic materials, functions, integration, and applications. *Chem. Rev.* **2013**, *113*, 2550–2583. [CrossRef] [PubMed]

70. Fiorini, G.S.; Yim, M.; Jeffries, G.D.M.; Schiro, P.G.; Mutch, S.A.; Lorenz, R.M.; Chiu, D.T. Fabrication improvements for thermoset polyester (TPE) microfluidic devices. *Lab Chip* **2007**, *7*, 923–926. [CrossRef] [PubMed]

71. Kim, J.-y.; deMello, A.J.; Chang, S.-I.; Hong, J.; O'Hare, D. Thermoset polyester droplet-based microfluidic devices for high frequency generation. *Lab Chip* **2011**, *11*, 4108–4112. [CrossRef] [PubMed]

72. Sollier, E.; Murray, C.; Maoddi, P.; Di Carlo, D. Rapid prototyping polymers for microfluidic devices and high pressure injections. *Lab Chip* **2011**, *11*, 3752–3765. [CrossRef] [PubMed]

73. Konstantinou, D.; Shirazi, A.; Sadri, A.; Young, E.W.K. Combined hot embossing and milling for medium volume production of thermoplastic microfluidic devices. *Sens. Actuators B Chem.* **2016**, *234*, 209–221. [CrossRef]

74. Stoller, M.A.; Konda, A.; Kottwitz, M.A.; Morin, S.A. Thermoplastic building blocks for the fabrication of microfluidic masters. *RSC Adv.* **2015**, *5*, 97934–97943. [CrossRef]

75. Vella, P.C.; Dimov, S.S.; Brousseau, E.; Whiteside, B.R. A new process chain for producing bulk metallic glass replication masters with micro- and nano-scale features. *Int. J. Adv. Manuf. Technol.* **2015**, *76*, 523–543. [CrossRef]

76. Li, G.; Parmar, M.; Lee, D.W. An oxidized liquid metal-based microfluidic platform for tunable electronic device applications. *Lab Chip* **2015**, *15*, 766–775. [CrossRef] [PubMed]

77. Ho, C.M.B.; Ng, S.H.; Li, K.H.H.; Yoon, Y.-J. 3D printed microfluidics for biological applications. *Lab Chip* **2015**, *15*, 3627–3637. [CrossRef] [PubMed]

78. Au, A.K.; Lee, W.; Folch, A. Mail-order microfluidics: Evaluation of stereolithography for the production of microfluidic devices. *Lab Chip* **2014**, *14*, 1294–1301. [CrossRef] [PubMed]

79. Gong, X.Q.; Yi, X.; Xiao, K.; Li, S.; Kodzius, R.; Qin, J.H.; Wen, W.J. Wax-bonding 3D microfluidic chips. *Lab Chip* **2010**, *10*, 2622–2627. [CrossRef] [PubMed]
80. Tsao, C.W.; Lee, Y.P. Magnetic microparticle-polydimethylsiloxane composite for reversible microchannel bonding. *Sci. Technol. Adv. Mater.* **2016**, *17*, 2–11. [CrossRef] [PubMed]
81. Tan, H.Y.; Loke, W.K.; Nguyen, N.T. A reliable method for bonding polydimethylsiloxane (PDMS) to polymethylmethacrylate (PMMA) and its application in micropumps. *Sens. Actuators B Chem.* **2010**, *151*, 133–139. [CrossRef]
82. Li, X.; Wu, N.Q.; Rojanasakul, Y.; Liu, Y.X. Selective stamp bonding of pdms microfluidic devices to polymer substrates for biological applications. *Sens. Actuators A Phys.* **2013**, *193*, 186–192. [CrossRef]
83. You, J.B.; Min, K.I.; Lee, B.; Kim, D.P.; Im, S.G. A doubly cross-linked nano-adhesive for the reliable sealing of flexible microfluidic devices. *Lab Chip* **2013**, *13*, 1266–1272. [CrossRef] [PubMed]
84. Yetisen, A.K.; Akram, M.S.; Lowe, C.R. Paper-based microfluidic point-of-care diagnostic devices. *Lab Chip* **2013**, *13*, 2210–2251. [CrossRef] [PubMed]
85. Xia, Y.N.; Whitesides, G.M. Soft lithography. *Annu. Rev. Mater. Sci.* **1998**, *28*, 153–184. [CrossRef]

micromachines

MDPI

Article

Large-Area and High-Throughput PDMS Microfluidic Chip Fabrication Assisted by Vacuum Airbag Laminator

Shuting Xie, Jun Wu, Biao Tang, Guofu Zhou, Mingliang Jin * and Lingling Shui *

Institute of Electronic Paper Displays, South China Academy of Advanced Optoelectronics & Joint International Research Laboratory of Optical Information of the Chinese Ministry of Education, South China Normal University, Guangzhou 510006, China; stxie@m.scnu.edu.cn (S.X.); 2006wujun1999@163.com (J.W.); tangbiao@scnu.edu.cn (B.T.); guofu.zhou@m.scnu.edu.cn (G.Z.)
* Correspondence: Jinml@m.scnu.edu.cn (M.J.); Shuill@m.scnu.edu.cn (L.S.); Tel.: +86-20-3931-4813 (M.J. & L.S.)

Received: 15 May 2017; Accepted: 29 June 2017; Published: 12 July 2017

Abstract: One of the key fabrication steps of large-area microfluidic devices is the flexible-to-hard sheet alignment and pre-bonding. In this work, the vacuum airbag laminator (VAL) which is commonly used for liquid crystal display (LCD) production has been applied for large-area microfluidic device fabrication. A straightforward, efficient, and low-cost method has been achieved for 400×500 mm^2 microfluidic device fabrication. VAL provides the advantages of precise alignment and lamination without bubbles. Thermal treatment has been applied to achieve strong PDMS–glass and PDMS–PDMS bonding with maximum breakup pressure of 739 kPa, which is comparable to interference-assisted thermal bonding method. The fabricated 152×152 mm^2 microfluidic chip has been successfully applied for droplet generation and splitting.

Keywords: large-area; microfluidic devices; fabrication; vacuum airbag laminator

1. Introduction

Microfluidics is also known by the names of microfluidic chip, lab on a chip (LOC) or micro-total-analysis-system (μ-TAS), which refers to building a micro-scale chemical or biological "lab" on a chip that is a few square centimeters in size [1,2]. A microfluidic chip contains a micro-channel network that allows the control of fluids to form a miniaturized chemical or biological "lab" where sample preparation, separation, detection, and other functions could be carried out [3,4]. Fabrication of stable and reliable microfluidic chips with various sizes is critical for their applications in different areas. Currently, many technologies aiming for functional chip fabrication in small size have been developed to fabricate microfluidic chips based on typical Micro-electromechanical Systems (MEMS) technologies [5–9]. When microfluidics is applied in the fields of emulsification [10], environmental analysis [11,12], and food safety [13,14], cheap and efficient technology for fabricating large area microfluidic chips becomes a bottleneck according to the limits of typical MEMS technologies. Therefore, it is necessary to find new technologies for large-area and high-throughput microfluidic chip fabrication, especially for industrialization.

Polydimethylsiloxane (PDMS) is the most popular material used to make microfluidic chips for both academic research (in biomedical, analytical, and biotechnological areas) and industrial production [8,15–17]. This is mainly attributed to the inherent properties of PDMS, such as optical transparency, biocompatibility, gas permeability, easy fabrication, low cost, and wide availability. On the other hand, soft-lithography technique has also enabled the widespread use of PDMS materials and opened up the era of PDMS-based microfluidics [18,19]. The fabrication of microfluidic devices typically includes structural design, mask fabrication, mold fabrication and replication, chip alignment, and bonding.

One of the key and difficult process challenges is how to align and bond the flexible PDMS sheet to another sheet of either hard glass or flexible PDMS to make a sealed chip. Precise alignment without bubble generation is mandatory for handling devices with flexible sheet. Up to now, there has been limited technical development focusing on such issues from the literature. Vacuum airbag laminator (VAL) is a commonly used piece of equipment in the liquid crystal display (LCD) production line. Two substrate sheets are respectively sucked onto top and bottom stages with vacuum of 0.01 bar. Typically, the alignment precision of 100 μm can be achieved for a 400 × 500 mm^2 display panel [20,21]. According to the vacuum environment, bubble generation is also avoided. So, we chose the VAL equipment to pre-bond a large-area PDMS sheet to another glass or PDMS sheet without bubble generation.

In addition, various issues like bonding strength, interfacial stress, microchannel fidelity, solvent compatibility, surface chemistry, and optical properties of the bonded microfluidic chips should all be taken into account for the selection of appropriate bonding methods. So far, a variety of bonding technologies have been developed to seal PDMS-based microfluidic chips, such as thermal bonding [22], solvent bonding [23], plasma-aided bonding [24–26], adhesive bonding [27,28], and ultrasonic welding [29,30]. However, the limitations of either fabrication size or materials selectivity are obvious for these methodologies. Thermal bonding has been frequently chosen according to its simple process and high bonding strength, which is also suitable for larger-area device processing.

In this work, the combination of VAL and thermal bonding has been applied for large-area PDMS microfluidic chip fabrication. Optimal parameters of PDMS components (mass ratio of the pre-polymer to the curing agent), VAL alignment and pre-bonding, and thermal annealing process were investigated. The bond strength was evaluated by the peel, leak, and burst tests. The fabricated 152 × 152 mm^2 microfluidic chips were successfully used to create and split microdroplets.

2. Materials and Methods

2.1. Materials

The glass substrate with thickness of 0.7 mm was purchased from Shenzhen Laibao Hi-tech Co. Ltd., Shenzhen, China. Deionized water (18.2 MΩ cm at 25 °C) was prepared by a water purification system (Water Purifier, Sichuan, China). The PDMS (Sylgard 184) package consisted of a base, and curing agent was purchased from Dow Corning Corporation (Midland, MI, USA). SU-8 3050 and its developer propylene glycol methyl ether acetate (PGMEA, 99%) were respectively purchased from MicroChem (Westborough, MA, USA) and Aladdin (Shanghai, China). N-hexadecane (99%) and sorbitane monooleate (Span 80, 99%) were purchased from Acros Organics (Geel, Belgium) and Aladdin (Shanghai, China), respectively. 1H,1H,2H,2H-perfluorodecyltrichlorosilane (FDTS, 96%) was purchased from Sigma Aldrich (Shanghai, China).

2.2. Fabrication of SU-8 Mold on Glass

Microfluidic channels on PDMS were fabricated using standard soft replication. According to the large-area device, the photomask and mold were also fabricated using glass substrate with the same size. Figure 1 shows the geometry of the microfluidic device. The microchannel shown in Figure 1a is a continuous Y-shaped microchannel network. It includes a vertically-tagged T-junction, and multiple tree-like Y-joints downstream microchannels. The width of the T-junction channel is 750.0 μm, and the width of the microchannel after each junction was sequentially decreased to half of the previous one. The smallest channel of the Y-shaped microchannel is about 11.7 μm. The microchannel length between two Y-joints is 2.7 mm. The whole length of the designed channel is 22.9 mm. The mold was fabricated on SU-8 3050 on a mother glass using standard lithography. Briefly, the photomask was obtained by transferring the CAD (computer-aided design) drawing onto a chrome plate. The mold glass substrate was chemically cleaned by immersing in piranha solution (H$_2$SO$_4$/H$_2$O$_2$ 3:1, *v/v*) for 10 min to remove organic impurities, and rinsed using deionized (DI)water and dried using nitrogen.

Subsequently, the negative photoresist of SU-8-3050 was spin-coated on the glass, soft baked at 95 °C for 2 min, exposed for 150 s using an aligner (URE-2000/35, Chinese Academy of Sciences, Beijing, China), post-baked on a hotplate (EH20B, LabTech, Beijing, China) at 95 °C for 3 min, developed in PGMEA for 3 min, and then hard-baked at 150 °C for 30 min to obtain the photoresist patterned glass mold. Afterwards, FDTS coating was formed by vapor deposition [31].

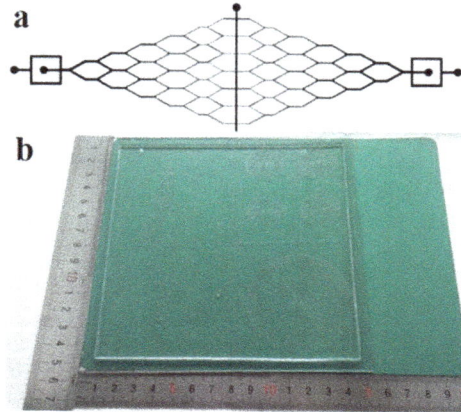

Figure 1. (**a**) Schematic drawing of the microfluidic device design; (**b**) Photograph of one fabricated large-area microfluidic plate with multiple chips.

2.3. Fabrication of PDMS Microfluidic Chips

The PDMS substrate was prepared by initially mixing the pre-polymer solution with the curing agent at a certain mass ratio. The mixture was then degassed in a vacuum chamber for 30 min. Subsequently, the mixture was poured onto the prepared mold and cured in an oven (EH20B, LabTech, Beijing, China) at 90 °C for 30 min to replicate the microfluidic patterns from mold to PDMS. After peeling the PDMS sheet from the mold, holes were punched through for the in-chip and out-chip fluidic interconnection. The depth of the obtained microchannel is approximately 38 μm.

2.4. Bonding Procedure

The bonding process contains two major steps: VAL aligning and pre-bonding, and thermal annealing. Figure 2 shows the schematic illustration of the VAL aligning and pre-bonding steps. Before pre-bonding, the glass substrate was chemically cleaned by immersing in piranha solution (H_2SO_4/H_2O_2 3:1, v/v) for 10 min, rinsed using DI water, and dried using N_2, and the PDMS sheet was also thoroughly rinsed using DI water and dried using N_2. Then, the PDMS sheet and glass substrate was positioned and attached on the top and bottom vacuum stages, respectively. The vacuum of top and bottom stages was respectively 0.01 and 0.008 bar, respectively. The top stage holding the PDMS sheet was then flipped (Figure 2a) and scrolled down to the appropriate position, and moved slowly to the right position for alignment from both *x*- and *y*-axis sides (Figure 2b). As soon as the alignment precision requirement was satisfied, two sheets were put together face-to-face (Figure 2c). The alignment precision was about 100 μm. Then, the PDMS sheet was released from the top plate by applying positive pressure to the top stage (Figure 2c). A pressure roller was then applied to the surface to press the PDMS and glass for pre-bonding (Figure 2d). The roller pressure was set at 0.15 MPa and the rolling velocity was 10 mm/s. The pre-bonded microfluidic chip was then released from the bottom stage by applying a positive pressure (Figure 2e). The prepared chip was then ready to be used directly or put into an oven for further thermal bonding.

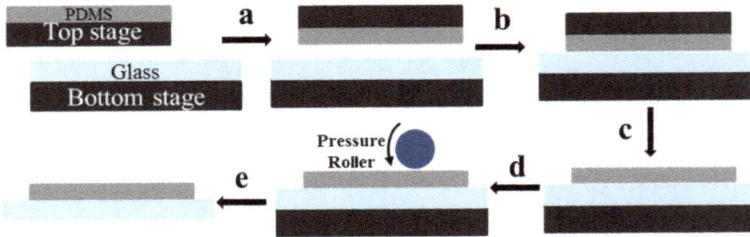

Figure 2. Schematic of the Vacuum airbag laminator (VAL) aligning and pre-bonding process: (**a**) flipping well-positioned top plate to the middle of the bottom plate; (**b**) aligning and laminating the flexible top sheet to the bottom plate; (**c**) releasing the polydimethylsiloxane (PDMS) sheet from the top stage; (**d**) pre-bonding the PDMS–glass chip by pressure roller; and (**e**) releasing the pre-bonded chip from the bottom stage.

2.5. Bond Strength Analysis

To characterize the bonding strength, peel and delamination tests were conducted. A home-made device was used to measure the force required to physically separate the patterned PDMS sheet and the glass substrate, as shown in Figure 3. Before testing, a PDMS substrate ($10 \times 10 \times 1.5$ mm^3) was ordinarily bonded with a glass slide by VAL pre-bonding with or without thermal bonding. For comparison, the other surface of the PDMS was bonded with another glass substrate by the normal plasma-aided bonding [15]. Figure 3a exhibits a PDMS chip sample. The force was measured using the method as described in references [32,33], using the home-made device with a digital tubular tensiometer (ALIYIOI, Wenzhou Yiding Instrument Manufacturing Co. Ltd., Wenzhou, China) as shown in Figure 3b. Figure 3c shows a typical pressure versus time curve during the measurement. Because the force to separate the plasma-aided bonding was higher than that of the thermal bonding, the thermally-bonded glass slide would be first peeled off. Figure 3c provides details of the pressure distribution applied on the contact surface between the two surfaces, provided by the software. The maximum value was counted as the bond strength of the PDMS and glass. In order to secure the reproducibility, the tests were performed in triplicate.

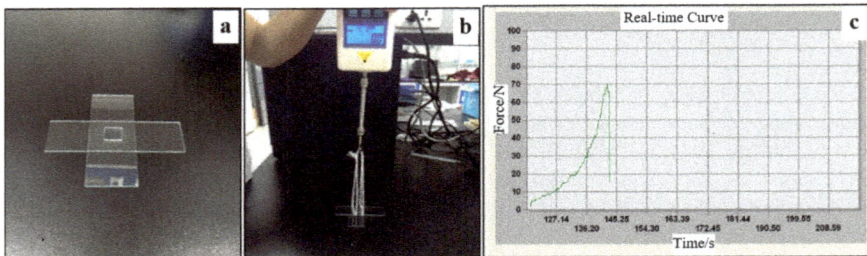

Figure 3. The bonding strength analysis: (**a**) the tested PDMS chip sample; (**b**) the home-made device with a digital tubular tensiometer; and (**c**) the software interface showing the pressure distribution applied on the contact surface between the two surfaces.

2.6. Leak Test

Leak test was conducted to assess the sealing efficiency of the large-area PDMS–glass chips. 0.5 w/w % Rhodamine B solution was pumped into the microchannels at different flow rates using a syringe pump (KDS 200, Kd Scientific, Hongkong, China). The flow rate at which the dyed liquid started to squeeze out of the microchannel is regarded as the beginning of the leakage.

2.7. Optical Measurement of Microfluidic Chip

To prove the validity of this method, the fabricated 152×152 mm^2 microfluidic chip was tested. Two immiscible liquid phases were introduced into the microchannels via the inlets through polytetrafluoroethylene (PTFE) tubes by two syringe pumps (KDS 200, Kd Scientific, Hongkong, China). The fluidic flow behavior was visualized and recorded using an inverted optical microscope (Olympus IX2, Tokyo, Japan) equipped with a high-speed camera (Phantom Miro M110, Vision Research Inc., Wayne County, NC, USA).

3. Results and Discussion

3.1. Thermal Bonding of PDMS–Glass

Bonding quality is a key factor in achieving reliable microfluidic chips. Therefore, optimal parameters of thermal bonding were investigated. Typically, thermal bonding is performed at a temperature higher than the glass transition temperature (T_g) of investigated materials. T_g of PDMS is ~120 °C [31], and the glass substrate can survive up to 600 °C without obvious property change. Therefore, the bonding temperature of 120–260 °C has been tested for optimizing the bonding process.

A PDMS substrate ($10 \times 10 \times 1.5$ mm^3) was manually put onto a glass slide surface and heated in an oven at different temperatures (T) for 4 h to achieve thermal bonding. At the same time, plasma-aided bonding was also carried out for comparison. Peel test was carried out to analyze the bonding strength and efficiency at different temperature. From the photos shown in Figure 4a–h, we can find that the quantity of residual PDMS on the glass surface bonded at 220 °C is the most after peeling off process, which is comparable to the plasma-aided bonding.

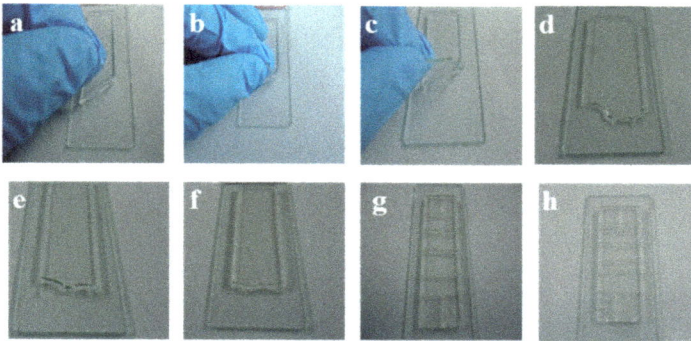

Figure 4. The peeling-off experiments at different temperatures: (**a**) 120 °C; (**b**) 140 °C; (**c**) 160 °C; (**d**) 180 °C; (**e**) 200 °C; (**f**) 220 °C; (**g**) 240 °C; and (**h**) 260 °C.

In addition, the force required to separate the bonded PDMS and glass sheets after thermal bonding was measured to evaluate the bond strength. The force (F) was obtained by averaging three measurements under the same conditions. Figure 5a presents the measured force as a function of the annealing temperature (T) (the mass ratio of the pre-polymer to the curing agent was 10:1). When the temperature was below 150 °C, F was maintained at about 51 kPa (5.1 N divided by 10×10 mm^2). With the increase of T, F increased when T was in the range of 150–200 °C; it decreased when T was higher than 200 °C. Thus, the maximum force was achieved at $T = 200$ °C, with the value of about 542 kPa.

The effect of holding time on the bonding efficiency was studied at $T = 200$ °C. The pre-bonded devices were put in the oven heated up to 200 °C and held for different periods of time (t_h). As shown in Figure 5b, the bonding strength increased with holding time when t_h was in the range of 0.5–4 h.

When $t_h > 4$ h, the adhesion force gradually reduced. The adhesion force was about 531 kPa when $t_h = 4$ h, which is consistent with the result of Figure 5a. The cracking and decreased adhesion force at higher temperature may be attributed to the oxidation or decomposition of PDMS.

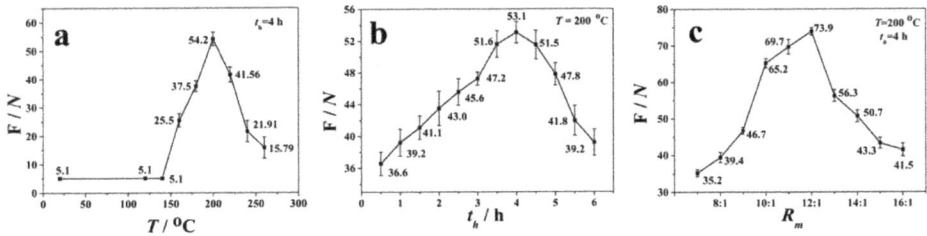

Figure 5. The variation of bonding force with (**a**) temperature (T); (**b**) holding time (t_h) at 200 °C; and (**c**) the mass ratio of pre-polymer to curing agent (R_m) of PDMS substrate.

Figure 5c shows the profile of the measured force between the PDMS–glass substrates with different PDMS compositions by varying the mass ratio (R_m) of the pre-polymer to the curing agent. The variation of mass ratios means different degrees of cross-linking of PDMS. The lower the mass ratio is, the lesser the degree to which the PDMS molecules are cross-linked, and the softer the PDMS sheet is. Varying the crosslink density in the polymer network allows us to tune the mechanical and optical properties of the obtained PDMS substrate to adapt to different applications. As seen in Figure 5c, the bonding strength increased with R_m when $R_m < 12$:1, then sharply reduced when R_m was changed from 12:1 to 13:1. After that, the force slowly decreased with R_m. Therefore, the maximum force of 739 kPa was obtained under the conditions of $R_m = 12$:1, $T = 200$ °C, and $t_h = 4$ h, which is higher than the plasma-aided bonding and interference-assisted thermal bonding [29,32]. Such parameters were then used commonly for the rest of experiments in this work.

3.2. Large-Area Microfluidic Device Fabrication Using VAL

The vacuum airbag laminator plays an important role in the LCD panel production line. Display devices with areas up to 400×500 mm^2 (G2.5 of LCD line) have been fabricated in our lab using VAL, with the alignment precision of about 100 μm. Both LCDs and microfluidic devices are composed of glass and flexible substrates. Therefore, the use of VAL has predetermined advantages for microfluidic device fabrication. The possible fabrication size can hereby be comparable to the maximum LCDs; for instance the G11 line (3000×3320 mm^2), which is highly valuable for industrialization.

VAL is a technology manipulated at room temperature, which can not only improve the alignment speed and accuracy but also avoid bubble generation, as described in the Materials and Methods session. To assess the sealing efficiency of the large-area PDMS–glass chips, leak tests were conducted by pumping 0.5 w/w % Rhodamine B solution into the whole microchannels from the inlet at different flow rates. The flow rate was increased from 100 μL·h^{-1} until the beginning of the leakage. When the dyed liquid started to squeeze out of the microchannel, we counted the flow rate as the beginning of the leakage. The pressure (ΔP, Pa) between inlet and leakage position of the channel was calculated by Poiseuille's law: $\Delta P = QR_{hy}$, where Q (m^3·s^{-1}) is the applied volume flow rate and R_{hy} (kg·s^{-1}·L^{-4}) is the hydrodynamic flow resistance. Assuming rectangular cross-section of the microchannels, R_{hy} could be approximately expressed as: $R_{hy} \approx \eta LC^2 A^{-3}$, where η (Pa·s) is the viscosity of the Rhodamine B solution, L (m) is the microchannel length, C (m) and A (m^2) are the perimeter and area of the cross-section. Based on the tree-like microchannel arrangement, the pressure at each Y-junction block was calculated and then added to obtain the total pressure drop over the microfluidic network, which was about 30.5 kPa.

Figure 6 shows the peel and leak tests of the fabricated PDMS–glass using only VAL pre-bonding (Figure 6a–c) and the combination of VAL pre-ponding and thermal bonding (Figure 6d,e). The only VAL pre-bonding could withstand a flow rate of 200 μL·h^{-1} (Figure 6b) with the maximum bonding strength of about 51 kPa (Figure 6a) which may not be enough for general microfluidic experiments. As seen in Figure 6c, the leakage started at the first joint where the sudden increase of flow velocity happened due to the large size change from 3 × 1.5 mm to 1 × 750 μm.

After thermal bonding treatment at 200 °C for 4 h, the average bonding strength reached 251 kPa, which is five times higher than those without thermal annealing, as seen from Figure 6d. The leakage flow rate was about 1000 μL·h^{-1}, as shown in Figure 6e,f. Obvious bonding strength improvement was obtained by the thermal annealing.

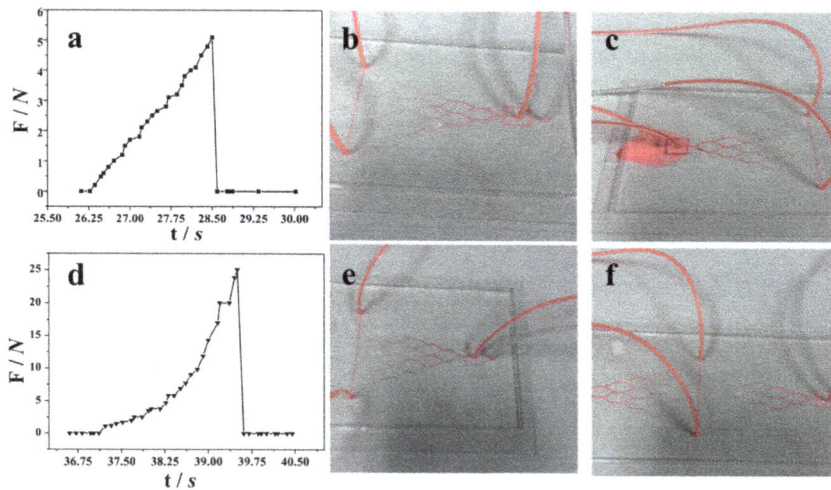

Figure 6. Comparison of adhesion force (**a,d**) and leakage flow rates (**b,c,e,f**) when PDMS–glass devices were bonded by VAL pre-bonding without thermal bonding (**a–c**), and the combination of VAL pre-bonding and thermal bonding at 200 °C for 4 h (**d–f**). The applied flow rates were: (**b**) 200 μL·h^{-1}; (**c**) 250 μL·h^{-1}; (**e**) 200 μL·h^{-1}; and (**f**) 1000 μL·h^{-1}, respectively.

After bonding, either one large-area multifunctional microfluidic chip or multiple small microfluidic chips cut from one large plate could be obtained at once. Therefore, the production speed is enhanced, and the price per chip is reduced.

3.3. Droplet Creation and Splitting in the Microfluidic Chip

To prove the validity of this method, the fabricated 152 × 152 mm^2 microfluidic chip was tested. Two immiscible liquid phases of DI water and hexadecane with 3.0 *w*/*w* % Span 80 were introduced into the microchannel via the inlets through PTFE tubes by two syringe pumps. The schematic illustration of the experimental set-up is shown in Figure 7a. The geometry of the microfluidic device is shown in Figure 1, which includes a vertically-tagged T-junction and multiple tree-like Y-joints downstream microchannel networks. When the two immiscible phases met at the T-junction, the water phase broke into droplets dispersed in the organic continuous phase (Figure 7b). When the generated larger droplets flew through the downstream, they were cut into smaller ones according to the flow resistance distribution caused by the obstacle of the Y-joint channel, as shown in Figure 7c and Mov.S1. The average diameter and standard deviation of microdroplets in the outlet channel were, respectively, 90.36 μm and 8.4% when the oil and water flow rates was 600 μL·h^{-1} and 200 μL·h^{-1}, respectively

(Figure 7d). The fabricated 152×152 mm^2 PDMS–glass microfluidic chip has been successfully applied for droplets creation and splitting.

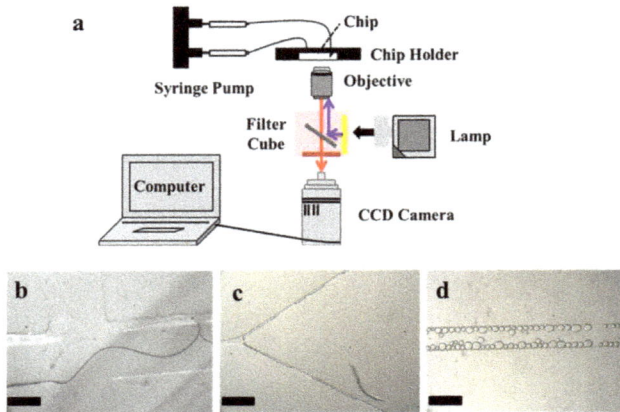

Figure 7. (**a**) Schematic illustration of the experimental set-up; (**b**) A snapshot of the droplet generation at the T-junction; (**c**) A snapshot of the droplet breakup at the Y-joint; (**d**) A snapshot of the droplets flow in the outlet channel after being broken at various ratio from the Y-joint. All scale bars in the figures are 500 μm.

4. Conclusions

In conclusion, a straightforward and efficient method for large-area PDMS–glass microfluidic chip fabrication has been proposed and verified. The vacuum airbag laminator plays a key role for flexible PDMS sheet aligning and pre-bonding with glass substrate without air bubble generation. The size of the bonded device is determined by the available size of the laminating machine. Thermal bonding has been applied to improve the bonding strength from 52 kPa to 739 kPa under the optimal conditions R_m = 12:1, T = 200 °C, and t_h = 4 h, which is comparable to those obtained from plasma-aided bonding and interference-assisted thermal bonding. The fabricated 152×152 mm^2 PDMS–glass microfluidic chip has been successfully applied for droplet creation and splitting in the same device. Either one large-area multifunctional microfluidic chip or multiple small microfluidic chips cut from one large plate could be obtained at once by this fabrication method. The VALs can realize large-area hard-to-hard, soft-to-hard, and soft-to-soft device fabrication. Such a method has borrowed and combined existed technologies from matured LCD production line to the microfluidic area, making it easier for industrialization and commercialization.

Supplementary Materials: The following are available online at www.mdpi.com/2072-666X/8/7/218/s1, Mov.S1: The process of generated bigger droplets cut into smaller ones when they flow through the downstream Y-junctions.

Acknowledgments: We appreciate the financial support from the National Natural Science Foundation of China (No. 61574065), the National Key Research & Development Program of China (2016YFB0401502), the Science and Technology Planning Project of Guangdong Province (2016B090906004), the Science and technology project of Guangdong Province (No. 2016A010101023). This work was also partially supported by the national 111 project, PCSIRT project (IRT13064) and Guangdong Innovative Research Team Program (No. 2011D039).

Author Contributions: Mingliang Jin and Lingling Shui designed this project. Jun Wu performed part of the experiments. Shuting Xie also carried out part of the experiments. Mingliang Jin especially conducted and supervised the VAL technology. The data analysis and writing of this article was completed together by Shuting Xie and Lingling Shui. Jun Wu, Biao Tang and Mingliang Jin contributed to the results discussion. Guofu Zhou gave suggestions on the project management and conducted helpful discussion on the experimental results.

Conflicts of Interest: The authors declare no conflict of interest.

References

1. DeMello, A.J. Control and detection of chemical reactions in microfluidic systems. *Nature* **2006**, *442*, 394–402. [CrossRef] [PubMed]
2. Manz, A.; Graber, N.; Widmer, H.M. Miniaturized total chemical analysis systems: A novel concept for chemical sensing. *Sens. Actuators B Chem.* **1990**, *1*, 244–248. [CrossRef]
3. Njoroge, S.K.; Witek, M.A.; Battle, K.N.; Immethun, V.E.; Hupert, M.L.; Soper, S.A. Integrated continuous flow polymerase chain reaction and micro-capillary electrophoresis system with bioaffinity preconcentration. *Electrophoresis* **2011**, *32*, 3221–3232. [CrossRef] [PubMed]
4. Chen, P.C.; Park, D.S.; You, B.H.; Kim, N.; Park, T.; Soper, S.A.; Nikitopoulos, D.E.; Murphy, M.C. Titer-plate formatted continuous flow thermal reactors: Design and performance of a nanoliter reactor. *Sens. Actuators B Chem.* **2010**, *149*, 291–300. [CrossRef] [PubMed]
5. Roy, S.; Yue, C.Y.; Venkatraman, S.S.; Ma, L.L. Fabrication of smart COC chips: Advantages of N-vinylpyrrolidone (NVP) monomer over other hydrophilic monomers. *Sens. Actuators B Chem.* **2013**, *178*, 86–95. [CrossRef]
6. Nguyen, N.T.; Shaegh, S.A.M.; Kashaninejad, N.; Phan, D.T. Design, fabrication and characterization of drug delivery systems based on lab-on-a-chip technology. *Adv. Drug Deliv. Rev.* **2013**, *65*, 1403–1419. [CrossRef] [PubMed]
7. Komuro, N.; Takaki, S.; Suzuki, K.; Citterio, D. Inkjet printed (bio)chemical sensing devices. *Anal. Bioanal. Chem.* **2013**, *405*, 5785–5805. [CrossRef] [PubMed]
8. Gao, C.L.; Sun, X.H.; Gillis, K.D. Fabrication of two-layer poly(dimethyl siloxane) devices for hydrodynamic cell trapping and exocytosis measurement with integrated indium tin oxide microelectrodes arrays. *Biomed. Microdevices* **2013**, *15*, 445–451. [CrossRef] [PubMed]
9. Oliveira, K.A.; De Oliveira, C.R.; Da Silveira, L.A.; Coltro, W.K.T. Laser-printing of toner-based 96-microzone plates for immunoassays. *Analyst* **2013**, *138*, 1114–1121. [CrossRef] [PubMed]
10. Utada, A.S.; Lorenceau, E.; Link, D.R.; Kaplan, P.D.; Stone, H.A.; Weitz, D.A. Monodisperse double emulsions generated from a microcapillary device. *Science* **2005**, *308*, 537–541. [CrossRef] [PubMed]
11. Dharmasiri, U.; Witek, M.A.; Adams, A.A.; Osiri, J.K.; Hupert, M.L.; Bianchi, T.S.; Roelke, D.L.; Soper, S.A. Enrichment and detection of escherichia coli O157:H7 from water samples using an antibody modified microfluidic chip. *Anal. Chem.* **2010**, *82*, 2844–2849. [CrossRef] [PubMed]
12. Eriksson, E.; Sott, K.; Lundqvist, F.; Sveningsson, M.; Scrimgeour, J.; Hanstorp, D.; Goksor, M.; Graneli, A. A microfluidic device for reversible environmental changes around single cells using optical tweezers for cell selection and positioning. *Lab Chip* **2010**, *10*, 617–625. [CrossRef] [PubMed]
13. Ferey, L.; Delaunay, N. Food analysis on electrophoretic microchips. *Sep. Purif. Rev.* **2016**, *45*, 193–226. [CrossRef]
14. Dong, Y.Y.; Liu, J.H.; Wang, S.; Chen, Q.L.; Guo, T.Y.; Zhang, L.Y.; Jin, Y.; Su, H.J.; Tan, T.W. Emerging frontier technologies for food safety analysis and risk assessment. *J. Integr. Agr.* **2015**, *14*, 2231–2242. [CrossRef]
15. Shiroma, L.S.; Piazzetta, M.H.; Duarte-Junior, G.F.; Coltro, W.K.; Carrilho, E.; Gobbi, A.L.; Lima, R.S. Self-regenerating and hybrid irreversible/reversible PDMS microfluidic devices. *Sci. Rep.* **2016**, *6*, 26032–26033. [CrossRef] [PubMed]
16. Vladisavljevic, G.T.; Khalid, N.; Neves, M.A.; Kuroiwa, T.; Nakajima, M.; Uemura, K.; Ichikawa, S.; Kobayashi, I. Industrial lab-on-a-chip: Design, applications and scale-up for drug discovery and delivery. *Adv. Drug Deliv. Rev.* **2013**, *65*, 1626–1663. [CrossRef] [PubMed]
17. Cai, D.K.; Neyer, A. Cost-effective and reliable sealing method for PDMS (polydimethylsiloxane)-based microfluidic devices with various substrates. *Microfluid. Nanofluid.* **2010**, *9*, 855–864. [CrossRef]
18. Xia, Y.N.; Whitesides, G.M. Soft lithography. *Angew. Chem. Int. Ed.* **1998**, *37*, 550–575. [CrossRef]
19. Qin, D.; Xia, Y.N.; Black, A.J.; Whitesides, G.M. Photolithography with transparent reflective photomasks. *J. Vac. Sci. Technol. B* **1998**, *16*, 98–103. [CrossRef]
20. Tseng, C.H.; Liu, Y.C.; Chen, M.S. Enhancement of Scribing Stability for Laminated TFT-LCD. Presented at 17th Annual International Display Workshops (IDW 10), Fukuoka, Japan, 1–3 December 2010; ITE/SID, INST Image Information & Television Engineers: Tokyo, Japan, 2010.

21. Hirakata, J.; Nagae, Y.; Kando, Y. Polarization analysis of a black and white supertwisted nematic liquid crystal display with two laminated birefringent films of different optical axes. *Jpn. J. Appl. Phys.* **1993**, *32*, 872. [CrossRef]

22. Carballo, V.M.B.; Melai, J.; Salm, C.; Schmitz, J. Moisture resistance of SU-8 and KMPR as structural material. *Microelectron. Eng.* **2009**, *86*, 765–768. [CrossRef]

23. Chen, P.C.; Duong, L.H. Novel solvent bonding method for thermoplastic microfluidic chips. *Sens. Actuators B Chem.* **2016**, *237*, 556–562. [CrossRef]

24. Carlier, J.; Chuda, K.; Arscott, S.; Thomy, V.; Verbeke, B.; Coqueret, X.; Camart, J.C.; Druon, C.; Tabourier, P. High pressure-resistant SU-8 microchannels for monolithic porous structure integration. *J. Micromech. Microeng.* **2006**, *16*, 2211–2219. [CrossRef]

25. Ye, M.Y.; Fang, Q.; Yin, X.F.; Fang, Z.L. Studies on bonding techniques for poly (dimethylsiloxane) microfluidic chips. *Chem. J. Chin. Univ.* **2002**, *23*, 2243–2246.

26. Igata, E.; Arundell, M.; Morgan, H.; Cooper, J.M. Interconnected reversible lab-on-a-chip technology. *Lab Chip* **2002**, *2*, 65–69. [CrossRef] [PubMed]

27. Chen, H.Y.; Mcclelland, A.A.; Chen, Z.; Lahann, J. Solventless adhesive bonding using reactive polymer coatings. *Anal. Chem.* **2008**, *80*, 4119–4124. [CrossRef] [PubMed]

28. Song, J.; Vancso, G.J. Effects of flame treatment on the interfacial energy of polyethylene assessed by contact mechanics. *Langmuir* **2008**, *24*, 4845–4852. [CrossRef] [PubMed]

29. Kreider, A.; Richter, K.; Sell, S.; Fenske, M.; Tornow, C.; Stenzel, V.; Grunwald, I. Functionalization of PDMS modified and plasma activated two-component polyurethane coatings by surface attachment of enzymes. *Appl. Surf. Sci.* **2013**, *273*, 562–569. [CrossRef]

30. Yussuf, A.A.; Sbarski, I.; Hayes, J.P.; Solomon, M.; Tran, N. Microwave welding of polymeric-microfluidic devices. *J. Micromech. Microeng.* **2005**, *15*, 1692–1699. [CrossRef]

31. Shui, L.L.; Van Den Berg, A.; Eijkel, J.C.T. Interfacial tension controlled W/O and O/W 2-phase flows in microchannel. *Lab Chip* **2009**, *9*, 795–801. [CrossRef] [PubMed]

32. Gong, Y.; Park, J.M.; Lim, J. An interference-assisted thermal bonding method for the fabricarion of thermoplastic microfluidic devices. *Micromachines* **2016**, *7*, 211–221.

33. Nguyen, T.P.O.; Tran, B.M.; Lee, N.Y. Thermally robust and biomolecule-friendly room-temperature bonding for the fabrication of elastmer-plastic hybrid microdevices. *Lab Chip* **2016**, *16*, 3251–3259. [CrossRef] [PubMed]

micromachines

MDPI

Article

Fabrication of All Glass Bifurcation Microfluidic Chip for Blood Plasma Separation

Hyungjun Jang [1], Muhammad Refatul Haq [1], Jonghyun Ju [1], Youngkyu Kim [1], Seok-min Kim [1,*] and Jiseok Lim [2,*]

[1] School of Mechanical Engineering, Chung-Ang University, Seoul 06974, Korea; janghj@cau.ac.kr (H.J.); refat@cau.ac.kr (M.R.H.); jhju@cau.ac.kr (J.J.); kykdes@cau.ac.kr (Y.K.)
[2] School of Mechanical Engineering, Yeungnam University, Gyeongsan-si, Gyeongsangbuk-do 38541, Korea
* Correspondence: smkim@cau.ac.kr (S.K.); jlim@yu.ac.kr (J.L.);
   Tel.: +82-2-820-5877 (S.K.); +82-53-810-2577 (J.L.)

Academic Editors: Weihua Li, Hengdong Xi and Say Hwa Tan
Received: 12 December 2016; Accepted: 20 February 2017; Published: 24 February 2017

Abstract: An all-glass bifurcation microfluidic chip for blood plasma separation was fabricated by a cost-effective glass molding process using an amorphous carbon (AC) mold, which in turn was fabricated by the carbonization of a replicated furan precursor. To compensate for the shrinkage during AC mold fabrication, an enlarged photoresist pattern master was designed, and an AC mold with a dimensional error of 2.9% was achieved; the dimensional error of the master pattern was 1.6%. In the glass molding process, a glass microchannel plate with negligible shape errors (~1.5%) compared to AC mold was replicated. Finally, an all-glass bifurcation microfluidic chip was realized by micro drilling and thermal fusion bonding processes. A separation efficiency of 74% was obtained using the fabricated all-glass bifurcation microfluidic chip.

Keywords: glass molding; amorphous carbon mold; microfluidics; blood plasma separation; bifurcation microfluidic chip

---

1. Introduction

Microfluidic chip technology has attracted much attention in the field of biological analysis because it decreases the amount of the sample and the time required for analysis [1–4]. Blood plasma provides plenty of vital information about health condition [5], and is one of the most widely used target fluids in microfluidic systems for disease diagnosis [6]. Blood plasma is commonly extracted using a centrifuge machine. However, the procedure is time-consuming and requires relatively large amounts of blood sample; hence, it is not suitable for use with the microfluidic chip technology. The blood plasma separation processes have been conducted on a microfluidic platform. The microfluidic methods for blood plasma separation can be categorized into two types: active and passive. The active methods use external forces such as centrifugal [7,8], acoustic [9], electric [10,11], and magnetic [12] forces, and the passive methods use hydrodynamics [13–16] or geometrical filter structures [17–19] for blood plasma separation. Although the centrifugal microfluidic platform has been effectively used for separating and analyzing blood plasma, it is not suitable for continuous real time blood separation and analysis processes, which are required in some medical treatments such as blood dialysis for renal insufficiency patients and cardiac surgery with cardiopulmonary bypass [20]. The bifurcation microfluidic chip [20–22], a passive chip using hydrodynamics, can be regarded as a suitable method for continuous blood plasma separation because it can separate blood plasma from the main blood stream and does not require any additional device for generating external force. In the bifurcation microfluidic chip, the blood plasma separation efficiency is mainly affected by the geometrical shape of the chip and the efficiency increases with increasing flow rate [23,24]. Therefore, dimensional stability at high flow rates is essential for the blood plasma separation microfluidic chip using bifurcation. In the fabrication of the bifurcation microfluidic chip, most of the researchers use a polydimethylsiloxane (PDMS) replication process because of its simplicity and cost-effectiveness. However, the PDMS microfluidic chip has some limitations regarding dimensional stability at high flow rates due to its elastic characteristics.

In this research, we fabricated a highly durable glass bifurcation microfluidic chip for blood plasma separation. To realize a glass microfluidic chip, various patterning techniques based on the material removal of glass substrate have been investigated, such as mechanical milling [25], laser ablation [26], and etching process [27]. However, these methods are time-consuming due to the high mechanical, chemical, and thermal resistances of the glass materials. As an alternative method for patterning the glass material at low cost, a glass molding process has been proposed. In this process, the glass substrate is heated up to higher than its glass transition temperature and pressed against a mold with the negative shape of the final structure. During the molding process, the cavity patterns of the mold are transferred onto the surface of the glass substrate. Since the processing temperature of glass molding (>600 °C) is much higher than that of the conventional polymer molding, the nickel mold commonly used in the micro polymer molding processes cannot be utilized for glass molding due to its low mechanical resistances at high temperatrue. Nickel alloy [28], silicon carbide [29], and amorphous carbon (AC) [30] have been used as mold materials for glass molding of microfluidic structures due to their high hot hardness. To prepare the cavity structure in these refractory materials, laser aberration was applied in the previous works [28–30]. However, the surface quality of the mold cavity pattern fabricated by laser aberration was not suitable for precise glass microfluidic chip applications. Furthermore, the laser aberration method is not suitable for fabricating many molds with the same design, which are required in the mass production of glass molded microfluidic devices, because it is a serial machining process and a long machining time is required due to the low material removal-rate of refractory materials and the large machining area to produce a protruding microchannel cavity. In our previous paper, we proposed a low-cost, parallel, and large-area fabrication method for an AC micro mold with high surface quality by the carbonization of a replicated furan precursor [31].

In this research, an all-glass bifurcation microfluidic chip for blood plasma separation was fabricated, using the glass molded microchannel plate prepared from an AC mold. Figure 1 shows the schematics of the steps of the fabrication process for the all-glass bifurcation microfluidic chip used in this study. Although various patterning techniques [32–35] can be applied as a master fabrication process, conventional photolithography was applied in this study. To compensate for the shape change during AC mold fabrication, an enlarged silicon master pattern was prepared. A furan precursor was obtained by thermal replication using a replicated PDMS mold from the master. After the carbonization of the furan precursor, an AC mold with microchannel cavities was obtained. Finally, the glass bifurcation microfluidic chip was obtained by implementing glass molding for the microchannel plate, micro-drilling for the inlet and outlet connecting holes, and thermal fusion bonding for sealing. To examine the feasibility of the all-glass bifurcation microfluidic chip, the performance of the blood plasma separation was evaluated.

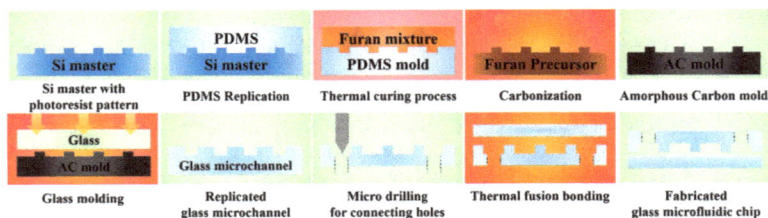

Figure 1. Schematic of the fabrication process for the AC mold by the carbonization of a replicated furan precursor and the all-glass bifurcation microfluidic chip by glass molding and thermal fusion bonding.

2. Fabrication of the All-Glass Microfluidic Chip Using a Glass Molded Microchannel

2.1. Design and Fabrication of the Silicon Master

A bifurcation microfluidic chip for blood plasma separation was examined for the feasibility of the glass molded microfluidic chip prepared using AC mold. Figure 2a shows the design of the blood plasma separation chip. The tortuous inlet microchannel was designed to improve the separation efficiency due to the centrifugal action [21]. Figure 2b shows the schematic of the main separation region of the bifurcation microfluidic chip. The designed widths of the blood inlet, blood outlet, and plasma extraction channels were 150, 450, and 150 μm, respectively, and the height of the channel was 55 μm. The performance of the bifurcation microfluidic chip can be examined using the channel resistance ratio between the plasma extraction and the blood outlet channels. The channel resistance (R) can be calculated by Equation (1) [36].

$$R = \frac{12\mu L}{\{1 - 0.63\left(\frac{h}{w}\right)\}} \frac{1}{h^3 w} \tag{1}$$

where μ is the viscosity of the blood (0.004 Pa·s) and L, w, and h are the length, width, and height of the channel, respectively. The lengths of the blood outlet and plasma extraction channels were 1.5 and 26.7 mm, respectively. The calculated channel resistance ratio between the plasma extraction and blood outlet channels was 63:1, which is an acceptable value for plasma extraction [36].

In the proposed AC mold fabrication process, a significant shape change inherently occurs during carbonization due to the material decomposition [31]. In addition, some shrinkages also occurred in PDMS and Furan replication processes due to the polymerization of material. To compensate the total shrinkage occurring in the AC mold fabrication process, we conducted a preliminary experiment using the master pattern having similar microchannel structures. In the preliminary experiment, the total shape change (shrinkage) of the AC mold compared to the master pattern was ~25%. To compensate for the shape change during AC mold fabrication, an enlarged (~133%) photoresist master with a positive microchannel structure (inverse shape of the final microchannel structure, positive channel structure) was fabricated. The dimensional targets for the master pattern were 200 μm for the width of the inlet and extraction channels, 600 μm for the outlet channel, and 70 μm for the height. About 5 g of negative photoresist (PR, SU-8 3050, Microchem Co. Ltd., Westborough, MA, USA) was poured onto a 4-inch Si wafer, and spin coated at a maximum rotation speed of 3000 rpm for 25 s to achieve a PR thickness of 70 μm. The coated photoresist was prebaked on a hot plate at 95 °C for 15 min and exposed to 20 mW/cm² UV light for 12 s using a mask aligner system (M-100, Prowin Co. Ltd., Daejeon, Korea). A post-exposure bake step was then conducted on a hot plate at 65 °C for 1 min and at 95 °C for 5 min. After the development process using the SU-8 developer for 8 min, the developed pattern was rinsed with isopropyl alcohol (IPA). Finally, a hard bake step was conducted on a hot plate at 100 °C for 1 h.

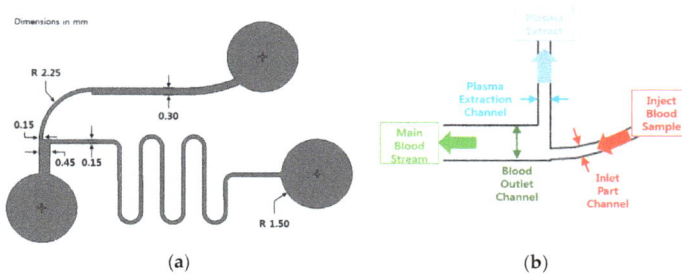

Figure 2. Schematics of (**a**) the designed bifurcation blood plasma separation chip and (**b**) the main separation region of the designed bifurcation blood plasma separation chip.

2.2. Fabrication of the AC Mold

An AC mold with microchannel cavities (positive channel structure) was fabricated by the carbonization of a replicated polymer precursor. A PDMS mold with the negative channel structure was replicated from the photoresist master pattern. An elastomer, Sylgard 184A (Dow Corning Co. Ltd., Auburn, MI, USA), was mixed with Sylgard 184B in a weight ratio of 10:1, poured on the photoresist master pattern, and cured on a hot plate at 150 °C for 1.5 h. Furan mixture—made of furan resin (Kangnam Chemical Co. Ltd., Seoul, Korea), *p*-toluenesulfonic acid (Kanto Chemical Co., Tokyo, Japan), and ethanol—was poured into the PDMS mold. To remove the air bubbles entrapped in the furan mixture during the mixing and pouring processes, a degassing process was conducted by placing the PDMS mold with the furan mixture into a vacuum chamber at room temperature for 3 h. After degassing, a two-step curing process was carried out to avoid the warpage of the replicated furan precursor. In the first curing phase, the furan mixture was cured under atmospheric conditions for five days, and in the second phase, it was cured in a convection oven for two days at ~100 °C. After curing, the furan precursor with positive microchannel structures was released from the PDMS mold and carbonized in a tube furnace at a maximum temperature of 1000 °C under nitrogen environment for five days. After carbonization, an AC mold with positive microchannel cavities was obtained [31]. Since the furan material is a carcinogen, the processes using furan was conducted in a fume hood.

2.3. Glass Molding of Microfluidic Channel Plate

A glass molding machine, comprising an insulated box furnace fitted with carbon heaters for heating up to 1000 °C at a heating rate of 5 °C/min, a hydraulic pressing unit for applying a compression pressure of ~30 ton during the glass molding, and a nitrogen purging unit for maintaining an inert gas environment during the glass molding, was designed and constructed as shown in Figure 3a. In the box furnace, the top and bottom graphite pressing jig structures (as shown in Figure 3b) were connected to the hydraulic unit and the load cell, respectively. A soda-lime glass plate (Low-Iron Glass, JMCGLASS Co. Ltd., Seoul, Korea) with a glass transition temperature (T_g) of 650 °C and a softening temperature (T_s) of 720 °C was used as the molding material. In the glass molding process, the fabricated AC mold was diced to a size of 20×15 mm^2 and placed on the bottom graphite jig. A glass plate with the same size as that of AC mold and a thickness of 3.2 mm was loaded onto the AC mold. To prevent the oxidization of the AC mold during the glass molding process, the box furnace was evacuated and purged with nitrogen gas. The glass molding process was divided into four stages: heating, holding, pressing, and cooling. Figure 3c shows the temperature and pressure history during the entire glass molding process. In the heating stage, the temperature of the furnace (ambient temperature of the AC mold and the glass plate) was increased up to the molding temperature of 700 °C. In the holding stage, the temperature was maintained for 15 min to achieve a stable temperature in the furnace (uniform temperature distribution between the AC mold and the glass plate). In the pressing stage, a compression pressure of 18 kPa was applied to the stack of the AC mold and glass plate for 15 min. In the cooling stage, the pressure was released and natural cooling was carried out. When the furnace temperature was reduced to the room temperature, the box furnace was opened and the replicated glass plate was detached from the AC mold. Figure 4 shows the SEM images of the fabricated (a) AC mold and (b) glass molded microchannel at blood separation region.

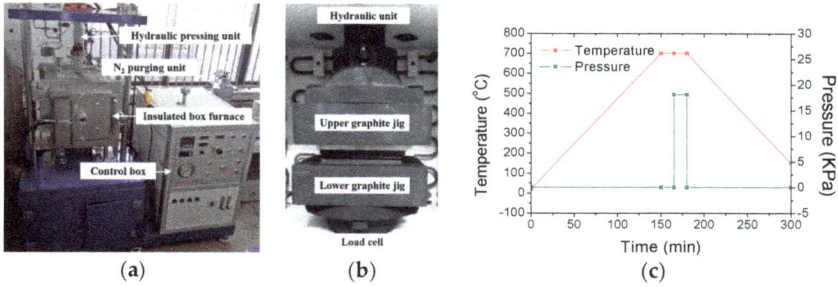

(a)   (b)   (c)

Figure 3. Photographs of the (**a**) constructed glass molding system; (**b**) graphite pressing jig in the box furnace; and (**c**) temperature and pressure history during the glass molding process.

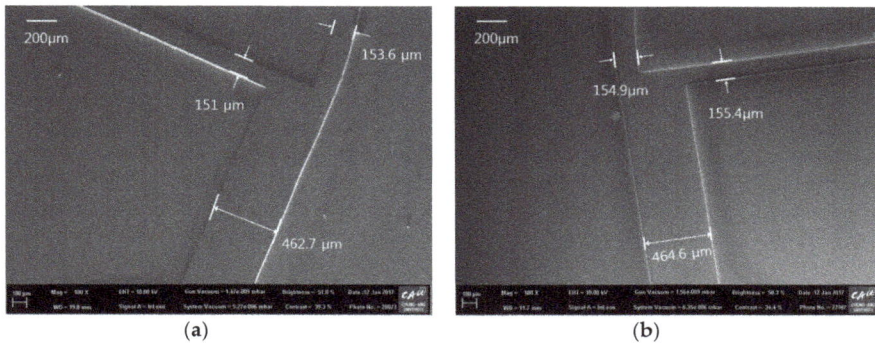

(a)   (b)

Figure 4. SEM images of fabricated (**a**) AC mold and (**b**) glass molded microchannel at blood separation region.

2.4. Preparation of the All-Glass Bifurcation Blood Plasma Separation Chip

To develop the blood plasma separation chip, three connection holes for blood in-flow, blood out-flow, and plasma out-flow were machined on the glass molded microchannel plate by the micro drilling process. A drill press (MM-180s, MANIX Co. Ltd., Pyeongtaek, Korea) with a carbide drill tool (ϕ1.5 mm, HAM precision Co. Ltd., Pewaukee, WI, USA) was used at a rotation speed of 2000 rpm. After drilling, the glass plate with microchannels was sufficiently cleaned by an ultrasonic cleaner using acetone and ethanol for 5 min each, because any particle or dust on the plate can cause failure of the subsequent glass sealing process. Finally, a planar glass plate was bonded onto the fabricated glass microchannel plate by a thermal fusion bonding process at a temperature of 630 °C and a pressure of 3 kPa for 3 h using the constructed glass molding system.

3. Analysis of the Geometrical Properties in Each Fabrication Step

To examine the geometrical properties of the fabricated samples, the surface profiles of each microfluidic channel structure were measured by a 3-dimensional (3D) confocal microscope (OLS-4100, Olympus Co. Ltd., Tokyo, Japan). Figure 5 shows the 3D surface profiles of the fabricated (a) silicon master; (b) PDMS mold; (c) furan precursor; (d) AC mold; and (e) molded glass microfluidic plate, obtained from the confocal microscopy results. It is clearly seen that the positive and negative microfluidic channel structures were uniformly transferred in each fabrication step. Figure 6 shows the comparisons of the cross-sectional surface profiles of the inlet microchannel (a) between the silicon master, PDMS mold (inverted), furan precursor, and AC mold, and (b) between the AC mold and the glass replica (inverted). Figure 6a shows that a small shape change (shrinkage) occurred during the PDMS and furan precursor replication processes due to polymerization, and a significant shape change took place during the AC mold fabrication because of material decomposition during carbonization. Although an inherent substantial shape change occurred during the AC mold fabrication process, the shape change during the glass molding process is negligible, as shown in Figure 6b.

For the quantitative analysis of the changes in the geometrical dimensions after each fabrication step, for a glass molded microfluidic chip with AC mold, the measured widths and heights of microchannels in each sample are summarized in Table 1. Although the target values of the silicon master pattern were 200, 600, 200, and 70 μm for the inlet width, outlet width, extraction width, and height, respectively, the fabricated photoresist master pattern had a dimensional error of 1.6% (average) due to the experimental errors in the spin coating and photolithography processes. The shape difference between the silicon master and the replicated furan precursor was 4.2% (dimensional change/dimension of master, average), and it might have been caused by the shrinkage during the polymerization of PDMS and furan materials. The shrinkage ratio due to the material decomposition in carbonization, which is the dimensional change ratio between the AC mold and the furan precursor (dimensional change/dimension of furan precursor), was ~22% (average). Therefore, the total shrinkage ratio between the AC mold and the silicon master (dimensional change/dimension of master) was 25.2% (average), which is almost the same as the values used for compensating the enlargement of the master pattern size. The standard deviation of the shrinkage rate was ~1.05% in the carbonization process and 1.42% in whole AC fabrication process. In this study, the designed values for the widths of the inlet, outlet, and extraction, and the height of the bifurcation fluidic chip were 150, 450, 150, and 55 μm, respectively. The average shape error between the design and the fabricated AC mold was 2.9%. Since the shape error in the conventional PDMS replication process, widely used in the microfluidic chip fabrication, was ~2%, the dimensional error of ~3% in the AC mold fabrication with our shrinkage compensation method is acceptable for the mold fabrication method, for microfluidic chip applications. To examine the repeatability of the proposed AC mold fabrication process, we fabricated five AC molds using the same master pattern. The average shrinkage ratio in whole AC mold fabrication process were 24.6%~25.7%. Although ~1% of shrinkage rate variation was existing in the repeated experiment, we believe that is acceptable in microfabrication process.

In the glass molding process, the widths and heights of microchannels in the glass molded plate were slightly greater than those of the AC mold (~1.5%). This difference might have been caused by the measurement errors of the confocal microscope while measuring the positive and negative shapes. Therefore, the shape change during the glass molding process can be neglected. Although the fabricated glass microfluidic channel has some dimensional errors compared to the designed values, the channel resistance ratio between the plasma extraction and the blood outlet channels calculated using the measured data was ~65:1, which is similar to the value obtained using the designed data.

We also measured the surface roughness of the fabricated samples at the bottom location of the microchannel. The measured root mean square (RMS) surface roughness values were 8, 34, 30, 12, and 8 nm on the silicon master, PDMS mold, furan precursor, AC mold, and molded glass plate samples, respectively. Although the surface roughness increased for the PDMS and the furan precursor due to the chain size of the polymer materials, it decreased during carbonization owing to the inherent shrinkage. In the glass molding process, the surface roughness also improved probably due to the surface energy of the softened glass material. It is clearly seen that the proposed AC mold fabrication process can provide a mold for glass molding with superior surface quality, which is difficult to obtain using the previous laser machined molds.

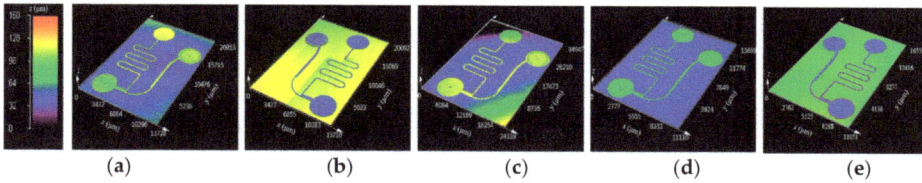

Figure 5. 3D microscope images of the (**a**) master pattern on Si wafer; (**b**) polydimethylsiloxane (PDMS) mold; (**c**) furan precursor; (**d**) AC mold; and (**e**) replicated glass microfluidic structure.

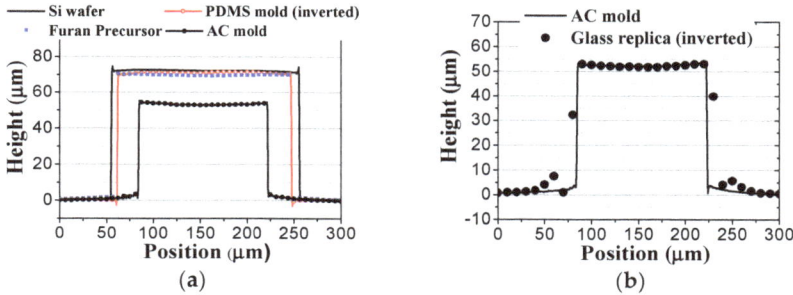

Figure 6. Comparison of cross-sectional surface profiles of the (**a**) silicon master, furan precursor, and AC mold; and (**b**) AC mold and replicated glass microfluidic structure.

Table 1. Summary of the measured widths and heights of the microchannels for each fabrication step.

Samples	Inlet Channel		Outlet Channel		Extraction Channel	
	Width (μm)	Height (μm)	Width (μm)	Height (μm)	Width (μm)	Height (μm)
Silicon master	204.5	71.4	600.4	71.9	200.3	71.6
polydimethylsiloxane (PDMS) mold	202.5	71.0	584.9	71.7	190.6	71.1
Furan precursor	196.4	67.8	585.9	68.2	190.8	68.9
amorphous carbon (AC) mold	153.6	51.6	462.7	53.7	151.0	53.4
Glass replica	154.9	52.9	464.6	54.4	155.4	53.9

4. Feasibility Analysis of the All-Glass Bifurcation Chip for Blood Plasma Separation

To examine the feasibility of the fabricated glass bifurcation microfluidic chip, we prepared a setup for blood plasma separation composed of an inverted microscope (CKX-41, Olympus Co., Tokyo, Japan) and a precision syringe pump (Legato 200, KD Scientific., Ringoes, NJ, USA). For the blood plasma separation experiment, diluted defibrinated sheep blood (MBcell, Los Angeles, CA, USA) in phosphate-buffered saline (PBS) with a hematocrit level of 25% [20] was infused into the glass bifurcation microfluidic chip using the precision syringe pump at flow rates of 0.1, 0.3, 0.5, and 0.7 mL/min. Figure 7a shows the experimental setup for imaging the blood plasma separation and Figure 7b shows the fabricated glass bifurcation microfluidic device with tubing during the separation process. Three Teflon tubes were connected to the micro-drilled inlet and outlet holes, and sealed using UV curable bonding material. To examine the effects of flow rates on blood plasma separation, the number of red blood cells in the inlet channel and the plasma extraction channels were measured using a hemacytometer, and the blood plasma separation efficiency (*E*) was calculated by Equation (2).

$$E = \frac{N_{inlet} - N_{plasma}}{N_{inlet}} \times 100 \ (\%) \tag{2}$$

where N_{inlet} and N_{plasma} were the measured number of red blood cells in the inlet channel (initial blood sample) and the plasma extraction channels (sample obtained from the tube attached to the plasma extraction reservoir), respectively. We fabricated five glass blood plasma separation chips and measured the blood plasma separation efficiencies twice at a flow rate of 0.1~0.7 mL/min for each chip. After the measurements with fixed flow rates, the chips were cleaned with a PBS solution for 3 min. Figure 8 shows the effects of the flow rate on the blood plasma separation efficiency. It was noted that the separation efficiency improved with increasing flow rate, and an efficiency of 74% was obtained at a flow rate of 0.7 mL/min. The relatively large variation in the measurement results might be mainly due to the large uncertainty in the blood cell measurement process using a hemacytometer [37].

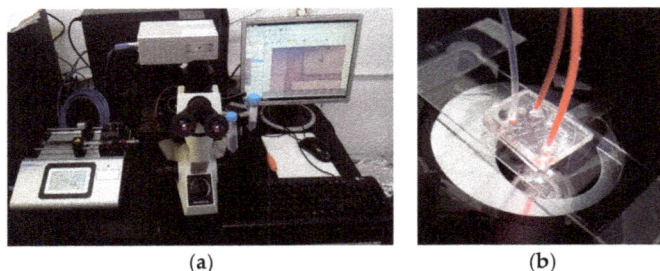

(a) (b)

Figure 7. Photographed images of the (**a**) experimental setup for blood plasma separation and (**b**) fabricated all-glass bifurcation microfluidic chip during blood plasma separation.

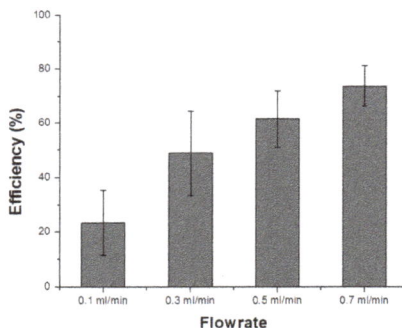

Figure 8. Blood plasma separation efficiency with various flow rates in the fabricated all-glass microfluidic chip with diluted and defibrinated sheep blood.

5. Conclusions

A durable all-glass bifurcation microfluidic chip was fabricated using a glass molded microchannel plate. An AC mold with microchannel cavities was fabricated by the carbonization of a replicated furan precursor, which is a low-cost and large-area patterning process for achieving high surface quality. During the AC mold fabrication process, a large and inevitable shrinkage (~22%) occurred in carbonization process due to the thermal decomposition process, and ~4% of shrinkage also occurred in PMDS and furan replication due to polymerization. To compensate the total shrinkage (~25%) during the fabrication of the AC mold, an enlarged photoresist master pattern was fabricated by photolithography on a silicon wafer. A PDMS mold was replicated from the silicon master and a furan precursor was replicated from the PDMS mold by the thermal curing process. The AC mold was obtained by the carbonization of the furan precursor. The shape difference between the AC mold and the silicon master was ~25.2%, which is similar to the value used for the compensation process. Although an inherent large shrinkage occurs in the AC mold fabrication process, the dimensional error between the designed value and the measured value from the fabricated AC mold was 2.9%, which is acceptable in a microfluidic device. Furthermore, almost half of the dimensional error (1.6%) occurred during the photolithography process. It clearly shows that the large amount of shrinkage occurring in the AC mold fabrication process is predictable and can be compensated. Finally, a glass microchannel plate with high fidelity was replicated using the glass molding system, and a flat glass plate was bonded to the replicated glass microchannel plate using the thermal fusion bonding method. The long cycle time issues of glass molding and thermal fusion bonding in this research are mainly due to our stand alone type glass molding system, and it can be solved by using a progressive type glass molding system [38]. Thus, we fabricated an all-glass bifurcation microfluidic chip using the glass molded microchannel plate and measured the blood plasma separation efficiency of 74% in the fabricated chip at a flow rate of 0.7 mL/min. Although the applied flow rate and the achieved plasma separation efficiency were not higher than those reported using the conventional PDMS chip because of the limitation of our injection system, it shows that the feasibility of the proposed all-glass chip fabrication process for microfluidics application. The application of the developed device to the whole blood plasma separation with high pressure injection system and real-time monitoring of the contents in the separated plasma using Raman analysis is the subject of our ongoing research.

Acknowledgments: This work was supported by the National Research Foundation of Korea (NRF) grant funded by the Korean Government (MSIP) (Nos. 2015R1A5A1037668, 216C000605), the Technology Innovation Program of the Korea Evaluation Institute of Industrial Technology (KEIT) granted financial resource from the Korean Government (MOTIE) (No. 10051636), the Korea Institute for Advancement of Technology (KIAT) grant funded by the Korean Government (MOTIE) (No. N0001075), and the Technological Innovation R&D Program funded by the SMBA, Korea (No. S2334634).

Author Contributions: Hyungjun Jang, Seok-min Kim, and Jiseok Lim wrote the main manuscript text. Hyungjun Jang, Muhammad Refatun Haq, Jonghyun Ju, and Youngkyu Kim conducted the experiments. Seok-min Kim and Jiseok Lim conceived the idea. All the authors reviewed the manuscript.

Conflicts of Interest: The authors declare no conflict of interest.

References

1. Chong, Z.Z.; Tor, S.B.; Loh, N.H.; Wong, T.N.; Gañán-Calvo, A.M.; Tan, S.H.; Nguyen, N.T. Acoustofluidic control of bubble size in microfluidic flow-focusing configuration. *Lab Chip* **2015**, *15*, 996–999. [CrossRef] [PubMed]

2. Song, C.; Nguyen, N.T.; Tan, S.H.; Asundi, A.K. A micro optofluidic lens with short focal length. *J. Micromech. Microeng.* **2009**, *19*, 085012. [CrossRef]

3. Song, C.; Nguyen, N.T.; Tan, S.H.; Asundi, A.K. A tuneable micro-optofluidic biconvex lens with mathematically predictable focal length. *Microfluid. Nanofluid.* **2010**, *9*, 889–896. [CrossRef]

4. Chong, Z.Z.; Tan, S.H.; Gañán-Calvo, A.M.; Tor, S.B.; Loh, N.H.; Nguyen, N.T. Active droplet generation in microfluidics. *Lab Chip* **2016**, *16*, 35–58. [CrossRef] [PubMed]

5. Kersaudy-Kerhoas, M.; Sollier, E. Micro-scale blood plasma separation: From acoustophoresis to egg-beaters. *Lab Chip* **2013**, *13*, 3323–3346. [CrossRef] [PubMed]

6. Choi, C.J.; Wu, H.-Y.; George, S.; Weyhenmeyer, J.; Cunningham, B.T. Biochemical sensor tubing for point-of-care monitoring of intravenous drugs and metabolites. *Lab Chip* **2012**, *12*, 574–581. [CrossRef] [PubMed]

7. Gorkin, R.; Park, J.; Siegrist, J.; Amasia, M.; Lee, B.S.; Park, J.; Kim, J.; Kim, H.; Madou, M.; Cho, Y. Centrifugal microfluidics for biomedical applications. *Lab Chip* **2010**, *10*, 1758–1773. [CrossRef] [PubMed]

8. Burger, R.; Reis, N.; Fonseca, J.G.D.; Ducree, J. Plasma extraction by centrifugo-pneumatically induced gating of flow. *J. Micromech. Microeng.* **2013**, *23*, 035035. [CrossRef]

9. Petersson, F.; Åberg, L.; Swärd-Nilsson, A.-M.; Laurell, T. Free Flow Acoustophoresis: Microfluidic-Based Mode of Particle and Cell Separation. *Anal. Chem.* **2007**, *79*, 5117–5123. [CrossRef] [PubMed]

10. Nakashima, Y.; Hata, S.; Yasuda, T. Blood plasma separation and extraction from a minute amount of blood using dielectrophoretic and capillary forces. *Sens. Actuators B* **2010**, *145*, 561–569. [CrossRef]

11. Jiang, H.; Weng, X.; Chon, C.H.; Wu, X.; Li, D. A microfluidic chip for blood plasma separation using electro-osmotic flow control. *J. Micromech. Microeng.* **2011**, *21*, 085019. [CrossRef]
12. Furlani, E. Magnetophoretic separation of blood cells at the microscale. *J. Phys. D Appl. Phys.* **2007**, *40*, 1313. [CrossRef]
13. Xiang, N.; Ni, Z. High-throughput blood cell focusing and plasma isolation using spiral inertial microfluidic devices. *Biomed. Microdevices* **2015**, *17*, 110. [CrossRef] [PubMed]
14. Sollier, E.; Rostaing, H.; Pouteau, P.; Fouillet, Y.; Achard, J.-L. Passive microfluidic devices for plasma extraction from whole human blood. *Sens. Actuators B* **2009**, *141*, 617–624. [CrossRef]
15. Tripathi, S.; Kumar, Y.B.V.; Prabhakar, A.; Joshi, S.S.; Agrawal, A. Performance study of microfluidic devices for blood plasma separation—A designer's perspective. *J. Micromech. Microeng.* **2015**, *25*, 084004. [CrossRef]
16. Maria, M.S.; Kumar, B.; Chandra, T.; Sen, A. Development of a microfluidic device for cell concentration and blood cell-plasma separation. *Biomed. Microdevices* **2015**, *17*, 1–19. [CrossRef] [PubMed]
17. Li, C.; Liu, C.; Xu, Z.; Li, J. Extraction of plasma from whole blood using a deposited microbead plug (DMBP) in a capillary-driven microfluidic device. *Biomed. Microdevices* **2012**, *14*, 565–572. [CrossRef] [PubMed]
18. Chen, X.; Liu, C.C.; Li, H. Microfluidic chip for blood cell separation and collection based on crossflow filtration. *Sens. Actuators B* **2008**, *130*, 216–221. [CrossRef]
19. Kim, Y.C.; Kim, S.-H.; Kim, D.; Park, S.-J.; Park, J.-K. Plasma extraction in a capillary-driven microfluidic device using surfactant-added poly(dimethylsiloxane). *Sens. Actuators B* **2010**, *145*, 861–868. [CrossRef]
20. Yang, S.; Undar, A.; Zahn, J.D. A microfluidic device for continuous, real time blood plasma separation. *Lab Chip* **2006**, *6*, 871–880. [CrossRef] [PubMed]
21. Prabhakar, A.; Kumar, Y.V.B.V.; Tripathi, S.; Agrawal, A. A novel, compact and efficient microchannel arrangement with multiple hydrodynamic effects for blood plasma separation. *Microfluid. Nanofluid.* **2015**, *18*, 995–1006. [CrossRef]
22. Rodriguez-Villarreal, A.I.; Arundell, M.; Carmona, M.; Samitier, J. High flow rate microfluidic device for blood plasma separation using a range of temperatures. *Lab Chip* **2010**, *10*, 211–219. [CrossRef] [PubMed]
23. Feketea, Z.; Nagya, P.; Huszkac, B.; Tolnerb, F.; Pongrácza, A.; Fürjesa, P. Performance characterization of micromachined particle separation system based on Zweifach–Fung effect. *Sens. Actuators B* **2012**, *162*, 89–94. [CrossRef]
24. Blattert, C.; Jurischka, R.; Tahhan, I.; Schoth, A.; Kerth, P.; Menz, W. Microfluidic blood/plasma separation unit based on microchannel bend structures. In Proceedings of the 3rd IEEE/EMBS Special Topic Conference on Microtechnology in Medicine and Biology, Oahu, HI, USA, 12–15 May 2005; IEEE: New York, NY, USA, 2005; pp. 38–41.
25. Lopes, R.; Rodrigues, R.O.; Pinho, D.; Garcia, V.; Schütte, H.; Lima, R.; Gassmann, S. Low cost microfluidic device for partial cell separation: Micromilling approach, Industrial Technology (ICIT). In Proceedings of the 2015 IEEE International Conference on Intelligent Transportation Systems, Canary Islands, Spain, 15–18 September 2015; IEEE: New York, NY, USA, 2015; pp. 3347–3350.
26. Nieto, D.; Delgado, T.; Flores-Arias, M.T. Fabrication of microchannels on soda-lime glass substrates with a Nd:YVO$_4$ laser. *Opt. Lasers Eng.* **2014**, *63*, 11–18. [CrossRef]
27. Bahadorimehr, A.; Majlis, B.Y. Fabrication of Glass-based Microfluidic Devices with Photoresist as Mask. *Elektron. Elektrotech.* **2011**, *116*, 45–48. [CrossRef]
28. Chen, Q.; Chen, Q.; Maccioni, G. Fabrication of microfluidics structures on different glasses by simplified imprinting technique. *Curr. Appl. Phys.* **2013**, *13*, 256–261. [CrossRef]
29. Huang, C.-Y.; Kuo, C.-H.; Hsiao, W.-T.; Huang, K.-C.; Tseng, S.-F.; Chou, C.-P. Glass biochip fabrication by laser micromachining and glass-molding process. *J. Mater. Process. Technol.* **2012**, *212*, 633–639. [CrossRef]
30. Tseng, S.-F.; Chen, M.-F.; Hsiao, W.-T.; Huang, C.-Y.; Yang, C.-H.; Chen, Y.-S. Laser micromilling of convex microfluidic channels onto glassy carbon for glass molding dies. *Opt. Lasers Eng.* **2014**, *57*, 58–63. [CrossRef]
31. Ju, J.; Lim, S.; Seok, J.; Kim, S.M. A method to fabricate low-cost and large area vitreous carbon mold for glass molded microstructures. *Int. J. Precis. Eng. Manuf.* **2015**, *16*, 287–291. [CrossRef]
32. Faustino, V.; Catarino, S.O.; Lima, R.; Minas, G. Biomedical microfluidic devices by using low-cost fabrication techniques: A review. *J. Biomech.* **2016**, *49*, 2280–2292. [CrossRef] [PubMed]
33. Pinto, V.C.; Sousa, P.J.; Cardoso, V.F.; Minas, G. Optimized SU-8 processing for low-cost microstructures fabrication without cleanroom facilities. *Micromachines* **2014**, *5*, 738–755. [CrossRef]

34. Pinto, E.; Faustino, V.; Rodrigues, R.O.; Pinho, D.; Garcia, V.; Miranda, J.M.; Lima, R. A rapid and low-cost nonlithographic method to fabricate biomedical microdevices for blood flow analysis. *Micromachines* **2014**, *6*, 121–135. [CrossRef]

35. Liu, Z.; Xu, W.; Hou, Z.; Wu, Z. A rapid prototyping technique for microfluidics with high robustness and flexibility. *Micromachines* **2016**, *7*. [CrossRef]

36. Tripathi, S.; Prabhakar, A.; Kumar, N.; Singh, S.G.; Agrawal, A. Blood plasma separation in elevated dimension T-shaped microchannel. *Biomed. Microdevices* **2013**, *15*, 415–425. [CrossRef] [PubMed]

37. Biggs, R.; Macmillan, R.L. The Error of the Red Cell Count. *J. Clin. Pathol.* **1948**, *1*, 288–291. [CrossRef] [PubMed]

38. Chang, S.-H.; Lee, Y.-M.; Shin, K.-H.; Heo, Y.-M. A study on the aspheric glass tens forming analysis in the progressive GMP process. *J. Opt. Soc. Korea* **2007**, *11*, 85–92. [CrossRef]

micromachines

MDPI

Article

Analysis of Liquid–Liquid Droplets Fission and Encapsulation in Single/Two Layer Microfluidic Devices Fabricated by Xurographic Method

Chang Nong Lim [1], Kai Seng Koh [2,*], Yong Ren [3,4], Jit Kai Chin [1], Yong Shi [3,4] and Yuying Yan [4,5]

[1] Department of Chemical and Environmental Engineering, University of Nottingham Malaysia Campus, Jalan Broga, 43500 Semenyih, Selangor, Malaysia; kebx4lcn@nottingham.edu.my (C.N.L.); Jit-Kai.Chin@nottingham.edu.my (J.K.C.)

[2] School of Engineering and Physical Sciences, Heriot-Watt University Malaysia, 62200 Putrajaya, Wilayah Persekutuan Putrajaya, Malaysia

[3] Department of Mechanical, Materials and Manufacturing Engineering, University of Nottingham Ningbo China, Ningbo 315100, China; yong.ren@nottingham.edu.cn (Y.R.); yong.shi@nottingham.edu.cn (Y.S.)

[4] Research Centre for Fluids and Thermal Engineering, University of Nottingham Ningbo China, Ningbo 315100, China; yuying.yan@nottingham.ac.uk

[5] Research Group of Fluids and Thermal Engineering, Faculty of Engineering, University of Nottingham, Nottingham NG7 2RD, UK

* Correspondence: k.koh@hw.ac.uk; Tel.: +60-3-8894-3999

Academic Editors: Weihua Li, Hengdong Xi and Say Hwa Tan
Received: 16 December 2016; Accepted: 4 February 2017; Published: 10 February 2017

Abstract: This paper demonstrates a low cost fabrication approach for microscale droplet fission and encapsulation. Using a modified xurography method, rapid yet reliable microfluidic devices with flexible designs (single layer and double layer) are developed to enable spatial control of droplet manipulation. In this paper, two different designs are demonstrated, i.e., droplet fission (single layer) and droplet encapsulation (double layer). In addition, the current fabrication approach reduces the overall production interval with the introduction of a custom-made polydimethylsiloxane (PDMS) aligner. Apart from that, the fabricated device is able to generate daughter droplets with the coefficient of variance (CV) below 5% and double emulsions with CV maintained within 10% without involvement of complex surface wettability modification.

Keywords: droplet manipulation; microfabrication; encapsulation; fission

1. Introduction

Droplet microfluidics, one subcategory under microfluidics, enable the manipulation of fluid in the discrete form such as micro-droplets that offers greater benefits than the conventional flow systems in terms of high throughput and scalability. The generated droplets can be manipulated according to the desired droplet operations such as droplet fusion, fission, mixing and sorting [1–4]. On the other hand, compartmentalization of functionalized fluid within a droplet is useful for formation of double emulsions, microcapsules and microbubbles [5–7]. Droplet fission is known as a simpler way to introduce multiple sample arrays for rapid mixing or chemical reaction while increasing the throughput for droplet production and digitizing the biological assays at the same time [8], whereas controlled micro-encapsulation or better known as double emulsions provides a high entrapment efficiency and desired particle size [9]. On top of that, the core–shell structure enables the encapsulation and protection of a specific small volume captive ingredients for controlled released applications [10].

Over the past decade, a range of materials [11–14] has been reported to successfully fabricate the microfluidic device with different fabrication techniques developed. Such examples include soft-lithography [15], deep reactive-ion etching (DRIE) [16], micromachining [17] and rapid prototyping [18]. In addition, crucial factors in fabrication, such as process parameters and material bonding were also developed and optimized [19]. As such, xurography [20] technique was introduced as a new rapid prototyping method capable to reproduce soft microfluidic devices and glass microchannel [21] with higher reproducibility, easier installation and comparable lower overall fabrication cost than other techniques [22].

While polydimethylsiloxane (PDMS) is versatile, comparable low cost fabrication and allows easy replication with the presence of a master mold [23], the spatial wettability modification required is relatively difficult to achieve especially when it involves droplet encapsulation in a single piece device [24]. Microfluidic device made of glass capillary tube [14] has recently aroused as an established device for droplet encapsulation due to its advantages of conveniently functioned spatially and the precision in controlling the number and structure of encapsulation. However, the current fabrication procedure involves intricate operations, expensive equipment and mostly requires clean room facilities, thus inappropriate for mass production and commercialization [25]. Such a handicap has hindered the further stretch of droplet microfluidic development, hence motivates the aim of this paper which is to provide an alternative low cost fabrication method, xurography technique for droplet manipulation. In particular, this paper proposes a modified xerographic method by which a PDMS-Glass hybrid microfluidic device is fabricated enabling the process of droplet encapsulation to be carried out under consistent wall surface properties. This fabrication technique circumvents use of the complex surface wettability modification method, i.e., sol-gel coating method [26] and oxygen plasma [27], hence achieves environmental friendliness through the reduction of chemical usage, equipment and energy usage.

Specifically, we demonstrate in this paper the capability of xurography technique in fabricating single layer and double layer microfluidic devices with flexibility in channel dimension for different purposes of droplet manipulation, i.e., droplet fission (single layer) and droplet encapsulation (double layer). While the usual photolithography technique tends to produce a master mold with uneven channel heights due to the ultra-violet (UV) light curable duration, xurography technique discussed in this paper provides a constant channel height which is fixed by the thickness of the adhesive vinyl film used to replace silicon wafer in the photolithography technique. This provides distinguished advantages for hybrid devices such that a connection between different materials with fixed dimension is required. On top of that, the disposable adhesive vinyl film is cheaper than the single usage of silicon wafer, thus imposing higher economic efficiency in our xurography-based microfabrication portfolio.

The optimization for the fabrication of the single layer microfluidic device for droplet fission mainly involves the cutting plotter settings such as cutting force and offset. In contrast, the optimization for the sealing of the double layer droplet encapsulation device introduces the use of a custom-made PDMS aligner. The different settings of the cutting plotter and the detailed operation of the custom-made PDMS aligner are discussed to enable sharp edge corner cutting while ensuring the consistency of the alignment and sealing of the microfluidic device. The experimental results show different flowrate ratios to attest the functionality of the microfluidic device in which the droplets generated are all well maintained within 10% of the coefficient of variance (CV) value.

2. Materials and Methods

The modified xurography process started with a drawing of the microchannel design features (Figure 1a–c) using conventional drawing software. The design was then sent to the cutting plotter (CE6000-60, Graphtec, Yokohama, Japan) to be plotted onto 100 μm thickness adhesive vinyl film (Oracal Intermediate Cal 651, Orafol, Oranienburg, Germany). The vinyl film was then transferred to a blank PDMS slab which was later adhered onto a paper mold (Figure 1d). The epoxy resin and hardener (CP362 A/B, Oriental Option Sdn Bhd, Penang, Malaysia) were prepared at a ratio of 2:1 (w/w) and poured into the paper mold to be cured overnight at the room temperature (25 °C). PDMS pre-polymer (Sylgard 184, Dow Corning, Midland, MI, USA) prepared at a ratio of 10:1 (w/w)

were poured into the master mold and left partial curing. The partially cured PDMS slabs were peeled from the master, aligned and sealed via a custom-made PDMS aligner. The double layer aligned PDMS layers was then heated for 2 h at 80 °C to complete.

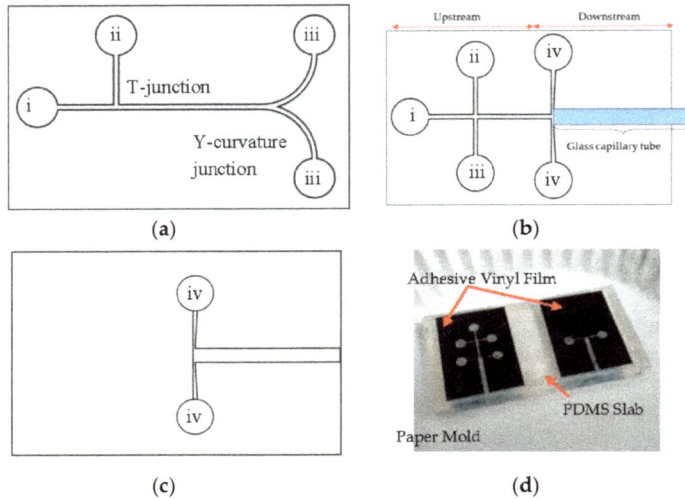

Figure 1. (**a**) Single layer microchannel design for droplet fission. The total length of the main channel is 28.10 mm with a 0.5 cm radii curve channel. (Label: i. cooking oil inlet; ii. DI water inlet; iii. Outlet). (**b**) Top layer design for droplet encapsulation. The channel height in the upstream is 0.2 mm. (Label: i. cooking oil inlet; ii. red dyed water inlet; iii. blue dyed water inlet; iv. yellow dyed water inlet). (**c**) Bottom layer design for droplet encapsulation to increase the downstream channel height to 1 mm for glass capillary tube fitting. (Label: iv. yellow dyed water inlet). (**d**) Layering for the fabrication of the epoxy master mold.

For liquid–liquid droplet fission experiments, a T-junction geometry was utilized for water-in-oil (W/O) droplet generation. The W/O droplets were then directed towards the Y-curvature junction via the main channel to split into two daughter droplets (Figure 1a). The channel height was maintained at 0.1 mm throughout the whole device. On the contrary, droplet encapsulation device consists of two channel heights one is 0.2 mm at the upstream and the other is 1 mm at the downstream to enhance the encapsulation process. The dual T-junction were used at the upstream of the device to produce two distinct W/O inner droplets differentiated by color, i.e., red and blue which was then being encapsulated at the downstream forming a water-in-oil-in-water (W/O/W) double emulsions (Figure 1b,c). Glass capillary tube with an inner diameter of 0.58 mm (World Precision Instrument, Inc., Sarasota, FL, USA) was inserted at the downstream of the encapsulation device with the aid of the tweezer to create a hydrophilic environment for W/O/W double emulsions formation.

Droplet fission experiments were carried out with blended cooking oil as the continuous phase and pure deionized (DI) water (18.2 MΩ-cm, Milli-Q, Millipore, Molsheim, France) as the dispersed phase to generate W/O droplets. Droplet encapsulation experiments on the other hand, utilized cooking oil as the middle continuous phase while the two inner aqueous phase and outer aqueous phase were composed of a mixture of DI water with 16.6% (v/v), 9.0% (v/v) and 11.8% (v/v) of Fortune Red, True Blue and Egg Yellow food coloring, respectively, for W/O/W double emulsions. The materials used for the experimental work were pure cooking oil and DI water with food grade dye without addition of any other chemical solvent including surfactants. The materials were all used as purchased.

All the aforementioned liquids were loaded individually into their respective inlets through polytetrafluoroethylene (PTFE) tubing with an outer diameter of 1/16″ (Omnifit® Labware, Diba Industries Ltd., Cambridge, UK), delivered by syringe pumps (Model KDS-200, KD Scientific Inc., Holliston, MA, USA and NE-4000, New Era Pump Systems Inc., Farmingdale, NY, USA). The process of the droplet manipulation was recorded by a high-speed camera (Phantom Miro M110, Vision Research, Wayne, NJ, USA) mounted onto a light inverted microscope (Olympus IX51, Olympus Corporation, Tokyo, Japan) which was connected to a desktop (Figure 2). The droplet images were then extracted from the videos for a size distribution analysis via an image processing software—ImageJ (1.50i, National Institutes of Health, Bethesda, MD, USA). Experiments were repeated at least 3 times for the result accuracy purpose.

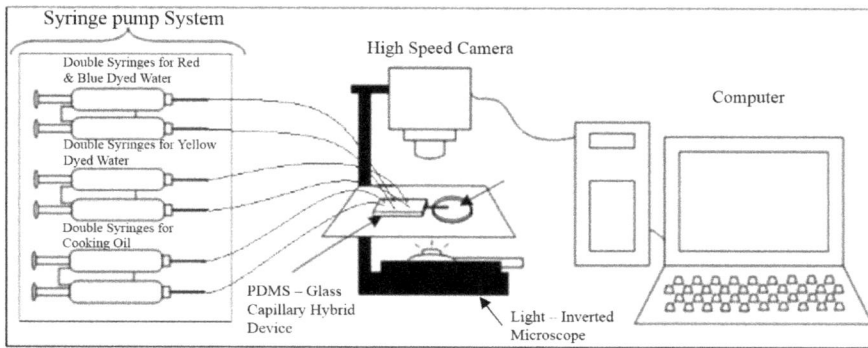

Figure 2. Schematic diagram of the experimental setup for droplet encapsulation and fission.

3. Results and Discussion

3.1. Fabrication Analysis

Previously, Pinto et al. [28] applied xurography to produce a biomedical microdevice. Since the xurography technique limits the minimum dimension of the microchannel width as 150 μm, they have compared the geometrical quality of the master mold with the corresponding microchannels using three sets of channel widths, i.e., 200, 300, and 500 μm. Their findings observe the largest inconsistence in channel width occurs at the smallest set of microchannels i.e., 200 μm with approximately of 50 μm difference. Therefore, in order to test on the increased accuracy and efficiency of the cutter plotter after optimization for xurography technique, the microchannel widths used in this study were mainly drawn in 200 μm. An orifice with a 150 μm channel width was also included for the optimization.

3.1.1. Cutter Plotter Optimization

Plotting quality can be directly affected by four main plotter settings: cutting force, offset value, offset angle and cutting mode. In this paper, cutter blade (CB09UB, Graphtec, Tokyo, Japan) was used to illustrate optimization of cutter setting on a 100 μm thickness vinyl film. New blade was used to carry out the optimization to prevent inconsistent results caused by blunted blade. The blade length was also adjusted and fixed to the optimum length using the built-in function. To tackle the problem systematically, the cutting force was first adjusted while keeping the offset and offset angle at default values.

The value of cutting force represents the pressure applied on the cutter blade to cut the design. Smooth cutting line can be obtained when the optimum cutting force is used. Starting with value 5, the setting was then gradually increased until traces of the blade were observed on the backing sheet.

Figure 3 shows the appearance of the film at undercut, optimal and overcut condition, respectively. The width of the channel was measured with Image J and summarized in Table 1.

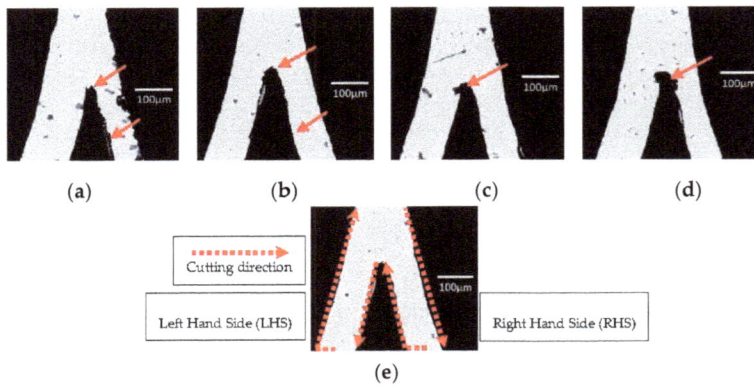

(a)　　　　　　　(b)　　　　　　　(c)　　　　　　　(d)

(e)

Figure 3. (**a**) Undercut vinyl film at the cutting force of 5; (**b**) vinyl film with the optimal cutting force of 12; (**c**) overcut vinyl film at the cutting force of 13; and (**d**) overcut vinyl film at the cutting force of 20, where the red arrows indicate the locations to be noticed on the sharpness of the cutting edge and the microchannel smoothness; and (**e**) the moving direction of the cutter blade when producing a microchannel.

Table 1. Relationship between the cutting force and the Y-curvature junction channel width.

Cutting Force	LHS Channel Width (µm) [1]	Error (%)	RHS Channel Width (µm) [1]	Error (%)
20	265.01	6.0	177.79	28.9
15	240.97	3.6	201.16	19.5
14	231.10	7.6	184.69	26.1
13	250.60	0.2	202.07	19.1
12	240.52	3.8	218.70	12.5
11	229.59	8.2	198.30	20.7
10	240.15	4.0	200.71	19.7
5	288.45	15.4	154.01	38.4

[1] Designed channel width for both the right hand side (RHS) and left hand side (LHS) are 250 µm respectively.

From visual inspection, the undercutting of the vinyl film was indicated by a rough channel surface with visible residue (Figure 3a). Since the pressure applied on the cutting blade was insufficient, the surface of the vinyl film was mostly dragged off by the blade resulting in a rough and torn pattern. On the contrary, excessive cutting force exerted onto the vinyl film affects the cutting area by indirectly ripping a larger shredded surface during the cutting process. At the optimal cutting force (Figure 3b), the channel surface presented is smooth and clean without visible residue as compared to the undercut and overcut vinyl films. The excessive cutting force applied has enlarged the inaccuracy caused by the offset and cutting mode, resulting in a blunt junction as indicated in Figure 3c,d. The moving direction of the blade when cutting this Y-curvature junction design further justifies the distortion towards the left (see Figure 3e). The measurement of the channel dimension is divided into two parts i.e., the left hand side (LHS) and right hand side (RHS) since a Y-curvature junction with 1:1 ratio was designed for this investigation purpose. The percentage error is calculated by |actual width − desired width|/(desired width) × 100%. We note that although the lowest percentage error is 0.2% at the value of 13 for the LHS, the cutting line is not as smooth (Figure 3c) as compared to the film cut at the value of 12 (Figure 3b). Hence, we conclude that the optimal cutting force is at the value of 12 with an error of 3.8% and 12.5% for LHS and RHS, respectively. Percentage error for the LHS falls under the range of 12%–38% due to the unequal width between both junctions. This can be resolved by further optimizing the offset or offset angle.

The offset of a cutting plotter is defined as a distance difference between the center of the plunger and the tip of the blade in the plunger. The tip of the cutting blade is usually located after the center point due to the design of the blade (Figure 4a). Hence, fine-tune of the offset value allows the plotter to adjust the center point to an optimized coordinate for sharp cutting, especially at the bending of channel dimension. For better illustration, the Y-curvature junction was not used for offset adjustment. Instead, a cross-junction geometry of 200 μm width, with an orifice of 150 μm width, was used (Figure 4b). Since the cutting force had been optimized previously, the cutting force was kept constant at value 12 while the offset values in the range of −2 to +3 were tested with the offset angle and cutting mode set at default values. The channel width for each offset value tested was then measured and compared with its respective reference line (Figure 4c).

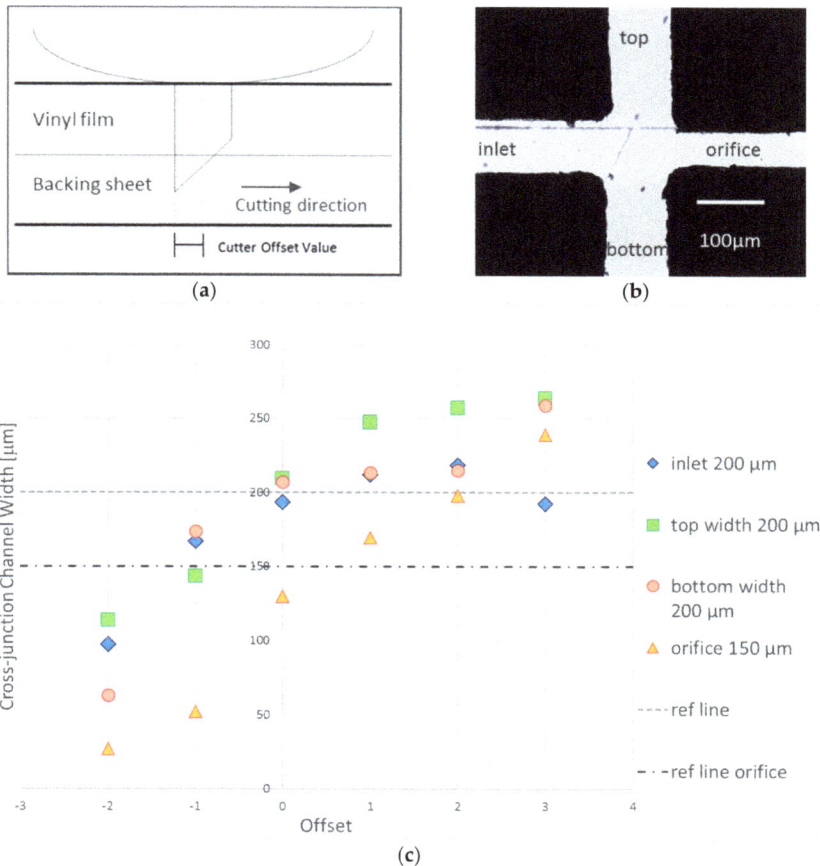

Figure 4. (**a**) Schematic illustration of the cutting mechanism of the cutter plotter; (**b**) vinyl film cutting at the optimal offset value 0; and (**c**) graph of cutting plotter offset versus cross-junction channel width. The reference (ref.) line refers to the desired width for the top, bottom and inlet channel i.e., 200 μm while the orifice width is at 150 μm. At the offset of 0 value, the errors of the channel widths produced towards the desired were below the 5% acceptable range.

As the offset values were set towards the negative value, the centerline moved backwards to the tip of the blade. As a result, the blade turned after the desired cutting path causing a narrower channel with a closed orifice. On the other hand, while the offset value increased towards the positive value, the centerline moved further from the tip of the blade to its front resulting in the blade to turn before reaching the desired cutting path. This caused a wider channel and rounded corner. At the optimal offset value 0, the channel widths produced were close to the desired cutting dimension with acceptable percentage errors below 5%, i.e., 3.25%, 4.75% and 3.5% for the inlet, top and bottom channel respectively. These results show better consistency as compared to the experiments at the channel width of 200 μm reported in [28]. Although we notice an exception at the orifice in which the percentage error goes up to 13.7%, it is still the lowest as compared with the other offset values where the range of errors falls within 13%–65%. It is worth mentioning that sharp edged cutting was also well observed at the optimal offset value (Figure 4b).

The third factor we investigated is the offset angle, which is applied when there is a larger angle change than the specified reference angle. A larger value of the reference angle can reduce the plotting time by reducing the blade control time, however, insufficient angle control on the blade can occur if the value is higher than the optimum. The offset angles in our experiments were manipulated within a range of 10°–60° (Figure 5). When the offset angle increases, we observe that the width deviation between the two channels reduces until it reaches an equal channel width at both sides at the offset angle of 60°.

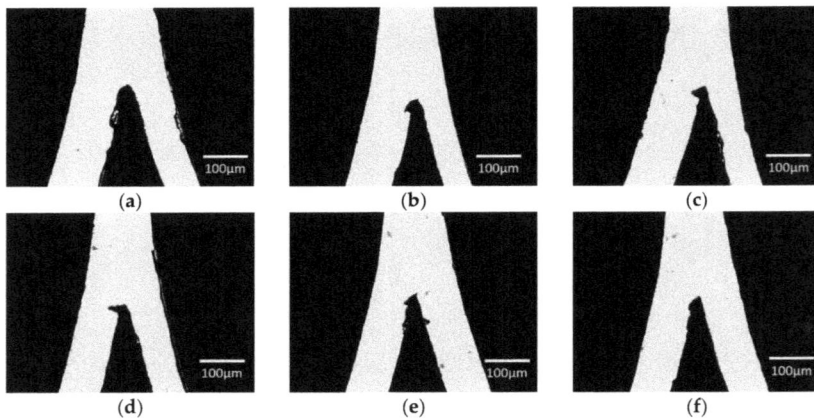

Figure 5. (**a**) Vinyl film cutting at the offset angle of 10°; (**b**) vinyl film cutting at the offset angle of 20°; (**c**) vinyl film cutting at the offset angle of 30°; (**d**) vinyl film cutting at the offset angle of 40°; (**e**) vinyl film cutting at the offset angle of 50°; and (**f**) vinyl film cutting at the offset angle of 60°.

Two cutting modes are available in the Graphtec cutting plotter CE6000-60, i.e., drag-knife and tangential mode. In the drag-knife mode, the blade cuts the vinyl film according to the path line, whereas, in the tangential mode, the blade is completely lifted from the film and rotates to a new position whenever the plotting involves a round curve and a sharp corner. The drag-knife mode (Figure 6a) is more appropriate when cutting a Y-curvature junction for a droplet fission device since the blade follows the cutting curve smoothly. This is different from the tangential mode where the blade lifts up whenever there is a bent. Therefore, the tangential mode can produce curves made up of lines with different angles, and provide a more precise and sharper edge cutting when straight channels are involved (Figure 6d). The drag-knife mode, on the other hand, gives a poor presentation when edge cutting is required (Figure 6c).

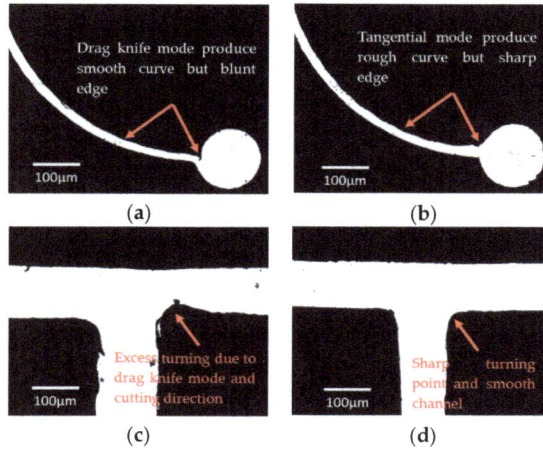

Figure 6. (**a**) Smooth channel curvature cut by the drag-knife mode. A clean cut near reservoir is harder to achieve in comparison to the tangential mode. (**b**) Rough channel curvature cut by the tangential mode with a cutting profile made of several straight lines with different angles. (**c**) Vinyl film with curvature at the corner by the drag-knife cutting even at the optimal offset value. (**d**) Clear edge cutting of the T-junction geometry by the tangential cutting.

In short, the cutting mode depends on the shape of the channel design. The drag-knife mode is more appropriate when the design involves curves and arcs whereas the tangential mode is suitable when the design consists mostly of straight channels with sharp edge turning such as cross-junction and T-junction geometry. For a 100 µm thickness vinyl film, the optimal cutting force is at the value of 12 while the optimal offset value and offset angle are at the value of 0° and 60°, respectively. The optimization for both the cutting force and offset value was then extended to a double layer vinyl film where the optimal cutting force for the 200 µm thickness vinyl film was set equal to 16 while the optimal offset value was −2. However, a quantitative analysis is unable to be obtained in the current experiments as: (1) the cutting range of the cutter blade, CB09UB, is limited to a maximum of 250 µm; and (2) the thickness of available adhesive vinyl films is fixed at 100 µm.

Another technical limitation of the cutting plotter that affects the cutting quality is the cutting direction of the blade. Although the optimal conditions were achieved, small defects such as a slightly narrower channel width or a small area of excessive film at the turning edge, were still observed in our experiments. Since the cutting direction is fixed by the plotter, the only way to alter the direction is to rotate the orientation of the design before sending to the plotter. Consequently, the defects are hard to be eliminated thoroughly. However, shifting the design to another position can give the minimal impacts on the channel geometry.

3.1.2. PDMS Aligner and Delta Analysis Results

The PDMS aligner (Figure 7) is made of stainless steel with 4 functional knots to adjust the PDMS layer in the x, y, z directions and allows rotation on the x-z plane by adjusting its skewness. In the alignment, the PDMS layer with a complex micro-channel design was placed onto the microscope stage as a base layer while the other PDMS layer was adjusted to the position by tuning the x-z knot until the edges of both layers were aligned accurately (Figure 8b–e). The entire process was completed under the light inverted microscope.

Figure 7. Image of the portable polydimethylsiloxane (PDMS) aligner with its functional knots that allow the PDMS layer to be moved in micro-meter (µm) intervals along the *x*, *y*, and *z* directions. (Labels: a. Left-right knot for PDMS slab's left-right adjustment along the *x*-coordinates; b. Up-down knot to move PDMS slab's up and down with micrometer intervals along the *y*-coordinates; c. Top-bottom knot to lower down the top layer of PDMS slab towards the bottom layer of PDMS slab along the *z*-coordinates and finally seal them together; d. PDMS-clip tightened knot for PDMS clip's position adjustment; e. Skew adjustment knot to rotate PDMS slab in the *x*-*z* plane; f. PDMS clip to secure the position of the PDMS slab; and g. Screw hole to temporarily secure the device on the microscopic stage).

The precision of the PDMS aligner was compared by measuring the channel width before and after the alignment process. The error and standard deviation were compared with bare-handed/manual alignment (Table 2). The channel widths of three locations (positions A, B and C in Figure 8a) were measured at the downstream of the encapsulation device, i.e., both the outer aqueous phase inlets and the distance between the glass capillary tube and the dispersion channel. To ensure repeatability and precision, aligner and manual alignment experiments were repeated by 10 times each and the average was calculated.

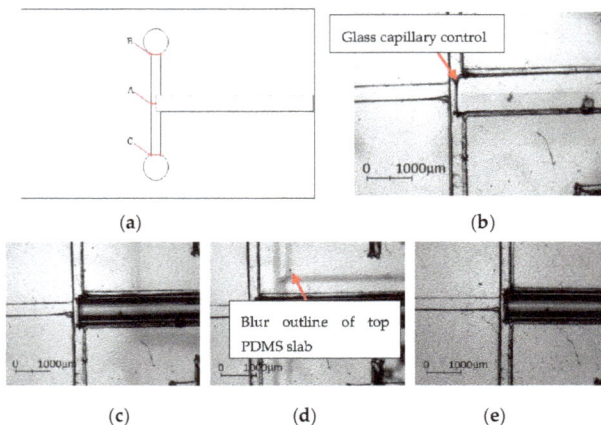

Figure 8. (a) Schematic diagram of the locations measured for the aligner precision analysis; (b) microscopic image of the bottom PDMS layer; (c) microscopic image of the bottom PDMS layer with glass capillary aligned on the channel; (d) microscopic image of aligning another layer of PDMS on the top; and (e) microscopic image of an aligned device.

Table 2. Comparisons of the channel width at each location before and after aligned using aligner and manual alignment.

Position	Default Channel Width (μm)	Channel Width ± Precision (μm, Aligner)	Error Percentage (%, Aligner)	Channel Width (μm, Manual)	Error Percentage (%, Manual)
A	151	151.13 ± 1.13	0.75	151.88 ± 1.13	0.72
B	310	309.60 ± 3.37	1.09	302.12 ± 1.63	2.34
C	310	310.28 ± 3.00	0.97	302.68 ± 1.39	2.83

In general, the error percentage of the alignment using the PDMS aligner is lower as compared to manual alignment except at Position A. However, this can be easily resolved by simply lifting the top PDMS piece without affecting the aligned coordinates to adjust the glass capillary tube back to its desired position with the aid of the tweezer. The position of the glass capillary is essential to ensure the success of the droplet encapsulation (Figure 9). For better demonstration, we use δ to indicate the channel width of the outer aqueous phase, i.e., the distance between the collection channel and the upstream dispersion channel. When the glass capillary tube was positioned at δ and 1.5δ, encapsulation was inconsistent as the pressure from the outer aqueous phase was insufficient to shear the incoming fluid. As the glass capillary tube moved towards the dispersion channel and was positioned at 0.5δ, the outer aqueous phase was forced to squeeze into the glass capillary tube. The pressure built up at the cross-junction allowed the formation of the encapsulated droplets.

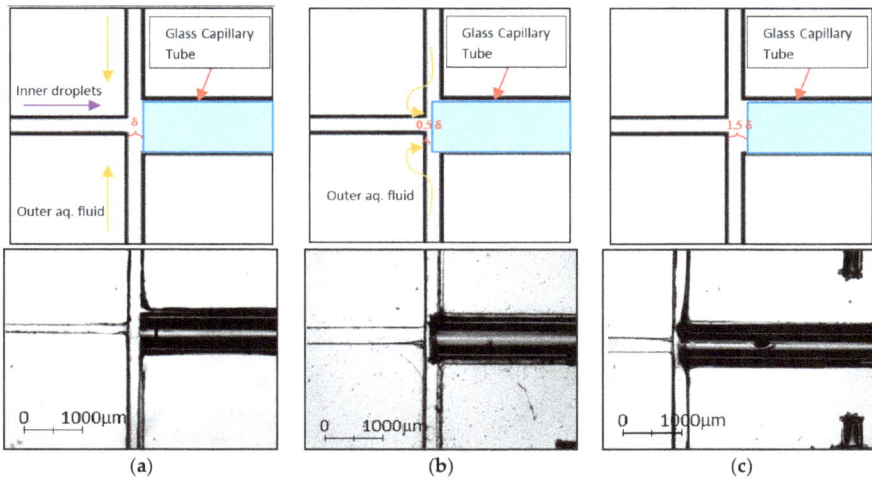

Figure 9. (**a**) Schematic illustration and the microscopic image of the glass capillary tube positioned at δ; (**b**) schematic illustration and the microscopic image of the glass capillary tube positioned at 0.5δ; and (**c**) schematic illustration and the microscopic image of the glass capillary tube positioned at 1.5δ. The glass capillary tube in the microscopic image dislocated due to expansion after heating. In the schematic diagram, yellow arrows indicate the flow of the outer aqueous phase according to different positions of the glass capillary tube while the purple arrow indicates the flow of the inner droplets.

3.2. Experimental Results—Droplet Encapsulation

The encapsulation experiments were conducted with four different flowrate ratios. The experimental data were grouped into two sets according to the manipulated flowrate ratios as shown in Table 3.

Table 3. Summary of the flowrate ratio and the manipulated parameter.

Group	Flowrate Ratio (μL/min)	Manipulated Flowrate
Group A	1:2:40 1:4:40 1:6:40	Middle phase—continuous cooking oil
Group B	1:6:40 1:6:60	Outer aqueous phase—continuous yellow dyed water

Relationship of Droplet Size with Flow Ratio, $Q_{disperse}/Q_{continuous}$

The flow regimes for the droplet generation are mainly distinguished by the capillary number of the continuous phase, Ca_c. Whilst $Ca_c < 0.002$ indicates the squeezing regime, $0.002 < Ca_c < 0.01$ indicates the transient regime and $0.01 < Ca_c < 0.3$ indicates the dripping regime [29]. The Ca_c at all flowrates were maintained at 0.0001 in our experiments indicating the squeezing regime for the downstream encapsulation process, while the droplet generation process at the upstream falls into the transient flow regime. Therefore, the droplet size depends predominantly on the flow ratio, $Q_{disperse}/Q_{continuous}$, instead of Ca_c [30]. Hence, Figure 10a shows the relationship between the droplet size and the flow ratio. Figure 10b illustrates the animation of droplet generation and encapsulation process of forming double emulsion droplets while Figure 10c shows the inconsistent droplet sizes formed due to the disproportion dominant force acting on droplet break up. The double emulsion droplets formed in the hybrid device were all discharged onto the petri dish as shown in Figure 10d–f.

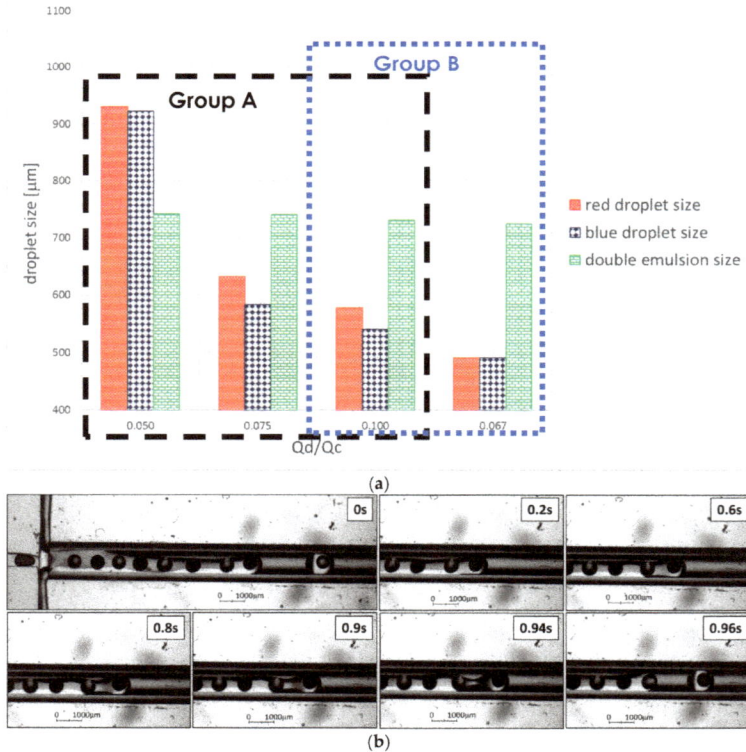

(a)

(b)

Figure 10. *Cont.*

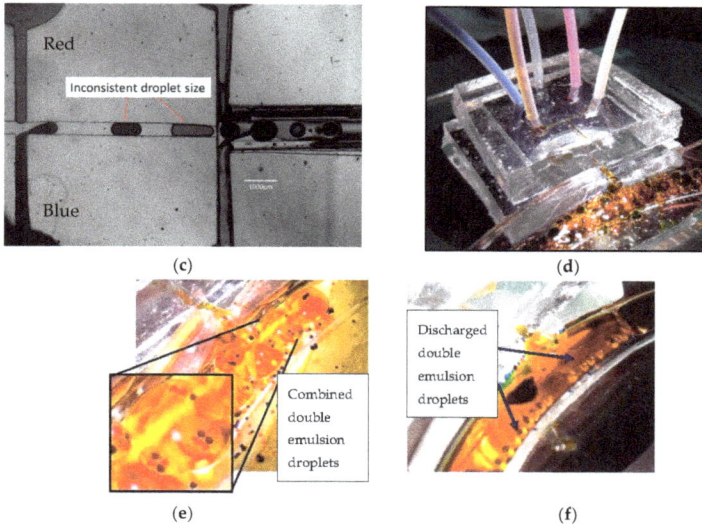

Figure 10. (**a**) Combination graph of droplet size versus flow ratio for Group A and Group B. Note that the values in the *y*-axis are not in an ascending trend since the sequence is in the ascending of flowrate ratio i.e., 1:2:40; 1:4:40; 1:6:40 and 1:6:60 µL/min. (**b**) Microscopic image of the droplet encapsulation process at the flowrate ratio of 1:6:40 µL/min. (**c**) Microscopic image of the droplet generation process at the flowrate ratio of 1:2:40 µL/min. (**d**) Image of the microfluidic device with the samples collected on the petri dish at the outlet. The double emulsion droplets will then slowly accumulate to form two layers of cooking oil and dyed water due to the absence of surfactant. (**e**) Image of the double emulsion droplets in the petri dish, two double emulsion droplets coalesced to form a larger droplet with four internal droplets due to the continuous flow disturbance from the outlet. (**f**) Image of the double emulsion droplets after discharged from the device with mostly single internal droplet encapsulated.

Since the droplet generation at flowrate ratio 1:2:40 µL/min falls under the transient regime, the dynamics of droplets break up would be dominated by both the shear force and interfacial force [29]. However, as Ca_c in these cases are very close to the squeezing regime, i.e., 0.003, the disproportion in both dominant forces i.e., shearing from the continuous phase and pressure built up across the inner droplets, causes the inconsistency of the internal droplet sizes (Figure 10c) with high values of CV i.e., 17.3% and 20.1% for red and blue droplets, respectively. From the experimental observations, a long water plug was formed when interfacial force dominated over the shear force and vice versa. Hence, this results in the formation of polydispersed internal droplets at the upstream of the device.

On the other hand, both inner droplet sizes show inconsistency at 1:4:40 (CV = 8% and 9.7%) and 1:6:40 µL/min (CV = 9.9% and 13.9%), even though the channel dimensions and inner flowrates were kept constant in these two cases. The experimental observations show a competition between both inner droplets squeezing out from their respective channel. This leads to two conditions: (1) the tip of the inner droplet was forced reverse to its channel by the opposite inner fluid before it could further elongate to the main channel and break off; and (2) the inner droplet had been sheared by the opposite inner fluid before it was fully developed to break up as a droplet. The first condition disrupts and randomizes the alternate sequencing of the droplets. The second condition, however, as the main reason influencing the consistency of the droplet production, results in a higher coefficient of variance (CV). The deviation between the two inner droplet sizes was calculated by |blue droplet diameter − red droplet diameter|/(red droplet diameter) × 100%. The results obtained are 0.8%, 7.5% and 6.4% respective to $Q_{disperse}/Q_{continuous}$ of 0.050, 0.075 and 0.100 (Figure 10a).

In group B, the inner droplet size of the 1:6:60 µL/min decreased significantly with higher similarity in size (CV = 5.9% and 9.1%). The competition between the droplets at the upstream was no longer observed once the system was stabilized. Only 0.1% difference between the red and blue inner droplet sizes was seen. This further evidences the influences and interactions among the interfacial forces of the inner, middle and outer fluids in the droplet encapsulation system. Current approach has produced monodispersed double emulsion with CV maintained at approximately or less than 10%. The double emulsion size remained at a range of 723–741 µm. While 1:6:60 µL/min possessed the smallest double emulsion size, flowrate ratio of 1:2:40 µL/min has the largest emulsion size. The CV percentages of all flowrate ratios were tabulated in Table 4.

Table 4. Coefficient of variance for different types of droplets at each flowrate ratio.

Group	Flowrate Ratio (µL/min)	Inner (Red) Droplets CV (%)	Inner (Blue) Droplets CV (%)	Double Emulsion CV (%)
Group A	1:2:40	17.3	20.1	10.5
	1:4:40	8.0	9.7	5.6
	1:6:40	9.9	13.9	6.7
Group B	1:6:40	9.9	13.9	6.7
	1:6:60	5.9	9.1	9.8

With different flowrate ratios, six different types of encapsulation are observed from the experimental work: (i) two distinct droplets with one red and one blue droplet encapsulated; (ii) two similar droplets with either two red; or (iii) two blue droplets encapsulated; (iv) single droplet encapsulation with either one red; or (v) one blue; and (vi) zero encapsulation which occurs whenever there is a flow disturbance from the syringe pumps. The encapsulation percentage/success rate of the double emulsion droplets formation are tabulated in Table 5.

Table 5. Success rate of the encapsulation for each combination.

Group	Flowrate Ratio (µL/min)	Percent of 2 Distinct Droplets Encapsulation (1R1B)	Percent of 2 Similar Droplets Encapsulation (2R or 2B)	Percent of Single Droplet Encapsulation (1R or 1B)	% of Zero Encapsulation
Group A	1:2:40	44.2	4.7	51.2	-
	1:4:40	19.1	5.6	69.7	5.6
	1:6:40	35.0	4.0	61.0	-
Group B	1:6:40	35.0	4.0	61.0	-
	1:6:60	49.6	11.9	38.5	0.6

Ultimately, distinct droplet encapsulation is preferred for better control and application wise. However, from an overall comparison, most of the flowrate ratios favors the single droplet encapsulation except for 1:6:60 µL/min which shows a higher total percentage in the double droplet encapsulation. Comparing different cases in group A, 1:4:40 µL/min shows the highest single encapsulation among the three flowrate ratios while 1:2:40 µL/min has the highest total percentage in double droplet encapsulation. At 1:2:40 µL/min, the internal droplets generated are in a longer plug size, i.e., 920 µm, with smaller spacing between the droplets. This narrow spacing between the droplets and low droplet generation frequency lead to a higher percentage in encapsulating two internal droplets. When the middle phase flowrate increased to 4 µL/min, the spacing between the droplets generated increased. This is because the middle continuous phase increases the flowing velocity of the generated droplet. However, the shear force was insufficient to increase the frequency of droplet production simultaneously, and thus leading to a higher percentage of single droplet encapsulation. As the middle phase flowrate further increased to 6 µL/min, the droplet spacing reduced with increasing droplet frequency. Therefore, we obtained a higher percentage of double droplets encapsulation at 1:6:40 µL/min.

The outer aqueous flowrate was then further increased to 60 µL/min with a constant inner and middle flowrates. Result shows the total percentage of double droplet encapsulation in this case is the highest at 61.5% with the lowest single droplet encapsulation at 38.5%. Until current stage of experiments, flowrate ratio 1:6:60 µL/min would be the optimum for encapsulating two distinct droplets. On top of that, no coalescence was observed between the two inner droplets though no surfactants were used. We attribute to the larger channel dimension of the glass capillary tube, where the droplet dimension is no longer constraint by the microchannel width and hence no coalescence occurs.

3.3. Experimental Results—Droplet Fission

Next, the droplet fission experiments will be further discussed. The droplet fission phenomenon started with a conventional upstream T-junction droplet formation [30] as shown in Figure 11a. Upon reaching the Y-curvature junction downstream, the mother droplet (Zone A, Figure 11b) experiences a competition among various forces subject to microchannel and liquid properties. The surface tension, originated from the dispersed water phase underwent stretching caused by uneven microchannel pressure distribution [30] and geometry change of the downstream. Eventually, the surface energy of a droplet will be surmounted by the combination forces, and the mother droplet will break into two equivalent daughter droplets (Figure 11) [31].

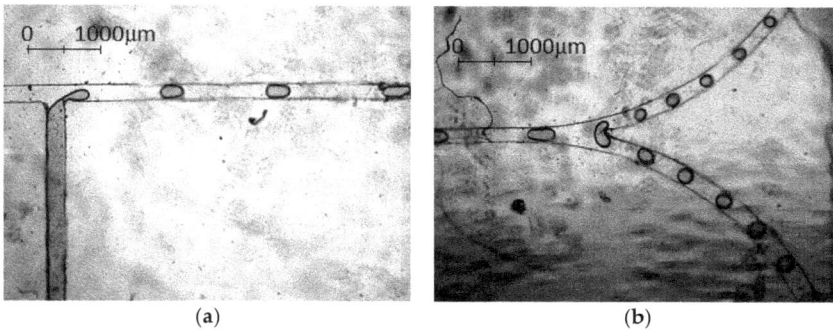

(a)   (b)

Figure 11. Droplet fission from the high speed camera of 1:1 curvature ratio at dispersed flowrate of 2 µL/min and continuous flowrate of 6 µL/min: (**a**) water droplet breakup at the T-junction; and (**b**) the droplet fission where a mother droplet breaks up into two daughter droplets. The blurriness of the images was caused by the cooking oil stain, which adhered to the top of the device surface during the experiment.

Effect of Flow Ratio

The flow ratio between the continuous and dispersed phase greatly affects the droplet formation as well as fission. With smaller droplets formed at the T-junction at the high velocity, the overall droplet fission size distribution shows an inversely proportional with the increment in flowrates of the dispersed and continuous phases at the constant flow ratio of 1:3. From Figure 12, the droplet fission size shows a direct relationship with respect to the alteration of flowrate ratio. As the flowrate increases, shearing effect is more effective as the continuous phase flow more easily overcomes the surface tension of the disperse phase at the T-junction. Such a phenomenon contributes to more uneven droplet fission distribution as pressure driven mechanism [30] at the splitting channel undergoes a strong competition with the geometry effect [31], which causes the splitting with a bigger droplet size variation.

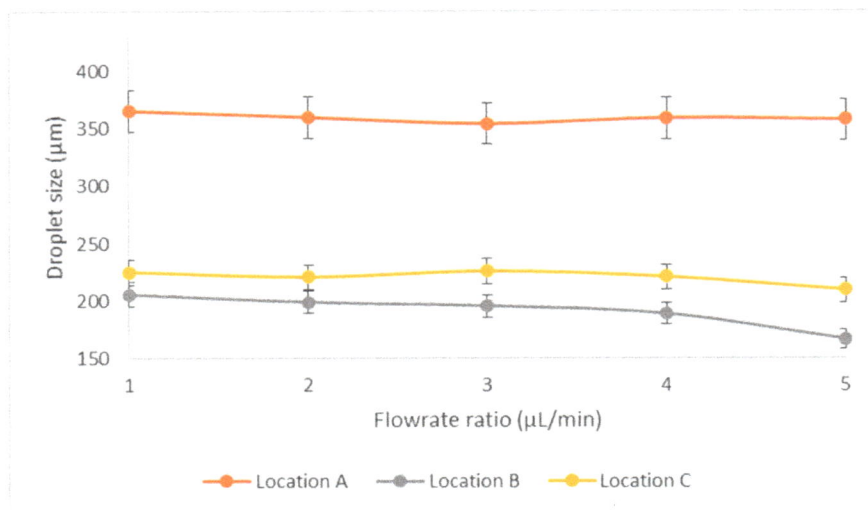

Figure 12. Droplet fission droplet profiles at different flow ratios. *x*-axis denotes the flowrate ratio pair of continuous and dispersed phase flowrate ranging from 2:6 to 6:18 µL/min.

4. Conclusions

In this paper, xurography was introduced as a promising fabrication method that has the capability of rapidly reproducing microfluidic device consisting of both single layer and double layer design with a comparable low installation and fabrication cost in bulk. Bartholomeusz et al. [20] shows that, although its resolution is lower than the standard lithography, the accuracy still falls within 10 µm of the drawn dimensions where the feature variability is less than 20%. However, our work in this paper proved that the error can be reduced to as low as 5% after appropriate optimization of the plotter settings. Utilizing custom-made PDMS aligner in alignment and sealing of two PDMS layers enable the alignment error to be dropped from 2.8% to 1.1%. This error can be further reduced to 0.75% if the glass capillary position control was implemented as we suggested. Experiments carried out using the xurography fabricated devices show convincing results in which all double emulsions produced were well maintained at CV below 10% indicating monodispersity, whereas the droplet fission produced daughter droplets with CV below 5%. We note that the monodispersity of the double emulsions produced is slightly lower as compared to the glass capillary device. Such limitations will be further improved in our future work through the addition of surfactants into the fluid phase. On top of that, flow phenomena studies will be further carried out in depth to better characterize the competition among different forces in droplets fission and break up.

Acknowledgments: This research was supported by Young Scientist Program from National Natural Science Foundation of China under Grant No. NSFC51506103/E0605; Zhejiang Provincial Natural Science Foundation of China under Grant Nos. Q15E090001 and LY16E060001; Ningbo Natural Science Foundation of China under Grant No. 2015A610281; and Ningbo Science and Technology Bureau Technology Innovation Team under Grant No. 2016B10010. Part of the experimental work for PDMS aligner analysis was carried out by the summer intern student, Lim Juin Yau.

Author Contributions: Chang Nong Lim and Kai Seng Koh conceived and designed the experiments while Jit Kai Chin supervised the whole process. Chang Nong Lim performed the experiments, analyzed and wrote the paper together with Kai Seng Koh and Yong Ren. Jit Kai Chin contributed the experimental materials and equipment. Yong Shi and Yuying Yan proofread the manuscript.

Conflicts of Interest: The authors declare no conflict of interest.

References

1. Chabert, M.; Dorfman, K.D.; Viovy, J.-L. Droplet fusion by alternating current (AC) field electrocoalescence in microchannels. *Electrophoresis* **2005**, *26*, 3706–3715. [CrossRef] [PubMed]
2. Link, D.R.; Anna, S.L.; Weitz, D.A.; Stone, H.A. Geometrically mediated breakup of drops in microfluidic devices. *Phys. Rev. Lett.* **2004**, *92*, 054503. [CrossRef]
3. Song, H.; Tice, J.D.; Ismagilov, R.F. A microfluidic system for controlling reaction networks in time. *Angew. Chem. Int. Ed. Engl.* **2003**, *42*, 768–772. [CrossRef] [PubMed]
4. Mazutis, L.; Gilbert, J.; Ung, W.L.; Weitz, D.A.; Griffiths, A.D.; Heyman, J.A. Single-cell analysis and sorting using droplet-based microfluidics. *Nat. Protoc.* **2013**, *8*, 870–891. [CrossRef] [PubMed]
5. Takeuchi, S.; Garstecki, P.; Weibel, D.B.; Whitesides, G.M. An axisymmetric flow-focusing microfluidic device. *Adv. Mater.* **2005**, *17*, 1067–1072. [CrossRef]
6. Datta, S.S.; Abbaspourrad, A.; Amstad, E.; Fan, J.; Kim, S.H.; Romanowsky, M.; Shum, H.C.; Sun, B.; Utada, A.S.; Windbergs, M.; et al. 25th anniversary article: Double emulsion templated solid microcapsules: Mechanics and controlled release. *Adv. Mater.* **2014**, *26*, 2205–2218. [CrossRef] [PubMed]
7. Drenckhan, W. Generation of superstable, monodisperse microbubbles using a pH-driven assembly of surface-active particles. *Angew. Chem. Int. Ed. Engl.* **2009**, *48*, 5245–5247. [CrossRef] [PubMed]
8. Day, P.; Manz, A.; Zhang, Y. *Microdroplet Technology: Principles and Emerging Applications in Biology and Chemistry*; Springer: Berlin, Germany, 2003.
9. Meng, F.T.; Ma, G.H.; Qiu, W.; Su, Z.G. W/O/W double emulsion technique using ethyl acetate as organic solvent: Effects of its diffusion rate on the characteristics of microparticles. *J. Control. Release* **2003**, *91*, 407–416. [CrossRef]
10. Kong, T.; Wu, J.; Yeung, K.W.K.; To, M.K.T.; Shum, H.C.; Wang, L. Microfluidic fabrication of polymeric core-shell microspheres for controlled release applications. *Biomicrofluidics* **2013**, *7*, 1–9. [CrossRef] [PubMed]
11. Okushima, S.; Nisisako, T.; Torii, T.; Higuchi, T. Controlled production of monodisperse double emulsions by two-step droplet breakup in microfluidic devices. *Langmuir* **2004**, *20*, 9905–9908. [CrossRef] [PubMed]
12. Deng, N.-N.; Meng, Z.-J.; Xie, R.; Ju, X.-J.; Mou, C.-L.; Wang, W.; Chu, L.-Y. Simple and cheap microfluidic devices for the preparation of monodisperse emulsions. *Lab Chip* **2011**, *11*, 3963–3969. [CrossRef] [PubMed]
13. Zhu, D.; Zhou, X.; Zheng, B. A double emulsion-based, plastic-glass hybrid microfluidic platform for protein crystallization. *Micromachines* **2015**, *6*, 1629–1644. [CrossRef]
14. Utada, A.S.; Lorenceau, E.; Link, D.R.; Kaplan, P.D.; Stone, H.A.; Weitz, D.A. Monodisperse double emulsions generated from a microcapillary device. *Science* **2005**, *308*, 537–541. [CrossRef] [PubMed]
15. Lei, K.F. Chapter 1 Materials and fabrication techniques for nano- and microfluidic devices. In *Microfluidics in Detection Science: Lab-on-a-Chip Technologies*; Labeed, F.H., Fatoyinbo, H.O., Eds.; RSC Detection Science, Royal Society of Chemistry: Cambridge, UK, 2014; pp. 1–28.
16. Marty, F.; Rousseau, L.; Saadany, B.; Mercier, B.; Français, O.; Mita, Y.; Bourouina, T. Advanced etching of silicon based on deep reactive ion etching for silicon high aspect ratio microstructures and three-dimensional micro- and nanostructures. *Microelectron. J.* **2005**, *36*, 673–677. [CrossRef]
17. Goral, V.N.; Hsieh, Y.-C.; Petzold, O.N.; Faris, R.A.; Yuen, P.K. Hot embossing of plastic microfluidic devices using poly(dimethylsiloxane) molds. In Proceedings of the 14th International Conference on Miniaturized Systems for Chemistry and Life Sciences, Groningen, The Netherlands, 3–7 October 2010; pp. 1214–1216.
18. Weigl, B.H.; Bardell, R.; Schulte, T.; Battrell, F.; Hayenga, J. Design and rapid prototyping of thin-film laminate-based microfluidic devices. *Biomed. Microdevices* **2001**, *3*, 267–274. [CrossRef]
19. Attia, U.M.; Marson, S.; Alcock, J.R. Micro-injection moulding of polymer microfluidic devices. *Microfluid. Nanofluid.* **2009**, *7*, 1–28. [CrossRef]
20. Bartholomeusz, D.A.; Boutté, R.W.; Andrade, J.D. Xurography: Rapid prototyping of microstructures using a cutting plotter. *J. Microelectromech. Syst.* **2005**, *14*, 1364–1374. [CrossRef]
21. Santana, P.P.; de Segato, T.P.; Carrilho, E.; Lima, R.S.; Dossi, N.; Kamogawa, M.Y.; Gobbi, A.L.; Piazzetaf, M.H.; Piccin, E. Fabrication of glass microchannels by xurography for electrophoresis applications. *Analyst* **2013**, *138*, 1660–1664. [CrossRef] [PubMed]
22. Bartholomeusz, D.A.; Boutté, R.W.; Gale, B.K. Xurography: Microfluidic prototyping with a cutting plotter. In *Lab on a Chip Technology: Fabrication and Microfluidics*; Caister Academic Press: Poole, UK, 2009; Volume 1, pp. 65–82.

23. Ren, K.; Zhou, J.; Wu, H. Materials for microfluidic chip fabrication. *Acc. Chem. Res.* **2013**, *46*, 2396–2406. [CrossRef] [PubMed]

24. Abate, A.R.; Weitz, D.A. High-order multiple emulsions formed in poly(dimethylsiloxane) microfluidics. *Small* **2009**, *5*, 2030–2032. [CrossRef] [PubMed]

25. Ren, Y.; Liu, Z.; Shum, H.C. Breakup dynamics and dripping-to-jetting transition in a Newtonian/shear-thinning multiphase microsystem. *Lab Chip* **2014**, *15*, 121–134. [CrossRef] [PubMed]

26. Roman, G.T.; Hlaus, T.; Bass, K.J.; Seelhammer, T.G.; Culbertson, C.T. Sol-gel modified poly(dimethylsiloxane) microfluidic devices with high electroosmotic mobilities and hydrophilic channel wall characteristics. *Anal. Chem.* **2005**, *77*, 1414–1422. [CrossRef] [PubMed]

27. Bhattacharya, S.; Datta, A.; Berg, J.M.; Gangopadhyay, S. Studies on surface wettability of poly(dimethyl) siloxane (PDMS) and glass under oxygen-plasma treatment and correlation with bond strength. *J. Microelectromech. Syst.* **2005**, *14*, 590–597. [CrossRef]

28. Pinto, E.; Faustino, V.; Rodrigues, R.O.; Pinho, D.; Garcia, V.; Miranda, J.M.; Lima, R. A rapid and low-cost nonlithographic method to fabricate biomedical microdevices for blood flow analysis. *Micromachines* **2015**, *6*, 121–135. [CrossRef]

29. Nunes, J.K.; Tsai, S.S.H.; Wan, J.; Stone, H.A. Dripping and jetting in microfluidic multiphase flows applied to particle and fiber synthesis. *J. Phys. D Appl. Phys.* **2013**, *46*, 1–6. [CrossRef] [PubMed]

30. Garstecki, P.; Fuerstman, M.J.; Stone, H.A.; Whitesides, G.M. Formation of droplets and bubbles in a microfluidic T-junction-scaling and mechanism of break-up. *Lab Chip* **2006**, *6*, 437–446. [CrossRef] [PubMed]

31. Ren, Y.; Koh, K.S. Droplet fission in non-Newtonian multiphase system using bilayer bifurcated microchannel. In Proceedings of the ASME 2016 5th Micro/Nanoscale Heat and Mass Transfer International Conference (MNHMT-16), Singapore, 4–6 January 2016.

micromachines

MDPI

Article

An Interference-Assisted Thermal Bonding Method for the Fabrication of Thermoplastic Microfluidic Devices

Yao Gong, Jang Min Park * and Jiseok Lim *

School of Mechanical Engineering, Yeungnam University, Daehakro 280, Gyeongsan, 38541 Gyeongbuk, Korea; manyjoy321@gmail.com
* Correspondence: jpark@yu.ac.kr (J.M.P.); jlim@yu.ac.kr (J.L.);
   Tel.: +82-53-810-2456 (J.M.P.); +82-53-810-2577 (J.L.)

Academic Editors: Weihua Li, Hengdong Xi and Say Hwa Tan
Received: 20 September 2016; Accepted: 17 November 2016; Published: 22 November 2016

Abstract: Solutions for the bonding and sealing of micro-channels in the manufacturing process of microfluidic devices are limited; therefore, further technical developments are required to determine these solutions. In this study, a new bonding method for thermoplastic microfluidic devices was developed by combining an interference fit with a thermal treatment at low pressure. This involved a process of first injection molding thermoplastic substrates with a microchannel structure, and then performing bonding experiments at different bonding conditions. The results indicated the successful bonding of microchannels over a wide range of bonding pressures with the help of the interference fit. The study also determined additional advantages of the proposed bonding method by comparing the method with the conventional thermal bonding method.

Keywords: lab-on-a-chip; thermal bonding; interference fit; injection molding

1. Introduction

A microfluidic device is a miniaturized device in which micrometer-scale components, such as microchannels, micromixers, and microincubators, are integrated to achieve high-throughput analysis with respect to chemical and biomedical reactions [1–6]. Recently, the market for lab-on-a-chips utilizing microfluidic devices has rapidly increased with increase in the need for point-of-care systems. Therefore, this has led to the availability of various types of commercial microfluidic devices in the market [7].

Microfluidic devices can be fabricated from various materials including glass, silicon, and polymeric materials [8,9]. Poly-dimethylsiloxane (PDMS) is the most common material used for the fabrication of microfluidic devices in the laboratory, as it allows simple replication and sealing procedures. Furthermore, advantages of PDMS, such as biocompatibility, high gas permeability, and superior optical transmittance, are required in many cases for biomedical applications. However, the issue of a long fabrication time is a disadvantage of PDMS; thus, it is difficult to use PDMS as a material for several practical applications.

An extensive review of commercial microfluidic devices [7,10] suggested that thermoplastic materials are the most commonly used materials in practical applications of microfluidic devices. The main advantages of thermoplastic materials include low cost and disposability of the device due to mass production.

The manufacturing process of thermoplastic microfluidic devices consists of the following major steps: (i) fabrication of a microfeatured substrate; and (ii) bonding the substrates and interconnecting the substrates with external devices [11]. In contrast to the first step, the second step continues to present a challenge, and thereby necessitates further technical developments as noted by current research [12].

Previous studies proposed various methods, such as adhesive bonding, solvent assisted bonding, local welding, and thermal bonding, for the development of bonding solutions for thermoplastic-based microfluidic devices [12,13]. A comprehensive review on the various bonding method can be found in the recent review paper by Temiz et al. [12]. Each proposed bonding method exhibits its own set of advantages and disadvantages when compared with other methods.

For example, adhesive bonding is a simple method in which strong bonding can be achieved at room temperature. However, the surface properties of adhesive bonding can be modified. Additionally, an adhesive can clog a microchannel. In contrast, local welding offers a fairly strong degree of bonding without changes in the surface chemistry, but requires special equipment for precise welding.

A thermal bonding process can be considered an appropriate solution. In this process, two substrates are first heated to a temperature approximately equal to the corresponding glass transition temperatures of the respective substrates, and this is followed by the application of pressure (p). Molecular diffusion occurs across the contact surface between the two substrates during this process. A bonded product is finally obtained after the cooling process. Figure 1a shows the schematic of a conventional thermal bonding method.

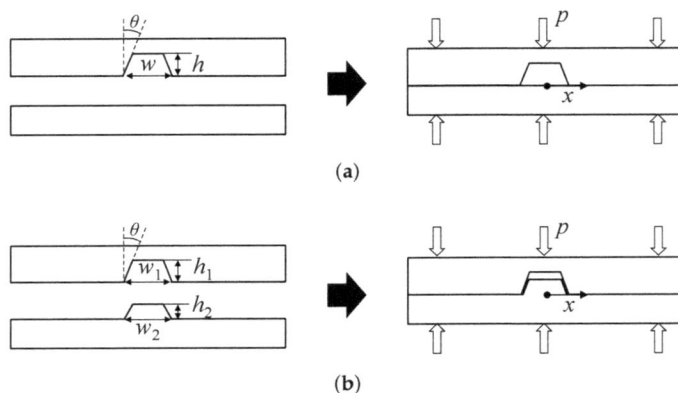

Figure 1. Schematics of (**a**) the conventional thermal bonding method; and (**b**) interference-assisted thermal bonding method.

The thermal bonding method offers certain advantages for microfluidic device assembly under optimized processing conditions. Specifically, it offers sufficiently high bonding strength, and the chemical properties of the thermoplastic surface remain unchanged [11,13,14]. However, the conventional thermal bonding method also entails issues limiting its application in practical manufacturing processes. In particular, it necessitates the uniform application of pressure over the entire substrate to seal the microchannels and other microfluidic components, and this is not trivial, especially in cases involving a large bonding area. Furthermore, given that the bonding area changes following the channel design, it is necessary to precisely calculate and control the applied force to avoid delamination or channel collapse during the bonding process [13].

As noted above, the most challenging part in the bonding of thermoplastic microfluidic device would be the uniformity of the applied pressure. In addition to this issue, warpage of the substrate is also a critical factor affecting the bonding uniformity [15]. To resolve these issues, Park et al. developed a pressure chamber system where pressure and temperature can be precisely controlled [16]. This method, however, still have limitations regarding the productivity. Since then, there has not been much progress made regarding the thermal bonding method.

This study involved the development of a new bonding method that combined interference fit and thermal bonding. This bonding method, termed the interference-assisted thermal bonding method, offered additional benefits when compared with the conventional thermal bonding method. In this method, the bonding performance was less sensitive to the applied pressure, since the bonding occurred due to the stress induced by mechanical interference between the substrates. The interference-assisted thermal bonding method for a simple microfluidic device was numerically analyzed and experimentally proved in the study.

2. Interference-Assisted Thermal Bonding Method

Figure 1b shows a schematic of the interference-assisted thermal bonding method. The upper substrate contains a microchannel of width w_1, while the lower one has a micro-rib of width w_2, and both the structures have a draft angle of θ. If $w_2 - w_1 > 0$, then interference between the microchannel and the micro-rib exists. This interference causes a pressure higher than the applied pressure (denoted by p) to develop along the sidewall of the micro-rib. This leads to strong bonding along the sidewall, and thus the microfluidic system is sealed.

Typically, the bonding strength increases with the applied pressure in the thermal bonding of thermoplastics [14]. Numerical simulation was first performed prior to the experimental implementation of the proposed bonding method to quantitatively analyze the pressure distribution along the contact surface between the two substrates. The numerical simulation involves two-dimensional models as shown in Figure 1. The microchannel width (w_1) and height (h_1) were fixed to 500 μm and 700 μm, respectively. Other geometrical parameters of draft angle (θ), micro-rib width (w_2) and height (h_2) were varied to observe their effect on the pressure distribution. The width and thickness of the substrate were 5 mm and 1 mm, respectively. The process was assumed to be isothermal at 95 °C, and ANSYS software was used to only perform the structural analysis. An elastic modulus of 533 MPa and a Poisson ratio of 0.4 were applied with respect to the material properties of the substrates. A constant pressure was applied at the upper boundary of the upper substrate, while the lower boundary of the lower substrate was fixed. The left and right boundaries of the substrates were constrained such that they did not exhibit any displacement in the lateral direction, and a constant friction coefficient of 0.2 was applied for the contact condition between the two substrates.

Figure 2 shows the stress distribution around the microchannel when a pressure p of 1 MPa was applied, and the interference is 10 μm. The overall stress level remained almost the same with the interference, but there was a significant change in the distribution especially at the corner region of the microchannel.

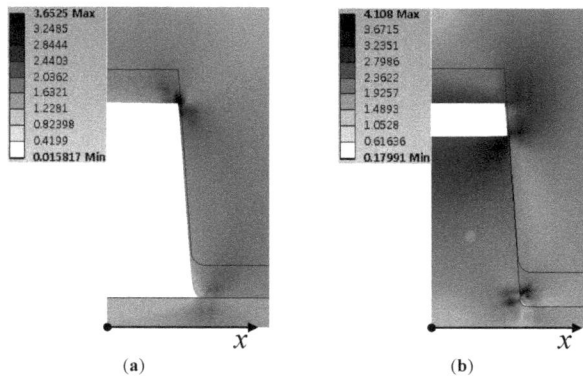

Figure 2. Simulation results of von Mises stress distribution around microchannel (**a**) without the interference fit and (**b**) with the interference fit. Applied pressure is 1 MPa. Interference height, width and draft angle are 500 μm, 10 μm and 4°, respectively.

Figure 3 provides further details as it shows the pressure distribution applied on the contact surface between the two substrates. In particular, effect of the interference width ($w_2 - w_1$), interference height (h_2) and draft angle (θ) was investigated to provide design guideline of the interference fit. Shown in Figure 3a is the effect of the interference on the pressure distribution when $h_2 = 500$ μm and $\theta = 4°$. In the conventional case, the pressure was close to 1 MPa in the overall region ($x > 0.27$ mm) except for the corner region (0.25 mm $< x <$ 0.27 mm) where the pressure was locally concentrated as expected. In the interference-assisted case, the pressure was slightly reduced in $x > 0.26$ mm, while it could be significantly increased along the interference region (0.22 mm $< x <$ 0.252 mm). The tendency grew more pronounced as the interference increased. This numerical result suggested that increased bonding strength could be achieved along the interference region while the remaining region was only slightly affected. However, it should be noted that one needs to apply the interference width of at least 10 μm in order to have the contact pressure along the interference region more than two times the applied pressure.

Figure 3b shows the effect of h_2 on the pressure distribution when $w_2 - w_1 = 10$ μm and $\theta = 4°$. The maximum pressure at the interference region was found to almost independent of h_2, while the average value is almost inversely proportional to h_2. Therefore, when h_2 is further increased as $h_2 \geq 400$ μm, the average pressure at the interference region is only slightly affected by h_2. From the view point of the process optimization, it would be better if the contact pressure is less sensitive to the geometrical parameter of the interference. In this regard, it is recommended to select the interference height (h_2) as at least two times the final microchannel depth ($h_1 - h_2$). In contrast, the draft angle θ had only minor effect on the pressure distribution as shown in Figure 3c where $w_2 - w_1 = 10$ μm and $h_2 = 500$ μm were applied. The maximum and average pressure at the interference region is slightly decreased as θ is decreased.

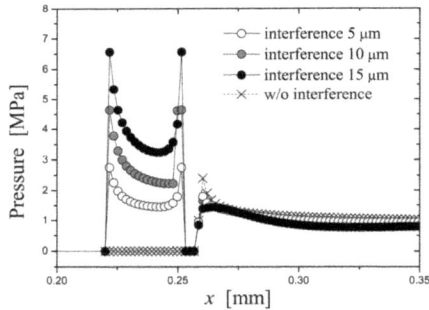

(a)

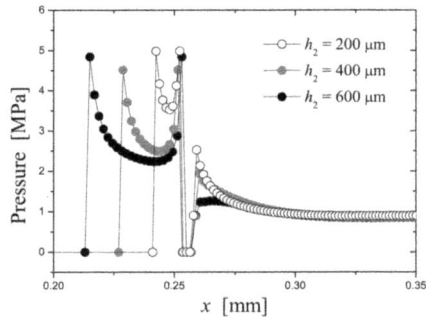

(b)

Figure 3. *Cont.*

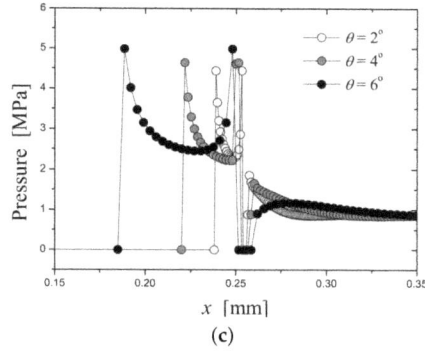

Figure 3. Simulation results of profiles of contact pressure between the two substrates: (**a**) Interference width is varied, while interference height (h_2) and draft angle (θ) are fixed as 500 μm and 4°; (**b**) Interference height is varied, while interferenc width and draft angle are fixed as 10 μm and 4°, respectively; (**c**) Draft angle is varied, while interference width and height are fixed as 10 μm and 500 μm, respectively. Applied pressure is 1 MPa.

According to the numerical simulation result, the interference height and draft angle had only minor effect on the pressure distribution in comparison to the interference width did. As a result, interference height and draft angle were fixed to 500 μm and 4°, respectively, in the following experiments. In order to determine appropriate value of the interference width, one would need more elaborate study than the simulation, since more complicated non-linear deformation and fracture phenomena could take place in practice, which were not included in the present simulation. Therefore, the interference width $w_2 - w_1$ were determined based on experimental tests, which will be discussed in the next chapter.

3. Experiments

In the study, the injection molding process replicated two substrates, which were used for the bonding experiments. Mold inserts were fabricated by micro-machining, and the thermoplastic material poly(methyl metacrylate) (PMMA, IH830H, LG Chem.) was used in the injection molding process. Table 1 shows the injection molding condition. It should be noted that the injection molding cycle time required to obtain a substrate was only 27 s.

Table 1. Injection molding condition.

Parameter	Value
Mold temperature	85 °C
Nozzle temperature	230 °C
Packing pressure	65 MPa
Packing time	3 s
Injection time	0.5 s
Cooling time	10 s

An in-house hot embossing equipment was used to perform the bonding experiments. The two substrates were first interference fitted and mounted on the hot embossing equipment. The specimen was then heated to the bonding temperature, and this was followed by the application of pressure for 3 min. The pressure was released after cooling the specimen to 50 °C, and the final product was obtained. In the study, the bonding temperature was varied between 95 °C, 97 °C and 100 °C, The bonding pressure was varied from 0.1 MPa to 3.0 MPa in order to observe the effect on the bonding quality. The processing condition is selected based on the previous work by Zhu et al. [14].

Figure 4 shows the schematics of the two substrates fabricated by the injection molding process. Each substrate had a thickness of 1 mm. With respect to the alignment between the two substrates, holes and rectangular pillars were located near the corners of the upper and lower substrates, respectively. These alignment structures of the holes and pillars also displayed interference fit in between the substrates. Each structure on the substrates had a draft angle of 4° to ease the demolding involved in the injection molding process. The upper substrate had two holes (with diameters of 3 mm and 5 mm) and a microchannel (with a width of 500 µm and a depth of 700 µm). The lower substrate exhibited protruded microstructures (with a height of 500 µm) with respect to the interference fit with the upper substrate. The lateral dimensions of the protruded microstructures were designed such that the interference was 10 µm, which was determined from the following experiments.

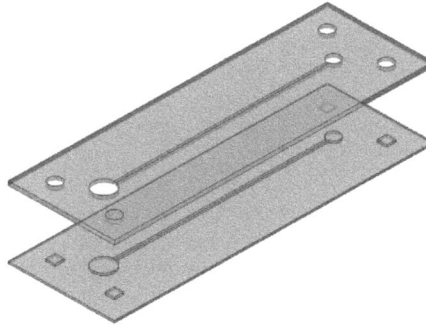

Figure 4. Schematics of the two substrates used in the bonding experiment. The upper substrate has two holes and a microchannel in the middle and four holes are located near the corner. The lower substrate has micro-rib and protrude structures which will result in interference fit with the upper substrate.

As discussed with Figure 3a, the interference width should be at least 10 µm to achieve significant increase of the contact pressure along the interference region. However, it should not be too large so that the interference does not deteriorate the original shape of the microfluidic system. In addition, it was necessary for the interference dimension to be within the resolution range of the manufacturing method employed to fabricate the mold inserts. In this study, mold inserts having different interference width of 10 µm to 200 µm were fabricated by micro-machining process, and bonding experiments were performed to determine the appropriate value of the interference. According to the test, specimens having the interference width larger than 20 µm had a fracture along the upper wall of the microchannel during the bonding process. This fracture was mainly due to a weldline of which details will be shown and explained in the next chapter. As a result, the interference width of 10 µm was applied in this study.

4. Results

Figure 5 shows photographs of the mold inserts, injection molded substrates and a micrograph of the microchannel. Several types of methods were available for the fabrication of the mold inserts [17]. In the case used in the present study, the minimum feature size of the microchannel was 500 µm, and the maximum aspect ratio was 1.4. The study involved obtaining an interference of 10 µm, and thus the feature tolerance of the manufacturing method could correspond to an order of micrometers. Additionally, fine roughness of the sidewalls was not required. Micro-machining satisfied these requirements and could provide suitable mold inserts. It might be mentioned that one should use other fabrication methods for the mold insert to produce devices having smaller features.

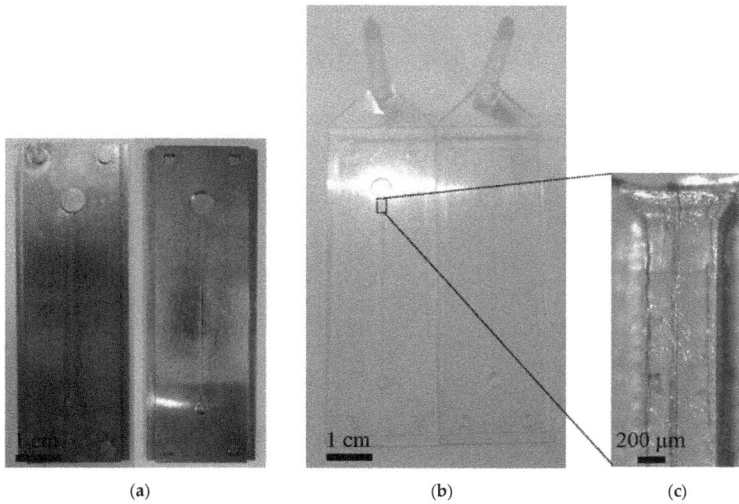

Figure 5. Photographs of mold inserts (**a**), injection molded substrates (**b**) and a micrograph of the microchannel (**c**). There is weldline along the center of the microchannel in the micrograph.

A fan gate was used to obtain uniform filling of the substrate. With respect to the injection molding of the substrate with holes and a microchannel (as shown in Figure 5), a weldline developed along the upper wall of the microchannel due to the presence of a hole. This weldline can be observed clearly in the magnified view of the microchannel in Figure 5. The weldline could result in a fracture of the upper wall during the interference-assisted thermal bonding process because of the weak mechanical properties of the welding. A mold temperature higher than that in the conventional molding condition of PMMA material was applied to avoid such defects (Table 1).

Figure 6 shows five different products obtained from bonding experiments with different bonding pressures. A dyed liquid was introduced at the larger well to confirm the sealing of the microchannel and wells. The products bonded at a pressure of 1 MPa and products bonded at higher pressure showed a distinct microchannel and two wells, thereby indicate complete sealing. Conversely, products bonded at low pressures of 0.1 MPa and 0.5 MPa exhibited leakage due to insufficient bonding.

Figure 6. Sealing test results of products bonded at different bonding pressures of 0.1 MPa, 0.5 MPa, 1.0 MPa, 1.5 MPa, and 2.0 MPa (from the left to right). The bonding temperature is 95 °C. Dyed water is injected through the upper chamber to visualize the sealing.

Cross-sections of the bonded products were observed to investigate the effect of the bonding condition on the final shape of the microchannel. Figure 7 shows the product bonded at 2.0 MPa and 95 °C. The micro-rib was retained securely without significant deformation. The microchannel only exhibited a slight deformation, and its depth was 682 µm. Figure 7b presents the scanning electron microscope (SEM) micrograph to confirm the bonding along the contact region. The other products that bonded at lower pressures of 1.0 MPa and 1.5 MPa also displayed approximately the same cross-sectional structures as shown in Figure 7 (although this was not shown in this study). However, the only notable difference was that the bonding in the plane became weaker as the pressure decreased.

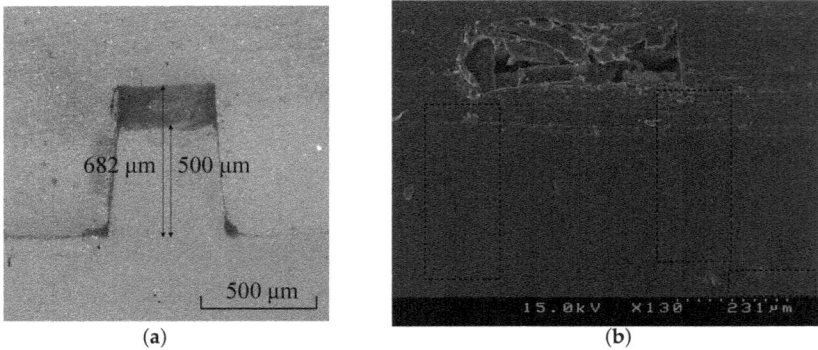

Figure 7. Cross-sectional micrographs of the microchannel observed by (**a**) optical microscope; and (**b**) scanning electron microscope (SEM). The dotted boxes in the SEM micrograph indicate the contact region. The bonding temperature and bonding pressure are 95 °C and 2 MPa, respectively.

Figure 8a,b show the cross-sections of the products bonded at higher temperatures of 97 °C and 100 °C, respectively. The results indicated that the microchannel collapsed further as the temperature increased. Additionally, the micro-rib was deformed when compared to that shown in Figure 7a. However, the height did not monotonically decrease with the temperature. This was attributed to the squeezing effect in the lateral direction, and this squeezing effect was reflected in the curved profile of the top surface.

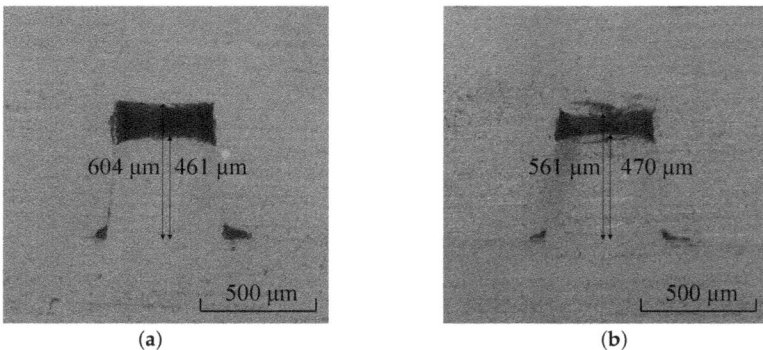

Figure 8. Cross-sectional micrographs of the microchannels bonded at temperatures of (**a**) 97 °C; and (**b**) 100 °C. The bonding pressure corresponds to 2 MPa.

5. Discussion and Conclusions

This study proposed a new bonding method that combined a thermal bonding method and an interference fit. The performance of the proposed method was experimentally investigated. This section discusses several features of the proposed bonding method and particularly focuses on the comparison of the proposed method with the conventional thermal bonding method.

In the present bonding method, the uniformity of the applied pressure did not affect the sealing of the microfluidic system to the same extent as that in the conventional thermal bonding method. The microfluidic system could be sealed first because of the interference fit between the microchannel and the micro-rib despite the potential lack of uniformity in the applied pressure over the entire substrate. This feature provides a substantial advantage. This is especially significant in cases where the substrate size increases such that the application of uniform pressure is more difficult to achieve.

In the proposed method, it is not necessary for the entire substrate to be completely bonded. It is only necessary for the sidewall of the microfluidic system to be strongly bonded by using the interference fit. This feature offers advantages in terms of process optimization and productivity. For example, as shown in Figure 6, it is possible to obtain well-sealed products with a bonding pressure ranging from 1 MPa to 2 MPa. This corresponded to a relatively wider processing window when compared with that of the conventional thermal bonding method. This indicated that process optimization was easier in the proposed method. Additionally, the proposed method reduced the probabilities of seal defects in the final product.

However, the proposed method also had certain disadvantages. First, the proposed method necessitates an additional mold insert, and this increases the overall manufacturing cost. In addition, the micro features should have extra height for the interference, which makes the mold insert fabrication more difficult. Also a possibility of bends or warpages with respect to a substrate after bonding was associated with the proposed method because of the presence of the interference. This was resolved in this study by introducing alignment structures near the corner of the substrate. As discussed previously, these alignment structures also displayed interference-assisted bonding, which prevented the bend or warpage of the substrate. Such structures could be introduced based on the design of the microfluidic system.

Despite these drawbacks, the interference-assisted thermal bonding method evidently offers certain advantages over the conventional thermal bonding method, especially with respect to productivity. Hence, it is expected that the proposed method will be widely used in practical manufacturing processes.

Acknowledgments: Y.G. and J.M.P. would like to thank the financial support from Basic Science Research Program through the National Research Foundation of Korea (NRF) Grant funded by the Ministry of Science, ICT & Future Planning (NRF 2015R1C1A1A02036960). J.L. would like to thank the financial support from the National Research Foundation of Korea (NRF) Grant funded by the Ministry of Science, ICT & Future Planning (NRF 215C000744). This research was also supported from National Research Foundation of Korea (NRF) Grant funded by the Korean Government (MSIP) (No. 2015R1A5A1037668). Finally, authors would like to thank Dong Sung Kim at POSTECH for providing hot embossing equipment for this research.

Author Contributions: J.M.P. and J.L. conceived and designed the experiments; Y.G. has performed the experiments and analyzed the results; Y.G., J.M.P., and J.L. wrote the paper.

Conflicts of Interest: The authors declare no conflict of interest.

References

1. Manz, A.; Graber, N.; Widmer, H.M. Miniaturized total chemical analysis systems: A novel concept for chemical sensing. *Sens. Actuators B Chem.* **1990**, *1*, 244–248.
2. Chow, A.W. Lab-on-a-Chip: Opportunities for chemical engineering. *AIChE J.* **2002**, *48*, 1590–1595.
3. Squires, T.M.; Quake, S.R. Microfluidics: Fluid physics at the nanoliter scale. *Rev. Mod. Phys.* **2005**, *77*, 977.
4. Whitesides, G.M. The origins and the future of microfluidics. *Nature* **2006**, *442*, 368–373.
5. Chin, C.D.; Linder, V.; Sia, S.K. Lab-on-a-chip devices for global health: Past studies and future opportunities. *Lab Chip* **2007**, *7*, 41–57.

6. Baret, J.C.; Beck, Y.; Billas-Massobrio, I.; Moras, D.; Griffiths, A.D. Quantitative cell-based reporter gene assays using droplet-based microfluidics. *Chem. Biol.* **2010**, *17*, 528–536.

7. Chin, C.D.; Linder, V.; Sia, S.K. Commercialization of microfluidic point-of-care diagnostic devices. *Lab Chip* **2012**, *12*, 2118–2134.

8. Iliescu, C.; Taylor, H.; Avram, M.; Miao, J.M.; Franssila, S. A practical guide for the fabrication of microfluidic devices using glass and silicon. *Biomicrofluidics* **2012**, *6*, 016505.

9. Becker, H.; Locascio, L.E. Polymer microfluidic devices. *Talanta* **2002**, *56*, 267–287.

10. Berthier, E.; Young, E.W.K.; Beebe, D. Engineers are from PDMS-land, Biologists are from Polystyrenia. *Lab Chip* **2012**, *12*, 1224–1237.

11. Kipling, G.D.; Haswell, S.J.; Brown, N.J. A considered approach to lab-on-a-chip fabrication. In *Lab-on-a-Chip Devices and Micro-Total Analysis Systes*; Castillo-León, J., Svendsen, W.E., Eds.; Springer: Berlin, Germany, 2015; pp. 53–82.

12. Temiz, Y.; Lovchik, R.D.; Kaigala, G.V.; Delamarche, E. Lab-on-a-chip devices: How to close and plug the lab? *Microelectron. Eng.* **2015**, *132*, 156–175.

13. Tsao, C.W.; DeVoe, D.L. Bonding of thermoplastic polymer microfluidics. *Microfluid. Nanofluid.* **2009**, *6*, 1–16.

14. Zhu, X.; Liu, G.; Guo, Y.; Tian, Y. Study of PMMA thermal bonding. *Microsyst. Technol.* **2007**, *13*, 403–407.

15. Chen, P.C.; Yen, Y.C. Warpage of embossed thermoplastic substrates and the effects on solvent bonding. *Microsyst. Technol.* **2016**, doi:10.1007/s00542-016-3035-8.

16. Park, T.; Song, I.H.; Park, D.S.; You, B.H.; Murphy, M.C. Thermoplastic fusion bonding using a pressure-assisted boiling point control system. *Lab Chip* **2012**, *12*, 2799–2802.

17. Giboz, J.; Copponnex, T.; Mélé, P. Microinjection molding of thermoplastic polymers: A review. *J. Micromech. Microeng.* **2007**, *17*, R96–R109.

micromachines

MDPI

Article

Rapid Capture and Analysis of Airborne *Staphylococcus aureus* in the Hospital Using a Microfluidic Chip

Xiran Jiang [1], Yingchao Liu [2], Qi Liu [3], Wenwen Jing [3], Kairong Qin [1] and Guodong Sui [3,*]

[1] Department of Biomedical Engineering, Dalian University of Technology, Dalian 116024, China; xrjiang@dlut.edu.cn (X.J.); krqin@dlut.edu.cn (K.Q.)
[2] Department of Neurosurgery, Provincial Hospital Affiliated to Shandong University, Jinan 250021, China; bxs103@sdu.edu.cn
[3] Department of Environmental Science & Engineering, Fudan University, Shanghai 200433, China; 13110740002@fudan.edu.cn (Q.L.); jingwenwen1983@gmail.com (W.J.)
* Correspondence: gsui@fudan.edu.cn; Tel.: +86-5566-4504

Academic Editors: Weihua Li, Hengdong Xi and Say Hwa Tan
Received: 3 August 2016; Accepted: 12 September 2016; Published: 15 September 2016

Abstract: In this study we developed a microfluidic chip for the rapid capture, enrichment and detection of airborne *Staphylococcus* (*S.*) *aureus*. The whole analysis took about 4 h and 40 min from airborne sample collection to loop-mediated isothermal amplification (LAMP), with a detection limit down to about 27 cells. The process did not require DNA purification. The chip was validated using standard bacteria bioaerosol and was directly used for clinical airborne pathogen sampling in hospital settings. This is the first report on the capture and analysis of airborne *S. aureus* using a novel microfluidic technique, a process that could have a very promising platform for hospital airborne infection prevention (HAIP).

Keywords: microfluidics; airborne bacteria; bioaerosols

1. Introduction

Staphylococcus (*S.*) *aureus*, which is one of the major community-acquired pathogens, has been reported to be responsible for various human diseases [1]. Further, *S. aureus* has the ability to colonize the human nose and wound area on the skin, causing hospital airborne infections [2]. Due to the lack of a technique for rapidly detecting airborne *S. aureus*, bacterial transfer by air is hard to prevent and has become a serious threat to public safety and patients in hospital wards. At present, few techniques are capable of rapid detection of airborne *S. aureus*.

In the field of airborne pathogen detection, several traditional techniques have been proposed, such as the Anderson sampler and all-glass impinger (AGI) sampler [3]. However, the pathogen concentrations in samples collected and recovered by these techniques are too low for direct bioanalysis [4,5]. Furthermore, these techniques all require an obligatory culturing step, which is the most time-consuming stage of the analysis, as this usually takes days to complete [6]. Moreover, most of the known pathogens that exist in the natural environment are in a viable but non-culturable (VBNC) state [7], thus they cannot be detected using a culture method. So rapid bioanalysis of airborne pathogen is very hard to perform, and this is the biggest bottleneck for current sampling techniques. As a result, the limited techniques have made it difficult to issue a pre-warning of airborne *S. aureus*–related diseases.

The efficient transfer of pathogens from air to favorable media is the most essential step [7] for a rapid analysis technique. Another vital factor is that the result of the analysis is easily visualized, without the need for extensive or elaborate processing such as conventional electrophoresis or hybridization. Loop-mediated isothermal amplification (LAMP) is an attractive technique with high sensitivity and selectivity, as well as less processing time [8] compared with traditional molecular analysis methods such as polymerase chain reaction (PCR). More importantly, the LAMP reaction can be performed under isothermal conditions and the result is visible to the naked eye under a 365 nm ultraviolet (UV) lamp, without the need for electrophoresis, which makes it a very convenient step for a downstream bioanalytical technique [9]. Although bioanalysis integrated with LAMP has been used for the detection of microorganisms, LAMP analysis of airborne samples has rarely been referred to.

Microfluidic-based techniques have recently drawn lots of attention because of their low reagent consumption, short analysis time and environmentally friendly process which can be used for developing portable biosensors [6]. Pathogen analysis using microfluidic chips has been frequently reported, but few studies have involved the airborne analysis of *S. aureus*. A microfluidic system with a staggered herringbone mixer (SHM) structure that is capable of rapid and efficient capture and enrichment of airborne bacteria was established previously in our lab [10]. Based on a similar technique, we report in this study the rapid capture and enrichment of airborne *S. aureus*, followed by direct LAMP analysis. Compared to the conventional Anderson sampler that has a collection and analysis time of several hours and days, respectively, the chip described in this study significantly decreased the collection (3 h and 30 min) and analysis time (40 min), and the device was much simpler and more portable. Further, for hospital airborne infection prevention (HAIP) purposes, we were the first to evaluate an airborne pathogen in a hospital environment using the microfluidic technique.

2. Materials and Methods

2.1. Bacteria and Reagents

Staphylococcus aureus was isolated from clinical samples taken from Neweast Hospital (Shanghai, China) and cultured in Luria-Bertani (LB) medium at 37 °C.

2.2. Chip Fabrication

The double-layer microfluidic chip was fabricated following the standard soft lithography described previously [11,12]. Two pieces of the silicon molds, including the upper fluidic layer and the bottom staggered herringbone layer, were constructed from 20 μm high SU-8 2025 photoresist (Microchem, Westborough, MA, USA). The staggered herringbone mixer (SHM) structure was designed based on previous work [10]. Two polydimethylsiloxane (PDMS) layers were bonded through baking at 80 °C for 12 h to give a radial chip with 18 s-shaped airborne bacteria capture channels. Access holes of 1.5 mm diameter were drilled along the edge of the round chip of each channel to be used as inlet. Meanwhile, a 3.5 mm diameter hole was drilled in the center of the chip for air flow and bacteria-capture outlet, connecting the 18 airborne bacteria-capture units.

2.3. Airborne Staphylococcus aureus Capture and Enrichment

An overnight culture of *S. aureus* suspension was diluted to different concentrations to generate a bioaerosol using an aerosol generator in a 125 L cube tank referred to in our previous studies [10,13]. The chip was placed in the tank and connected to a pump to facilitate the airborne bacteria capture and enrichment.

For the limit of detection (LOD) evaluation, two chips were used in a parallel manner in the experiment. One of the chips was for collecting and counting the captured airborne bacteria, whereas the other one was used for bacterial capture and analysis. An LB culture dish was placed on the bottom of the tank as a parallel control. The pump process has been referred to in our previous work [10,13], but the chip vacuum time being extended to 3 h and 30 min. Following the process of enrichment, the counting chip was washed with ddH$_2$O to flush the SHM channels. The washed bacterial cells were then collected with a pipette and transferred to a 1.5 mL tube for counting using the dilution-plate counting method [10]. As for the capture chip, it was washed with 0.5 μL lysis buffer (DEAOU Biotechnology,

Dalian, China) per channel in the same manner as for the counting chip, with the difference being that the collected suspension was maintained at room temperature for 30 min to allow lysis of bacterial cells to occur. The cell lysate was then used for direct LAMP analysis without any purification process.

2.4. LAMP Reaction System for Nuc Gene Detection

The lysed bacterial suspension was mixed with LAMP reagent (DEAOU Co., Dalian, China) consisting of 0.8 µM each of the inner primer (FIP and BIP), 0.4 µM each of the loop primer (LF and BF), 0.2 µM each of outer primer (F3 and B3), 8U *Bst* DNA Polymerase and 12.5 µL Reaction Mix provided in the kit. The species-specific primer sequences (a total of four primers) of the *nuc* gene (Gene Accession no. V01281) were as described in our previous work [14], and synthesized by Invitrogen (Shanghai, China). LAMP amplification was performed for 40 min in a 65 °C water bath, followed by fluorescence detection under UV excitation at 365 nm.

2.5. Clinical Airborne Staphylococcus aureus Analysis

Clinical airborne samples were obtained from six different settings in Shandong Hospital, including the intensive care unit (ICU), surgery room, emergency room, surgical ward, outpatient service hall and doctor's office. The radial chip was placed on a well-ventilated site for airborne sample capture. The vacuum time was set as 3 h and 30 min, followed by washing of the channels and LAMP analysis.

3. Results and Discussion

3.1. Chip Design

The microfluidic chip with 18 SHM channels is shown in Figure 1. The chip had a diameter of about 7 cm, and each of the channels had a height of 40 µm and a width of 600 µm. To facilitate the capture and enrichment of airborne bacteria, the 18 channels within the chip in the bottom layer contained the SHM structure for capturing the airborne bacteria, while the upper layer contained 18 flow channels at the corresponding position. The chip was assembled from the SHM layer and flow layer, both fabricated from polydimethylsiloxane (PDMS). Eighteen access holes along the edge of the round chip were used as airflow inlets. The central hole was used for air flow and bacteria-capture outlet.

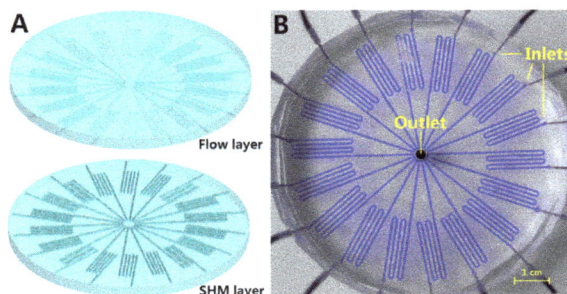

Figure 1. Schematic illustration of the radial airborne bacteria capture chip. (**A**) Schematic illustration of the assembly of the microfluidic chip; (**B**) Top view of the radial chip, showing 18 inlets along the outer edge of each channel for air inlet and a 3.5 mm diameter hole in the center for air and a washed solution outlet. The channels were loaded with blue dye to show the clear structure.

To facilitate the capture of airborne bacteria, a vacuum was connected to the center of the radial chip via a 3.5-mm-diameter stainless steel tube, drawing the bacterial aerosol into the channels from the holes along the outer edge of the chip under the negative force created by the vacuum. By breaking the laminar flow to twisted air flow inside the channel, the contact opportunity between the channel wall and the bacteria in the airflow is increased, so the chip can collect airborne bacteria with very high efficiency [7,10]. The number of SHM channels within the chip was increased to 18 so as to increase the total amount of air that would pass through the channels.

3.2. Detection Limit of Staphylococcus aureus in the Microfluidic Chip

We evaluated the utility of the microfluidic method using standard bacterial bioaerosols in a 125 L cube tank. *S. aureus* was chosen because it is frequently associated with hospital airborne infectious diseases. The sensitivity of the microfluidic chip was validated by *S. aureus* bioaerosols using overnight cultured bacteria ranging from 1.92×10^2 to 1.92×10^{-1} CFU/mL. The plate sedimentation method was also performed in parallel for comparison with the chip method.

The airborne *S. aureus* in the bioaerosol was successfully captured by using the microfluidic chips containing the SHM channels. Figure 2 compares the number of *S. aureus* that was collected by the radial chip and the plate sedimentation method. Each experiment was performed three times and the average number was used.

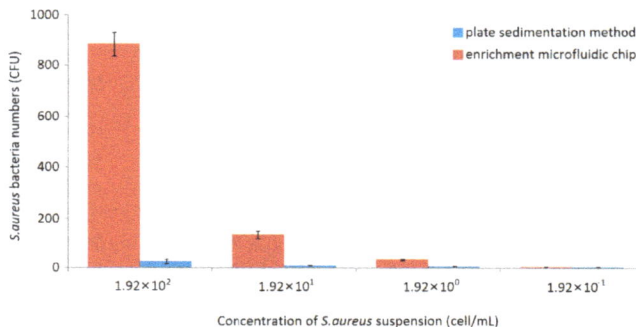

Figure 2. *Staphylococcus (S.) aureus* cells collected by the plate sedimentation method or captured by the microfluidic chip.

Compared to our previous data [5], the extent of the vacuum time significantly increased the ratio of the number of chip-captured bacteria to the number of plate-collected bacteria. The reason might be the increased number of the capture channels, which greatly increased the air flow passing through the SHM structures, thus capturing more bacteria within the chip.

When the concentration of the *S. aureus* suspension was 10^2 cell/mL, approximately 885 cells (average number) were captured by the chip, which was about 42 times higher than that collected by the plate sedimentation method (21 cells). When the bacterial concentration was decreased to 10 cell/mL, 142 cells were captured by the chip compared to about five cells collected by the plate sedimentation method. Further reduction of the *S. aureus* concentration to 1 cell/mL, although a very low cell concentration, still resulted in 27 cells being captured by the chip, compared to one cell collected by the plate sedimentation method. When the bacterial suspension was used at a concentration of 10^{-1} cell/mL, only about one cell was captured both by the chip method and the plate sedimentation method.

The chip-captured bacteria were washed and used as samples for direct LAMP analysis. The *nuc* gene was chosen for the LAMP analysis, because the gene encodes a thermostable nuclease that is found only in *S. aureus*, and thus can be used as a specific target for the detection of *S. aureus* [5,15]. Positive LAMP results were obtained when the numbers of chip-captured bacteria were equivalent to approximately 885, 142 and 27 cells (Figure 3). Further reduction of the collected cells to about one cell did not yield any detectable signal. Therefore, the detection limit for the system was about 27 CFU for *S. aureus*.

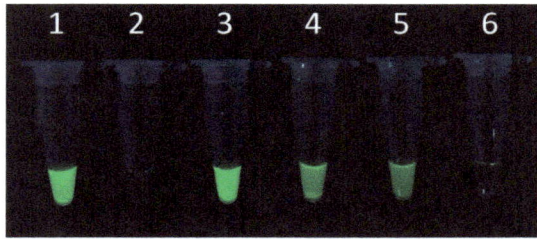

Figure 3. Loop-mediated isothermal amplification (LAMP) analysis of the system sensitivity using diluted *S. aureus*. (**1**) *S. aureus* DNA; (**2**) ddH$_2$O; (**3**) LAMP product of about 885 cells collected by the chip; (**4**) LAMP product of about 142 cells; (**5**) LAMP product of about 27 cells; (**6**) LAMP product of about one cell.

3.3. Clinical Airborne Staphylococcus aureus Analysis

Six different wards in Shandong Hospital were chosen for airborne sample collections. The collected samples were directly subjected into LAMP assay without a conventional DNA extraction step. As shown in Figure 4, the LAMP results for the samples collected from the hospital were negative, indicating that the number of bacteria collected was lower than our LOD (about 27 cells).

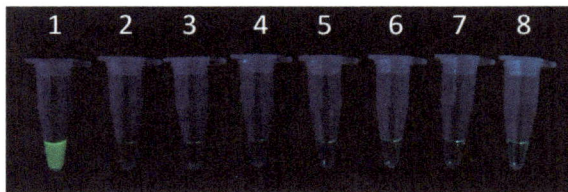

Figure 4. LAMP analysis of airborne bacteria collected in Shandong Hospital. (**1**) *S. aureus* DNA; (**2**) ddH$_2$O; (**3**) Airborne bacteria samples from intensive care unit (ICU); (**4**) Surgery room; (**5**) Emergency room; (**6**) Surgical ward; (**7**) Outpatient service hall; (**8**) Doctor's office.

This is the first report to describe the utilization of a microfluidic chip for airborne bacteria analysis in the hospital. Although traditional techniques for the detection of airborne bacteria have been used since 1881 [7], there is still a gap between collecting the bacterial sample and direct diagnosis, and the reason is that the bacterial concentration of the collected and recovered samples is too low for direct bioanalysis, such as PCR or immunoanalysis. Current sampling techniques could not solve the problem of bacterial concentration, while the culturing step needed for the identification of the microorganisms usually takes days to complete, which is not ideal for the early diagnosis of diseases.

Furthermore, there is still no technique capable of accurately evaluating the distribution of pathogens in the air, which is a critical piece of data for addressing the concern of hospital infection as it enables the indoor air quality in a hospital to be assessed. Based on the published data using an Anderson sampler, the airborne bacterial concentration is approximately 1×10^3 to 6×10^3 CFU/m^3 in an ordinary hospital [16]. The total number of viable cells was greatly underestimated, considering that some of the bacteria might have been killed by the strong impact at which the cells strike the culture plate during the Anderson sampling process. The collected dead bacteria that were distributed in the air would not grow on a culture plate. Besides, many pathogen species are in a viable but non-culturable state (VBNC) in the natural environment [4]. Thus, the number of bacteria that would finally grow on the culture plate could not really account for the actual number of bacteria originally present in the air. Although the Anderson sampling method is limited as discussed above, there has been no other technique that can be used in parallel to verify the amount of bacteria in the air.

The microfluidic chip that we presented in this paper could capture both living and dead bacteria, as well as DNA fragments suspended in the air. During the wash step, the bacteria along with the DNA fragments were all mixed in the solution. Thus, the microfluidic chip can collect a huge number of bacteria within a couple of microliters of aqueous medium, so that the bacterial concentration would be high enough for direct bioanalysis. By obtaining samples of airborne bacteria with a high concentration, molecular bioanalysis can be carried out as soon as the sampling process is completed, and this was considered as a great progress in the field of the rapid analysis of airborne pathogens in the hospital.

However, having an effective amount of air to pass through the channels is still the bottleneck of the microfluidic chip technique. In our study, to increase the amount of air passing through the chip channels, the air flow was set at 79.2 mL/min, and this was almost the highest air flow (mL/min) that could be obtained for a normal-sized microfluidic chip, given the necessary channel length should be longer than 17.4 cm to ensure no airborne cells would leak from the chip [10]. After 3 h and 30 min of the vacuum process, the amount of air flow that passed through the chip channels was estimated to be about 16.63 L, a very large amount of air for a normal-sized microfluidic chip. In order to further increase the air flow, a longer vacuum time and more pumps are needed.

The negative result obtained for the hospital airborne sample analysis showed that the number of captured *S. aureus* was lower than the LOD of the microfluidic chip (27 cells). Nevertheless, the microfluidic chip showed that the concentration of *S. aureus* in the hospital air was lower than 1.6 cells per liter of air (27 cells divided by 16.63 L), a result that was not possible to obtain with any existing traditional techniques. On the basis of this, our work may provide a potential platform for airborne pathogen sampling and bioanalysis, especially in the prevention of hospital airborne pathogen infections.

4. Conclusions

Few studies have been conducted to evaluate airborne *S. aureus*. We have successfully demonstrated a radial airborne bacteria capture and enrichment chip for fast assessment of airborne *S. aureus*. The system could perform airborne *S. aureus* capture, enrichment, and rapid LAMP analysis. The bacteria were collected by the radial chip with an SHM structure. They were washed and then directly subjected to LAMP analysis without the need for DNA purification. Standard *S. aureus* bioaerosol was used to validate the system. A detection limit down to approximately 27 cells was achieved for *S. aureus*. The presented microfluidic technique for rapid capture and analysis of airborne *S. aureus* could have a huge potential in disease control and clinical applications, making it a promising for future point-of-care tests in the field, especially given its unique properties compared to traditional techniques.

Acknowledgments: This work was funded by NFSC (81501833) and the Natural Science Foundation of Liaoning Province (2015020574).

Author Contributions: X.J. and G.S. conceived and designed the experiments; X.J., Y.L. and Q.L. performed the experiments; X.J., Y.L. and W.J. analyzed the data; K.Q. gave scientific support and conceptual advices; X.J. and Q.L. wrote the paper. All authors discussed the results and implications and commented on the manuscript.

Conflicts of Interest: The authors declare no conflict of interest.

References

1.   Aires, S.M.; Lencastre, H. Bridges from hospitals to the laboratory: Genetic portraits of methicillin-resistant *Staphylococcus aureus* clones. *FEMS Immunol. Med. Microbiol.* **2004**, *40*, 101–111. [CrossRef]
2.   Rodriguez, E.A.; Correa, M.M.; Ospina, S.; Atehortua, S.L.; Jimenez, J.N. Differences in Epidemiological and Molecular Characteristics of Nasal Colonization with *Staphylococcus aureus* (MSSA-MRSA) in Children from a University Hospital and Day Care Centers. *PLoS ONE* **2014**, *9*, e101417. [CrossRef] [PubMed]

3. Glasgow, H.B.; Burkholder, J.M.; Schmechel, D.E.; Tester, P.A.; Rublee, P.A. Insidious effects of a toxic estuarine dinoflagellate on fish survival and human health. *Toxicol. Environ. Health* **1995**, *46*, 501–522. [CrossRef] [PubMed]
4. Heidelberg, J.F.; Eisen, J.A.; Nelson, W.C.; Clayton, R.A.; Gwinn, M.L.; Dodson, R.J.; Haft, D.H.; Hickey, E.K.; Peterson, J.D.; Umayam, L.; et al. DNA sequence of both chromosomes of the cholera pathogen *Vibrio cholerae*. *Nature* **2000**, *406*, 477–483. [PubMed]
5. Jiang, X.R.; Jing, W.W.; Zheng, L.L.; Liu, S.X.; Wu, W.J.; Sui, G.D. A Continuous-flow high-throughput microfluidic device for airborne bacteria PCR detection. *Lab Chip* **2014**, *14*, 671–676. [CrossRef] [PubMed]
6. Thorsen, T.; Maerkl, S.J.; Quake, S.R. Microfluidic large-scale integration. *Science* **2002**, *298*, 580–584. [CrossRef] [PubMed]
7. Sui, G.D.; Cheng, X.J. Microfluidics for detection of airborne pathogens: What challenges remain. *Bioanalysis* **2014**, *6*, 5–7. [CrossRef] [PubMed]
8. Tomita, N.; Mori, Y.; Kanda, H.; Notomi, T. Loop-mediated isothermal amplification (LAMP) of gene sequences and simple visual detection of products. *Nat. Protoc.* **2008**, *3*, 877–882. [CrossRef] [PubMed]
9. Mori, Y.; Notomi, T. Loop-mediated isothermal amplification (LAMP): A rapid, accurate, and cost-effective diagnostic method for infectious diseases. *J. Infect. Chemother.* **2009**, *15*, 62–69. [CrossRef] [PubMed]
10. Jing, W.W.; Zhao, W.; Liu, S.X.; Li, L.; Tsai, C.T.; Fan, X.Y.; Wu, W.J.; Li, J.Y.; Yang, X.; Sui, G.D. Microfluidic Device for efficient airborne bacteria capture and enrichment. *Anal. Chem.* **2013**, *85*, 5255–5262. [CrossRef] [PubMed]
11. Whitesides, G.M.; Ostuni, E.; Takayama, S.; Jiang, X.Y.; Ingber, D.E. Soft lithography in biology and biochemistry. *Annu. Rev. Biomed. Eng.* **2001**, *3*, 335–373. [CrossRef] [PubMed]
12. Liu, C. Rapid fabrication of microfluidic chip with three-dimensional structures using natural lotus leaf template. *Microfluid. Nanofluid.* **2010**, *9*, 923–931. [CrossRef]
13. Jing, W.W.; Jiang, X.R.; Zhao, W.; Liu, S.X.; Cheng, X.J.; Sui, G.D. Microfluidic platform for direct capture and analysis of airborne *Mycobacterium tuberculosis*. *Anal. Chem.* **2014**, *86*, 5815–5821. [CrossRef] [PubMed]
14. Yuan, H.; Liu, Y.C.; Jiang, X.R.; Xu, S.C.; Sui, G.D. Microfluidic chip for rapid analysis of cerebrospinal fluid infected with *Staphylococcus aureus*. *Anal. Methods* **2014**, *6*, 2015–2019. [CrossRef]
15. Brakstad, O.G.; Aasbakk, K.; Maeland, J.A. Detection of *Staphylococcus aureus* by Polymerase Chain Reaction Amplification of the *nuc* gene. *J. Clin. Microbiol.* **1992**, *30*, 1654–1660. [PubMed]
16. Lin, A.H.; Zhang, S.H. Analysis of airborne bacteria concentration and distribution in hospital wards. *Hubei J. Prev. Med.* **2002**, *4*, 29–36.

![micromachines logo] *micromachines*

MDPI

Article

A Reconfigurable Microfluidics Platform for Microparticle Separation and Fluid Mixing

Young Ki Hahn [1], Daehyup Hong [2], Joo H. Kang [3],* and Sungyoung Choi [2],*

[1] Samsung Electronics, 4 Seocho-daero 74-gil, Seocho-gu, Seoul 06620, Korea; hahnv79@gmail.com
[2] Department of Biomedical Engineering, Kyung Hee University, Yongin-si, Gyeonggi-do 17104, Korea; hdh9080@gmail.com
[3] Department of Biomedical Engineering, School of Life Science, Ulsan National Institute of Science and Technology (UNIST), 50 UNIST-gil, Ulsan 44919, Korea
*   Correspondence: jookang@unist.ac.kr (J.H.K.); s.choi@khu.ac.kr (S.C.);
    Tel.: +82-52-217-2595 (J.H.K.); +82-31-201-3496 (S.C.)

Academic Editors: Weihua Li, Hengdong Xi and Say Hwa Tan
Received: 1 July 2016; Accepted: 3 August 2016; Published: 8 August 2016

Abstract: Microfluidics is an engineering tool used to control and manipulate fluid flows, with practical applications for lab-on-a-chip, point-of-care testing, and biological/medical research. However, microfluidic platforms typically lack the ability to create a fluidic duct, having an arbitrary flow path, and to change the path as needed without additional design and fabrication processes. To address this challenge, we present a simple yet effective approach for facile, on-demand reconfiguration of microfluidic channels using flexible polymer tubing. The tubing provides both a well-defined, cross-sectional geometry to allow reliable fluidic operation and excellent flexibility to achieve a high degree of freedom for reconfiguration of flow pathways. We demonstrate that microparticle separation and fluid mixing can be successfully implemented by reconfiguring the shape of the tubing. The tubing is coiled around a 3D-printed barrel to make a spiral microchannel with a constant curvature for inertial separation of microparticles. Multiple knots are also made in the tubing to create a highly tortuous flow path, which induces transverse secondary flows, Dean flows, and, thus, enhances the mixing of fluids. The reconfigurable microfluidics approach, with advantages including low-cost, simplicity, and ease of use, can serve as a promising complement to conventional microfabrication methods, which require complex fabrication processes with expensive equipment and lack a degree of freedom for reconfiguration.

Keywords: reconfigurable microfluidics; inertial microfluidics; microparticle separation; fluid mixing

1. Introduction

Microfluidics is a technology capable of controlling and transferring small quantities of liquids, ranging from nanoliters to microliters, which enables multiple biological assays and high-throughput screening. Particle separation and fluid mixing are important in biochemical and clinical applications for the identification and analysis of specific target molecules or cells. To achieve this, microfluidic channels have been integrated into miniaturized (lab-on-a-chip) and point-of-care testing devices as indispensable components with various advantages, such as ensuring precise control of a fluid and reducing the time and costs associated with routine biological analysis [1]. In particular, microfluidic technologies play a key role in the separation of cells or microparticles and the rapid mixing of fluids, which have been demonstrated with microfabricated, complicated structures, such as a chaotic mixer, serpentine channels, herringbone structures, and a contraction-expansion array channel [2–8].

Inertial microfluidics is a useful technique that exploits a unique physical phenomenon in microchannels, which utilizes the inertia of fluid flows at intermediate ranges of the Reynolds number (*Re*) (1 < *Re* < 100). This technique enables high throughput separation of microparticles and the mixing of fluids without mechanical or electrical assistance. Inertial microfluidics-based separation and mixing have shown unique utility in various applications, such as inertial focusing, ordering, and separation of microparticles and blood cells [9], DNA [10], bacterial cells [11,12], and tumor cells [13]. Most microfluidic devices, including devices for inertial microfluidics, are fabricated by conventional photolithography and soft lithography processes based on polydimethylsiloxane (PDMS) [14]. Although the fabrication process is now well established and widely used, there are several limitations to the process in terms of design flexibility and user-friendliness. First, changing the device application inevitably accompanies the modification of microfluidic devices with new designs and dimensions. For example, microfluidic mixers typically require sudden changes in flow paths to achieve efficient mixing [15], and repetitive microfluidic patterns often have to be formed for sorting microparticles [16]. Microfluidic methods, thus, require adjustment of channel geometries, according to changing target applications, for different sizes of microparticles and cells [17]. Although microfluidic devices fabricated using 3D printing technologies have been recently introduced, with the advantages of rapid prototyping and easy fabrication, these devices have a major limitation in that channel dimensions are restricted by the minimum resolution of a 3D printer [18]. The minimum resolution of 3D printers is typically in the range of tens to hundreds of micrometers, which does not permit microfabrication of fluidic channels smaller than 100 μm [19]. Recently, considerable efforts have been also made to improve the design flexibility of microfluidic platforms using modular microfluidic components [19] and punch-card-based microfluidics [20]. However, materials fabricating the channels used above are rigid, which do not allow us to adapt or change their configurations as needed.

To overcome these limitations, we propose a new approach to reconfigurable microfluidics to achieve facile, on-demand reconfiguration of microfluidic channels using flexible polymer tubing (Figure 1). Because flexible tubing is commonly used in most microfluidics laboratories, it is easy to access tubing components and to obtain a specific geometry of microchannels by configuring or knotting tubing without microfabrication processes. To evaluate the performance of the reconfigurable microfluidics platform, we demonstrated continuous separation of microparticles and the rapid mixing of fluids. First, we separated 15-μm and 25-μm particles with the flexible tubing coiled on a 3D-printed barrel by utilizing differential inertial focusing of particles of different sizes. Second, we confirmed that the mixing efficiency of fluids significantly improves when flowing fluids through a series of knots of flexible tubing. Thus, we successfully carried out particle separation and fluid mixing in the reconfigurable microfluidics platform, which is intended to complement existing fabrication methods for microfluidic channels and also improve the flexibility of microchannels for reconfiguration.

2. Materials and Methods

2.1. Materials

Tygon® tubing with an inner diameter (I.D.) of 190 μm and an outer diameter of 2 mm was purchased from Cole-Parmer Inc. (Vernon Hills, IL, USA). For microparticle separation, 15-μm, 20-μm, and 25-μm particles were obtained from Polysciences Inc. (Warrington, PA, USA). For fluid mixing, 100-nm fluorescent particles, purchased from Life Technologies Corp. (Carlsbad, CA, USA), were used. The particles were re-suspended in 0.1% bovine serum albumin (BSA) solution (Sigma-Aldrich Co., St. Louis, MO, USA) at a concentration of 10^4 to 3×10^4 beads/mL. BSA was used to passivate PDMS surfaces from non-specific adsorption. PDMS for microchannel fabrication was purchased from Dow Corning Inc. (Midland, MI, USA).

2.2. Microfluidic Setup and Analysis

A syringe pump (KD Scientific, Holliston, MA, USA) was used to flow sample solutions at a constant flow rate. Microfluidic behaviors were observed using a charge coupled device (CCD) camera (Nikon, Tokyo, Japan) and a high-speed camera (CASIO, Tokyo, Japan) equipped with a fluorescence microscope (Nikon) (Figure 2a). For analysis of spatial distributions of microparticles, the lateral positions of the microparticles were measured using the image analysis software, ImageJ (National Institutes of Health, Bethesda, MD, USA). After microparticle separation, the samples collected from outlet 1 and outlet 2 were analyzed using a flow cytometer (BD Biosciences, San Jose, CA, USA) (Figure 2b).

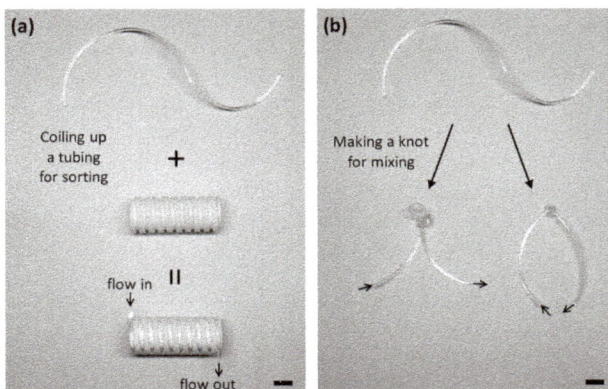

Figure 1. Photographs of reconfiguring flexible tubing for microparticle separation and fluid mixing. (**a**) A spiral microchannel is configured for microparticle separation by coiling the tubing onto a 3D-printed grooved barrel. (**b**) The tubing knots generate abrupt changes in a flow direction to mix laminar flows through the tubing. It shows two different ways of knotting tubing (knot with or without an inner loop) and presents that even one without microfluidic experience can construct various channel geometries by simply making knots of tubing. (scale bar: 1 cm)

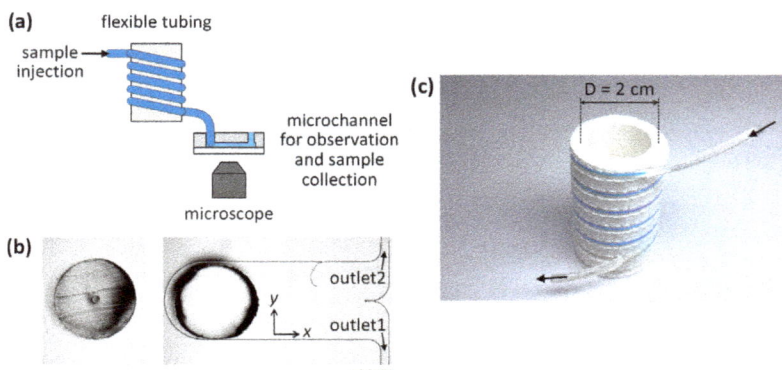

Figure 2. A reconfigurable microfluidics platform for microparticle separation. (**a**) Schematic showing the experimental setup for observation of the spatial distributions of microparticles after passing through the tubing. (**b**) Photographs of (**left**) a cross-section of a Tygon® tubing with an inner diameter of 190 μm and an outer diameter of 2 mm, and (**right**) a microchannel for measurement of the lateral distributions of the microparticles and for the collection of separated microparticles. (**c**) Photograph of the assembled tubing (35 cm in length) onto the grooved barrel that defines a fixed radius (1 cm) of the coiled tubing. (scale bar: 1 mm)

2.3. Fabrication of Microchannels for the Validation of a Reconfigurable Platform

A master mold for microchannels to provide fluidic access for the tubing was fabricated by standard photolithography processes, including spin-coating of photoresist, ultraviolet (UV) exposure through a photomask, development for the removal of the unexposed area of photoresist, and baking for the removal of residual solvents and solidification of the photoresist structures. Then, microfluidic channels were made by PDMS molding processes, including the mixing of PDMS with a curing agent, hardening the mixture for 1 h at 75 °C, punching to form inlet and outlet holes, and irreversible bonding between each PDMS microchannel and a glass slide. The PDMS microchannels presented in this work are intended for use in characterization of particle separation and fluid mixing and are unnecessary when used for practical applications using the reconfigurable microfluidics because other commercial tubing accessories, such as Y-shape connectors, can be simply substituted for the PDMS devices.

2.4. Fabrication of a Grooved Barrel

A grooved barrel was printed using a Mojo 3D printer (Stratasys, Eden Prairie, MN, USA). The barrel has a diameter of 2.2 cm and a spiral groove with a width of 2 mm and a depth of 2 mm, and a pitch of 6 mm (Figure 2c). The barrel is used for geometric guidance, which ensures that the tubing has a fixed radius of curvature (~1 cm) and coiling pitch.

3. Results and Discussion

3.1. Microparticle Separation Using Coiled Flexible Tubing

A reconfigurable microfluidics platform was applied to microparticle separation using the flexible tubing coiled onto the 3D-printed barrel, as shown in Figure 2c. Prior to the separation of microparticles, we performed a computational simulation (COMSOL Multiphysics®, 5.1, COMSOL Inc., Burlington, MA, USA) to predict a secondary flow induced by Dean flow in the coiled tubing, wherein rotating flows in a cross-section of the tubing were induced. The Dean flow was enhanced as Re increased (Figure 3a). Re is defined as $\rho UL/\mu$, where ρ is the density of the fluid, μ is its viscosity, U is its average velocity, and L is a characteristic dimension of a channel cross-section. From the simulation results, we predicted that microparticles flowing through the coiled tubing would laterally circulate, following the Dean flow, and would be confined at a certain position where the two secondary flows bifurcate next to the inner wall surface (Figure 3b, upper left). The stability of maintaining the equilibrium position can be determined by balancing between inertial lift and Dean drag forces. Microparticles smaller than a critical size will keep the flow path following the lateral secondary flows (Dean flow), as opposed to being gathered in the equilibrium position of the coiled tubing (Figure 3b, upper right). To corroborate our prediction, 20-μm and 25-μm fluorescent particles were, respectively, injected into the coiled tubing, and their lateral position in the tubing was observed, as shown in Figure 3b. The obtained images confirmed that 25-μm particles were aligned at the equilibrium lateral position, whereas 20-μm particles were dispersed throughout the tubing. The alignment of the larger microparticles (25 μm in diameter) is attributed to the interplay between inertial lift forces and Dean drag forces. In the circular cross-section of the tubing, microparticles can migrate to the periphery of the tubing by shear-induced and wall-induced lift forces at high Re, which is known as the "tubular pinch" effect [21]. The coiled tubing configuration generates centrifugal forces directed outward that induce counter-rotating vortices (Dean flow). As a result of the interplay between the inertial lift and Dean drag forces, the equilibrium position of the larger microparticles can be reduced in the position right next to the inner wall of the tubing. While straight microchannels typically form multiple equilibrium positions for particles influenced by inertial lift forces at high Re, additional inertial forces, such as centrifugal forces and Dean drag forces, in a curved microchannel are superposed with the lift forces and can reduce particle focusing into a single stream. Because the magnitudes of the inertial lift forces and Dean drag forces highly depend on the diameter of microparticles, microparticles of different diameters can occupy different lateral positions. Due to the smaller diameter of the microparticles (20 μm in diameter), the microparticles tend to position either above or below the center line, which results in the continuous circulation of the microparticles, following the secondary lateral flows in the coiled tubing at high *Re* over 22. We note that the polystyrene microparticles ($\rho = 1.04$ g/cm^3) used in this experiment have a density that is slightly higher than the surrounding medium. Their sedimentation velocity due to the gravitational force is less than 1 μm/s while their residence time in the tubing is typically less than a second. Thus, the behavior of particles is determined, not by the gravitational force, but by inertial forces.

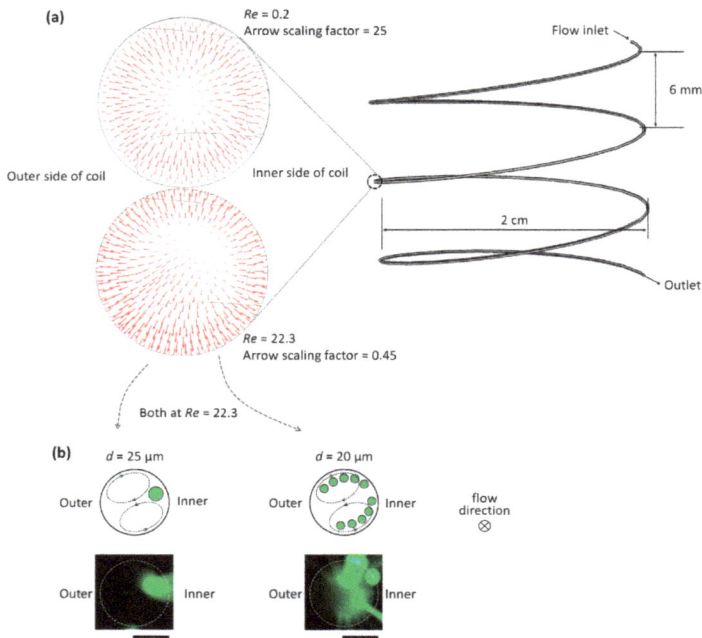

Figure 3. Distributions of secondary flows and the focused microparticles in a cross-section of the coiled tubing. (**a**) A computational simulation result predicts lateral flows induced by Dean flow in the coiled tubing, which results in bifurcating circulation of microparticles smaller than a critical diameter or focusing of microparticles larger than a critical diameter. Each arrow indicates a lateral flow velocity with a scale factor of 0.45 (at high *Re* of ~22) and 25 (at low *Re* of 0.2), respectively, supporting that the lateral flow is enhanced as *Re* increases. (**b**) A hypothetical diagram of the equilibrium positions of 25-μm and 20-μm particles in a cross-section of the tubing (**top**) and the corresponding experimental results (**bottom**). The photographs were taken at the end of the tubing plugged in the inlet of the microchannel for observation of microparticle distribution. Inner and outer denote the inner and outer sides of the spiral channel, respectively. The applied flow rate was 200 μL/min. (scale bars: 100 μm)

To quantitatively validate the separation capability, we connected the tubing with a microfluidic channel (2 mm in width) and measured the lateral positions of microparticles in the microfluidic channel at different flow rates ranging from 50 to 300 μL/min, which correspond to *Re* values of 5.5–33. Inertial migration of particles was first demonstrated by Segré and Silberberg in macroscale circular tubing [22]. In recent years, inertial fluidic behaviors in intermediate flow rates ($1 < Re < 100$) and microscale channels (from tens of microns to hundreds of microns in diameter) have been extensively explored in which deterministic and controllable motions of particles and fluids were observed. We also observed similar behaviors of particle migration at the intermediate *Re* regime. The equilibrium position was aligned to the right sidewall of the microchannel. The microparticles of three different diameters (15, 20, and 25 μm) were respectively flowed through the coiled tubing. As predicted, 25-μm particles coming out from the outlet of the microfluidic channels were aligned near the inner sidewall of the tubing and exited along the right sidewall of the microchannel, and the alignment of 25-μm particles was improved due to enhanced inertial lift forces as the flow rate increased to 300 μL/min (Figure 4, left-row panels). Because microparticles smaller than the critical size are more affected by the secondary flow, the particles were not confined at a certain lateral position of the tubing (Figure 4, right-row panels). The ratio of the particle diameter (a_p) to the channel diameter (D) is a critical factor for inertial focusing in a straight pipe and needs to be $a_p/D \geqslant 0.07$ [23]. For $D = 190$ μm, a_p needs to be larger than 13.3 μm to satisfy the criterion. In a curved microchannel, the secondary flow, Dean flow can perturb flowing particles and affect their equilibrium position. The ratio of inertial lift and Dean drag ($R_f = 2a_p{}^2R/D^3$) considerably increase with a_p [24] and so relatively strong Dean forces for 15-μm particles likely result in disruption of particle focusing, where R is a radius of curvature of the coiled tubing. In a curved microchannel, inertial particle focusing depends on the balance between inertial lift and Dean drag forces. The focusing behavior of 20 μm-particles is significantly affected by a flow rate (Figure 4) and this can be explained by increased Dean drag forces which can destabilize a preferred focusing position at high *Re* over 22.

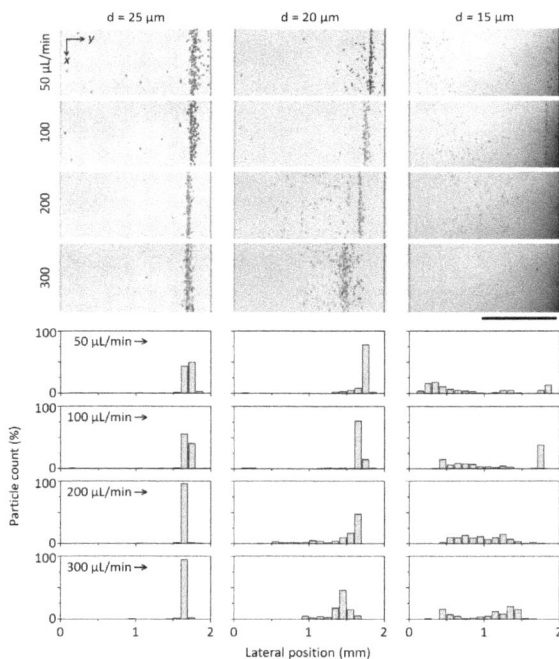

Figure 4. Stacked images and distribution plots show the distributions of microparticles according to the particle diameter and the flow rate. The flow direction is along the *x*-axis. (scale bar: 1 mm)

In addition to measuring the lateral positions of microparticles, we demonstrated the separation and collection of microparticles using the microchannel with two outlets connected to the coiled tubing (Figure 2b). The populations of microparticles collected from outlet 1 and outlet 2, respectively, were analyzed by a flow cytometry. The flow cytometry results support that two different sizes of microparticles (15 and 25 μm in diameter) mixed in a solution can be separated and significantly enriched when flowing through the coiled tubing with a throughput of 4.6×10^3 microparticles/min. A solution containing the two different sizes of microparticles (15 μm and 25 μm with an initial ratio of 38.6% and 61.4%, respectively) was injected into the coiled tubing at a flow rate of 200 μL/min. The purity of 15-μm particles in the sample collected from outlet 1 was significantly improved to $98.5 \pm 2.7\%$, whereas the purity of 25-μm particles in the sample from outlet 2 was $75.6 \pm 4.4\%$ ($n = 3$). This is most likely due to the continuous circulation of smaller microparticles (15 μm) by the balance between inertial lift and Dean drag forces. Additionally, the recoveries of 15-μm and 25-μm microparticles were determined to be $50.9 \pm 5.3\%$ in outlet 1 and $99.5 \pm 0.9\%$ in outlet 2 (Figure 5) ($n = 3$). Recovery is defined as the number of sorted target particles divided by the total number of the particles injected.

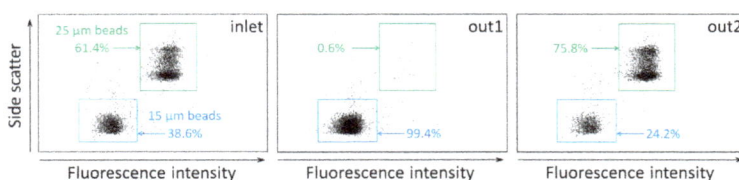

Figure 5. Flow cytometry results showing the separation of 15-μm and 25-μm particles. The purity of separated microparticles was analyzed using a flow cytometer. Purity is defined as the percentage of the number of target microparticles in each sorted population.

3.2. Mixing of Laminar Flows Using Knots of Tubing

We then demonstrated that a reconfigurable microfluidics platform could be applied to achieve high mixing efficiency with a simple configuration of tubing with a series of knots (Figure 6). Mixing efficiency was defined as a standard deviation, σ, of the fluorescence distribution. A value of 0 corresponds to complete mixing and 0.5 to complete segregation. For the mixing of fluids, multiple knots (trials with 3 and 10 knots) were made within a tubing (Figure 6c). We did not observe any structural changes in the knots during experiments. A solution containing fluorescent nanoparticles (100 nm in diameter) and the buffer solution were injected through a Y-shaped microchannel with two inlets, and the intersection where the two fluids met was directly connected to the tubing inlet without premixing in the microchannel (Figure 6b). Because the diffusion coefficient (D_{diff}) of nanoparticles (100 nm, ~10^{-8} cm^2/s) is about three order magnitudes lower than small fluorescent molecules (FITC, $D_{diff} = 0.64 \times 10^{-5}$ cm^2/s), the Peclet number (Pe) of the mass transport across the two laminar flows driven by diffusion is not effective. Thus, the mixing of the nanoparticles in laminar flows requires a certain mixing component to achieve complete homogenization in fluids. Efficient mixing can be achieved by knots of tubing because the direction of the secondary lateral flow (Dean flow) is irregular and dramatically changes at every turn of knotted tubing, which causes non-uniform lateral circulatory flows, and subsequently causes effective mixing in the knotted tubing. In addition, mixing efficiency is predicted to be improved when a flow rate increases because the Dean number is proportional to the Reynolds number [25]. At low Re (1.1) herein, efficient mixing was not achieved, even after passing through 10 serial knots of tubing. However, when we increased Re to 22.3, the complete mixing of 100-nm fluorescent particles was achieved after flowing through 10 serial knots of tubing (Figure 6d). The standard deviation (σ) of the fluorescence intensity across the microchannel is plotted in Figure 6e, revealing that the number of knots of tubing is more critical to achieve complete mixing of fluids than influences of Re. This is because the mixing efficiency affected by Re plateaued above a certain flow rate. The cross-sectional deformation of the knotted tubing may occur and affect the mixing performance; however, the tubing was adequately (not too much tightly) knotted to minimize such deformation. Thus, the mixing of fluids is mainly affected by abrupt changes in a flow direction rather than subtle changes in a channel cross-section. These results support that the proposed reconfigurable microfluidics platform can simply replace conventional microfluidic components and facilitate facile integration of the components as shown in Figure 7.

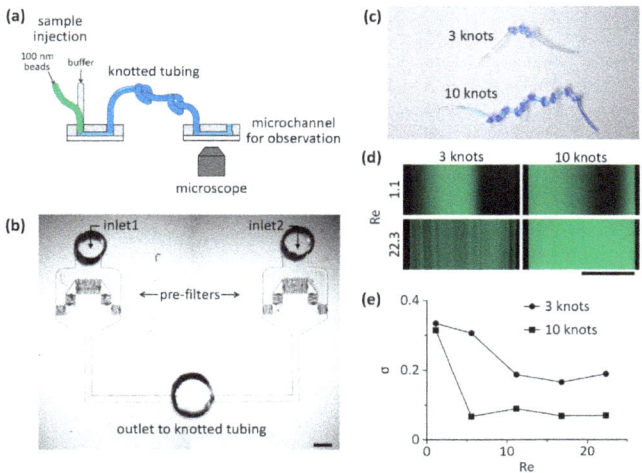

Figure 6. Mixing of fluids using knots of flexible tubing. (**a**) Schematic showing the experimental setup for observation of the spatial distribution of fluorescent nanoparticles after mixing. (**b**) Photograph of a microchannel for the infusion of two types of fluids to assess the mixing efficiency (one with 100 nm fluorescent nanoparticles and the other one without the particles). The outlet is directly connected to the flexible tubing with serial knots without premixing. (**c**) Photograph of knots in the tubing, which induce abrupt changes in a flow direction. (**d**) Fluorescence images after the mixing of the buffer solution (no fluorescent nanoparticles) and the solution containing 100-nm green-fluorescent nanoparticles. (**e**) The degree of mixing efficiency can be obtained from the standard deviation of the fluorescent intensity across the channel. The 100-nm fluorescent particles in laminar flows were completely mixed when flowing through 10 serial knots of tubing, even at low Reynolds numbers, in comparison to the tubing with 3 knots. (scale bars: 1 mm)

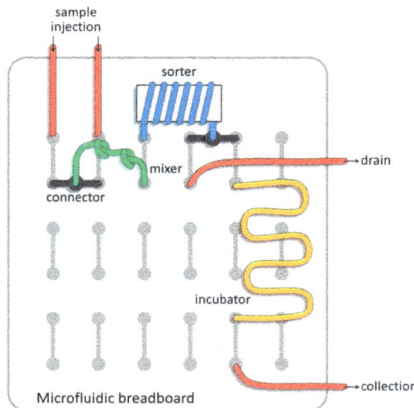

Figure 7. Schematic showing the potential utility of the reconfigurable microfluidics approach for a microfluidic breadboard that enables rapid testing of a prototype microfluidic circuit with modular microfluidic components.

4. Conclusions and Perspectives

We demonstrated that size-based separation of microparticles and mixing of laminar flows at low *Re* can be achieved by coiled and knotted flexible tubing. We constructed coiled tubing using a 3D-printed barrel that fixes tubing to a coiled configuration. Microparticles were laterally circulated, following the secondary flow (Dean flow) generated by the curvature of the coiled tubing, and the lateral positions of the microparticles could be confined at a certain position when the particle size was greater than 20 μm. The enrichment of microparticles was successfully demonstrated using a mixture of microparticles with diameters of 15 μm and 25 μm. The purity of 15-μm particles in outlet 1 was 98.5 ± 2.7% while the recovery rate of 25-μm particles from outlet 2 was 99.5 ± 0.9%. Because the focusing and separation of microparticles in the coiled tubing is attributed to the balance between the secondary lateral flow (Dean flow) and inertial lift forces, the critical size for separation can be determined by the relative difference between the particle diameter, the inner diameter of the coiled tubing, and the Dean number of the flow. To evaluate the versatility of the reconfiguration approach, we uncoiled the tubing and tied it to make serial knots to induce abrupt changes in a lateral flow direction and improve the mixing of fluids. The knots of tubing were made to form repeating tortuous flow paths that yield efficient mixing of laminar flows even at low *Re* (5.6–22.3). These knots of tubing induced the secondary circulating flows in irregular directions, resulting in local agitation and effective mixing of laminar flows in the tubing.

The strong point of the proposed reconfigurable microfluidics platform using flexible tubing is that it allows one even without microfabrication experiences to construct microfluidic channels for particle separation and fluid mixing, and that the shape of flexible tubing can be easily adapted to a certain configuration when required. In this study, we demonstrated the utility of the reconfigurable flexible tubing-based microfluidics platform as a substitution for the existing microfluidic components for microparticle separation and fluid mixing. More importantly, our proposed approach can be applied to the research field of microfluidic chemical synthesis, which requires proficiency in mixing and separating product particles, as well as robust reliability in enduring harsh chemical conditions (nearly impossible to achieve with conventional PDMS-based microfluidic devices due to the inherent characteristics of PDMS) [26]. The construction of a reconfigurable microfluidics system using chemical-resistant flexible tubing, such as Teflon tubing, would make the design and fabrication of a microfluidic system much more straightforward, in which a series of chemical mixing, synthesis, and size-based separation can simultaneously take place, as shown in Figure 7. The proposed platform will easily extend its applicability to chemical screening and synthesis if a microfluidic breadboard is printed in thermoplastic materials with chemical resistance using a 3D printer [19].

Acknowledgments: This research was supported by the Pioneer Research Center Program through the National Research Foundation (NRF) of Korea, funded by the Ministry of Science, ICT & Future Planning (MSIP) (2013M3C1A3064777), an NRF grant funded by the Korea government (MSIP) (2014R1A2A2A09052449), a NRF grant funded by the Korea government (2015R1C1A1A01053990), and the research fund (Project Number: 1.160053.01) of UNIST (Ulsan National Institute of Science and Technology), Korea.

Author Contributions: S.C. and J.H.K. conceptualized and directed the research project. S.C., Y.K.H., D.H., and J.H.K. prepared the manuscript. Y.K.H. and D.H. performed the experiments and data analysis with S.C. and J.H.K. All authors discussed the results and commented on the manuscript.

Conflicts of Interest: The authors declare no conflict of interest.

References

1. Mitchell, P. Microfluidics-downsizing large-scale biology. *Nat. Biotechnol.* **2001**, *19*, 717–721. [CrossRef] [PubMed]
2. Stroock, A.D.; Dertinger, S.; Ajdari, A.; Mezić, I.; Stone, H.A.; Whitesides, G.M. Chaotic mixer for microchannels. *Science* **2002**, *295*, 647–651. [CrossRef] [PubMed]
3. Kim, D.S.; Lee, S.H.; Kwon, T.H.; Ahn, C.H. A serpentine laminating micromixer combining splitting/recombination and advection. *Lab Chip* **2005**, *5*, 739–747. [CrossRef] [PubMed]

4. Choi, S.; Karp, J.M.; Karnik, R. Cell sorting by deterministic cell rolling. *Lab Chip* **2012**, *12*, 1427–1430. [CrossRef] [PubMed]
5. Yamada, M.; Seki, M. Hydrodynamic filtration for on-chip particle concentration and classification utilizing microfluidics. *Lab Chip* **2005**, *5*, 1233–1239. [CrossRef] [PubMed]
6. Mazutis, L.; Gilbert, J.; Ung, W.L.; Weitz, D.A.; Griffiths, A.D.; Heyman, J.A. Single-cell analysis and sorting using droplet-based microfluidics. *Nat. Protoc.* **2013**, *8*, 870–891. [CrossRef] [PubMed]
7. Dykes, J.; Lenshof, A.; Åstrand-Grundström, I.-B.; Laurell, T.; Scheding, S. Efficient removal of platelets from peripheral blood progenitor cell products using a novel micro-chip based acoustophoretic platform. *PLoS ONE* **2011**, *6*, e23074. [CrossRef] [PubMed]
8. Lee, M.G.; Choi, S.; Park, J.-K. Rapid laminating mixer using a contraction-expansion array microchannel. *Appl. Phys. Lett.* **2009**, *95*, 051902. [CrossRef]
9. Di Carlo, D.; Irimia, D.; Tompkins, R.G.; Toner, M. Continuous inertial focusing, ordering, and separation of particles in microchannels. *Proc. Natl. Acad. Sci. USA* **2007**, *104*, 18892–18897. [CrossRef] [PubMed]
10. Choi, S.; Song, S.; Choi, C.; Park, J.-K. Hydrophoretic sorting of micrometer and submicrometer particles using anisotropic microfluidic obstacles. *Anal. Chem.* **2009**, *81*, 50–55. [CrossRef] [PubMed]
11. Hou, H.W.; Bhattacharyya, R.P.; Hung, D.T.; Han, J. Direct detection and drug-resistance profiling of bacteremias using inertial microfluidics. *Lab Chip* **2015**, *15*, 2297–2307. [CrossRef] [PubMed]
12. Wu, Z.; Willing, B.; Bjerketorp, J.; Jansson, J.K.; Hjort, K. Soft inertial microfluidics for high throughput separation of bacteria from human blood cells. *Lab Chip* **2009**, *9*, 1193–1199. [CrossRef] [PubMed]
13. Ozkumur, E.; Shah, A.M.; Ciciliano, J.C.; Emmink, B.L.; Miyamoto, D.T.; Brachtel, E.; Yu, M.; Chen, P.I.; Morgan, B.; Trautwein, J.; et al. Inertial focusing for tumor antigen-dependent and -independent sorting of rare circulating tumor cells. *Sci. Transl. Med.* **2013**, *5*, 179ra47. [CrossRef] [PubMed]
14. McDonald, J.C.; Duffy, D.C.; Anderson, J.R.; Chiu, D.T.; Wu, H.; Schueller, O.J.; Whitesides, G.M. Fabrication of microfluidic systems in poly(dimethylsiloxane). *Electrophoresis* **2000**, *21*, 27–40. [CrossRef]
15. Liu, R.H.; Stremler, M.A.; Sharp, K.V.; Olsen, M.G.; Santiago, J.G.; Adrian, R.J.; Aref, H.; Beebe, D.J. Passive mixing in a three-dimensional serpentine microchannel. *J. Microelectromech. Syst.* **2000**, *9*, 190–197. [CrossRef]
16. Huang, L.R.; Cox, E.C.; Austin, R.H.; Sturm, J.C. Continuous particle separation through deterministic lateral displacement. *Science* **2004**, *304*, 987–990. [CrossRef] [PubMed]
17. Lee, M.G.; Choi, S.; Park, J.-K. Inertial separation in a contraction-expansion array microchannel. *J. Chromatogr. A* **2011**, *1218*, 4138–4143. [CrossRef] [PubMed]
18. Lee, W.; Kwon, D.; Choi, W.; Jung, G.Y.; Jeon, S. 3D-printed microfluidic device for the detection of pathogenic bacteria using size-based separation in helical channel with trapezoid cross-section. *Sci. Rep.* **2015**, *5*, 7717. [CrossRef] [PubMed]
19. Bhattacharjee, N.; Urrios, A.; Kang, S.; Folch, A. The upcoming 3D-printing revolution in microfluidics. *Lab Chip* **2016**, *16*, 1720–1742. [CrossRef] [PubMed]
20. Korir, G.; Prakash, M. Punch card programmable microfluidics. *PLoS ONE* **2015**, *10*, e0115993. [CrossRef] [PubMed]
21. Clime, L.; Morton, K.J.; Hoa, X.D.; Veres, T. Twin tubular pinch effect in curving confined flows. *Sci. Rep.* **2015**, *5*, 9765. [CrossRef] [PubMed]
22. Segré, G.; Silberberg, A. Radial particle displacements in Poiseuille flow of suspensions. *Nature* **1961**, *189*, 209–210. [CrossRef]
23. Bhagat, A.A.S.; Kuntaegowdanahalli, S.S.; Papautsky, I. Inertial microfluidics for continuous particle filtration and extraction. *Microfluid. Nanofluid.* **2008**, *7*, 217–226. [CrossRef]
24. Amini, H.; Lee, W.; Di Carlo, D. Inertial microfluidic physics. *Lab Chip* **2014**, *14*, 2739–2761. [CrossRef] [PubMed]
25. Dean, W.R. Fluid motion in a curved channel. *Proc. R. Soc. A* **1928**, *121*, 402–420. [CrossRef]
26. Lee, J.N.; Park, C.; Whitesides, G.M. Solvent compatibility of poly(dimethylsiloxane)-based microfluidic devices. *Anal. Chem.* **2003**, *75*, 6544–6554. [CrossRef] [PubMed]

micromachines

MDPI

Article

Thermoplastic Micromodel Investigation of Two-Phase Flows in a Fractured Porous Medium

Shao-Yiu Hsu [1], Zhong-Yao Zhang [2] and Chia-Wen Tsao [2,*]

[1] Department of Bioenvironmental System Engineering, National Taiwan University, Taipei 10617, Taiwan; syhsu@ntu.edu.tw

[2] Department of Mechanical Engineering, National Central University, Taoyuan 32001, Taiwan; stargold1224@gmail.com

* Correspondence: cwtsao@ncu.edu.tw; Tel.: +886-3-426-7343

Academic Editors: Weihua Li, Hengdong Xi, Say Hwa Tan and Nam-Trung Nguyen
Received: 16 September 2016; Accepted: 19 January 2017; Published: 26 January 2017

Abstract: In the past few years, micromodels have become a useful tool for visualizing flow phenomena in porous media with pore structures, e.g., the multifluid dynamics in soils or rocks with fractures in natural geomaterials. Micromodels fabricated using glass or silicon substrates incur high material cost; in particular, the microfabrication-facility cost for making a glass or silicon-based micromold is usually high. This may be an obstacle for researchers investigating the two-phase-flow behavior of porous media. A rigid thermoplastic material is a preferable polymer material for microfluidic models because of its high resistance to infiltration and deformation. In this study, cyclic olefin copolymer (COC) was selected as the substrate for the micromodel because of its excellent chemical, optical, and mechanical properties. A delicate micromodel with a complex pore geometry that represents a two-dimensional (2D) cross-section profile of a fractured rock in a natural oil or groundwater reservoir was developed for two-phase-flow experiments. Using an optical visualization system, we visualized the flow behavior in the micromodel during the processes of imbibition and drainage. The results show that the flow resistance in the main channel (fracture) with a large radius was higher than that in the surrounding area with small pore channels when the injection or extraction rates were low. When we increased the flow rates, the extraction efficiency of the water and oil in the mainstream channel (fracture) did not increase monotonically because of the complex two-phase-flow dynamics. These findings provide a new mechanism of residual trapping in porous media.

Keywords: thermoplastic; micromodel; cyclic olefin copolymer; UV/Ozone bonding; residual trapping; flow visualization

1. Introduction

When applying engineering processes to porous media involving two-phase flows, e.g., enhanced oil recovery, geological carbon sequestration, water infiltration, and groundwater remediation, a fractured pore structure is commonly observed. To understand the fluid interactions in a fractured porous material, a uniform pore-pattern embedded with a single macropore channel is commonly used to represent the pore geometry of the fractured porous medium [1–4]. A thorough conceptual and quantitative understanding of the flow behavior of fluids in pore spaces forms the basis of multiphase flow in porous materials. These subsurface multiphase flows are of interest to hydrologists, agricultural/environmental engineers, and petroleum engineers. Nevertheless, since most soils and geomaterials are not transparent, visualizing their flow behavior in porous materials is a long-term persistent challenge.

Micromodels are one of the commonly used tools for investigating and visualizing physical, chemical, and biological processes at a small scale in two or three dimensions [5]. Karadimitriou and

Hassanizadeh [5] defined a micromodel as "an artificial representation of a porous medium made of a transparent material." A micromodel usually comprises an artificial pore structure of connected pores, whose shapes are designed to represent the simplified geometry of a geomaterial. The geometries of the early micromodels were mostly simple and regular [6,7]. Since the 1980s, micromodel geometry has become more complicated, and various sizes of pores and channels have been computer-generated based on statistical distributions and fractal patterns [8,9]. Some irregular patterns that have properties similar to real porous media have been generated using the Delaunay triangulation [10]. Such patterns can also be designed for studying flow behaviors in pore geometries that are present either in natural porous materials or in theoretical models [1,3,11].

Micromodels for porous media are usually fabricated using glass plates packed with glass beads [12,13]. In addition, glass [14], silicon wafer [15], and photoresist [16] surfaces have been used to construct micromodels. The material cost and, in particular, the microfabrication-facility cost for making a glass or silicon-based micromodel are high; this creates obstacles for researchers investigating two-phase-flow behavior in porous media. Polymers are low-cost materials and are therefore an attractive alternative materials for microscale geomaterial models. Various polymer materials with different mechanical, chemical, and optical properties are available in the market; more importantly, the facility costs for making polymer micromodels are much lower than those for glass or silicon micromodels. Polymer materials have been proposed in the microfluidics community since the late 1990s [17] and are being increasingly used since then owing to their low cost and disposability. Polymer materials such as thermoplastics, polycarbonate (PC), polystyrene (PS), polymethylmethacrylate (PMMA), and cyclic olefin copolymer (COC) or polydimethylsiloxane (PDMS) elastomers are commonly used in microfluidics. Rigid thermoplastic materials are preferable for porous micromodels because they have better resistance to surface infiltration and deformation than PDMS [18]. An artificial pore structure can be fabricated via replication using a thermoplastic material. Polymer-replication processes such as injection molding [19,20], hot embossing [21–23], and roller printing [24,25] require a master micromold to produce the inverted polymer replicas for mass production. After replication, bonding is necessary with another cover plate to create an enclosed microchannel. Various thermoplastic bonding methods have been reported for sealing the microchannel [26]. Thermoplastic bonding is one of the most simple and straightforward thermoplastic processes [23]; however, it imparts a low bonding strength. Therefore, high-strength bonding techniques using materials such as solvents [27], adhesives [28], UV/Ozone [29], O_2 plasma [30], as well as welding [31] have been reported. Each polymer-replication or bonding technique presents specific advantages as well as process limitations for different applications. Therefore, appropriate selection of the polymer microfabrication process is important while developing micromodels for geo-fluid experimental usage.

Two-phase-flow experiments in a well-manufactured micromodel with a fractured pattern are still limited. A challenge in using polymer materials for representing natural porous materials is the polymers' surface wettability. The water contact angle of soil cracks is approximately $20°–30°$; however, it varies under different conditions [32–34]. Generally, natural porous materials are hydrophilic. Unlike glass or silicon substrates, polymer substrates such as PC, PMMA, and COC usually exhibit weak hydrophobicity, which may not be suitable for representing hydrophilic geomaterials. A hydrophilic surface coating may be used in a porous polymer micromodel. Normally, this coating is temporary, and the aqueous-phase coating reagent may not be uniformly filled and removed from the microchannel for homogeneous coating, especially in our high-density porous micromodel design. In this study, we demonstrate the use of COC, a thermoplastic material, in geo-fluid experimental micromodel investigations. In addition, we fabricate an artificial pore structure via hot embossing and use UV/Ozone for bonding as well as for controlling surface wettability. Furthermore, we perform two-phase-flow micromodel experiments in a fractured pattern and investigate the spatial distribution of oil and groundwater subjected to injection and extraction rates.

2. Experimental Section

2.1. Materials and Reagent

The materials used for fabricating the micromodels include COC, silicon wafer, SU-8 resist, surgical needle, acetone (ACE), and isopropyl alcohol (IPA). COC 8007 granules were purchased from TOPAS Advanced Polymers Inc. (Florence, KY, USA). The COC granules were injected into custom-made stainless-steel molds to fabricate COC substrates with a diameter of 7 cm and a thickness of 2 mm. A P-type (100) silicon wafer with a diameter of 10 cm was purchased from Summit-Tech Resource Corp. (Hsinchu, Taiwan). SU-8 3050 and SU-8 developer were purchased from MicroChem Corp. (Westborough, MA, USA). Surgical needles (SC20/15, LS20) were purchased from Instech Laboratories Inc. (Plymouth Meeting, PA, USA). Medical-grade tubing (Tygon S-50-HL) was purchased from Saint-Gobain Performance Plastics Corp. (Akron, OH, USA). ACE and IPA were purchased from Sigma-Aldrich (St. Louis, MO, USA).

2.2. Contact Angle Measurement

The contact angle of the COC surface was measured using a custom-made optical goniometer comprising a high-resolution digital camera (Canon EOS 450D/TAMRON Macro 90 mm F2.8 lens), a precision stage, and an optical light source. To measure the contact angle, a 5-μL water droplet was placed on the COC surface using a pipette; then, a side-view image of the microdroplet was taken using a digital camera. The contact angle was determined by measuring the angle formed between the microdroplet tangent line and the COC surface on the basis of microdroplet images obtained using AutoCAD 2013 (Autodesk, Inc., Mill Valley, CA, USA).

2.3. Micromold Fabrication

Fabrication of the SU-8 micromold started with cleaning silicon water with ACE, IPA, deionized (DI) water, and N_2 blow-drying followed by hot baking at 130 °C for 15 min (Super-Nuova, Thermo Scientific Inc., Ocala, FL, USA). Spin-coating with a thin hexamethyldisilazane (HMDS) was applied to remove the moisture from the silicon surface; subsequently, spin-coating was performed with SU-8 3050 at 700 rpm for 30 s and at 1200 rpm for 40 s (SPC-703, Yi Yang Co., Taoyuan, Taiwan) to apply a 100-μm SU-8 layer on the silicon substrate. Then, soft baking was performed at 110 °C for 15 min. The mask was aligned and then exposed to UV (AGL100 UV Light source, M&R Nano Technology Co., Taoyuan, Taiwan) for 90 s. Then, the UV-exposed SU-8 substrate was hard-baked at 95 °C for 5 min, and the SU-8 micromold was developed in the SU-8 developer within 2–3 min. The SU-8 micromold was rinsed clean with IPA, DI water, and N_2 blow drying. Finally, a hard-baking cure at 130 °C was applied for 8 h to enhance the SU-8 micromold's mechanical strength for the hot-embossing process.

2.4. Micromodel Design and Experimental Setup

The artificial porous micromodel design is shown in Figure 1. To mimic a natural fractured porous material, the micromodel device comprises a 2.6-mm-wide and 50-mm-long microchannel (mainstream microchannel) in the middle of a uniform 30 mm × 30 mm pore pattern containing 1332 pieces of microcavities. Each microcavity was 500 μm in diameter and 100 μm in depth. The microcavities were separated by a 200-μm distance (see Figure 1a).

The artificial pattern represents a two-dimensional cross-section of a fractured pore material. The inlet and outlet of the mainstream microchannel represent the injection and extraction points of water and oil during groundwater remediation or enhanced oil recovery, respectively. The secondary outlets represent another parallel fracture connected to air or another extraction well away from the main fracture. They allow the displaced fluid to flow in the direction perpendicular to the main flow direction.

The experimental setup of the micromodel experiment comprised a COC micromodel, a syringe pump (KD Scientific Legato 100), a high-speed camera (Point Grey Flea3, CMOS sensor, maximum resolution: 1280 × 1024 with maximum frame rate of 150 fps), and a background light source (Figure 1b). Images of the entire micromodel were taken with the high-speed camera to record the changes in the two fluid distributions during both the imbibition and drainage processes. The left inlet of the mainstream of the micromodel was connected to the syringe pump, which provided a constant-flow-rate boundary condition. The secondary microchannel outlets connected to atmospheric air provided a constant-pressure boundary condition. The secondary outlets represented another parallel fracture connected to air or another extraction well away from the main fracture.

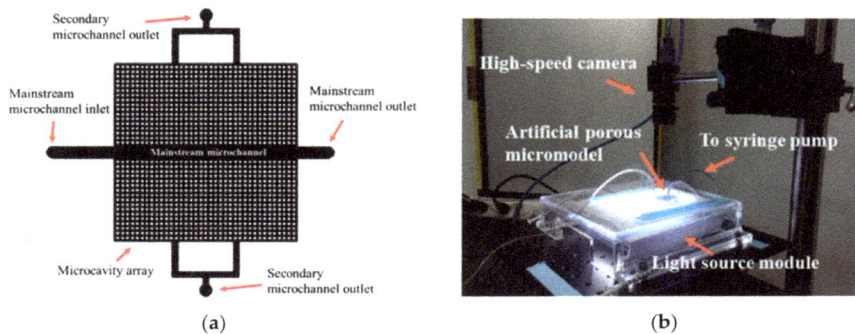

Figure 1. (a) Artificial porous micromodel design to mimic a natural fractured porous material. (b) Photograph of the porous micromodel experimental setup.

3. Results and Discussion

3.1. Polymer Micromodel Fabrication

Polymer selection is the first important step in determining the success of microfluidic-chip fabrication for an artificial porous micromodel. In our design, because the micromodel comprises 1332 pieces and 500-μm microcavities, the hydraulic resistance of the microchannel is high. Thus, a high pumping pressure is required to drive the fluid flow. Moreover, a good mechanical strength is required; therefore, an elastomer material such as PDMS is not preferred in this study. A thermoplastic material TOPAS COC 8007 is used herein because of its high mechanical rigidity, low water absorption (0.01%), and good optical transmissivity (91%) for optical observation [35]. The COC-based micromodel fabrication includes three major processes: micromold fabrication, hot embossing, and bonding. As shown in Figure 2a, the micromold is first fabricated using the standard SU-8 photolithography process. Figure 2b shows the SEM image of the microcavity arrays on the SU-8 micromold.

The SU-8 micromold is assembled to a custom-machined micromold holder and fixed to a hot embosser (Ray Cheng Enterprise Co. Ltd., Taoyuan, Taiwan). Because the glass-transition temperature (T_g) of COC is 78 °C, the COC substrate is placed above the SU-8 micromold and close to the hot embosser to preheat the COC substrate at 77 °C for 5 min. Then, a hot-embossing temperature of 83 °C (5 °C above T_g) and an embossing pressure of 1.45 MPa are applied for 7 min. After cooling down to 70 °C for 15 min, the hot embosser is opened and the inversely embossed COC replicas are removed when one hot-embossing process has been completed. The embossed COC microchannel layer is bonded to another COC cover layer to enclose the microchannel. Because of a high-porosity micromodel layout, the microchannel hydraulic resistance is high; therefore, a high bonding-strength polymer-bonding method is required. In this paper, we select the high bonding-strength UV/Ozone method to seal the COC substrate [29]. The bonding procedure starts with predrilling another COC cover layer with 1.5-mm-diameter reservoir holes prior to bonding. The embossed COC microchannel is placed and the cover layers are treated in the UV/Ozone cleaner (PSD-UV, Novascan Technologies, Ames, IA, USA) for 300 s. The top surface of the COC layers is activated by UV/Ozone treatment for consequent low-temperature UV/Ozone bonding. The COC pair is aligned and put in the hot embosser press-bond at 3.7 MPa and 70 °C for 110 min. After UV/Ozone bonding, surgical needles are inserted into the cover plate reservoirs as the inlet and outlet connections. Figure 2c shows the porous micromodel after bonding and the surgical-needle insertion; all microchannels and microcavity arrays are well-bonded without de-bonding or fluid leakage.

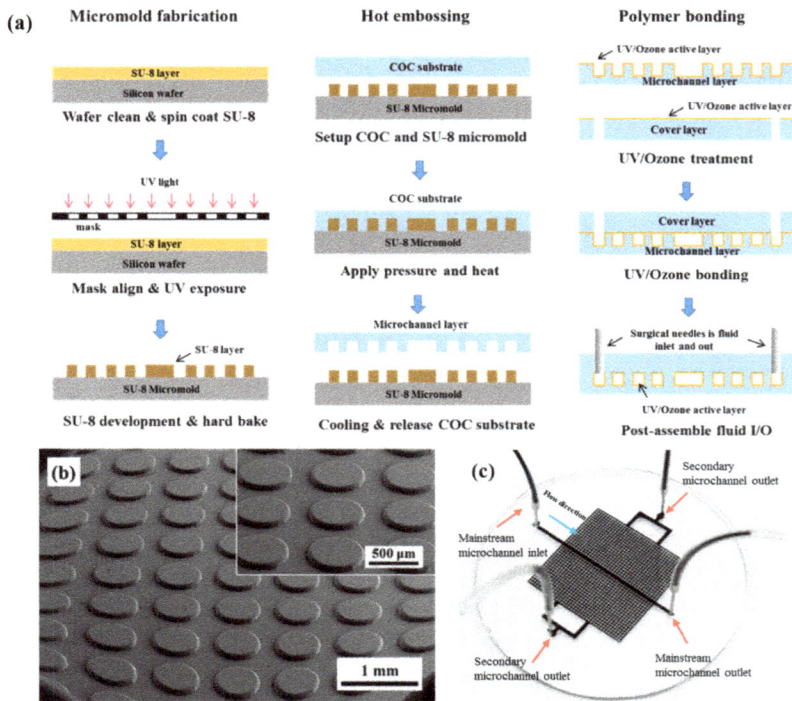

Figure 2. (**a**) Schematic of the micromodel fabrication flow chart. (**b**) 40× fisheye SEM image of the SU-8 micromold showing microcavity arrays. The top-right corner shows an enlarged 100× SEM image. (**c**) Micromodel with black-food-dye injection.

3.2. Microchannel Surface Wettability Characterization

One of the major challenges in using a polymer micromodel for two-phase-flow studies on natural porous materials is the wettability of the polymer. It is important to have a uniform and stable wettability throughout the pore space. In this study, we use UV/Ozone to physically treat the COC surface in order to enhance the bonding as well as to render the COC surface hydrophilic. UV/Ozone is widely used in dry cleaning aimed at removing organic contaminants from silicon surface. The mercury lamp outputs a 184.9 nm wavelength generating ozone in the chamber and a 253.7 nm wavelength that decomposes the ozone molecules. The continuous process of formation and decomposition of ozone molecules leads to surface oxidation and changes the COC-surface wettability. After UV/Ozone treatment, the COC surface becomes super hydrophilic. As shown in Figure 3a, the native COC surfaces exhibit weak hydrophobicity. After UV/Ozone surface treatment, the COC surface becomes hydrophilic. In the UV/Ozone bonding process, as shown in Figure 2a, the top, bottom, and side wall surfaces of the unbonded microchannel remain hydrophilic after UV/Ozone binding. Therefore, we used UV/Ozone treatment to make the micromodel's surface wettability similar to that of natural porous materials.

The wettability of a UV/Ozone-treated surface is a dynamic process that may vary with time when subjected to an aqueous environment. Figure 3b,c show the contact angle of the UV/Ozone-treated COC surface after immersion in water from 30 min to 60 days for 100 s and 300 s UV/Ozone treatment, respectively. The native COC-surface water contact angle is approximately 90.5° ± 2.8°. After 100 and 300 s of UV/Ozone treatment, the water contact angle on the COC surface changes drastically, thereby making the COC surface superhydrophilic. After UV/Ozone-treatment of the COC surface in water for 30 min, it still shows high hydrophilicity in both the 100-s (21.7° ± 1.4°) and 300-s (7.8° ± 1.6°) UV/Ozone-treatment cases. The water contact angle increases with the water-immersion time until it reaches a plateau. For the 100-s UV/Ozone treatment (Figure 3b), the water contact angle increases from 21.7° to 48.2° within the first week (seven days) and then remains constant at 45°–50° between Day 7 and 60. For the 300-s UV/Ozone treatment (Figure 3c), the water contact angle increases to 20.2° and 24.2° after 2 and 8 h of water immersion, respectively; furthermore, the COC-surface water contact angle reaches a plateau (27°–31°) after 8 h. Therefore, in the following porous micromodel experiments, the COC surfaces were treated for 300 s with UV/Ozone to bond the COC chip and enhance the bonding strength. After bonding, the microfluidic channels were rinsed with water for more than 8 h in order to ensure good contact-angle stability—i.e., a water contact angle of approximately 24°—which is analogous to those of natural porous materials. In our experiments, we made the micromodel hydrophilic and uniformly maintained it in the same state for a long period.

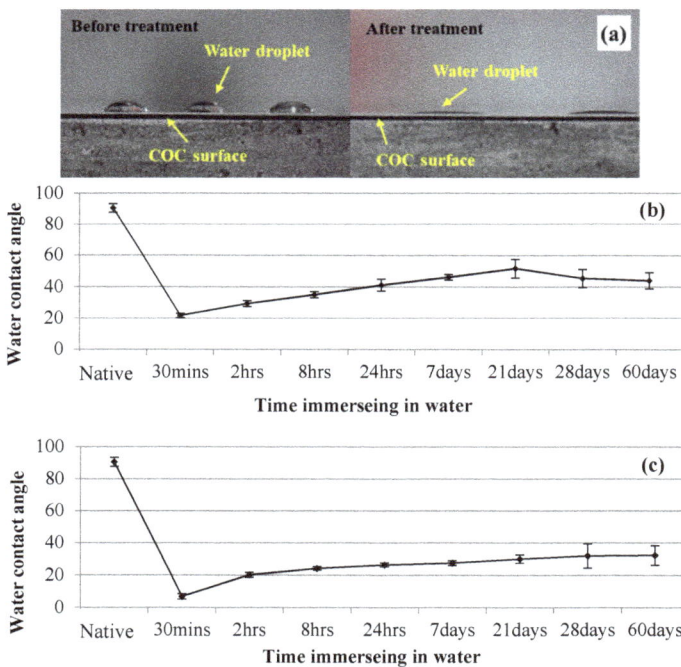

Figure 3. (**a**) Change in the COC-surface wettability after UV/Ozone treatment. The COC-surface water contact angle at different water-immersion times after (**b**) 100 s and (**c**) 300 s of UV/Ozone surface treatments.

3.3. Results of the Two-Phase-Flow Micromodel Experiments

We performed the experiments in micromodels comprising mainstream channels with different widths and different injection and extraction rates. The two-phase-flow experimental setup shown in Figure 4 involves imbibition and drainage processes. The experimental fluids are water and oil (gasoline). DI water is dyed blue to enhance visual observation and facilitate image-analysis of saturation. The dynamic viscosities of water and oil are 0.089 and 0.06 Pa·s, respectively. The surface tensions of the water and oil are 71.5 and 22.4 dyne/cm, respectively. Although the surface tension between oil and water is not measured directly, it is around 30–40 dyne/cm based on previous studies [36,37]. In the drainage process, the micromodel with a hydrophilic surface is first filled with dyed water. We drain the water out of the micromodel using two different methods: (1) injecting air into the model and (2) extracting water from the model. Since our syringe pump can provide a constant extraction rate, the extraction rates of both water and air are constant. In the imbibition process, the micromodel is first filled with oil. Then, the oil is extracted by injecting dyed water into the model. The constant injection and extraction rates of oil and water for both drainage and imbibition are 50, 100, 200, and 400 µL/min. During the drainage and imbibition processes, the high-speed camera records the dynamics of the fluid distributions. Injection or extraction is terminated when the fluid distribution reaches equilibrium.

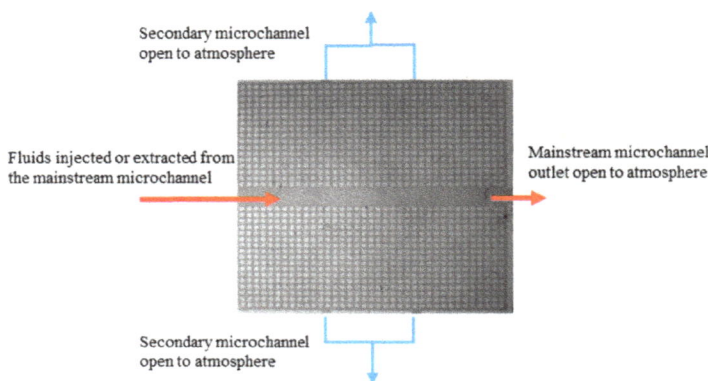

Figure 4. The inlet and outlets of the micromodel. The fluids are injected or extracted from the left inlet of the mainstream microchannel, which provides a constant-flow-rate boundary condition. The five outlets are connected to atmospheric air, thereby providing constant-pressure boundary conditions. The outlet on the right end of the micromodel is directly connected to the mainstream microchannel, and the other four outlets are connected to the top and bottom of the secondary microchannels (microarrays). The direction shown in the figure is for the injection experiments, and its opposite direction is for the extraction experiments.

3.3.1. Water–Air Displacement: Air Injection and Water Extraction

The water-drainage experiments are performed in two different processes: air injection and water extraction. Figures 5–8 show the changes in the water–air distributions during the drainage process at air-injection and water-extraction rates of 50, 100, 200, and 400 µL/min. As shown in Figure 5a–c, the air gradually pushes out the dyed water in the bottom microarray. At the end of the experiment, the injected air bypasses most of the water and directly flows out of the model via the bottom outlet close to the left inlet, as shown in Figure 5c. For water extraction from the left inlet, the air first moves into the micromodel via the right outlet and quickly forms an air path connecting the right and bottom outlets, as shown in Figure 5d. In the middle of water extraction, air flows into the micromodel from the right, top, and bottom outlets. At the end of the experiment, all the air from the outlets flows out via the water extraction inlet, and the water distribution in the porous micromodel becomes steady, as shown in Figure 5f.

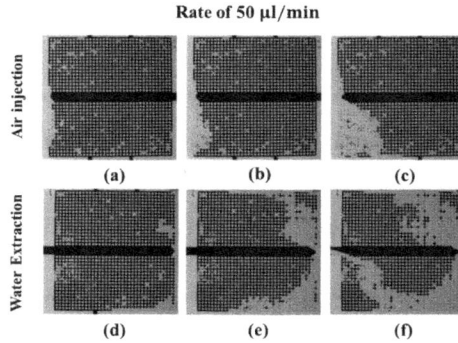

Figure 5. Dynamics of the water (dark)–air (transparent) distributions during the drainage process at an air-injection (**a**–**c**) and a water-extraction (**d**–**f**) rate of 50 µL/min; (**a,d**) correspond to the beginning, (**b,e**) to the middle, and (**c,f**) to the end of the drainage process.

In Figure 6a–c, air pushes out the dyed water in the upper microarray. When the injected air flows out via the right outlet, the water distribution in the upper microarray quickly reaches a steady state. At the end of the experiment, the injected air pushes out most of the water in the upper microarray, as shown in Figure 6c. For water extraction from the left inlet, the air first moves into the micromodel via the right outlet and drains the upper microarray and the lower microarray, as shown in Figure 6d. In the middle of water extraction, the air uniformly drains the water in the upper microarray and forms a preferential pathway connecting the right and bottom outlets in the lower microarray. At the end of the experiment, all the air flows out of the micromodel via the water extraction inlet, and the water distribution in the micromodel becomes steady, as shown in Figure 6f.

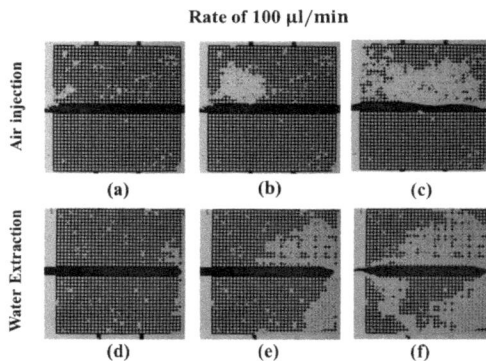

Figure 6. Dynamics of the water (dark)–air (transparent) distributions during the drainage process with an air-injection (**a**–**c**) and a water-extraction (**d**–**f**) rate of 100 µL/min; (**a,d**) correspond to the beginning, (**b,e**) to the middle, and (**c,f**) to the end of the drainage process.

In Figure 7a–c, the air pushes out the dyed water in the upper microarray. When the injected air flows out via the right outlet, the water distribution in the upper microarray quickly reaches a steady state. At the end of the experiment, the injected air pushes out most of the water in the upper microarray, as shown in Figure 7c. For water extraction from the left inlet, the air first moves into the micromodel via the right outlet and drains the upper microarray and the lower microarray, as shown in Figure 7d. In the middle of the water extraction, the air uniformly drains the water in the upper microarray, and the air forms a preferential pathway connecting the right and bottom outlets in the lower microarray. At the end of the experiment, all the air from the outlets flows out via the water extraction inlet, and the water distribution in the micromodel becomes steady, as shown in Figure 7f.

Rate of 200 μl/min

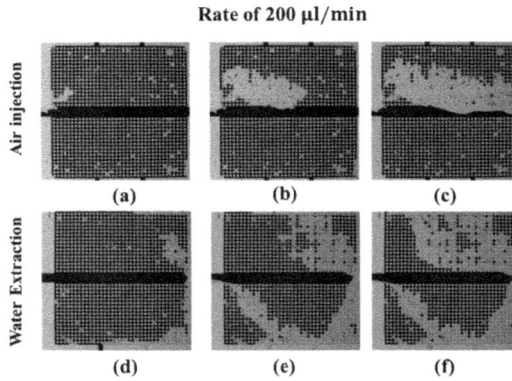

Figure 7. The dynamics of the water (dark)-air (transparent) distributions during the drainage process at an air injection (**a–c**) and water extraction (**d–f**) rate of 200 μL/min. (**a,d**) correspond to the beginning of the drainage process, (**b,e**) the middle of the drainage process, and (**c,f**) the end of the drainage process.

In Figure 8a–c, the air pushes out the dyed water in the upper microarray and then the lower microarray. When the injected air flows out via the right outlet, the water distribution in the upper microarray quickly reaches a steady state; then, the air pushes water in the bottom microarray. At the end of the experiment, the injected air pushes out more than 50% of the water in both the upper and lower microarrays, as shown in Figure 8c. For water extraction from the left inlet, the air first moves into the micromodel via the right outlet and drains the upper microarray and the lower microarray, as shown in Figure 8d. In the middle of the water extraction, the air uniformly drains the water in the upper microarray and forms a preferential pathway connecting the right and bottom outlets in the lower microarray. At the end of the experiment, all the air from the outlets flows out via the water extraction inlet, and the water distribution in the micromodel becomes steady, as shown in Figure 8f.

Rate of 400 μl/min

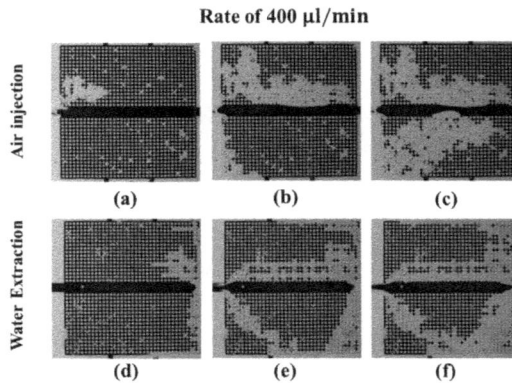

Figure 8. Dynamics of the water (dark)–air (transparent) distributions during the drainage process at an air-injection (**a–c**) and water-extraction (**d–f**) rate of 400 μL/min; (**a,d**) correspond to the beginning, (**b,e**) to the middle, and (**c,f**) to the end of the drainage process.

Figure 9 shows the evolution of water saturation versus the volume ratio of the injected air, which is the ratio of the total injected air volume to the total pore volume of the micromodel. Because during the water extraction, air also moves out of the micromodel via the inlet, the amount of water extracted cannot be estimated only from the extraction rate of the syringe pump. Nevertheless, we can estimate the injected-air volume during the water-extraction experiments. Therefore, we use the volume ratio of the injected air for all experiments, as shown in Figure 9. For air injection, the water-saturation-decreasing rate increases with the air-injection rate. At injection rates higher than 100 µL/min, the residual-water saturations all converge to approximately 70%. For water extraction, the water-saturation-decreasing rate increases with the air-injection rate. Nevertheless, at the extraction rate of 400 µL/min, the water distribution reaches an equilibrium quickly, with water saturation of approximately 80%. This is because the preferential air path is readily formed at the beginning of the experiments.

Both the air-injection and water-extraction experiments indicate that at injection/extraction rates of less than 200 µL/min, the minimum water-saturation rate decreases with increasing injection or extraction rate, as shown in Figure 9. However, at injection/extraction rates of 400 µL/min, our experimental results show that an increase in these flow rates does not significantly reduce the minimum water-saturation rate. Conversely, an increase in the water-extraction rate increases the minimum water-saturation rate. Therefore, we conclude that an increase in the air-injection rate can lead to a decrease in the minimum water-saturation rate; however, this is not a monotonic trend in water-extraction experiments because of the preferential path.

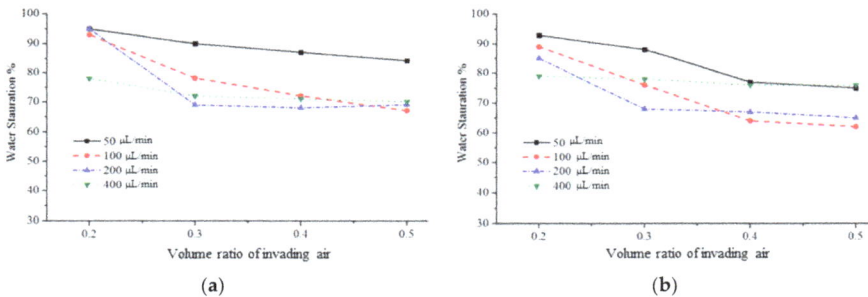

(a)　　　　　　　　　　　　　(b)

Figure 9. Water saturation versus the volume ratio of invading air in the models with (**a**) different air injection rates and (**b**) different water extraction rates.

Figure 10 shows the changes in the water–air distributions during the drainage process at an air-injection rate of 400 µL/min in the micromodels comprising narrow mainstream channels. In Figure 10a–c, the air uniformly pushes out the dyed water in the upper and the lower microarrays. When the injected air flows out via the left upper and bottom outlets, the water distribution in the microarray gradually reaches a steady state. At the end of the experiment, the injected air pushes out one-third of the water in both the upper and lower microarrays, as shown in Figure 10c.

The effect of the width of the mainstream channel on the water drainage process is significant, as indicated by a comparison of Figures 8 and 10. In the model comprising a narrow mainstream channel (Figure 10), the water is pushed out from the mainstream channel and the upper and lower microarrays at the same time because of the small difference between the widths of the mainstream and the microchannel. Since the water is displaced first in the narrow mainstream, the injected air can be connected in the upper and lower microarrays that leads to a uniform sweep. Therefore, the injected air is connected in the upper and lower microarrays. On the other hand, the air pushes water from the upper microarray first and then from the bottom microarray in the model comprising a wide mainstream channel (Figure 8). Nevertheless, the water in the wide mainstream channel is not displaced by air. The air in the upper and lower microarrays is disconnected; therefore, the air–water distribution patterns are totally different in the upper and lower microarrays. For water extraction at a rate of 400 µL/min, the air is pulled via all outlets into the micromodel comprising the wide mainstream channel. Once the air from the bottom is connected to the left inlet, the extraction process reaches equilibrium. In the narrow mainstream channel, the air does not flow through the main channel and less oil is extracted.

Rate of 400 µl/min

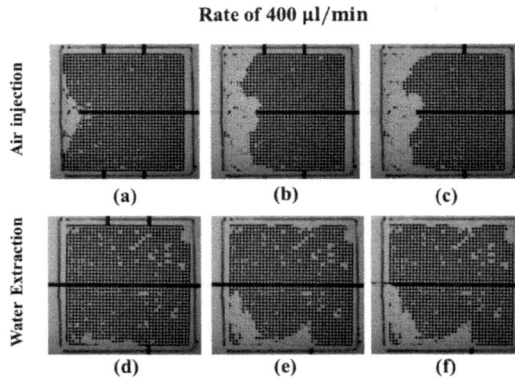

Figure 10. Dynamics of the water (dark)–air (transparent) distributions during the drainage process at an air-injection (**a–c**) and water-extraction (**d–f**) rate of 400 µL/min in a micromodel comprising a narrow mainstream channel with a width of 1 mm; (**a,d**) correspond to the, (**b,e**) to the middle, and (**c,f**) to the end of the drainage process.

3.3.2. Oil-Water Displacement: Water Injection

In the water-imbibition experiments, water is injected into the oil-filled micromodel. Figure 11 shows the oil–water distribution dynamics at different injection rates. The amount of oil extracted increases with the water-injection. Figure 11a–c shows that at an injection rate of 50 µL/min, a clear fingering water flow is formed in the lower microarray close to the mainstream channel. Water only flows in the newly formed preferential channel toward the upper and right outlets, and no more oil is extracted. When the injection rate increases to 100 µL/min, as shown in Figure 11d–f, the oil in the upper and lower microarrays is partially extracted but that in the mainstream channel mostly remains. As the water-injection rate increases to 200 µL/min, the injected water first pushes the oil in the mainstream channel and then pushes it to the upper and lower microarrays, as shown in Figure 11g–i. At a water-injection rate of 400 µL/min, the injected water pushes out most of the oil in the mainstream channel and the upper and lower microarrays, as shown in Figure 11j–l. However, the amount of residual oil in the mainstream channel is greater than that obtained at a water-injection rate of 200 µL/min. The oil-extraction efficiency in the mainstream channel (fracture) non-monotonically increases with the water-injection rate.

From the air–water and oil–water experiments, we found that the injected air and water bypass the fracture, especially at the beginning of the experiments. The major reason for this bypass flow is the additional secondary outlets. In the classical pattern of the pore structure with fracture, the only outlet exists at the end of the fracture. In our new model, we added four secondary outlets to create an additional flow direction. Since the distance between the mainstream inlet and the secondary outlet is closer to the mainstream outlet, the pressure gradient is larger between the mainstream inlet and the secondary outlet than that between the mainstream inlet and the mainstream outlet. Therefore, even though the flow-resistance is larger in the secondary microchannel, the injected air tends to flow toward the secondary microchannel (top or bottom) first. At low-flow rates in the air-injection experiments, air flows out via the secondary outlet. However, when the air-injection rate increases, all the air flows toward the secondary microchannel first and then toward the mainstream outlet instead of flowing out via the secondary outlet, which becomes a flow pathway with lower resistance. In the water extraction experiments, because of the high pressure gradient between the mainstream inlet and the secondary outlet, the air flows into the secondary microchannel via the secondary outlet first and then through the mainstream outlet. A similar phenomenon occurs in the oil–water experiments.

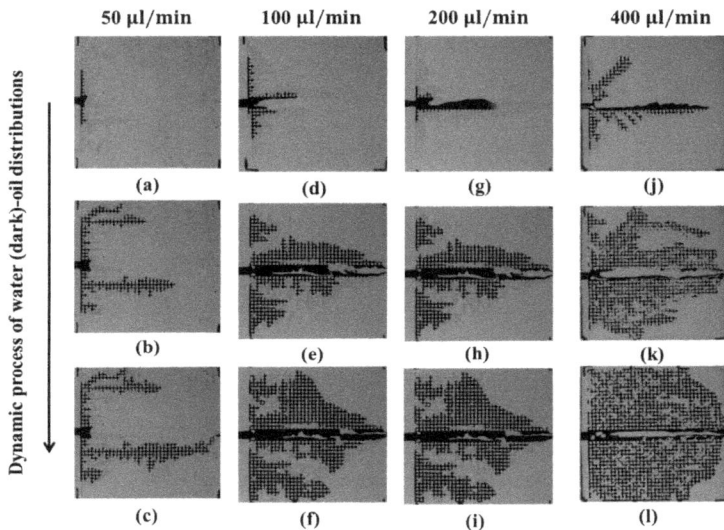

Figure 11. Dynamics of the water (dark)–oil (transparent) distributions during the imbibition process at water-injection rates of 50 μL/min (**a–c**), 100 μL/min (**d–f**), 200 μL/min (**g–i**), and 400 μL/min (**j–l**) in a micromodel. (**a,d,g,j**) correspond to the beginning; (**b,e,h,k**) to the middle; and (**c,f,i,l**) to the end of the imbibition process.

4. Conclusions

The development of porous micromodels is important for visualizing two-phase-flow phenomena in natural porous materials. Compared with glass-based or other silicon-based micromodels, the thermoplastic micromodel demonstrated in this study provides the advantages of design flexibility, precision scales, low cost, and low fabrication complexity. The COC porous micromodels were herein fabricated via hot embossing followed by bonding. When using thermoplastics in a porous micromodel, the issues of high hydraulic resistivity and surface hydrophobicity need to be considered. In this research, we used the UV/Ozone bonding method to increase the bonding strength and to render the surface hydrophilic (analogous to the surface wettability of natural geomaterials) for successful porous micromodel preparation.

The porous micromodel two-phase-flow experiments conducted herein show that, during the drainage experiment, the extraction efficiency upon direct water-extraction is higher than that obtained via air-injection. At low air-injection rates, the drainage process quickly reaches equilibrium (no more water is extracted) once the preferential channel connects the inlet and outlet. When the air-injection rate increases, the water-extraction rate increases and the saturation rate of the residual water decreases. The same situation occurs when we inject water to extract oil. Our experimental results show that the water- and oil-extraction efficiency in a fractured porous medium depends on the boundary conditions, the injection and extraction rates, and the width of the mainstream channel. In addition, the dynamics of the displaced water and oil in the mainstream channel are more complicated than those in the microarrays. Because of the complex two-phase-flow dynamics in the mainstream channel, the extraction efficiency in the mainstream channel does not monotonically increase with the injection rate. The common assumption that water or oil is extracted first in the fracture during the drainage and imbibition processes is not supported by our micromodel experiments.

Acknowledgments: The authors thank the Ministry of Science and Technology, Taiwan, for financially supporting this project under Grant No. MOST 105-2116-M-002-028.

Author Contributions: S.Y.H. and C.W.T. conceived and designed the experiments; Z.Y.Z. performed the experiments; S.Y.H. and C.W.T. analyzed the data and wrote the paper; S.Y.H. and C.W.T. contributed equally to this work.

Conflicts of Interest: The authors declare no conflict of interest.

References

1. Rangel-German, E.; Kovscek, A. A micromodel investigation of two-phase matrix-fracture transfer mechanisms. *Water Resour. Res.* **2006**, *42*. [CrossRef]
2. Conn, C.A.; Ma, K.; Hirasaki, G.J.; Biswal, S.L. Visualizing oil displacement with foam in a microfluidic device with permeability contrast. *Lab Chip* **2014**, *14*, 3968–3977. [CrossRef] [PubMed]
3. Wan, J.; Tokunaga, T.K.; Tsang, C.-F.; Bodvarsson, G.S. Improved glass micromodel methods for studies of flow and transport in fractured porous media. *Water Resour. Res.* **1996**, *32*, 1955–1964. [CrossRef]
4. Long, J.; Witherspoon, P.A. The relationship of the degree of interconnection to permeability in fracture networks. *J. Geophys. Res. Solid Earth* **1985**, *90*, 3087–3098. [CrossRef]
5. Karadimitriou, N.K.; Hassanizadeh, S.M. A review of micromodels and their use in two-phase flow studies. *Vadose Zone J.* **2012**, *11*. [CrossRef]
6. Chatenever, A.; Calhoun, J.C., Jr. Visual examinations of fluid behavior in porous media-part I. *J. Pet. Technol.* **1952**, *4*, 149–156. [CrossRef]
7. Nuss, W.; Whiting, R. Technique for reproducing rock pore space: Geological notes. *AAPG Bull.* **1947**, *31*, 2044–2049.
8. Tsakiroglou, C.; Avraam, D. Fabrication of a new class of porous media models for visualization studies of multiphase flow processes. *J. Mater. Sci.* **2002**, *37*, 353–363. [CrossRef]
9. Cheng, J.T.; Pyrak-Nolte, L.J.; Nolte, D.D.; Giordano, N.J. Linking pressure and saturation through interfacial areas in porous media. *Geophys. Res. Lett.* **2004**, *31*. [CrossRef]
10. Blunt, M.; King, P. Relative permeabilities from two-and three-dimensional pore-scale network modelling. *Transp. Porous Media* **1991**, *6*, 407–433. [CrossRef]
11. Chatzis, I.; Dullien, F. Dynamic immiscible displacement mechanisms in pore doublets: Theory versus experiment. *J. Colloid Interface Sci.* **1983**, *91*, 199–222. [CrossRef]
12. Cuenca, A.; Chabert, M.; Morvan, M.; Bodiguel, H. Axisymmetric drainage in hydrophobic porous media micromodels. *Oil Gas Sci. Technol.* **2012**, *67*, 953–962. [CrossRef]
13. Corapcioglu, M.Y.; Fedirchuk, P. *Glass Bead Micromodel Study of Solute Transport*; FAO: Rome, Italy, 1999.
14. Mohammadi, S.; Maghzi, A.; Ghazanfari, M.H.; Masihi, M.; Mohebbi, A.; Kharrat, R. On the control of glass micro-model characteristics developed by laser technology. *Energy Sources Part A* **2013**, *35*, 193–201. [CrossRef]
15. Liu, Y.Y.; Zhang, C.Y.; Hilpert, M.; Kuhlenschmidt, M.S.; Kuhlenschmidt, T.B.; Nguyen, T.H. Transport of cryptosporidium parvum oocysts in a silicon micromodel. *Environ. Sci. Technol.* **2012**, *46*, 1471–1479. [CrossRef] [PubMed]
16. Cheng, J.T.; Giordano, N. Fluid flow through nanometer-scale channels. *Phys. Rev. E* **2002**, *65*. [CrossRef] [PubMed]
17. Becker, H.; Gartner, C. Polymer microfabrication technologies for microfluidic systems. *Anal. Bioanal. Chem.* **2008**, *390*, 89–111. [CrossRef] [PubMed]
18. Roh, C.; Lee, J.; Kang, C. The deformation of polydimethylsiloxane (PDMS) microfluidic channels filled with embedded circular obstacles under certain circumstances. *Molecules* **2016**, *21*, 798. [CrossRef] [PubMed]
19. Attia, U.M.; Marson, S.; Alcock, J.R. Micro-injection moulding of polymer microfluidic devices. *Microfluid. Nanofluid.* **2009**, *7*, 1–28. [CrossRef]

20. Attia, U.M.; Marson, S.; Alcock, J.R. Design and fabrication of a three-dimensional microfluidic device for blood separation using micro-injection moulding. *J. Eng. Manuf.* **2014**, *228*, 941–949. [CrossRef]

21. Jeon, J.S.; Chung, S.; Kamm, R.D.; Charest, J.L. Hot embossing for fabrication of a microfluidic 3D cell culture platform. *Biomed. Microdevices* **2011**, *13*, 325–333. [CrossRef] [PubMed]

22. Huang, M.S.; Chiang, Y.C.; Lin, S.C.; Cheng, H.C.; Huang, C.F.; Shen, Y.K.; Lin, Y. Fabrication of microfluidic chip using micro-hot embossing with micro electrical discharge machining mold. *Polym. Adv. Technol.* **2012**, *23*, 57–64. [CrossRef]

23. Li, Y.; Buch, J.S.; Rosenberger, F.; DeVoe, D.L.; Lee, C.S. Integration of isoelectric focusing with parallel sodium dodecyl sulfate gel electrophoresis for multidimensional protein separations in a plastic microfludic network. *Anal. Chem.* **2004**, *76*, 742–748. [CrossRef] [PubMed]

24. Tsao, C.W.; Chen, T.Y.; Woon, W.Y.; Lo, C.J. Rapid polymer microchannel fabrication by hot roller embossing process. *Microsyst. Technol.* **2012**, *18*, 713–722. [CrossRef]

25. Yeo, L.P.; Ng, S.H.; Wang, Z.F.; Xia, H.M.; Wang, Z.P.; Thang, V.S.; Zhong, Z.W.; de Rooij, N.F. Investigation of hot roller embossing for microfluidic devices. *J. Micromech. Microeng.* **2010**, *20*, 015017. [CrossRef]

26. Tsao, C.-W.; DeVoe, D.L. Bonding of thermoplastic polymer microfluidics. *Microfluid. Nanofluid.* **2008**, *6*, 1–16. [CrossRef]

27. Koesdjojo, M.T.; Koch, C.R.; Remcho, V.T. Technique for microfabrication of polymeric-based microchips from an SU-8 master with temperature-assisted vaporized organic solvent bonding. *Anal. Chem.* **2009**, *81*, 1652–1659. [CrossRef] [PubMed]

28. Riegger, L.; Strohmeier, O.; Faltin, B.; Zengerle, R.; Koltay, P. Adhesive bonding of microfluidic chips: Influence of process parameters. *J. Micromech. Microeng.* **2010**, *20*, 087003. [CrossRef]

29. Tsao, C.W.; Hromada, L.; Liu, J.; Kumar, P.; DeVoe, D.L. Low temperature bonding of PMMA and COC microfluidic substrates using UV/Ozone surface treatment. *Lab Chip* **2007**, *7*, 499–505. [CrossRef] [PubMed]

30. Sunkara, V.; Park, D.K.; Hwang, H.; Chantiwas, R.; Soper, S.A.; Cho, Y.K. Simple room temperature bonding of thermoplastics and poly(dimethylsiloxane). *Lab Chip* **2011**, *11*, 962–965. [CrossRef] [PubMed]

31. Truckenmuller, R.; Ahrens, R.; Cheng, Y.; Fischer, G.; Saile, V. An ultrasonic welding based process for building up a new class of inert fluidic microsensors and -actuators from polymers. *Sens. Actuators A Phys.* **2006**, *132*, 385–392. [CrossRef]

32. Hirasaki, G.; Zhang, D.L. Surface chemistry of oil recovery from fractured, oil-wet, carbonate formations. *SPE J.* **2004**, *9*, 151–162. [CrossRef]

33. Carrillo, M.; Yates, S.; Letey, J. Measurement of initial soil-water contact angle of water repellent soils. *Soil Sci. Soc. Am. J.* **1999**, *63*, 433–436. [CrossRef]

34. Letey, J.; Osborn, J.; Pelishek, R. Measurement of liquid-solid contact angles in soil and sand. *Soil Sci.* **1962**, *93*, 149–153. [CrossRef]

35. Technical Datasheet. Available online: http://www.topas.com/tech-center/datasheets (accessed on 19 Feburary 2017).

36. Peters, F.; Arabali, D. Interfacial tension between oil and water measured with a modified contour method. *Colloids Surf. A Physicochem. Eng. Asp.* **2013**, *426*, 1–5. [CrossRef]

37. McDowell, C.J.; Powers, S.E. Mechanisms affecting the infiltration and distribution of ethanol-blended gasoline in the vadose zone. *Environ. Sci. Technol.* **2003**, *37*, 1803–1810. [CrossRef] [PubMed]

MDPI

Article

Dynamic Electroosmotic Flows of Power-Law Fluids in Rectangular Microchannels

Cunlu Zhao [1,*], Wenyao Zhang [1] and Chun Yang [2]

[1] Key Laboratory of Thermo-Fluid Science and Engineering of MOE, School of Energy and Power Engineering, Xi'an Jiaotong University, Xi'an 710049, China; wyalbertzhang@foxmail.com
[2] School of Mechanical and Aerospace Engineering, Nanyang Technological University, 50 Nanyang Avenue, Singapore 639798, Singapore; MCYang@ntu.edu.sg
* Correspondence: mclzhao@mail.xjtu.edu.cn; Tel.: +86-29-82663222

Academic Editors: Weihua Li, Hengdong Xi and Say Hwa Tan
Received: 12 December 2016; Accepted: 18 January 2017; Published: 24 January 2017

Abstract: Dynamic characteristics of electroosmosis of a typical non-Newtonian liquid in a rectangular microchannel are investigated by using numerical simulations. The non-Newtonian behavior of liquids is assumed to obey the famous power-law model and then the mathematical model is solved numerically by using the finite element method. The results indicate that the non-Newtonian effect produces some noticeable dynamic responses in electroosmotic flow. Under a direct current (DC) driving electric field, it is found that the fluid responds more inertly to an external electric field and the steady-state velocity profile becomes more plug-like as the flow behavior index decreases. Under an alternating current (AC) driving electric field, the fluid is observed to experience more significant acceleration and the amplitude of oscillating velocity becomes larger as the fluid behavior index decreases. Furthermore, our investigation also shows that electroosmotic flow of power-law fluids under an AC/DC combined driving field is enhanced as compared with that under a pure DC electric field. These dynamic predictions are of practical use for the design of electroosmotically-driven microfluidic devices that analyze and process non-Newtonian fluids such as biofluids and polymeric solutions.

Keywords: microfluidics; flow enhancement; electroosmosis; non-Newtonian fluids

1. Introduction

Nowadays, microfluidic devices find promising applications in a variety of fields, including chemical analysis, medical diagnostics and material synthesis etc. The ultimate goal of microfluidics is to replace conventional large-scale laboratories with single, disposable microchips. This leads to some distinctive advantages, such as fast analyses, low sample consumption and cost as well as minimum personnel requirements etc. Usually, a microfluidic device has to perform multiple types of liquid sample manipulation to finish one single analysis, for example, pumping, mixing, injection, dispensing, just to name a few, among which pumping is the most fundamental one [1–3]. In order to achieve an optimal design and a better control of microfluidic devices, one needs to have a fundamental understanding of the liquid pumping in microchannels. In general, two popular ways to pump liquids in microfluidic devices are pressure-driven flow and electroosmotic flow. The problem for pressure driven flow is that it becomes increasingly difficult to pump liquids as channel size reduces to micron and submicron range. Electroosmotic flow however does not suffer from this problem, thereby providing an efficient way of pumping in microfluidic devices. In addition, electroosmotic flow possesses other advantages over pressure-drive flow, such as ease of fabrication and control, no need for moving parts, and easy integration with electronic controlling circuits for automation etc.

Particularly, the plug-like velocity profile in electroosmotic flow minimizes the sample dispersion, which is essential for high-resolution separation in capillary electrophoresis [4].

Viscous relaxation or diffusion is the characteristic time scale for microscale liquid flow to develop to steady-state, and typically it is in the order of milliseconds [5]. An increasing number of practical microfluidic applications involving electroosmotic flows are in the sub-millisecond range, such as high-speed electrophoretic separation [6–8], the decoupling of particle velocity and background electroosmotic flow velocity with pulsed electric fields [9] as well as microfluidic pumping and mixing with alternating current (AC) or modulated direct current (DC) fields [10–12]. Therefore, understanding the dynamic characteristics of electroosmotic flows is highly important for these microfluidic applications. Previous theoretical studies of dynamic characteristics of electroosmosis have been focusing on Newtonian fluids under various modes of suddenly applied external fields [9,13–20]. In addition, experimental investigation of the dynamics of electroosmotic flows of Newtonian liquids was performed by utilizing state-of-the-art micro-particle image velocimetry (micro-PIV) techniques [21].

However, microfluidic devices are practically used to process biofluids (such as solutions of blood and DNA) which cannot be treated as Newtonian fluids and are usually characterized with viscosities dependent on the rate of shear. Therefore, the more general Cauchy momentum equation with a proper constitutive law, instead of the Navier–Stokes equation, should be used to describe flow characteristics of such fluids. Among various constitutive laws for non-Newtonian fluids, power-law constitutive law is the simplest yet most popular one. It has been shown to be suitable for the description of pressure-driven flows of various non-Newtonian fluids, such as polymeric solutions [22,23] and blood solutions [24–26]. A number of recent investigations [27–37] already showed that electroosmotically driven flows of non-Newtonian fluids behave differently from those of Newtonian fluids. However, their attention was unanimously focused on the steady-state characteristics and the dynamic aspects were missing from these investigations.

At present, the dynamic characteristics of electroosmosis of non-Newtonian fluids were mainly investigated for viscoelastic fluids. The existing studies have analyzed the dynamic electroosmosis of viscoelastic fluids in slit channel [38,39], circular channel [40,41], semi-circular channel [42] and rectangular channel [43]. It was revealed that the presence of the viscoelasticity can essentially affect dynamic aspects of electroosmosis. For power-law fluids, the study is however quite rare. The most relevant work at the moment is by Deng et al. [44] who analyzed the unsteady electroosmotic flow in a rectangular microchannel. Yet, their investigation is limited to low channel zeta potential and a pure DC driving electric field. In the present study, we report numerical analyses of transient electroosmotic flows of power-law fluids in a rectangular microchannel driven by three modes of electric field, namely a pure DC electric field, a pure AC electric field and a combination of AC and DC electric field. Besides, our analyses are valid for arbitrary channel zeta potential. The numerical simulations are carried out by using the finite element method which is verified through a comparison with the exact solution available for Newtonian fluids. Parametric studies are performed to examine the effects of fluid rheology (fluid behavior index) on the dynamics of electroosmosis of power-law fluids.

2. Problem Formulation

Figure 1 shows the dimensions of the microchannel and the coordinate system adopted in the present work. The channel is filled with a liquid solution having a dielectric constant of ε_r. It is assumed that all channel walls are uniformly charged with a zeta potential of ψ_w, and the liquid solution exhibits a typical non-Newtonian behavior which is described by the well-known power-law model. As soon as an external dynamic electric field $E_0 f(t)$ is imposed along the x-axis direction, the fluid in the microchannel is set in motion due to electroosmosis. $f(t)$ is a time-dependent function characterizing the dynamic behavior of the applied electric field. In this study, we consider three different modes of electric fields: a DC driving electric field with $f(t) = 1$, an AC driving electric field with $f(t) = \sin(\omega t)$ and a combination of AC and DC electric fields with $f(t) = 1 + \varepsilon\sin(\omega t)$, in which ω

and ε are the frequency and the amplitude of the AC component in the combined electric field. Because of the geometrical symmetry, the analysis would be restricted in the first quadrant of z-y plane.

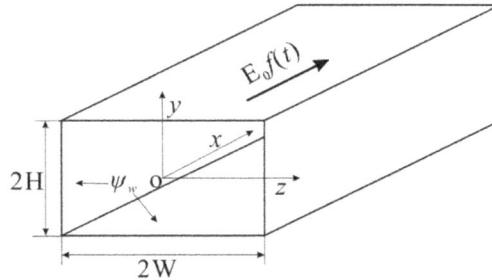

Figure 1. Electroosmotic flow system in a rectangular microchannel. The width of the channel is $2W$ and the depth of the channel is $2H$. All the walls are uniformly charged with a zeta potential ψ_w, and the dynamic electric field $E_0 f(t)$ is applied along the axial direction of the microchannel. The zeta potential on the walls induces a near-wall electric double layer (EDL) which has a non-zero charge density. Then, interaction of the external electric field with the non-zero charge density induces a driving force for electroosmosis.

2.1. Electric Field in the EDL

As aqueous solution in the microchannel contacts the charged wall, a thin charged solution layer forms near the wall to neutralize the surface charge on the channel wall. This layer is commonly referred to as the electric double layer (EDL). According to the electrostatic theory, electric potential distribution in the EDL region is governed by Poisson equation which can be expressed as

$$\frac{\partial^2 \psi}{\partial y^2} + \frac{\partial^2 \psi}{\partial z^2} = -\frac{\rho_e}{\varepsilon_0 \varepsilon_r} \tag{1}$$

where ε_0 is the electric permittivity of vacuum, ρ_e is the net charge density in the *EDL* region, and can be related to the EDL potential via (by invoking assumptions of Boltzmann distribution and z_v:z_v symmetric electrolyte) [45]

$$\rho_e = (n_+ - n_-)z_v e = -2z_v e n_\infty \sinh\left(\frac{z_v e \psi}{k_B T}\right) \tag{2}$$

where n_+ and n_- are respectively number of concentrations of cations and anions in the EDL region. n_∞ and z_v are the bulk number concentration and the valence of ions, respectively. e is the elementary charge, k_B is the Boltzmann constant, and T is the absolute temperature.

Introducing dimensionless groups: $\overline{y} = y/D_h$, $\overline{z} = z/D_h$, $K = \kappa D_h$, and $\overline{\psi} = z_v e \psi/(k_B T)$, then substituting Equation (2) into Equation (1), one can show that electrical potential profile in the EDL is governed by the so-called Poisson–Boltzmann equation

$$\frac{\partial^2 \overline{\psi}}{\partial \overline{y}^2} + \frac{\partial^2 \overline{\psi}}{\partial \overline{z}^2} = K^2 \sinh \overline{\psi} \tag{3}$$

which is subject to the following boundary conditions

$$\overline{\psi}\big|_{\overline{y}=H/D_h} = \overline{\psi}_w, \quad \overline{\psi}\big|_{\overline{z}=W/D_h} = \overline{\psi}_w \tag{4}$$

$$\frac{\partial \overline{\psi}}{\partial \overline{y}}\bigg|_{\overline{y}=0} = 0, \quad \frac{\partial \overline{\psi}}{\partial \overline{z}}\bigg|_{\overline{z}=0} = 0 \tag{5}$$

In the above equations, D_h represents the hydrodynamic diameter of the rectangular microchannel and is defined as $D_h = 4HW/(H + W)$, the dimensionless wall zeta potential is given by $\overline{\psi}_w = z_v e\psi_w/(k_B T)$, and the Debye length κ^{-1} is defined as $\kappa^{-1} = [\varepsilon_0\varepsilon_r k_B T/(2e^2 z_v^2 n_\infty)]^{1/2}$.

2.2. Electroosmotic Flow of Power-Law Fluids

When an external electric field is applied, the flow of an incompressible power-law liquid induced by electroosmosis is jointly governed by the general Cauchy momentum equation and the continuity equation, i.e.,

$$\rho\left[\frac{\partial \mathbf{V}}{\partial t} + (\mathbf{V} \cdot \nabla)\mathbf{V}\right] = -\nabla p + \nabla \cdot [2\mu(\Gamma)\mathbf{\Gamma}] + \mathbf{F} \tag{6}$$

$$\nabla \cdot \mathbf{V} = 0 \tag{7}$$

where $\mathbf{V}$ is the velocity vector, ρ is the density of the liquid, p is the pressure, $\mathbf{F}$ is the body force vector, $\mathbf{\Gamma}$ is the rate of strain tensor and is given by $\mathbf{\Gamma} = [\nabla\mathbf{V} + (\nabla\mathbf{V})^T]/2$. $\mu(\Gamma)$ is the dynamic viscosity and generally is a function of the magnitude of $\mathbf{\Gamma}$ tensor, Γ. The present work considers a power-law non-Newtonian fluid, and its dynamic viscosity is given by

$$\mu(\Gamma) = m(2\Gamma)^{n-1} \tag{8}$$

where m is the flow consistency index, and n is the flow behavior index. Shear-thinning (also termed as pseudoplastic) behavior is defined by $n < 1$, and it indicates that the fluid viscosity decreases with the increasing rate of shear. The pseudoplastic effect commonly exists in polymeric fluids which are subject to the high rate of shear, as is developed in microchannels and nanochannels. Newtonian behavior is defined by $n = 1$. Shear-thickening (also termed as dilatant) behavior is defined by $n > 1$, and it shows that the fluid viscosity increases with the increasing rate of shear. The dilatant effect is unusual and rarely encountered in practical applications.

For the unidirectional electroosmotic flow considered here, the velocity vector can be simplified as

$$\mathbf{V} = u(y,z,t)\,\mathbf{e}_x \tag{9}$$

where u is the x-component of velocity and $\mathbf{e}_x$ the is unit vector along the x-direction. Clearly, the continuity Equation (7) is automatically satisfied for the velocity field given by Equation (9). Furthermore, for electroosmotic flow, the only driving force is due to the interaction of the applied electrical field $E_0 f(t)$ with the net charge density ρ_e in the EDL region. In the present system shown in Figure 1, such body force acts only along the x direction, and is expressed as

$$\mathbf{F} = \rho_e E_0 f(t)\mathbf{e}_x \tag{10}$$

For an open-end, horizontally placed channel, there is no induced pressure gradient along the channel and hence the pressure gradient term in the Cauchy momentum equation can be neglected.

Besides the nondimensional groups used in the previous subsection, we introduce additional nondimensional parameters

$$\bar{t} = \frac{\mu_0}{\rho D_h^2}t, \quad \bar{u} = \frac{u}{u_0}, \quad \overline{m} = \frac{m(2n_\infty k_B T)^{n-1}}{\mu_0^n}, \quad \overline{E}_0 = \frac{z_v e D_h E_0}{k_B T} \tag{11}$$

and take into account the aforementioned simplifications, then the nondimensional version of Equation (6) reads

$$\frac{K^{2(n-1)}}{\overline{m}}\frac{\partial \overline{u}}{\partial \overline{t}} = \frac{\partial}{\partial \overline{y}}\left[\overline{\mu}(\overline{\Gamma})\frac{\partial \overline{u}}{\partial \overline{y}}\right] + \frac{\partial}{\partial \overline{z}}\left[\overline{\mu}(\overline{\Gamma})\frac{\partial \overline{u}}{\partial \overline{z}}\right] - \frac{K^{2n}}{\overline{m}}\overline{E}_0\overline{f}(\overline{t})\sinh(\overline{\psi}) \tag{12}$$

where $\bar{\mu}(\bar{\Gamma})$ can be formulated as

$$\bar{\mu}(\bar{\Gamma}) = (2\bar{\Gamma})^{n-1} = \left[\left(\frac{\partial\bar{u}}{\partial\bar{y}}\right)^2 + \left(\frac{\partial\bar{u}}{\partial\bar{z}}\right)^2\right]^{\frac{n-1}{2}} \tag{13}$$

In Equation (11), μ_0 denotes the viscosity of Newtonian fluids and it has the same magnitude as the flow consistency index m. u_0 then can be viewed as the Helmholtz–Smoluchowski velocity for Newtonian liquids over a solid surface with zeta potential being equal to the thermal voltage $(k_BT/(z_ve))$ under an electric field strength of $k_BT/(z_veD_h)$, and is given by

$$u_0 = \frac{\varepsilon_0\varepsilon_r}{\mu_0}\frac{k_BT}{z_ve}\frac{k_BT}{D_hz_ve} \tag{14}$$

It is worth mentioning that reference quantities for time and velocity are independent of the rheological properties of fluids (n and m). Choosing reference quantities in such a manner is convenient when discussing the effect of fluid rheology on magnitudes of both electroosmotic velocity and transient start-up time in Section 4.

The initial and boundary conditions applicable to Equation (12) are

$$\bar{u}|_{\bar{t}=0} = 0 \tag{15}$$

$$\frac{\partial\bar{u}}{\partial\bar{y}}\bigg|_{\bar{y}=0} = 0, \quad \frac{\partial\bar{u}}{\partial\bar{z}}\bigg|_{\bar{z}=0} = 0 \tag{16}$$

$$\bar{u}|_{\bar{y}=H/D_h} = 0, \quad \bar{u}|_{\bar{z}=W/D_h} = 0 \tag{17}$$

3. Numerical Method and Model Validation

In the present analysis, both EDL potential field and electroosmotic flow field are solved in the partial differential equation (PDE) module of finite element numerical analysis package COMSOL Multiphysics 5.1 (COMSOL, Inc., Stockholm, Sweden). In the PDE module, the general form of PDE is given in terms of a series of coefficients and a source term which are left for the users to specify for formulating their models. These coefficients and the source term can be either constants or variables, thereby generating high flexibility for handling PDEs. In our work, a PDE governing the EDL potential (Equation (3)) and a PDE governing electroosmotic flow field (Equation (12)) are both constructed from the general form of PDE in Comsol. Through the source term $\sinh(\bar{\psi})$ in Equation (12), these two PDEs are coupled together.

In order to check the validity of the present model, we compared our numerical result with the exact result [46] derived for the starting electroosmotic flow of Newtonian fluids in a rectangular microchannel. However, their result was obtained under the Debye–Hückel linear approximation which assumes a small zeta potential on the channel wall. Thus, in the numerical validation, a small zeta potential ($\psi_w = -k_BT/(z_ve) \approx -25$ mV for monovalent electrolytes at room temperature) was prescribed for the channel walls and geometric dimensions of the microchannel were chosen as $2H = 10$ μm and $2W = 15$ μm. The working fluid flowing in the microchannel is the Newtonian solution (a special power-law fluid with flow behavior index $n = 1$) of a symmetric electrolyte ($z_v:z_v$), say NaCl. The bulk ionic number concentration was set to $n_\infty = 6.022 \times 10^{20}/\text{m}^3$. The dielectric constant of electrolytic solution was taken to be the same as that of room-temperature water, namely $\varepsilon_r = 78.5$. In electroosmotic flows, the velocity experiences steep changes in the EDL region near the channel walls. Therefore, in the present analysis, the mesh near the channel wall is finest to ensure that the velocity change in the EDL can be captured, and at least ten cells are positioned inside the EDL region. The mesh size increases towards the center region of the cross-section with mesh ratios of 1.04 and 1.03 in y and z directions, respectively. The maximal cell has dimensions of $\Delta y/D_h = \Delta z/D_h = 1.72 \times 10^{-2}$.

The time step used in this study is controlled to satisfy $\Delta t/T \leq 2 \times 10^{-3}$, where T is the start-up time for DC-driven electroosmosis or the period of the AC electric field. The calculated solutions were carefully validated to exclude both mesh dependency and time-step dependency. Mesh-independence was examined for two different mesh systems whose total cell numbers are 15,000 (150×100) and 60,000 (300×200), respectively. Two different time steps, i.e., 1×10^{-3} and 5×10^{-4}, were also examined. It was found that calculated flow rate differences under two examinations were both less than 1%. Therefore, mesh independence and time-step independence were confirmed, and then the mesh system with 15,000 cells (150×100) and a time step of 1×10^{-3} were applied in the study. The UMFPACK solver was used to solve the system with relative tolerances of spatial and temporal solutions both being 10^{-6}.

Figure 2 shows the velocity profiles at $\bar{z} = 0$ for three different time instants computed with the analytical formula [46] and our Comsol model. It can be seen from this plot that the numerical results of velocity distributions obtained from the Comsol model at three different time instants agree perfectly well with those obtained from the existing analytical model, which validates the high robustness and accuracy of the Comsol model.

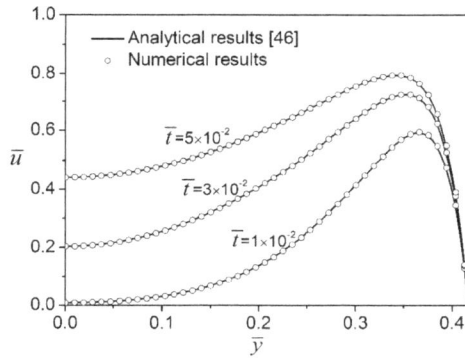

Figure 2. Transient velocity profiles along the $\bar{y}$ axis when $\bar{z} = 0$ at three different time instants for Newtonian fluids ($n = 1.0$) under a DC electric field $\bar{E}_0 = 1$ and a zeta potential $\bar{\psi}_w = -1$.

4. Results and Discussion

To predict dynamic behaviors of electroosmotic flows of power-law fluids under various modes of electric fields, we take values of some parameters to be the same as those in Section 3, i.e., $2H = 10$ μm, $2W = 15$ μm, $n_\infty = 6.022 \times 10^{20}$ m^{-3}. The dynamic viscosity of Newtonian fluids is set to be $\mu_0 = 9 \times 10^{-4}$ Pa·s (the same as room-temperature water) and flow consistency index of power-law fluids is taken as $m = 9 \times 10^{-4}$ Pa·s^n (the same magnitude as dynamic viscosity of Newtonian fluids). The corresponding dimensionless electrokinetic parameter $K = 40$, which makes sure that the microchannel has a moderately thin EDL, and thus the dynamic momentum transfer from the EDL to the bulk flow can be identified.

4.1. Transient Electroosmotic Flows of Power-Law Fluids under DC Electric Fields

Figure 3 shows electroosmotic velocity profiles of a power-law fluid with $n = 0.8$ at different time instants. Initially, the liquid in the whole microchannel is quiescent (not shown in Figure 3). As soon as the electric field is applied, the liquid within the EDL starts to flow immediately, but the bulk liquid in the middle portion of microchannel remains stationary. As the dimensionless time evolves to 10^{-3}, at $\bar{y} = 0$, the velocity reaches a local maximum near the vertical side wall (inside the EDL of side wall), and then drops gradually to zero as the distance is away from the side wall. At $\bar{y} = 0.35$ (very near the top wall), there is similarly a local maximal velocity near the side wall. However, the liquid far from the vertical side wall in this case is already in motion because of the EDL of the top wall. Moreover, the maximal velocity at $\bar{y} = 0.35$ is higher than that at $\bar{y} = 0$. This is because at $\bar{y} = 0.35$ near the vertical side wall (around the right upper corner of channel cross-section), the liquid is actuated by the electrostatic body force due to EDLs on both the top and side walls; while at $\bar{y} = 0$ near the vertical side wall, the liquid is actuated by the electrostatic body force due to the EDL on the side wall alone. In electroosmotic flows, the driving force is only present in the EDL region, and the generation of momentum is then also limited in the EDL region. As time evolves, the fluid velocity within the EDL continues to increase; at the same time, the bulk fluid starts moving due to the gradual transfer of momentum from the EDL to bulk liquid (see velocity profiles at $\bar{t} = 10^{-2}$ and $\bar{t} = 10^{-1}$). When the flow develops to the steady state ($\bar{t} \to \infty$), the velocity distribution exhibits a plug-like profile. It is also observed that the velocity profile at $\bar{y} = 0.35$ develops faster than that at $\bar{y} = 0$, which is peculiar to rectangular channels. This effect can be ascribed to the fact that at $\bar{y} = 0.35$ the driving force is present along the entire $\bar{z}$ axis, while at $\bar{y} = 0$ the driving force is present only near the vertical side wall.

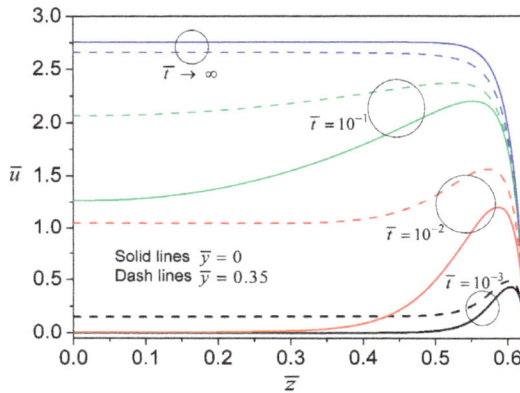

Figure 3. Transient evolution of axial velocity profiles for a power-law fluid with $n = 0.8$ and $\overline{\psi}_w = -1$ due to an applied DC electric field $\overline{E}_0 = 1$.

Usually, the fluid behavior index (n) is varied by the addition of polymers into the solutions which also changes the value of zeta potential. This indicates that the zeta potential is practically a function of n. However, at present, the quantitative relation between zeta potential and n is unclear and remains to be investigated. Therefore, for convenience, the zeta potential is assumed to be an independent variable which is not influenced by n in the current study. Such an assumption is widely adopted in the literature for study of electroosmotic flow of non-Newtonian fluids [27–37]. Figure 4 characterizes the transient development of electroosmosis of power-law fluids with different fluid behavior indices. The velocity in the whole channel domain is zero at $t = 0$ for all values of the fluid behavior index (not shown in Figure 4). As shown in Figure 4a, when dimensionless time evolves to 10^{-3}, the fluids with smaller fluid behavior indices acquire higher velocities inside the EDL region near the channel wall. The velocity inside the EDL becomes higher as time evolves, and at the same time the momentum generated inside the EDL gradually diffuses to the bulk. At $\bar{t} = 10^{-2}$, the fluids with smaller fluid behavior indices still have higher velocities inside the EDL region, while outside the EDL the velocity for a larger fluid behavior index surpasses that for a smaller fluid behavior index. This feature indicates that the momentum transfer is faster for a larger fluid behavior index due to the stronger viscous coupling between the EDL and the bulk liquid. At the steady state ($\bar{t} \to \infty$), normalized velocities in the bulk flow for four fluid behavior indices all increase to their corresponding constant values, which are typical for electroosmotically-driven flows. Furthermore, at the steady state, the velocity profiles for smaller fluid behavior indices become more plug-like and also the magnitude of bulk velocity is larger for a smaller fluid behavior index. As is the case for $n = 0.7$, the steady sate bulk liquid velocity is more than five times higher than that of Newtonian fluids ($n = 1$), which implies that the electroosmotic pumping of shearing-thinning fluids is far more efficient than that of Newtonian fluids. For situations where we have large-sized channels or thin EDLs (i.e., $K \gg 1$), it can be expected that the power-law fluid in the entire microchannel moves with a uniform bulk velocity. Consequently, the constant bulk velocities for various values of flow behavior index can be effectively seen as the so-called Helmholtz–Smoluchowski velocities in electrokinetics of power-law fluids (i.e., electrophoresis of particles in power-law fluids and electroosmosis of power-law fluids). In addition, it is shown in Figure 4b that the fluids with larger fluid behavior indices approach the steady state more quickly. This is because the fluids with larger fluid behavior indices are more viscous and then the momentum generated inside the EDL can be transferred more promptly to the center portion of the channel.

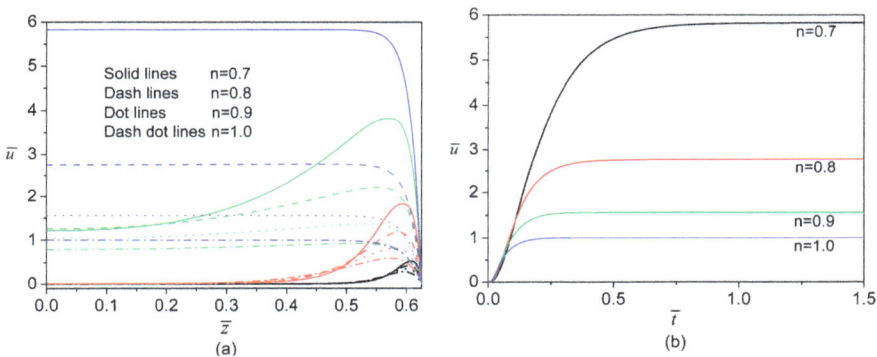

Figure 4. Comparison of transient development of electroosmosis for four different fluid behavior indices ($n = 0.7, 0.8, 0.9$ and 1.0) when $\bar{E}_0 = 1$ and $\bar{\psi}_w = -1$. (**a**) Transient velocity profiles along $\bar{z}$ axis at $\bar{y} = 0$. There are four groups of velocity profiles in this plot and each group represents the velocity profiles of four flow behavior indices at a specific time instant: the group with black color is at $\bar{t} = 10^{-3}$, the group with red color is at $\bar{t} = 10^{-2}$, the group with green color is at $\bar{t} = 10^{-1}$ and the group with blue color is at the steady state ($\bar{t} \to \infty$); (**b**) Time evolution of velocity at the center of cross-section ($\bar{z} = \bar{y} = 0$).

The effects of DC field strength and wall zeta potential on the transient development of electroosmosis are shown in Figure 5. The transient start-up time during which the velocity develops from zero to the steady state becomes shorter when the strength of electric field/zeta potential is decreased, and the magnitude of steady-state velocity increases nonlinearly with the increase of the strength of external electric field/zeta potential. These characteristics clearly differ from electroosmotic flows of Newtonian fluids for which the transient start-up time is independent of the strength of electric field/zeta potential, and also the magnitude of steady-state velocity increases *linearly* with the increasing strength of electric field/zeta potential [17,46,47].

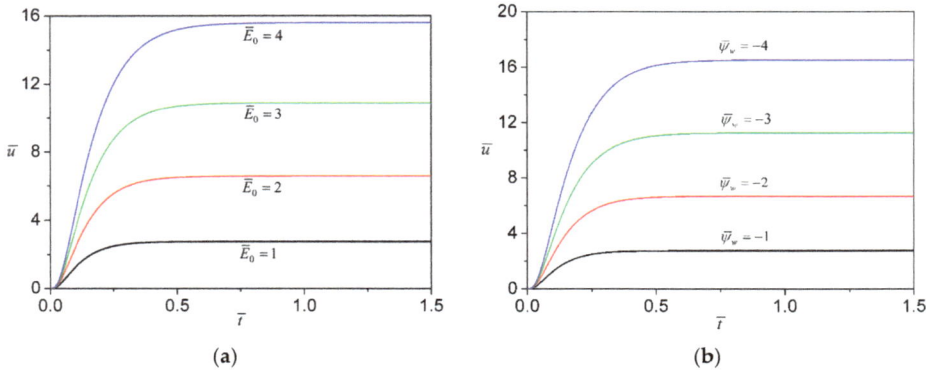

Figure 5. (a) Comparison of transient development of the axial velocity at the transverse center ($\bar{z} = \bar{y} = 0$) under different magnitudes of external electric field strength when $n = 0.8$ and $\bar{\Psi}_w = -1$; (b) Comparison of transient development of the axial velocity at the transverse center ($\bar{z} = \bar{y} = 0$) for different magnitudes of zeta potential when $n = 0.8$ and $\bar{E}_0 = 1$.

4.2. Transient Electroosmotic Flows of Power-Law Fluids under AC Electric Fields

In this particular investigation, the electroosmotic flow is driven by a pure AC electric field. Then, in the simulation, we choose $\bar{f}(\bar{t}) = \sin(\bar{\omega}\bar{t})$ and the corresponding dimensionless frequency to be $\bar{\omega} = \omega \rho D_h^2/\mu_0 = \pi$. Figure 6 presents the steady-state development of the axial velocity profile in the transverse section for a half period (from phase $\bar{\omega}\bar{t} = 0$ to phase $\bar{\omega}\bar{t} = \pi$) when $\bar{E}_0 = 1$ and $\bar{\Psi}_w = -1$. At $\bar{\omega}\bar{t} = 0$, although the electric field strength is zero, the flow field lags behind the electric field and the preceding negative electric field strength causes liquid in the microchannel to move along the negative x direction (negative velocity). As time elapses, the liquid within the EDL is rapidly driven to the positive x direction. Then, at the same time, the momentum transfer from the EDL to the bulk flow progresses, leading to the expansion of the positive-velocity region from the EDL towards the central region of the microchannel. Until phase $\bar{\omega}\bar{t} = \pi/5$, the positive-velocity region already expands to occupy the entire microchannel. From phase $\bar{\omega}\bar{t} = \pi/5$ to phase $\bar{\omega}\bar{t} = \pi/2$, the momentum transfer from the EDL to the bulk flow is enhanced by the increasing electric field strength, and thus the velocity in the whole channel domain continues to grow. After phase $\bar{\omega}\bar{t} = \pi/2$, the strength of the electric field begins to decrease, and the liquid the within EDL responds instantaneously to such change. Therefore, there is a slight reduction in the positive axial velocity near the walls. Nevertheless, the positive axial velocity in the microchannel center still increases due to the inertial acceleration (see profiles at phase $\bar{\omega}\bar{t} = 3\pi/5$). After phase $\bar{\omega}\bar{t} = 3\pi/5$ (such as phase $\bar{\omega}\bar{t} = 4\pi/5$), the decrement of momentum inside the EDL expands towards the central region of channel, which makes the axial velocity in the bulk flow decrease. At phase $\bar{\omega}\bar{t} = \pi$, it is noted that the axial flow velocity profiles strongly resemble those at phase $\bar{\omega}\bar{t} = 0$ in terms of their shapes. However, the direction of axial velocity is opposite to that at phase $\bar{\omega}\bar{t} = 0$. During the second half period (from $\bar{\omega}\bar{t} = \pi$ to $\bar{\omega}\bar{t} = 2\pi$), since the variation of an AC driving electric field is a mirror image of that during the preceding half period (from $\bar{\omega}\bar{t} = 0$ to $\bar{\omega}\bar{t} = \pi$), it is quite understandable that the corresponding evolution of axial velocity profiles is also a mirror image (symmetric with respect to the dot line $\bar{u} = 0$ in Figure 6) of the preceding half period.

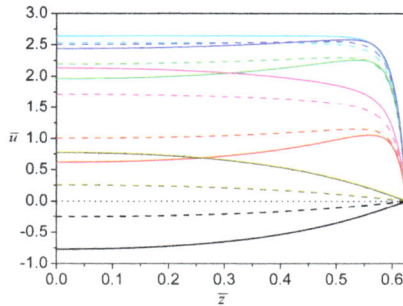

Figure 6. Steady-state oscillating axial velocity profiles of a power-law fluid with *n* = 0.8 at different phases in a half period from $\overline{\omega}\overline{t} = 0$ to $\overline{\omega}\overline{t} = \pi$ when $\overline{E}_0 = 1$, $\overline{\Psi}_w = -1$ and $\overline{\omega} = \pi$. There are seven groups of velocity profiles differentiated by different colors and each group represents the profiles at a given AC phase. In each group, the solid line is the velocity profile at $\overline{y} = 0$ and the dash line is the profile at $\overline{y} = 0.35$. The group with black color represents the profiles at $\overline{\omega}\overline{t} = 0$; the group with red color represents the profiles at $\overline{\omega}\overline{t} = \pi/5$; the group with green color represents the profiles at $\overline{\omega}\overline{t} = 2\pi/5$; the group with blue color represents the profiles at $\overline{\omega}\overline{t} = \pi/2$; the group with cyan color represents the profiles at $\overline{\omega}\overline{t} = 3\pi/5$; the group with magenta color represents the profiles at $\overline{\omega}\overline{t} = 4\pi/5$ and the group with dark-yellow color represents the profiles at $\overline{\omega}\overline{t} = \pi$. On the straight dot line, $\overline{u} = 0$.

Figure 7 presents the comparison of transient velocity development for different values of fluid behavior index at both the start-up stage and the steady-state oscillation. After turning on the AC electric field at $\overline{t} = 0$, it is seen from Figure 7a that the fluid at the transverse center ($\overline{y} = \overline{z} = 0$) remains quiescent for a very short period of time. At this moment, the momentum generated inside the EDL is still limited to the regions near the channel walls and therefore needs time to diffuse to the bulk liquid. After a certain amount of time, the momentum is gradually transferred to the bulk liquid and then the liquid starts to move. The fluid with a larger fluid behavior index responds more promptly to the applied AC field and then reaches the peak velocity more quickly. This is consistent with the case of the DC electric field in which the fluid with a larger fluid behavior index reaches the steady state more quickly. When the flow attains the steady-state oscillation (Figure 7b), the velocity generally lags behind the applied AC electric field, and the phase lag increases with the decrease of fluid behavior index, as is indicated in Figure 7c. We also note from Figure 7b that from one peak to its corresponding trough, power-law fluids with a smaller fluid behavior index experience more significant acceleration. Furthermore, the amplitude of oscillating velocity increases as the fluid behavior index decreases.

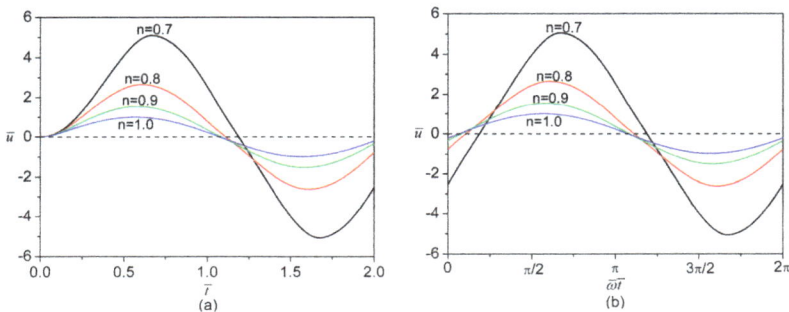

Figure 7. *Cont.*

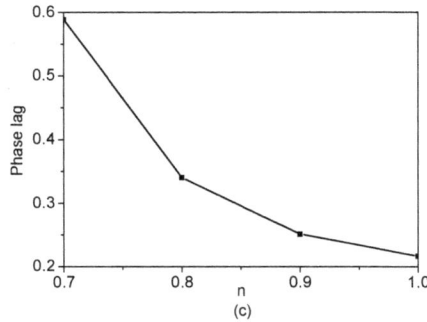

(c)

Figure 7. Comparison of the velocity evolution at the transverse center ($\bar{z} = \bar{y} = 0$) of the microchannel for different values of flow behavior index (n = 0.7, 0.8, 0.9, 1.0) under an AC electric field with $\bar{E}_0 = 1$, $\bar{w} = \pi$ and $\bar{\Psi}_w = -1$. (**a**) Start-up characteristics of the electroosmotic velocity; (**b**) Steady-state oscillation of the electroosmotic velocity; (**c**) Variation of the phase lag between velocity and AC field with the fluid behavior index at the steady-state oscillation.

4.3. Enhancement of Electroosmotic Flows of Power-Law Fluids by AC/DC Combined Electric Fields

For pressure-driven flows of power-law fluids, it is known that the flows can be enhanced by introducing one pulsatile pressure gradient to a constant pressure gradient [22,48]. Generally, this flow enchantment arises from the nonlinear relationship between the stress and the rate of strain which reduces effective viscosity of the liquids. Our investigation here proves that a similar concept can be used to enhance the electroosmotically-driven flow of power-law fluids by adding one AC electric field to a DC electric field. Particularly, the time characteristics of an AC/DC combined electric field is characterized by $\bar{f}(\bar{t}) = 1 + \varepsilon \sin(\bar{w}\bar{t})$, where ε defines the amplitude of the AC component of the electric field. In addition, a percentile, q = 100% × $(Q_\varepsilon - Q_0)/Q_0$, is defined to quantify the flow enhancement due to the AC electric field. In the definition of this percentile, Q_ε is the flow rate due to an AC/DC combined electric field $\bar{E}_0[1 + \varepsilon \sin(\bar{w}\bar{t})]$, and Q_0 is the flow rate due to a DC electric field $\bar{E}_0$ alone. The higher the q is, the more significant the flow enhancement is. Figure 8 shows the effects of AC amplitude and flow behavior index on q. It is clear that q increases with the increase of AC amplitude or the decrease of flow behavior index. These predictions are similar to the case of pressure-driven power-law fluid flow [22,48] in which the flow enhancement is amplified by increasing the amplitude of pulsatile pressure gradient or decreasing the flow behavior index.

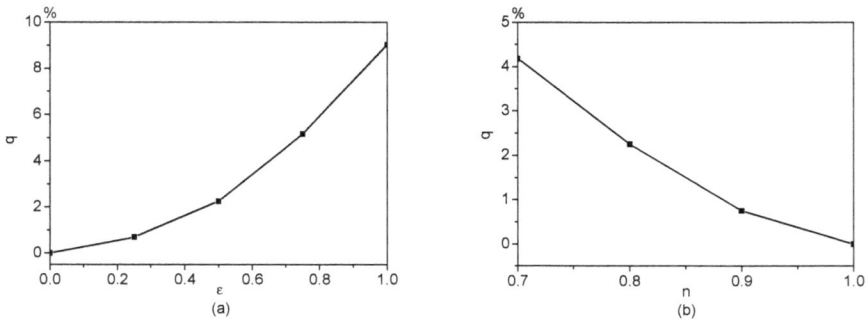

Figure 8. Flow enhancement of electroosmotic flows of power-law fluids by an AC/DC combined electric field with $\bar{E}_0 = 1$, $\bar{w} = \pi$ and $\bar{\Psi}_w = -1$. (**a**) Variation of q with ε when n = 0.8; (**b**) Variation of q with the flow behavior index n when ε = 0.5.

5. Conclusions

We have presented a comprehensive numerical analysis of dynamic electroosmotic flows of power-law fluids in rectangular microchannels under three modes of electric fields. For the case of transient electroosmotic flow driven by a pure DC electric field, initially, the DC electric field drives the liquid within the EDL immediately to move in the axial direction. Then the momentum generated in the EDL gradually transfers to the bulk region of channel, which leads to a plug-like velocity profile at the steady state. Generally, the non-Newtonian nature of fluids complicates the transient dynamics of electroosmosis. It is observed that the flow with a higher fluid behavior index responds more promptly to the external DC electric field and reaches the steady state more quickly. Another prominent feature is that the transient start-up time becomes dependent on the strength of the electric field/zeta potential for power-law fluids.

For the case of a pure AC electric field, the results show that the flow in the microchannel initially shows a transient start-up after the immediate application of the electric field and finally attains a steady-state oscillation. The electroosmosis of fluid with a larger fluid behavior index demonstrates a faster response to the external AC electric field and consequently has a smaller phase lag behind the applied AC electric field. At last, for the case of an AC/DC combined electric field, it is shown that the flow is enhanced as compared to a pure DC electric field. This feature is similar to the flow enhancement in non-Newtonian fluid flows driven by a pulsatile pressure gradient. The results show that increasing the amplitude of AC field component or decreasing the flow behavior index can intensify the electroosmotic flow enhancement of power-law fluids. These conclusions are of practical significance because they can be of potential use in guiding the design of microfluidic analytical devices which involve electroosmotic flows of non-Newtonian fluids.

Acknowledgments: C.Z. is supported by the "Top Young Talent Support Plan" from Xi'an Jiaotong University.

Author Contributions: C.Z. formulated the model and performed the numerical simulations; C.Z. and W.Z. analyzed the results and co-wrote the paper; C.Y. gave advice on the numerical simulations. All authors discussed the results.

Conflicts of Interest: The authors declare no conflict of interest.

References

1. Harrison, D.J.; Fluri, K.; Seiler, K.; Fan, Z.; Effenhauser, C.S.; Manz, A. Micromachining a miniaturized capillary electrophoresis-based chemical analysis system on a chip. *Science* **1993**, *261*, 895–897. [CrossRef] [PubMed]
2. Bousse, L.; Cohen, C.; Nikiforov, T.; Chow, A.; Kopf-Sill, A.R.; Dubrow, R.; Parce, J.W. Electrokinetically controlled microfluidic analysis systems. *Annu. Rev. Biophys. Biomol. Struct.* **2000**, *29*, 155–181. [CrossRef] [PubMed]
3. Whitesides, G.M. The origins and the future of microfluidics. *Nature* **2006**, *442*, 368–373. [CrossRef] [PubMed]
4. Manz, A.; Effenhauser, C.S.; Burggraf, N.; Harrison, D.J.; Seiler, K.; Fluri, K. Electroosmotic pumping and electrophoretic separations for miniaturized chemical analysis systems. *J. Micromech. Microeng.* **1994**, *4*, 257–265. [CrossRef]
5. Yossifon, G.; Frankel, I.; Miloh, T. Macro-scale description of transient electro-kinetic phenomena over polarizable dielectric solids. *J. Fluid Mech.* **2009**, *620*, 241–262. [CrossRef]
6. Fan, Z.H.; Harrison, D.J. Micromachining of capillary electrophoresis injectors and separators on glass chips and evaluation of flow at capillary intersections. *Anal. Chem.* **1994**, *66*, 177–184. [CrossRef]
7. Jacobson, S.C.; Culbertson, C.T.; Daler, J.E.; Ramsey, J.M. Microchip structures for submillisecond electrophoresis. *Anal. Chem.* **1998**, *70*, 3476–3480. [CrossRef]
8. Jacobson, S.C.; Hergenröder, R.; Koutny, L.B.; Ramsey, J.M. High-Speed Separations on a Microchip. *Anal. Chem.* **1994**, *66*, 1114–1118. [CrossRef]
9. Söderman, O.; Jönsson, B. Electro-osmosis: Velocity profiles in different geometries with both temporal and spatial resolution. *J. Chem. Phys.* **1996**, *105*, 10300–10311. [CrossRef]
10. Ajdari, A. Pumping liquids using asymmetric electrode arrays. *Phys. Rev. E* **2000**, *61*, R45–R48. [CrossRef]

11. González, A.; Ramos, A.; Green, N.G.; Castellanos, A.; Morgan, H. Fluid flow induced by nonuniform AC electric fields in electrolytes on microelectrodes. II. A linear double-layer analysis. *Phys. Rev. E* **2000**, *61*, 4019–4028. [CrossRef]

12. Ramos, A.; Morgan, H.; Green, N.G.; Castellanos, A. AC electric-field-induced fluid flow in microelectrodes. *J. Colloid Interface Sci.* **1999**, *217*, 420–422. [CrossRef] [PubMed]

13. Hanna, W.T.; Osterle, J.F. Transient electro-osmosis in capillary tubes. *J. Chem. Phys.* **1968**, *49*, 4062–4068. [CrossRef]

14. Ivory, C.F. Transient electroosmosis: The momentum transfer coefficient. *J. Colloid Interface Sci.* **1983**, *96*, 296–298. [CrossRef]

15. Keh, H.J.; Tseng, H.C. Transient electrokinetic flow in fine capillaries. *J. Colloid Interface Sci.* **2001**, *242*, 450–459. [CrossRef]

16. Kang, Y.; Yang, C.; Huang, X. Dynamic aspects of electroosmotic flow in a cylindrical microcapillary. *Int. J. Eng. Sci.* **2002**, *40*, 2203–2221. [CrossRef]

17. Yang, C.; Ng, C.B.; Chan, V. Transient analysis of electroosmotic flow in a slit microchannel. *J. Colloid Interface Sci.* **2002**, *248*, 524–527. [CrossRef] [PubMed]

18. Yang, J.; Kwok, D.Y. Time-dependent laminar electrokinetic slip flow in infinitely extended rectangular microchannels. *J. Chem. Phys.* **2003**, *118*, 354–363. [CrossRef]

19. Campisi, M.; Accoto, D.; Dario, P. AC electroosmosis in rectangular microchannels. *J. Chem. Phys.* **2005**, *123*, 204724. [CrossRef] [PubMed]

20. Mishchuk, N.A.; González-Caballero, F. Nonstationary electroosmotic flow in open cylindrical capillaries. *Electrophoresis* **2006**, *27*, 650–660. [CrossRef] [PubMed]

21. Yan, D.; Yang, C.; Nguyen, N.T.; Huang, X. Diagnosis of transient electrokinetic flow in microfluidic channels. *Phys. Fluids* **2007**, *19*, 017114. [CrossRef]

22. Sundstrom, D.W.; Kaufman, A. Pulsating Flow of Polymer Solutions. *Ind. Eng. Chem. Process Des. Dev.* **1977**, *16*, 320–325. [CrossRef]

23. Phan-Thien, N. On pulsating flow of polymeric fluids. *J. Non-Newton. Fluid Mech.* **1978**, *4*, 167–176. [CrossRef]

24. Mazumdar, J.N. *Biofluid Mechanics*; World Scientific: Singapore, 1992.

25. Tu, C.; Deville, M. Pulsatile flow of non-Newtonian fluids through arterial stenoses. *J. Biomech.* **1996**, *29*, 899–908. [CrossRef]

26. Buchanan, J.R.; Kleinstreuer, C.; Comer, J.K. Rheological effects on pulsatile hemodynamics in a stenosed tube. *Comput. Fluids* **2000**, *29*, 695–724. [CrossRef]

27. Zimmerman, W.; Rees, J.; Craven, T. Rheometry of non-Newtonian electrokinetic flow in a microchannel T-junction. *Microfluid. Nanofluid.* **2006**, *2*, 481–492. [CrossRef]

28. Das, S.; Chakraborty, S. Analytical solutions for velocity, temperature and concentration distribution in electroosmotic microchannel flows of a non-Newtonian bio-fluid. *Anal. Chim. Acta* **2006**, *559*, 15–24. [CrossRef]

29. Zhao, C.; Zholkovskij, E.; Masliyah, J.H.; Yang, C. Analysis of electroosmotic flow of power-law fluids in a slit microchannel. *J. Colloid Interface Sci.* **2008**, *326*, 503–510. [CrossRef] [PubMed]

30. Park, H.M.; Lee, W.M. Effect of viscoelasticity on the flow pattern and the volumetric flow rate in electroosmotic flows through a microchannel. *Lab Chip* **2008**, *8*, 1163–1170. [CrossRef] [PubMed]

31. Berli, C.L.A. Output pressure and efficiency of electrokinetic pumping of non-Newtonian fluids. *Microfluid. Nanofluid.* **2009**, *8*, 197–207. [CrossRef]

32. Zhao, C.; Yang, C. Analysis of Power-Law Fluid Flow in a Microchannel with Electrokinetic Effects. *Int. J. Emerg. Multidiscip. Fluid Sci.* **2009**, *1*, 37–52. [CrossRef]

33. Zhao, C.; Yang, C. Nonlinear Smoluchowski velocity for electroosmosis of Power-law fluids over a surface with arbitrary zeta potentials. *Electrophoresis* **2010**, *31*, 973–979. [CrossRef] [PubMed]

34. Zhao, C.; Yang, C. An exact solution for electroosmosis of non-Newtonian fluids in microchannels. *J. Non-Newton. Fluid Mech.* **2011**, *166*, 1076–1079. [CrossRef]

35. Zhao, C. Electro-osmotic mobility of non-Newtonian fluids. *Biomicrofluidics* **2011**, *5*, 014110. [CrossRef] [PubMed]

36. Vakili, M.A.; Sadeghi, A.; Saidi, M.H.; Mozafari, A.A. Electrokinetically driven fluidic transport of power-law fluids in rectangular microchannels. *Colloids Surf. A* **2012**, *414*, 440–456. [CrossRef]

37. Chen, S.; He, X.; Bertola, V.; Wang, M. Electro-osmosis of non-Newtonian fluids in porous media using lattice Poisson-Boltzmann method. *J. Colloid Interface Sci.* **2014**, *436*, 186–193. [CrossRef] [PubMed]
38. Jian, Y.; Su, J.; Chang, L.; Liu, Q.; He, G. Transient electroosmotic flow of general Maxwell fluids through a slit microchannel. *Z. Angew. Math. Phys.* **2014**, *65*, 435–447. [CrossRef]
39. Bandopadhyay, A.; Ghosh, U.; Chakraborty, S. Time periodic electroosmosis of linear viscoelastic liquids over patterned charged surfaces in microfluidic channels. *J. Non-Newton. Fluid Mech.* **2013**, *202*, 1–11. [CrossRef]
40. Zhao, M.; Wang, S.; Wei, S. Transient electro-osmotic flow of Oldroyd-B fluids in a straight pipe of circular cross section. *J. Non-Newton. Fluid Mech.* **2013**, *201*, 135–139. [CrossRef]
41. Liu, Q.-S.; Jian, Y.-J.; Chang, L.; Yang, L.-G. Alternating current (AC) electroosmotic flow of generalized Maxwell fluids through a circular microtube. *Int. J. Phys. Sci.* **2012**, *7*, 5935–5941.
42. Bao, L.-P.; Jian, Y.-J.; Chang, L.; Su, J.; Zhang, H.-Y.; Liu, Q.-S. Time Periodic Electroosmotic Flow of the Generalized Maxwell Fluids in a Semicircular Microchannel. *Commun. Theor. Phys.* **2013**, *59*, 615–622. [CrossRef]
43. Zhao, C.; Yang, C. Exact solutions for electro-osmotic flow of viscoelastic fluids in rectangular micro-channels. *Appl. Math. Comput.* **2009**, *211*, 502–509. [CrossRef]
44. Deng, S.Y.; Jian, Y.J.; Bi, Y.H.; Chang, L.; Wang, H.J.; Liu, Q.S. Unsteady electroosmotic flow of power-law fluid in a rectangular microchannel. *Mech. Res. Commun.* **2012**, *39*, 9–14. [CrossRef]
45. Li, D. *Electrokinetics in Microfluidics*; Elsevier Academic Press: London, UK, 2004.
46. Chang, C.C.; Wang, C.Y. Starting electroosmotic flow in an annulus and in a rectangular channel. *Electrophoresis* **2008**, *29*, 2970–2979. [CrossRef] [PubMed]
47. Marcos Yang, C.; Wong, T.N.; Ooi, K.T. Dynamic aspects of electroosmotic flow in rectangular microchannels. *Int. J. Eng. Sci.* **2004**, *42*, 1459–1481. [CrossRef]
48. Phan-Thien, N.; Dudek, J. Pulsating flow of a plastic fluid. *Nature* **1982**, *296*, 843–844. [CrossRef]

Article

A Y-Shaped Microfluidic Device to Study the Combined Effect of Wall Shear Stress and ATP Signals on Intracellular Calcium Dynamics in Vascular Endothelial Cells

Zong-Zheng Chen [1], Zheng-Ming Gao [1], De-Pei Zeng [1], Bo Liu [1], Yong Luan [2,*] and Kai-Rong Qin [1,*]

[1] Department of Biomedical Engineering, Dalian University of Technology, Dalian 116024, China; zongzheng@mail.dlut.edu.cn (Z.-Z.C.); gaozhengming@mail.dlut.edu.cn (Z.-M.G.); 1925044995@mail.dlut.edu.cn (D.-P.Z.); lbo@dlut.edu.cn (B.L.)

[2] Department of Anesthesiology, The First Affiliated Hospital of Dalian Medical University, Dalian 116011, China

* Correspondence: cclyyly@hotmail.com (Y.L.); krqin@dlut.edu.cn (K.-R.Q.); Tel.: +86-8363-5963-2054 (Y.L.); +86-411-8470-9690 (K.-R.Q.)

Academic Editors: Weihua Li, Hengdong Xi and Say Hwa Tan
Received: 13 October 2016; Accepted: 18 November 2016; Published: 23 November 2016

Abstract: The intracellular calcium dynamics in vascular endothelial cells (VECs) in response to wall shear stress (WSS) and/or adenosine triphosphate (ATP) have been commonly regarded as an important factor in regulating VEC function and behavior including proliferation, migration and apoptosis. However, the effects of time-varying ATP signals have been usually neglected in the past investigations in the field of VEC mechanobiology. In order to investigate the combined effects of WSS and dynamic ATP signals on the intracellular calcium dynamic in VECs, a Y-shaped microfluidic device, which can provide the cultured cells on the bottom of its mixing micro-channel with stimuli of WSS signal alone and different combinations of WSS and ATP signals in one single micro-channel, is proposed. Both numerical simulation and experimental studies verify the feasibility of its application. Cellular experimental results also suggest that a combination of WSS and ATP signals rather than a WSS signal alone might play a more significant role in VEC Ca^{2+} signal transduction induced by blood flow.

Keywords: Y-shaped microfluidic device; wall shear stress; adenosine triphosphate (ATP) signal; combined effect; vascular endothelial cells; calcium dynamics

---

1. Introduction

Vascular endothelial cells (VECs) lining the innermost layer of vessel walls directly contact the flowing blood and thus are exposed to both wall shear stress (WSS) induced by blood flow and adenosine triphosphate (ATP) contained in the blood. A number of investigations have revealed that this shear stress, either alone or along with the presence of ATP, activates the dynamic response of intracellular calcium ion (Ca^{2+}) signaling system in VECs [1–10]. From systemic dynamic point of view, the VEC intracellular Ca^{2+} signaling system can be considered as a dynamic system, with the WSS and ATP stimuli as the input signals and the intracellular Ca^{2+} dynamic response as the output signal of the dynamic system [11]. The intracellular Ca^{2+} dynamics in VECs motivated by WSS and/or ATP have been commonly regarded as a critical factor in regulating VEC function and behavior including proliferation, migration and apoptosis [12–14]. Therefore, it is of significance to experimentally investigate the VEC intracellular Ca^{2+} dynamics induced by WSS and/or ATP from systemic dynamic point of view.

In blood circulatory systems in vivo, the WSS and ATP signals are very complicated because they are influenced by many interference factors. In vitro experimental studies could exclude these interference factors existing in vivo. Since the 1980s, an in vitro flow shear device, namely parallel-plate flow chamber (PPFC), of which the height is far smaller than the width and the length, has been usually adopted to quantitatively simulate the WSS and ATP in the extracellular microenvironments and investigate the intracellular Ca^{2+} dynamics in VECs in response to WSS and/or ATP [1–10]. Using the PPFC, it has been demonstrated that the dynamic behavior of the VEC intracellular Ca^{2+} in response to WSS could be modulated by extracellular ATP in a dose-dependent manner [1–9]. However, all these excellent investigations have focused on either WSS alone or WSS together with a constant ATP concentration. The effects of dynamic ATP signals, which would be pivotal in the VEC intracellular Ca^{2+} signaling system, have been totally neglected in these researches by PPFCs.

In recent years, with the development of micro- and nano-technology, microfluidics has been becoming an emerging bioengineering approach to study cellular dynamics with the ability of precisely controlling the spatial and/or temporal distribution of biochemical factors in one microfluidic channel [15–20]. Using a microfluidic device, Bibhas et al. characterized the spatiotemporal evolution of intracellular calcium "flickers" in response to steady, pulsatile, or oscillatory WSS through a frequency controlled solenoid valve [15]; however, the effects of the flowing media containing biochemical signals (e.g., ATP signal) were missing in their studies. A number of investigations by microfluidics have studied the effects of dynamic biochemical signals on the function and behavior of biological cells [16–20]. For instance, Yamada et al. invented a Y-shaped microfluidic device for rapidly switching ATP solution or no ATP solution on HEK293 cells, and studied the intracellular Ca^{2+} response following dynamic ATP signal [16]. However, all these studies have not fully considered the influence of WSS signal on the biological cells [17–20].

In order to efficiently investigate the combined effect of WSS and ATP signals on the VEC intracellular calcium dynamics, particularly the effect of dynamic ATP signals together with WSS signal, a Y-shaped microfluidic device, which possesses an inlet A with inflowing static or dynamic ATP signal and an inlet B without ATP signal, is designed based upon the principles of fluid mechanics and mass transfer (Figure 1a). In this microfluidic device, the flow rate in the inlet A is constant but that in the inlet B is time-varying. Therefore, different types of stimulating signals, including static or dynamic WSS alone, dynamic WSS signal together with static ATP signal, static WSS signal together with dynamic ATP signal, could be implemented in the mixing micro-channel C. The implementation of the different types of dynamic biochemical signals was experimentally validated by fluorescent signals which could be easily observed by a fluorescence microscope with a charge-coupled device (CCD) camera (DS126431, Canon Inc., Tokyo, Japan). Finally, the dynamic responses of the intracellular Ca^{2+} in human umbilical vein endothelial cells (HUVECs) in exposure to the different kinds of stimulating signals were detected using the proposed microfluidic device.

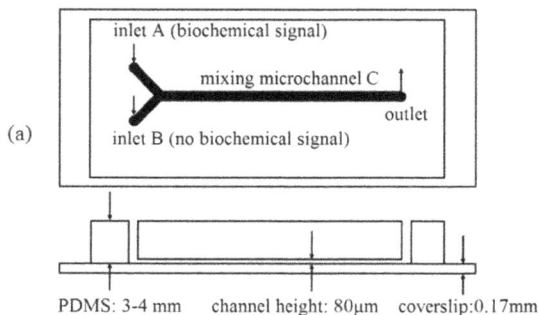

Figure 1. *Cont.*

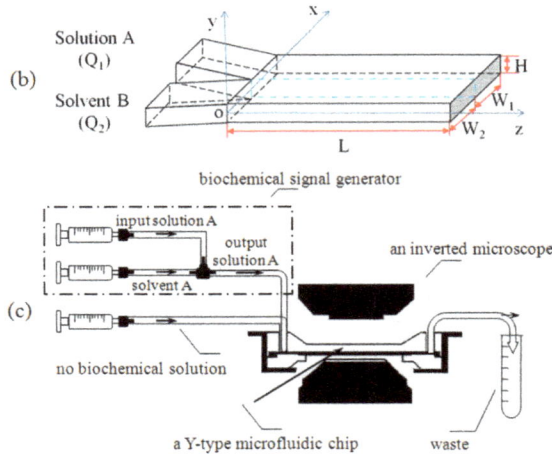

Figure 1. Schematic diagram of a Y-shaped microfluidic device. (**a**) Polydimethylsiloxane (PDMS)-glass structure; (**b**) coordinate system; (**c**) the integrated experimental system.

2. Materials and Methods

2.1. Equations Governing Pulsatile Flow and Mass Transfer

The geometry and the rectangular coordinate system *oxyz* of the shallow Y-shaped microfluidic chip used in this study are illustrated in Figure 1b. It is assumed that the height, *H*, is far smaller than the width, *W*, and the length, *L*, of the mixing micro-channel. A solution A with dynamic biochemical signal and a solution B without biochemical signal were driven into the inlet A and B by two programmable syringe pumps, respectively.

2.1.1. Equation Governing Pulsatile Flow in the Mixing Micro-Channel C

It is assumed that both of the two solutions are Newtonian fluids with identical viscosity, the pulsatile flow in the mixing micro-channel C driven by the programmable syringe pumps is a fully developed laminar flow. By neglecting the boundary effects from the lateral sides and the ends, the equation governing the pulsatile flow in the mixing micro-channel can be simplified as,

$$\frac{\partial u}{\partial t} = -\frac{1}{\rho}\frac{\partial p}{\partial z} + \frac{\eta}{\rho}\frac{\partial^2 u}{\partial y^2} \tag{1}$$

where, $u = u(y,t)$ is the fluid velocity along *z*-direction, $p = p(z,t)$ is the pressure, *t* is the time, η is the fluid viscosity, ρ is the fluid density.

With the assumption of quasi-steady flow, the velocity profile $u(y,t)$, the height-wise averaging velocity $\bar{u}$ and the shear stress $\tau_w(t)$ can be given by [21],

$$u(y,t) = \frac{3}{2WH}\left[1 - \left(\frac{2y}{H}\right)^2\right]Q(t) \tag{2}$$

$$\bar{u}(t) = \frac{Q(t)}{WH} \tag{3}$$

$$\tau_w(t) = \frac{6\mu Q(t)}{WH^2} \tag{4}$$

where $Q(t)$ is the total flow rate through the mixing micro-channel C.

2.1.2. Control of Two-Stream Flow Widths in the Mixing Micro-Channel C

It is assumed that solution A has a constant volume flow rate Q_A, and solvent B has a dynamically changing volume flow rate Q_B. The flow velocity u_A of solution A is the same as the flow velocity u_B of solvent B, which satisfies [22],

$$u_A(y,t) = u_B(y,t) = \frac{3}{2WH} \left[1 - \left(\frac{2y}{H} \right)^2 \right] Q(t) \tag{5}$$

where $Q(t) = Q_A + Q_B(t)$, the volume flow rates Q_A and $Q_B(t)$ satisfy [22],

$$\frac{W_1}{W_2} = \frac{Q_A}{Q_B(t)} \tag{6}$$

where W_1 and W_2 are the widths of the solution A and solvent B, respectively, $W = W_1 + W_2$. Equation (6) shows that the ratio of the widths W_1/W_2 of two streams in the mixing channel is uniquely determined by the externally controlled flow rate ratio $Q_A/Q_B(t)$. In this work, the flow rate ratio is set to be $Q_A/Q_B(t) = (1 - \varepsilon(t))/\varepsilon(t)$. Hence, the width of the solvent B in the Y-shaped channel is εW, which primarily determines the inlet boundary ($z = 0$) of biochemical flow in the mixing micro-channel C.

2.1.3. Taylor-Aris Dispersion in the Mixing Micro-Channel C

In the mixing micro-channel C where molecules are mixed by diffusion, the concentration ϕ of a biochemical substance is governed by [22],

$$\frac{\partial \phi}{\partial t} + u(y,t)\frac{\partial \phi}{\partial z} = D \left(\frac{\partial^2 \phi}{\partial x^2} + \frac{\partial^2 \phi}{\partial y^2} + \frac{\partial^2 \phi}{\partial z^2} \right) \tag{7}$$

where $\phi = \phi(x,y,z,t)$ is the concentration of biochemical substance, D is molecular diffusivity coefficient. Because the height of the micro-channel is smaller, a uniform concentration distribution of biochemical substance is easily formed in the y direction. Therefore, in this study, we only consider the height-wise averaging concentration, $\overline{\phi} = \overline{\phi}(x,z,t)$, defined as

$$\overline{\phi}(x,z,t) = \frac{1}{H} \int_{-H/2}^{H/2} \phi(x,y,z,t) dy \tag{8}$$

The transportation of the height-wise averaging concentration $\overline{\phi}$ in the mixing micro-channel C is governed by the following Taylor-Aris dispersion equation [22],

$$\frac{\partial \overline{\phi}}{\partial t} + \overline{u}\frac{\partial \overline{\phi}}{\partial z} = D\frac{\partial^2 \overline{\phi}}{\partial x^2} + D_{\text{eff}}\frac{\partial^2 \overline{\phi}}{\partial z^2} \tag{9}$$

In the Equation (9), D_{eff} is referred to the effective dispersion coefficient, which is superposed by molecular diffusion coefficient D and Taylor dispersion coefficient D_T as [22],

$$D_{\text{eff}} = D + D_T = D \left[1 + \frac{1}{210} \left(\frac{\overline{u}H}{D} \right)^2 \right] \tag{10}$$

Suppose the solution A with a biochemical factor concentration $\overline{\phi}_A(t)$, and solvent B with no biochemical factor, the boundary conditions for the Equation (9) are as,

$$\text{B.C.1}: \quad \overline{\Phi}(x,z,t)\big|_{z=0} = \overline{\Phi}_A(t), \quad \varepsilon W < x \le W$$

$$\text{B.C.2}: \quad \overline{\Phi}(x,z,t)\big|_{z=0} = 0, \quad 0 \le x \le \varepsilon W \tag{11}$$

$$\text{B.C.3}: \quad \partial\overline{\Phi}/\partial z\big|_{z\to\infty} = 0, \quad \partial\overline{\Phi}/\partial x\big|_{x=0} = 0, \quad \partial\overline{\Phi}/\partial x\big|_{x=W} = 0$$

The initial condition is:

$$\overline{\Phi}(x, z > 0, t)\big|_{t=0} = 0 \tag{12}$$

2.2. Microfluidic Device Fabrication and Experimental Setup

A polydimethylsiloxane (PDMS)-glass Y-shaped microfluidic device is designed as shown in Figure 1a. The height H of all the micro-channels is 80 μm, the width W and the length L of the mixing micro-channel C is 1000 μm and 4 cm, respectively. All the micro-channels were patterned in PDMS (Sylgard 184, Dow Corning, Midland, MI, USA) by replica molding. The mold was prepared by spin coating a thin layer of negative photoresist (SU-8, MicroChem, Westborough, MA, USA) onto a single side polishing silicon wafers and patterned with ultraviolet (UV) exposure. Next, the micro-channel layer was obtained by pouring PDMS with 10:1 (w/w) base: crosslinker ratio onto the mold yielding a thickness of 3 mm roughly. After curing the elastomer for 2 h at 80 °C, the PDMS slab was peeled from the mold, punched and hermetically bonded to a coverslip by plasma oxidation.

As shown in Figure 2, the fabricated Y-shaped microfluidic chip (Figure 2a) connected with three syringe pumps (NE-1000, New Era Pump Systems, Inc., Farmingdale, NY, USA) for controlling the dynamic biochemical signal and the magnitude of WSS in mixing micro-channel C by regulating the flow rates from the three syringe pumps. More specifically, the inlet A was connected to two syringe pumps with a T-bend and silicone tubes (Figure 2b). The dynamic biochemical signal was generated by controlling the flow rates of the solution A and the solvent A from two syringe pumps, respectively. The inlet B was connected to the third syringe pump to generate time-varying laminar flow without biochemical factor. An inverted microscope (CKX41, Olympus Corporation, Tokyo, Japan) equipped with a CCD camera (DS126431, Canon Inc., Tokyo, Japan) was adopted to observe the biochemical signal and the intracellular calcium signal in vascular endothelial cells cultured on the bottom of the mixing channel C in real time (Figure 2c).

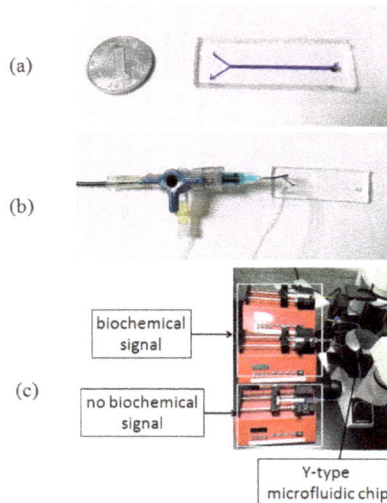

Figure 2. An actual Y-shaped microfluidic chip. (**a**) PDMS-glass microfluidic chip; (**b**) generator of dynamic biochemical signals; (**c**) the actual experimental setup.

2.3. Generation of Dynamic Biochemical Signals

As shown in Figure 1c, the dynamic biochemical signal flowing into the inlet A was generated by controlling the flow rates of two syringes connected by a T-bend and two silicone tubes. The syringes were driven by two syringe pumps, respectively. Suppose that Q_{A1}, Q_{A2} and Q_A were the flow rates of input solution A, solvent A and output solution A respectively, and ϕ_{A1} and ϕ_A were the concentrations of input solution A and output solution A, the mass conservation law led to

$$Q_{A1} + Q_{A2} = Q_A$$
$$Q_{A1}\phi_{A1} = Q_A\phi_A \tag{13}$$

and then Q_{A1}, Q_{A2} were expressed as

$$Q_{A1} = \frac{Q_A\phi_A}{\phi_{A1}} \tag{14}$$

and

$$Q_{A2} = Q_A \left(1 - \frac{\phi_A}{\phi_{A1}}\right) \tag{15}$$

From Equations (14) and (15), a desired biochemical signal with the flow rate Q_A and the concentration ϕ_A were implemented by controlling the syringe pumps by setting the flow rates Q_{A1} and Q_{A2} of syringes filled with the solution A and solvent A, respectively. After solution A and solvent A was fully mixed, the biochemical signal was generated, and then delivered to the inlet A of the Y-shaped microfluidic chip through a short silicone tube (Figure 2b). In this delivering process, because the signal frequency is very low (~1/60 Hz), the attenuation of the signal could be ignored before it reached the inlet A of the Y-shaped microfluidic device [23].

2.4. Transport of Dynamic Biochemical Signals in the Mixing Micro-Channel C

Before the dynamic biochemical signals input at the inlet A reached the mixing micro-channel C, they transported through a fully mixing microfluidic channel A which acts as a low-pass filter [24]. As the signal frequency in this study was as low as 1/60 Hz, the effect of this fully mixing micro-channel A on the signal transportation was not considered. As the dynamic biochemical signals transported along the mixing micro-channel C, the spatiotemporal concentration profiles of biochemical signals were described by Equation (9) together with the boundary conditions (11). This subsection presents the numerical and experimental simulation studies about the transport of dynamic biochemical signals in the mixing micro-channel C.

2.4.1. Numerical Simulation

For numerical simulation studies, Equation (9) was solved by a finite difference method. An Euler explicit discretization was used for the temporal derivation. The first-order and second-order central differences were adopted to approximate the first-order and second-order spatial derivation, respectively. Given the boundary conditions (11), i.e., the flow rates Q_A and $Q_B(t)$ satisfying that $Q_A/Q_B(t) = (1 - \varepsilon(t))/\varepsilon(t)$, and the input signal $\overline{\phi}_A(t)$, the spatiotemporal dynamic biochemical signal in the mixing micro-channel C were numerically simulated using MATLAB (Version R2009b, The Math Works, Inc., Natick, MA, USA). All the simulation results were normalized to a constant reference value. In the numerical simulations, all the parameters for the Y-shaped microfluidic device and the solutions are listed in Table 1.

Table 1. Default parameters used in the model.

Parameters	Values
L (z-direction)	4 cm
H (y-direction)	80 μm
W (x-direction)	1000 μm
η	0.001 Pa·s
$D_{fluorescent}$	8.2×10^{-10} m^2/s
D_{ATP}	2.36×10^{-10} m^2/s

2.4.2. Experimental Validation

For actual experimental validation, the fluorescent solution (Rhodamine-6, Sigma-Aldrich, St. Louis, MO, USA) with time-dependent concentration was used to simulate the dynamic ATP signal. The fluorescent signal $\overline{\phi}_A(t)$ was input through the inlet A of the Y-shaped microfluidic chip at a constant volumetric flow rate ($Q_A = 3.6$ mL/h). The volume flow rate $Q_B(t)$ of solvent B changes as a square wave with a period of 60 s between Q_A (3.6 mL/h) and $2Q_A$ (7.2 mL/h). The dynamic biochemical signal $\overline{\phi}_A(t)$ from the inlet A synchronizes with the flow rate $Q_B(t)$ from the inlet B. The time-varying images for dynamic fluorescent signals at any positions in mixing channel C could be observed and detected with the fluorescence microscope with the CCD camera (Figure 2c). The dynamic fluorescent intensities were then extracted from the images using MATLAB (The Math Works R2009b, Inc.). While the fluorescent intensities at each time point were calculated, the grey-values from the image background were subtracted. All the experimental results were normalized to a constant reference value.

2.5. Cell Culture and Intracellular Calcium Dynamic Response

HUVECs (derived from Dalian Medical University, Dalian, China) were cultured in Dulbecco's Modified Eagle's Medium (DMEM) (Invitrogen, Carlsbad, CA, USA) supplemented with 10% Fetal Bovine Serum (FBS) (Gibco, Thermo Fisher Scientific, Waltham, MA, USA) and were maintained at 37 °C with 5% CO$_2$ in culture flask. 0.25% Trypsin/EDTA (Gibco) was used to detach cells from plates and transfer them to the microfluidic chip as shown in Figure 2a. To ensure cell adhesion, the chip was subsequently filled with 100 mg/mL fibronectin (Sigma) and allowed to incubate at 37 °C for two hours. Afterwards, the chip was flushed and refilled with DMEM supplemented with 10% FBS. HUVECs cells were then seeded in the mixing micro-channel C from the outlet. The cells were then allowed to attach overnight. When HUVECs cells had been inoculated in the microfluidic chip for 4 days, the cytosolic calcium ions in cells were stained with 5 nM Fluo-3 AM for 45 min in a culture medium at 37 °C. The cells were then rinsed with dye-free medium twice. The entire operation was performed with extreme caution to minimize the response of cells to early agitations. When microfluidic device was placed under microscope with a CCD camera, the syringe pumps would start up according to the designed program. The time-varying fluorescent images for the intracellular calcium response in HUVECs at the regions of interest were recorded in a sampling frequency (one frame per 4 s) with the CCD for 3 min at room temperature. The calcium fluorescent intensities were then extracted from the dynamic images using the same method as described for dynamic fluorescent images in previous Section 2.4.2.

3. Results

3.1. Spatiotemporal Profiles of Static and Dynamic Fluorescent Signals in the Mixing Micro-Channel C

Figure 3 shows the spatiotemporal profiles of a static fluorescent signal transporting in a dynamic flow in the mixing micro-channel C. For this case, the input concentration $\overline{\phi}_A(t)$ of solution A is a constant of 5 μmol/mL, the volume flow rate Q_A is a constant of 3.6 mL/h while the volume flow rate $Q_B(t)$ is a dynamic signal as a square wave with a period of 60 s between 3.6 and 7.2 mL/h, and thus the WSS signal changes as a square wave with a period of 60 s between 1.875 and 2.813 Pa as well (Figure 3a). The spatial distribution of the fluorescent signal concentration at $t = 15$ s and at $t = 45$ s are exhibited in Figure 3b. It can be clearly seen from Figure 3b that at any position along the length of the channel (z-direction), the fluorescent signal concentration keeps at 0 μmol/mL while x is close to 0 mm and at a constant value 5 μmol/mL while x is close to 1 mm. The signal concentration will dramatically increase from 0 to 5 μmol/mL along x-direction at the region around the interface between two streams from the inlet A and B. Besides, because the volume flow rate $Q_B(t)$ dynamically changes as a square-wave like signal, this interface also dynamically changes its position along x-direction. These numerical simulation results can be validated by experimental results as shown in Figure 3c.

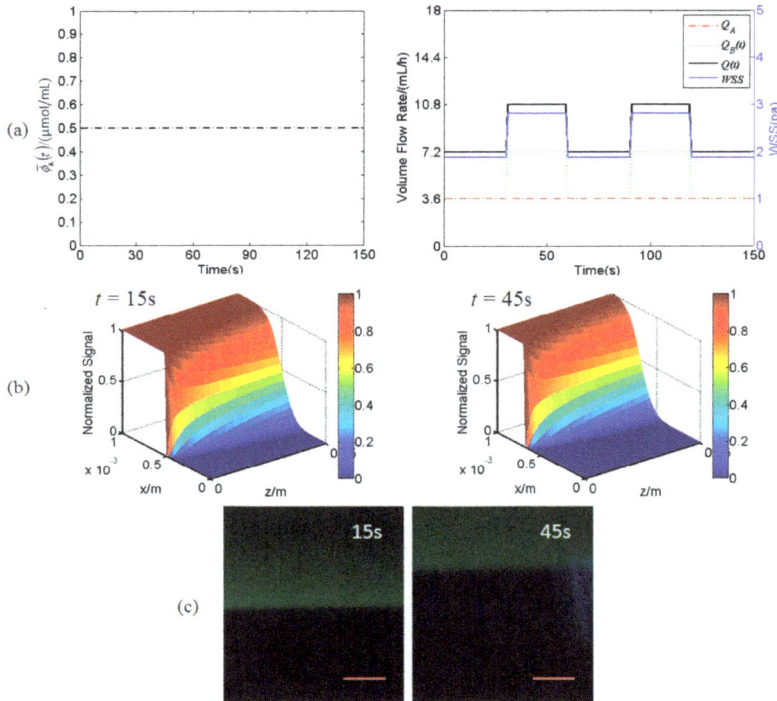

Figure 3. An input static fluorescent signal and its spatiotemporal concentration profile in a dynamic flow in the mixing micro-channel C. (**a**) The static fluorescent signal $\overline{\phi}_A(t)$, the volume flow rates Q_A, $Q_B(t)$, $Q(t)$ and wall shear stress (WSS); (**b**) numerically simulated concentration profile at $t = 15$ s and $t = 45$ s in x-z plane, respectively; (**c**) experimental concentration profiles at $t = 15$ s and $t = 45$ s in x-z plane, respectively. Scale bar is 100 μm.

The spatiotemporal concentration profiles of a square-wave-like fluorescent signal transporting in a steady flow in the mixing micro-channel C are illustrated in Figure 4. Under the steady flow, the concentration $\overline{\phi}_A(t)$ of solution A is a dynamic square wave with a period of 60 s between 5 μmol/mL and 0 μmol/mL (Figure 3a). As shown in Figure 4a, for the steady flow, the volume flow rates Q_A and $Q_B(t)$ are the same as 3.6 mL/h and thus the WSS is constant at 1.875 Pa. Figure 4b shows the concentration profile at $z = 2$ cm in x-t plane under steady flow. It is obvious in Figure 4b that the dynamic fluorescent signal keeps square-wave-like at the region that x is close to 1 mm, but the amplitude of the signal at the region around the interface between two streams from the inlet A and B decreases due to transverse molecular diffusion. Furthermore, the dynamic fluorescent signal has no significant amplitude attenuation and phase delay while it is transporting along the mixing micro-channel (data not shown). In addition, this interface keeps its position along x-direction under steady flow. All these numerical simulation results can also be reproduced by experimental results as shown in Figure 4c.

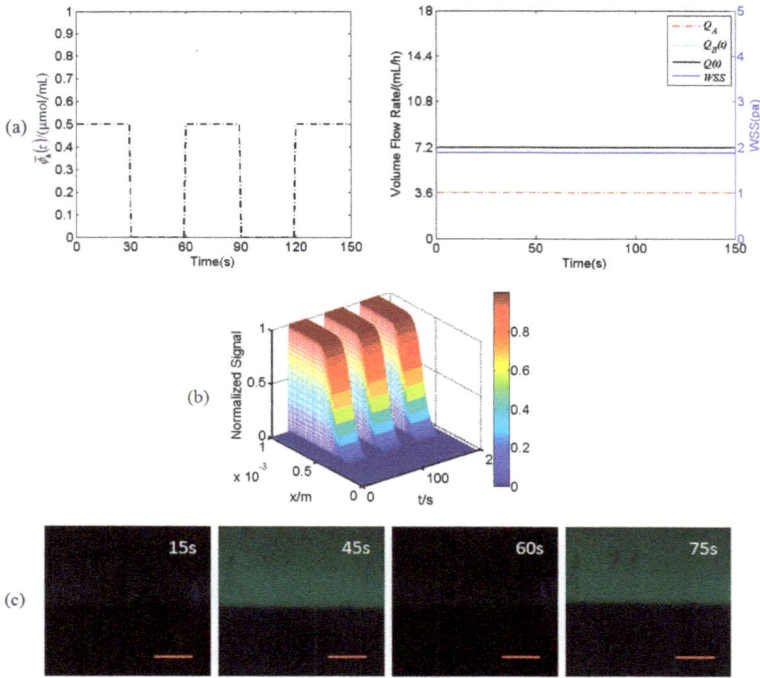

Figure 4. An input dynamic fluorescent signal and its spatiotemporal concentration profile in steady flow in the mixing micro-channel C. (**a**) The dynamic fluorescent signal $\overline{\Phi}_A(t)$ and the volume flow rates Q_A, $Q_B(t)$, $Q(t)$ and WSS; (**b**) numerically simulated spatiotemporal concentration profiles at $z = 2$ cm in x-t plane under steady flow; (**c**) experimental spatiotemporal concentration profiles at the region around $x = 2$ cm in x-z plane under steady flow. Scale bar is 100 μm.

3.2. Combination of WSS and Fluorescent Signals at the Central Regime Along the Mixing Micro-Channel C

It can be clearly observed from Figures 3 and 4 that at any z position along the mixing micro-channel C, there exist different regimes along x-direction where the transporting WSS and/or fluorescent signals are different. Figure 5 shows the comparison between simulation results and experimental data of combined WSS and fluorescent signals at three ($x = W/8$, $W5/8$ and $W7/8$ in Figure 5a) or two ($x = W/8$ and $W7/8$ in Figure 5b) different locations, at the central regime ($z = 2$ cm) in the mixing micro-channel C. It can be easily seen form Figure 5 that different combinations of WSS and fluorescent signals are produced in the mixing micro-channel. More specifically, Figure 5a exhibits the dynamic WSS alone at $x = W/8$, the combination of dynamic WSS and dynamic fluorescent signal at $x = W5/8$, and the combination of dynamic WSS and static fluorescent signal at $x = W7/8$; Figure 5b illustrates the static WSS alone at $x = W/8$ and the combination of static WSS and dynamic fluorescent signal at $x = W7/8$, respectively. All the fluorescent signals in Figures 4 and 5 were replaced by ATP signals instead in the HUVECs calcium dynamic response experiments as shown in Figures 6 and 7.

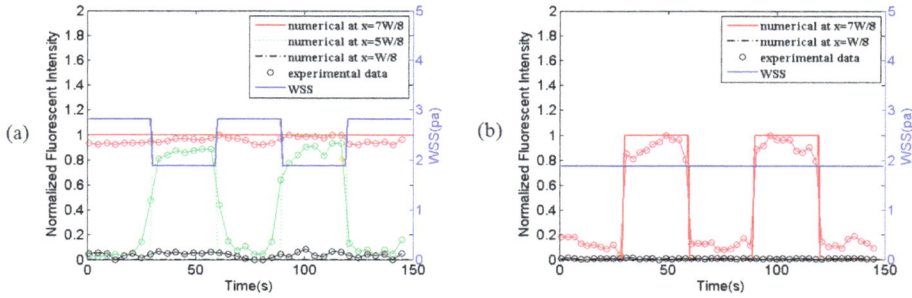

Figure 5. Comparison between simulation results and experimental data of combined WSS and fluorescent signals at different locations at $z = 2$ cm in the mixing micro-channel C. (**a**) Static fluorescent signal under dynamic flow; (**b**) dynamic fluorescent signal under steady flow. All the signals are normalized to a constant reference value.

3.3. Intracellular Ca^{2+} Dynamics in Huvecs in Response to Combined Effects of WSS and ATP Signals

Figure 6a shows the images of the HUVECs were cultured on the bottom of mixing micro-channel for 4 days. After incubated with 5 nM Fluo3-AM for 45 min, the dynamic responses of the intracellular Ca^{2+} concentration induced by static or dynamic WSS alone, as well as different combinations of WSS and ATP signals were carefully measured. As a specific case, Figure 6b exhibits the intracellular Ca^{2+} intensity at 3 s, 24 s, 69 s, and 111 s, respectively, in the HUVECs at $z = 2$ cm in the mixing micro-channel C under the stimulation of a combination of static ATP signal (1 µmol/L) and dynamic WSS (with the period of 60 s).

Different combined effects of WSS and ATP signals on the intracellular Ca^{2+} dynamics in HUVECs are shown in Figure 7. Figure 7a demonstrates that intracellular Ca^{2+} response dynamics under the stimulation of static WSS signal alone at $x = W/8$ under the condition that volume flow rate is constant. It can be seen from Figure 7a that after motivated by this static WSS signal alone, as increases in time, the intracellular Ca^{2+} concentration increases to a maximum and then deceases to the original value. Only single peak Ca^{2+} response is observed for this case. However, once this static WSS signal co-works with a dynamic ATP signal, a second dynamic response of intracellular Ca^{2+} concentration with the same frequency as that of ATP signal can be observed at $x = W7/8$ of the micro-channel C as shown in Figure 7b. Similar dynamic response of the intracellular Ca^{2+} concentration can also be found in HUVECs activated by a dynamic WSS alone (Figure 7c) and a combination of static ATP signal and dynamic WSS (Figure 7d) at the religion $x = W/8$ and $x = W7/8$ respectively.

It is easy to observe that the intracellular Ca^{2+} dynamic responses in HUVECs in exposure to combined effect of WSS and ATP synchronizes with the stimulating dynamic WSS or dynamic ATP signal, however, dynamic WSS or dynamic ATP signal alone cannot induce much more transient Ca^{2+} responses. These interesting results demonstrate that the synergistic effect of WSS and ATP signals might play a critical role in the HUVEC signal transduction. Further investigations will be required to understand this phenomenon and the underlying molecular mechanism.

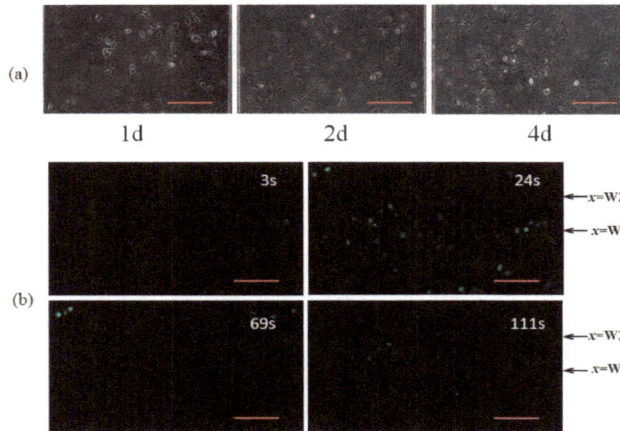

Figure 6. The intracellular Ca^{2+} response in the human umbilical vein endothelial cells (HUVECs) culture on the bottom of the mixing micro-channel C at $z = 2$ cm. (**a**) HUVECs cultured for 1d, 2d and 4d, respectively; (**b**) the intracellular Ca^{2+} intensity at $t = 3$ s, 24 s, 69 s, and 111 s, respectively. Scale bar is 100 μm.

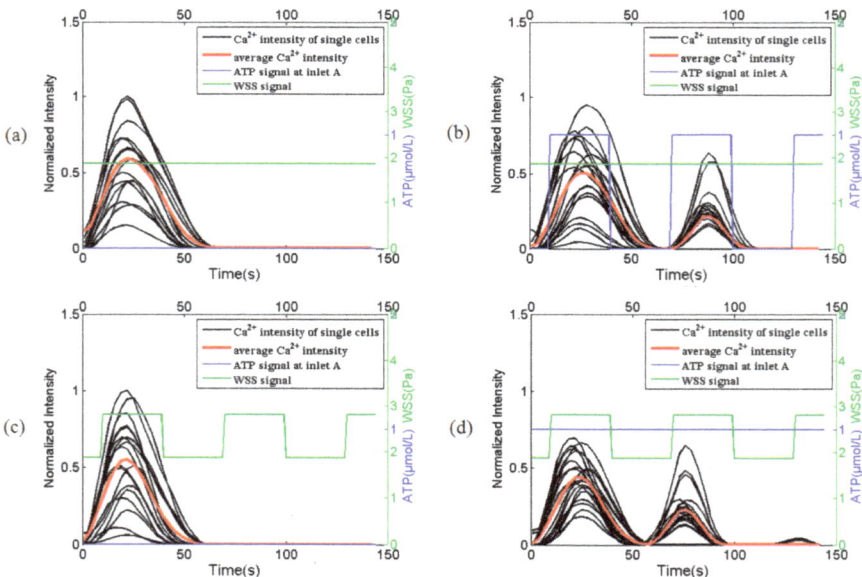

Figure 7. The intracellular Ca^{2+} dynamics in HUVECs in response to WSS signal alone and combined WSS and adenosine triphosphate (ATP) signals. (**a**) static WSS signal alone and (**b**) static WSS signal with dynamic ATP signal at the religion of $x = W/8$ or $x = W7/8$ in mixing micro-channel C under the condition that volume flow rate is constant, respectively; (**c**) dynamic WSS signal alone and (**d**) dynamic WSS signal with static ATP signal at the religion of $x = W/8$ or $x = W7/8$ in mixing micro-channel C under the condition that volume flow rate is dynamic changing, respectively. Experimental data in (**a,b**) are measured from one group of HUVECs while those in (**c,d**) are measure from another group of HUVECs. The average Ca^{2+} intensity is the average value of 20 single cells. All the data are normalized to a constant reference value.

4. Discussion

A Y-shaped microfluidic device, which provides different combinations of WSS and ATP signals in one single micro-channel, is proposed based upon the principles of fluid mechanics and mass transfer. This Y-shaped microfluidic device has an inlet A to input static or dynamic ATP signal, an inlet B to input static or dynamic laminar flow, and a mixing micro-channel for cell culture. The combinations of WSS and ATP signals are generated by controlling the volume flow rates of three programmable syringe pumps. To the best of our knowledge, such a Y-shaped microfluidic device is firstly used to study the combined effects of WSS and ATP signals, especially the impact of dynamic ATP signals, on intracellular Ca^{2+} dynamic in VECs although this type of Y-shaped microfluidic chip has been commonly adopted in biochemical mixing [25,26], cell sorting [27–29] and rapid biochemical switching for analysis at the cellular level [16,17].

It is of importance to understand the transport characteristics of the dynamic biochemical signals transporting in fluid flows in the micro-channels of the Y-shaped microfluidic device in order to precisely load biochemical signals on the desired cells cultured on the bottom of the mixing micro-channel as these microfluidic channels act as low-pass filters [22]. The WSS signals and the spatiotemporal concentration profiles of biochemical signals in the mixing micro-channel are carefully analyzed by numerically solving the equations governing the dynamic laminar flow and time-dependent Taylor-Aris dispersion (see Figures 3–5). Numerical simulation results demonstrate that the WSS signals on the bottom of the mixing micro-channel are the same. The biochemical signals with low frequency (e.g., $1/60$ Hz) have very little amplitude attenuation and phase delay while transporting in steady flow in the silicone tube or the micro-channels (data not shown), which is consistent with the conclusion in the literature [22–24]. Thus, the concentration profiles of a static or dynamic biochemical signal at any locations along the length of the mixing micro-channel (z-direction) are almost the same, which ensures that the cultured cells around the central regime ($z \approx 2$ cm) along z-direction are in exposure to very similar and stable biochemical signals (see Figures 3c and 4c). However, at any z positions and along x-direction, the biochemical signal switches dramatically from zero to a stationary signal around the interface between two streams from the inlet A and B, resulting in two (Figure 5b) or three (Figure 5a) regimes with WSS signal alone or different combinations of WSS and biochemical signals as shown in Figure 5. All these numerical simulation results can be validated by using fluorescent simulation experiments, demonstrating the Y-shaped microfluidic device can provide the culture cells on the bottom of the mixing channel with desired WSS and ATP signals as well as their combinations.

Using the proposed Y-shaped microfluidic device, the different combined effects of WSS and ATP signal on the intracellular Ca^{2+} dynamics in HUVECs culture on the bottom of the mixing micro-channel are measured. The experimental results suggest that the intracellular Ca^{2+} dynamic responses in HUVECs in exposure to combined effect of WSS and ATP synchronizes with the stimulating dynamic WSS or dynamic ATP signals, however, dynamic WSS or dynamic ATP signal alone cannot induce much more transient Ca^{2+} responses. These interesting results demonstrate that the synergistic effect of WSS and ATP signals, but not WSS or ATP signal alone, might play a critical role in the HUVEC signal transduction. A potential mechanism for this synergistic effect is that the inflowing ATP together with endogenously released ATP from HUVECs by wall shear stress activated P2X/P2Y signaling pathways, which in turn induced calcium release from calcium stores [5–7,30]; meanwhile, the mechano-sensitive channels on cell membrane directly activated by wall shear stress, leading to extracellular calcium influx across the cell membrane [11,31]; these two signaling pathways would interplay with each other. Further studies are necessary to confirm this novel phenomenon and figure out the underlying molecular mechanism.

5. Conclusions

A Y-shaped microfluidic device is proposed to investigate the combined effects of WSS and ATP signals on the intracellular Ca^{2+} dynamics in VECs. The Y-shaped microfluidic devices can provide the cultured cells on the bottom of its mixing micro-channel with stimuli of WSS signal alone and different combinations of WSS and ATP signals in one single micro-channel, which are validated by both numerical and experimental simulation studies. Cellular Ca^{2+} dynamic response experiments also verify the feasibility of application of the device. Preliminary experimental results of intracellular Ca^{2+} dynamics show that a combination of WSS and ATP signals rather than a WSS signal alone might play a more important role in VEC Ca^{2+} signal transduction induced by blood flow.

Acknowledgments: The authors would like to appreciate Wenyu Liu from Dalian University of Technology for English Editing. The project is, in part, supported by the National Natural Science Foundation of China (Nos. 11172060, 31370948, 11672065), State Natural Science Foundation of Liaoning (2015020303), and Science & Technology Foundation of Dalian (2015E12SF167).

Author Contributions: Z.-Z.C. and K.-R.Q. conceived and designed the experiments; Z.-Z.C. performed the experiments; Z.-Z.C., Z.-M.G. and D.-P.Z. analyzed the data; Y.L. and B.L. gave scientific support and conceptual advices; Z.-Z.C. and K.-R.Q. wrote the paper. All authors discussed the results and implications and commented on the manuscript.

Conflicts of Interest: The authors declare no conflict of interest.

Abbreviations

WSS	Wall shear stress
ATP	Adenosine triphosphate
VECs	Vascular endothelial cells
D	Diffusivity of solute (m^2/s)
D_{eff}	Effective diffusivity coefficient of solute (m^2/s)
H	Height of the mixing micro-channel (m)
p	Pressure (Pa)
Q	Total flow rate in the mixing micro-channel C (mL/h)
Q_A	Flow rate at inlet A (mL/h)
$Q_B(t)$	Flow rate at inlet B (mL/h)
u	Velocity of fluid in z direction (m/s)
W	Width of the mixing micro-channel (m)
W_1	Width of solution A in the mixing micro-channel (m)
W_2	Width of solvent B in the mixing micro-channel (m)
$\varepsilon(t)$	Ratio of W_2 to W
η	Viscosity of fluid (Pa·s)
$\overline{\phi}_A(t)$	Concentration of solution in the micro-channel ($\mu mol/mL$)
$\overline{\phi}(t)$	Height-wise averaging concentration of solution ($\mu mol/mL$)
$\overline{\phi}_A(t)$	Height-wise averaging concentration of solution at inlet A ($\mu mol/mL$)
$\tau_w(t)$	Wall shear stress (Pa)

References

1. Ando, J.; Ohtsuka, A.; Korenaga, R.; Kamiya, A. Effect of extracellular ATP level on flow-induced Ca^{++} response in cultured vascular endothelial cells. *Biochem. Biophys. Res. Commun.* **1991**, *179*, 1192–1199. [CrossRef]
2. Ando, J.; Kamiya, A. Cytoplasmic calcium response to fluid shear stress in cultured vascular endothelial cells. *In Vitro Cell. Dev. Biol.* **1988**, *24*, 871–877. [CrossRef] [PubMed]
3. Dull, R.O.; Davies, P.F. Flow modulation of agonist (ATP)-response (Ca^{2+}) coupling in vascular endothelial cells. *Am. J. Physiol.* **1991**, *261*, 149–154.
4. Mo, M.; Eskin, S.G.; Schilling, W.P. Flow-induced changes in Ca^{2+} signaling of vascular endothelial cells: Effect of shear stress and ATP. *Am. J. Physiol.* **1991**, *260*, 1698–1707.

5. Yamamoto, K.; Korenaga, R.; Kamiya, A.; Ando, J. Fluid shear stress activates Ca^{2+} influx into human endothelial cells via P2X4 purinoceptors. *Circ. Res.* **2000**, *87*, 385–391. [CrossRef] [PubMed]
6. Yamamoto, K.; Korenaga, R.; Kamiya, A.; Qi, Z.; Sokabe, M.; Ando, J. P2X4 receptors mediate ATP-induced calcium influx in human vascular endothelial cells. *Am. J. Physiol. Heart Circ.* **2000**, *279*, 285–292.
7. Yamamoto, K.; Sokabe, T.; Ohura, N.; Nakatsuka, H.; Kamiya, A.; Ando, J. Endogenously released ATP mediates shear stress-induced Ca^{2+} influx into pulmonary artery endothelial cells. *Am. J. Physiol. Heart Circ.* **2003**, *285*, 793–803. [CrossRef] [PubMed]
8. Shen, J.; Luscinskas, F.W.; Gimbrone, M.A.; Dewey, C.F. Fluid flow modulates vascular endothelial cytosolic calcium responses to adenine nucleotides. *Microcirculation* **1994**, *1*, 67–78. [CrossRef] [PubMed]
9. James, N.L.; Harrison, D.G.; Nerem, R.M. Effects of shear on endothelial cell calcium in the presence and absence of ATP. *FASEB J.* **1995**, *9*, 968–973. [PubMed]
10. Helmlinger, G.; Berk, B.C.; Nerem, R.M. Pulsatile and steady flow-induced calcium oscillations in single cultured endothelial cells. *J. Vasc. Res.* **1996**, *33*, 360–369. [CrossRef] [PubMed]
11. Li, L.F.; Xiang, C.; Qin, K.R. Modeling of TRPV4-C1-mediated calcium signaling in vascular endothelial cells induced by fluid shear stress and ATP. *Biomech. Model. Mechanbiol.* **2015**, *14*, 979–993. [CrossRef] [PubMed]
12. Davies, P.F. Flow-mediated endothelial mechanotransduction. *Physiol. Rev.* **1995**, *75*, 519–560. [PubMed]
13. Berridge, M.J.; Lipp, P.; Bootman, M.D. The versatility and universality of calcium signaling. *Nat. Rev. Mol. Cell Biol.* **2000**, *1*, 11–21. [CrossRef] [PubMed]
14. Clapham, D.E. Calcium signaling. *Cell* **1995**, *80*, 259–268. [CrossRef]
15. Bibhas, R.; Tamal, D.; Debasish, M.; Maiti, T.K.; Chakraborty, S. Oscillatory shear stress induced calcium flickers in osteoblast cells. *Integr. Biol.* **2014**, *6*, 289–299.
16. Yamada, A.; Katanosaka, Y.; Mohri, S.; Naruse, K. A rapid microfluidic switching system for analysis at the single cellular level. *IEEE Trans. Nanobiosci.* **2009**, *8*, 306–311. [CrossRef] [PubMed]
17. Kuczenski, B.; Ruder, W.C.; Messner, W.C.; Leduc, P.R. Probing cellular dynamics with a chemical signal generator. *PLoS ONE* **2009**, *4*, e4847. [CrossRef] [PubMed]
18. Zhang, X.; Grimley, A.; Bertram, R.; Roper, M.G. Microfluidic system for generation of sinusoidal glucose waveforms for entrainment of islets of Langerhans. *Anal. Chem.* **2010**, *82*, 6704–6711. [CrossRef] [PubMed]
19. Kim, Y.T.; Joshi, S.D.; Messner, W.C.; LeDuc, P.R.; Davidson, L.A. Detection of dynamic spatiotemporal response to periodic chemical stimulation in a Xenopus Embryonic tissue. *PLoS ONE* **2011**, *6*, e14624. [CrossRef] [PubMed]
20. Shin, H.; Mahto, S.K.; Kim, J.H.; Rhee, S.W. Exposure of BALB/3T3 fibroblast cells to temporal concentration profile of toxicant inside microfluidic device. *Biochip J.* **2011**, *5*, 214–219. [CrossRef]
21. Wang, Y.X.; Xiang, C.; Liu, B.; Zhu, Y.; Luan, Y.; Liu, S.T.; Qin, K.R. A multi-component parallel-plate flow chamber system for studying the effect of exercise-induced wall shear stress on endothelial cells. *Biomed. Eng.* **2016**. submitted for publication.
22. Li, Y.J.; Li, Y.Z.; Cao, T.; Qin, K.R. Transport of dynamic biochemical signals in steady flow in a shallow Y-shaped microfluidic channel: Effect of transverse diffusion and longitudinal dispersion. *J. Biomech. Eng.* **2013**, *135*, 121011–121019. [CrossRef] [PubMed]
23. Qin, K.R.; Xiang, C.; Ge, S.S. Generation of dynamic biochemical signals with a tube mixer: Effect of dispersion in an oscillatory flow. *Heat Mass Transf.* **2010**, *46*, 675–686. [CrossRef]
24. Xie, Y.; Wang, Y.; Mastrangelo, C.H. Fourier microfluidics. *Lab Chip* **2008**, *8*, 779–785. [CrossRef] [PubMed]
25. Capretto, L.; Cheng, W.; Hill, M.; Zhang, X. Micromixing within microfluidic devices. *Top. Curr. Chem.* **2011**, *304*, 27–68. [PubMed]
26. Nguyen, N.T.; Wu, Z. Micromixers—A review. *J. Micromech. Microeng.* **2005**, *15*, 1–16. [CrossRef]
27. Furdui, V.I.; Harrison, D.J. Immunomagnetic T cell capture from blood for PCR analysis using microfluidic systems. *Lab Chip* **2004**, *4*, 614–618. [CrossRef] [PubMed]
28. Mazutis, L.; Gilbert, J.; Ung, W.L.; Weitz, D.A.; Griffiths, A.D.; Heyman, J.A. Single-cell analysis and sorting using droplet-based microfluidics. *Nat. Protoc.* **2013**, *8*, 870–891. [CrossRef] [PubMed]
29. Yamada, M.; Nakashima, M.; Seki, M. Pinched flow fractionation: Continuous size separation of particles utilizing a laminar flow profile in a pinched microchannel. *Anal. Chem.* **2004**, *76*, 5465–5471. [CrossRef] [PubMed]

30. Malek, A.M.; Zhang, J.; Jiang, J.; Alper, S.L.; Izumo, S. Endothelin-1 gene suppression by shear stress: Pharmacological evaluation of the role of tyrosine kinase, intracellular calcium, cytoskeleton, and mechanosensitive channels. *J. Mol. Cell. Cardiol.* **1999**, *31*, 387–399. [CrossRef] [PubMed]
31. Kanai, A.J.; Strauss, H.C.; Truskey, G.A.; Crews, A.L.; Grunfeld, S.; Malinski, T. Shear stress induces ATP-independent transient nitric oxide release from vascular endothelial cells, measured directly with a porphyrinicmicrosensor. *Circ. Res.* **1995**, *77*, 284–293. [CrossRef] [PubMed]

MDPI AG

St. Alban-Anlage 66

4052 Basel, Switzerland

Tel. +41 61 683 77 34

Fax +41 61 302 89 18

http://www.mdpi.com

Micromachines Editorial Office

E-mail: micromachines@mdpi.com

http://www.mdpi.com/journal/micromachines